Graduate Texts in Mathematics

Graduate Texts in Mathematics bridge the gap between passive study and creative understanding, offering graduate-level introductions to advanced topics in mathematics. The volumes are carefully written as teaching aids and highlight characteristic features of the theory. Although these books are frequently used as textbooks in graduate courses, they are also suitable for individual study.

More information about this series at http://www.springer.com/series/136

Albert N. Shiryaev

Probability-1

Third Edition

Translated by R.P. Boas[†] and D.M. Chibisov

 Springer

Albert N. Shiryaev
Department of Probability Theory
  and Mathematical Statistics
Steklov Mathematical Institute and
  Lomonosov Moscow State University
Moscow, Russia

Translated by R.P. Boas[†] and D.M. Chibisov

ISSN 0072-5285   ISSN 2197-5612   (electronic)
Graduate Texts in Mathematics
ISBN 978-1-4939-7905-9   ISBN 978-0-387-72206-1   (eBook)
DOI 10.1007/978-0-387-72206-1

Springer New York Heidelberg Dordrecht London

Printed on acid-free paper

Springer Science+Business Media LLC New York is part of Springer Science+Business Media (www.
springer.com)

Preface to the Third English Edition

The present edition is the translation of the fourth Russian edition of 2007, with the previous three published in 1980, 1989, and 2004. The English translations of the first two appeared in 1984 and 1996. The third and fourth Russian editions, extended compared to the second edition, were published in two volumes titled "Probability-1" and "Probability-2". Accordingly, the present edition consists of two volumes: this vol. 1, titled "Probability-1," contains chapters 1 to 3, and chapters 4 to 8 are contained in vol. 2, titled "Probability-2," to appear in 2016.

The present English edition has been prepared by translator D. M. Chibisov, Professor of the Steklov Mathematical Institute. A former student of N. V. Smirnov, he has a broad view of probability and mathematical statistics, which enabled him not only to translate the parts that had not been translated before, but also to edit both the previous translation and the Russian text, making in them quite a number of corrections and amendments. He has written a part of Sect. 13, Chap. 3, concerning the Kolmogorov–Smirnov tests.

The author is sincerely grateful to D. M. Chibisov for the translation and scientific editing of this book.

Moscow, Russia   A.N. Shiryaev
2015

Preface to the Fourth Russian Edition

The present edition contains some new material as compared to the third one. This especially concerns two sections in Chap. 1, "Generating Functions" (Sect. 13) and "Inclusion–Exclusion Principle" (Sect. 14).

In the elementary probability theory, dealing with a discrete space of elementary outcomes, as well as in the discrete mathematics in general, the method of generating functions is one of the powerful tools of algebraic nature applicable to diverse problems. In the new Sect. 13, this method is illustrated by a number of probabilistic-combinatorial problems, as well as by the problems of discrete mathematics like counting the number of integer-valued solutions to linear relations under various constraints on the solutions or writing down the elements of sequences satisfying certain recurrence relations.

The material related to the principle (formulas) of inclusion–exclusion is given undeservedly little attention in textbooks on probability theory, though it is very efficient in various probabilistic-combinatorial problems. In Sect. 14, we state the basic inclusion–exclusion formulas and give examples of their application.

Note that after publication of the third edition in two volumes, "Probability-1" and "Probability-2," we published the book, "Problems in Probability Theory," [90] where the problems were arranged in accordance with the contents of these two volumes. The problems in this book are not only "problems-exercises," but are mostly of the nature of "theory in problems," thus presenting large additional material for a deeper study of the probability theory.

Let us mention, finally, that in "Probability-1" and "Probability-2" some corrections of editorial nature have been made.

Moscow, Russia   A.N. Shiryaev
November 2006

Preface to the Third Russian Edition

Taking into account that the first edition of our book "Probability" was published in 1980, the second in 1989, and the present one, the third, in 2004, one may say that the editions were appearing once in a decade. (The book was published in English in 1984 and 1996, and in German in 1988.)

Time has shown that the selection of the topics in the first two editions remained relevant to this day. For this reason, we retained the structure of the previous editions, having introduced, though, some essential amendments and supplements in the present books "Probability-1" and "Probability-2."

This is primarily pertinent to the last, 8th, chapter (vol. 2) dealing with the theory of Markov chains with discrete time. This chapter, in fact, has been written anew. We extended its content and presented the detailed proofs of many results, which had been only sketched before. A special consideration was given to the strong Markov property and the concepts of stationary and ergodic distributions. A separate section was given to the theory of stopping rules for Markov chains.

Some new material has also been added to the 7th chapter (vol. 2) that treats the theory of martingales with discrete time. In Sect. 9 of this chapter, we state a discrete version of the K. Ito formula, which may be viewed as an introduction to the stochastic calculus for the Brownian motion, where Ito's formula for the change of variables is of key importance. In Sect. 10, we show how the methods of the martingale theory provide a simple way of obtaining estimates of ruin probabilities for an insurance company acting under the Cramér–Lundberg model. The next Sect. 11 deals with the "Arbitrage Theory" in stochastic financial mathematics. Here we state two "Fundamental Theorems of the Arbitrage Theory," which provide conditions in martingale terms for absence of arbitrage possibilities and conditions guaranteeing the existence of a portfolio of assets, which enables one to achieve the objected aim. Finally, Sect. 13 of this chapter is devoted to the general theory of optimal stopping rules for arbitrary random sequences. The material presented here demonstrates how the concepts and results of the martingale theory can be applied in the various problems of "Stochastic Optimization."

There are also a number of changes and supplements made in other chapters.

We point out in this respect the new material concerning the theorems on monotonic classes (Sect. 2 of Chap. 2), which relies on detailed treatment of the concepts and properties of "π-λ" systems, and the fundamental theorems of mathematical statistics given in Sect. 13 of Chap. 3.

The novelty of the present edition is also the "Outline of historical development of the mathematical probability theory," placed at the end of "Probability-2."

In a number of sections new problems have been added.

The author is grateful to T. B. Tolosova for her laborious work over the scientific editing of the book and thanks the Publishing House of the Moscow Center for Continuous Mathematical Education for the offer of the new edition and the fast and efficient implementation of the publication project.

Moscow, Russia   A.N. Shiryaev
2003

Preface to the Second Edition

In the Preface to the first edition, originally published in 1980, we mentioned that this book was based on the author's lectures in the Department of Mechanics and Mathematics of the Lomonosov University in Moscow, which were issued, in part, in mimeographed form under the title "Probability, Statistics, and Stochastic Processes, I, II" and published by that University. Our original intention in writing the first edition of this book was to divide the contents into three parts: probability, mathematical statistics, and theory of stochastic processes, which corresponds to an outline of a three-semester course of lectures for university students of mathematics. However, in the course of preparing the book, it turned out to be impossible to realize this intention completely, since a full exposition would have required too much space. In this connection, we stated in the Preface to the first edition that only *probability theory* and the *theory of random processes with discrete time* were really adequately presented.

Essentially all of the first edition is reproduced in this second edition. Changes and corrections are, as a rule, editorial, taking into account comments made by both Russian and foreign readers of the Russian original and of the English and German translations [88, 89]. The author is grateful to all of these readers for their attention, advice, and helpful criticisms.

In this second English edition, new material also has been added, as follows: in Chap. 3, Sect. 5, Sects. 7–12; in Chap. 4, Sect. 5; in Chap. 7, Sect. 8. The most important additions are in the third chapter. There the reader will find expositions of a number of problems connected with a deeper study of themes such as the distance between probability measures, metrization of weak convergence, and contiguity of probability measures. In the same chapter, we have added proofs of a number of important results on the rate of convergence in the central limit theorem and in Poisson's theorem on the approximation of the binomial by the Poisson distribution. These were merely stated in the first edition.

We also call attention to the new material on the probabilities of large deviations (Chap. 4, Sect. 5), and on the central limit theorem for sums of dependent random variables (Chap. 7, Sect. 8).

During the last few years, the literature on probability published in Russia by Nauka has been extended by Sevastyanov [86], 1982; Rozanov [83], 1985; Borovkov [12], 1986; and Gnedenko [32], 1988. In 1984, the Moscow University Press published the textbook by Ya. G. Sinai [92]. It appears that these publications, together with the present volume, being quite different and complementing each other, cover an extensive amount of material that is essentially broad enough to satisfy contemporary demands by students in various branches of mathematics and physics for instruction in topics in probability theory.

Gnedenko's textbook [32] contains many well-chosen examples, including applications, together with pedagogical material and extensive surveys of the history of probability theory. Borovkov's textbook [12] is perhaps the most like the present book in the style of exposition. Chapters 9 (Elements of Renewal Theory), 11 (Factorization Identities) and 17 (Functional Limit Theorems), which distinguish [12] from this book and from [32] and [83], deserve special mention. Rozanov's textbook contains a great deal of material on a variety of mathematical models which the theory of probability and mathematical statistics provides for describing random phenomena and their evolution. The textbook by Sevastyanov is based on his two-semester course at the Moscow State University. The material in its last four chapters covers the minimum amount of probability and mathematical statistics required in a 1-year university program. In our text, perhaps to a greater extent than in those mentioned above, a significant amount of space is given to set-theoretic aspects and mathematical foundations of probability theory.

Exercises and problems are given in the books by Gnedenko and Sevastyanov at the ends of chapters, and in the present textbook at the end of each section. These, together with, for example, the problem sets by A. V. Prokhorov and V. G. and N. G. Ushakov (*Problems in Probability Theory*, Nauka, Moscow, 1986) and by Zubkov, Sevastyanov, and Chistyakov (*Collected Problems in Probability Theory*, Nauka, Moscow, 1988), can be used by readers for independent study, and by teachers as a basis for seminars for students.

Special thanks to Harold Boas, who kindly translated the revisions from Russian to English for this new edition.

Moscow, Russia   A.N. Shiryaev

Preface to the First Edition

This textbook is based on a three-semester course of lectures given by the author in recent years in the Mechanics–Mathematics Faculty of Moscow State University and issued, in part, in mimeographed form under the title *Probability, Statistics, Stochastic Processes, I, II* by the Moscow State University Press.

We follow tradition by devoting the first part of the course (roughly one semester) to the elementary theory of probability (Chap. 1). This begins with the construction of probabilistic models with finitely many outcomes and introduces such fundamental probabilistic concepts as sample spaces, events, probability, independence, random variables, expectation, correlation, conditional probabilities, and so on.

Many probabilistic and statistical regularities are effectively illustrated even by the simplest random walk generated by Bernoulli trials. In this connection we study both classical results (law of large numbers, local and integral De Moivre and Laplace theorems) and more modern results (for example, the arc sine law).

The first chapter concludes with a discussion of dependent random variables generated by martingales and by Markov chains.

Chapters 2–4 form an expanded version of the second part of the course (second semester). Here we present (Chap. 2) Kolmogorov's generally accepted axiomatization of probability theory and the mathematical methods that constitute the tools of modern probability theory (σ-algebras, measures and their representations, the Lebesgue integral, random variables and random elements, characteristic functions, conditional expectation with respect to a σ-algebra, Gaussian systems, and so on). Note that two measure-theoretical results—Carathéodory's theorem on the extension of measures and the Radon–Nikodým theorem—are quoted without proof.

The third chapter is devoted to problems about weak convergence of probability distributions and the method of characteristic functions for proving limit theorems. We introduce the concepts of relative compactness and tightness of families of probability distributions, and prove (for the real line) Prohorov's theorem on the equivalence of these concepts.

The same part of the course discusses properties "with probability 1" for sequences and sums of independent random variables (Chap. 4). We give proofs of the "zero or one laws" of Kolmogorov and of Hewitt and Savage, tests for the convergence of series, and conditions for the strong law of large numbers. The law of the iterated logarithm is stated for arbitrary sequences of independent identically distributed random variables with finite second moments, and proved under the assumption that the variables have Gaussian distributions.

Finally, the third part of the book (Chaps. 5–8) is devoted to random processes with discrete time (random sequences). Chapters 5 and 6 are devoted to the theory of stationary random sequences, where "stationary" is interpreted either in the strict or the wide sense. The theory of random sequences that are stationary in the strict sense is based on the ideas of ergodic theory: measure preserving transformations, ergodicity, mixing, etc. We reproduce a simple proof (by A. Garsia) of the maximal ergodic theorem; this also lets us give a simple proof of the Birkhoff-Khinchin ergodic theorem.

The discussion of sequences of random variables that are stationary in the wide sense begins with a proof of the spectral representation of the covariance function. Then we introduce orthogonal stochastic measures and integrals with respect to these, and establish the spectral representation of the sequences themselves. We also discuss a number of statistical problems: estimating the covariance function and the spectral density, extrapolation, interpolation and filtering. The chapter includes material on the Kalman–Bucy filter and its generalizations.

The seventh chapter discusses the basic results of the theory of martingales and related ideas. This material has only rarely been included in traditional courses in probability theory. In the last chapter, which is devoted to Markov chains, the greatest attention is given to problems on the asymptotic behavior of Markov chains with countably many states.

Each section ends with problems of various kinds: some of them ask for proofs of statements made but not proved in the text, some consist of propositions that will be used later, some are intended to give additional information about the circle of ideas that is under discussion, and finally, some are simple exercises.

In designing the course and preparing this text, the author has used a variety of sources on probability theory. The Historical and Bibliographical Notes indicate both the historical sources of the results and supplementary references for the material under consideration.

The numbering system and form of references is the following. Each section has its own enumeration of theorems, lemmas and formulas (with no indication of chapter or section). For a reference to a result from a different section of the same chapter, we use double numbering, with the first number indicating the number of the section (thus, (2.10) means formula (10) of Sect. 2). For references to a different chapter we use triple numbering (thus, formula (2.4.3) means formula (3) of Sect. 4 of Chap. 2). Works listed in the References at the end of the book have the form [Ln], where L is a letter and n is a numeral.

The author takes this opportunity to thank his teacher A. N. Kolmogorov, and B. V. Gnedenko and Yu. V. Prokhorov, from whom he learned probability theory and whose advices he had the opportunity to use. For discussions and advice, the author also thanks his colleagues in the Departments of Probability Theory and Mathematical Statistics at the Moscow State University, and his colleagues in the Section on probability theory of the Steklov Mathematical Institute of the Academy of Sciences of the U.S.S.R.

Moscow, Russia   A.N. Shiryaev

Contents

Introduction

The subject matter of probability theory is the mathematical analysis of random events, i.e., of those empirical phenomena which can be described by saying that:

They do not have *deterministic regularity* (observations of them do not always yield the same outcome) whereas at the same time:

They possess some *statistical regularity* (indicated by the *statistical stability of their frequencies*).

We illustrate with the classical example of a "fair" toss of an "unbiased" coin. It is clearly impossible to predict with certainty the outcome of each toss. The results of successive experiments are very irregular (now "head," now "tail") and we seem to have no possibility of discovering any regularity in such experiments. However, if we carry out a large number of "independent" experiments with an "unbiased" coin we can observe a very definite statistical regularity, namely that "head" appears with a frequency that is "close" to $\frac{1}{2}$.

Statistical stability of frequencies is very likely to suggest a hypothesis about a possible quantitative estimate of the "randomness" of some event A connected with the results of the experiments. With this starting point, probability theory postulates that corresponding to an event A there is a definite number $P(A)$, called the probability of the event, whose intrinsic property is that as the number of "independent" trials (experiments) increases the frequency of event A is approximated by $P(A)$.

Applied to our example, this means that it is natural to assign the probability $\frac{1}{2}$ to the event A that consists in obtaining "head" in a toss of an "unbiased" coin.

There is no difficulty in multiplying examples in which it is very easy to obtain numerical values intuitively for the probabilities of one or another event. However, these examples are all of a similar nature and involve (so far) undefined concepts such as "fair" toss, "unbiased" coin, "independence," etc.

Having been invented to investigate the quantitative aspects of "randomness," probability theory, like every exact science, became such a science only at the point when the concept of a probabilistic model had been clearly formulated and axiomatized. In this connection it is natural for us to discuss, although only briefly, the fundamental steps in the development of probability theory. A detailed "Outline of the history of development of mathematical probability theory" will be given in the book "Probability-2."

Probability calculus originated in the middle of the seventeenth century with Pascal (1623–1662), Fermat (1601–1655), and Huygens (1629–1695). Although special calculations of probabilities in games of chance had been made earlier, in the fifteenth and sixteenth centuries, by Italian mathematicians (Cardano, Pacioli, Tartaglia, etc.), the first general methods for solving such problems were apparently given in the famous correspondence between Pascal and Fermat, begun in 1654, and in the first book on probability theory, *De Ratiociniis in Aleae Ludo* (*On Calculations in Games of Chance*), published by Huygens in 1657. It was at this time that the fundamental concept of "mathematical expectation" was developed and theorems on the addition and multiplication of probabilities were established.

The real history of probability theory begins with the work of Jacob[1] Bernoulli (1654–1705), *Ars Conjectandi* (*The Art of Guessing*) published in 1713, in which he proved (quite rigorously) the first limit theorem of probability theory, the law of large numbers; and of de Moivre (1667–1754), *Miscellanea Analytica Supplementum* (a rough translation might be *The Analytic Method* or *Analytic Miscellany*, 1730), in which the so-called central limit theorem was stated and proved for the first time (for symmetric Bernoulli trials).

J. Bernoulli deserves the credit for introducing the "classical" definition of the concept of the *probability* of an event as the *ratio* of the number of possible outcomes of an experiment, which are favorable to the event, to the number of possible outcomes.

Bernoulli was probably the first to realize the importance of considering infinite sequences of random trials and to make a clear distinction between the probability of an event and the frequency of its realization.

De Moivre deserves the credit for defining such concepts as *independence, mathematical expectation, and conditional probability.*

In 1812 there appeared Laplace's (1749–1827) great treatise *Théorie Analytique des Probabilités* (*Analytic Theory of Probability*) in which he presented his own results in probability theory as well as those of his predecessors. In particular, he generalized de Moivre's theorem to the general (asymmetric) case of Bernoulli trials thus revealing in a more complete form the significance of de Moivre's result.

Laplace's very important contribution was the application of probabilistic methods to errors of observation. He formulated the idea of considering errors of observation as the cumulative results of adding a large number of independent elementary errors. From this it followed that under rather general conditions the distribution of errors of observation must be at least approximately normal.

The work of Poisson (1781–1840) and Gauss (1777–1855) belongs to the same epoch in the development of probability theory, when the center of the stage was held by limit theorems. In contemporary probability theory the name of Poisson is attributed to the probability distribution which appeared in a limit theorem proved

---

[1] Also known as James or Jacques.

by him and to the related stochastic process. Gauss is credited with originating the theory of errors and, in particular, justification of the fundamental *method of least squares.*

The next important period in the development of probability theory is connected with the names of P. L. Chebyshev (1821–1894), A. A. Markov (1856–1922), and A. M. Lyapunov (1857–1918), who developed effective methods for proving limit theorems for sums of independent but arbitrarily distributed random variables.

The number of Chebyshev's publications in probability theory is not large—four in all—but it would be hard to overestimate their role in probability theory and in the development of the classical Russian school of that subject.

> "On the methodological side, the revolution brought about by Chebyshev was not only his insistence for the first time on complete rigor in the proofs of limit theorems, . . . but also, and principally, that Chebyshev always tried to obtain precise estimates for the deviations from the limiting laws that are available for large but finite numbers of trials, in the form of inequalities that are certainly valid for any number of trials."
>
> (A. N. KOLMOGOROV [50])

Before Chebyshev the main interest in probability theory had been in the calculation of the probabilities of random events. He, however, was the first to realize clearly and exploit the full strength of the concepts of random variables and their mathematical expectations.

The leading exponent of Chebyshev's ideas was his devoted student Markov, to whom there belongs the indisputable credit of presenting his teacher's results with complete clarity. Among Markov's own significant contributions to probability theory were his pioneering investigations of limit theorems for sums of *dependent* random variables and the creation of a new branch of probability theory, the theory of dependent random variables that form what we now call a Markov chain.

> "Markov's classical course in the calculus of probability and his original papers, which are models of precision and clarity, contributed to the greatest extent to the transformation of probability theory into one of the most significant branches of mathematics and to a wide extension of the ideas and methods of Chebyshev."
>
> (S. N. BERNSTEIN [7])

To prove the central limit theorem of probability theory (the theorem on convergence to the normal distribution), Chebyshev and Markov used what is known as the method of moments. Under more general conditions and by a simpler method, the method of characteristic functions, the theorem was obtained by Lyapunov. The subsequent development of the theory has shown that the method of characteristic functions is a powerful analytic tool for establishing the most diverse limit theorems.

The modern period in the development of probability theory begins with its axiomatization. The first work in this direction was done by S. N. Bernstein (1880–1968), R. von Mises (1883–1953), and E. Borel (1871–1956). A. N. Kolmogorov's book *Foundations of the Theory of Probability* appeared in 1933. Here he presented the axiomatic theory that has become generally accepted and is not only applicable to all the classical branches of probability theory, but also provides a firm foundation for the development of new branches that have arisen from questions in the sciences and involve infinite-dimensional distributions.

The treatment in the present books "Probability-1" and "Probability-2" is based on Kolmogorov's axiomatic approach. However, to prevent formalities and logical subtleties from obscuring the intuitive ideas, our exposition begins with the elementary theory of probability, whose elementariness is merely that in the corresponding probabilistic models we consider only experiments with finitely many outcomes. Thereafter we present the foundations of probability theory in their most general form ("Probability-1").

The 1920s and 1930s saw a rapid development of one of the new branches of probability theory, the theory of stochastic processes, which studies families of random variables that evolve with time. We have seen the creation of theories of Markov processes, stationary processes, martingales, and limit theorems for stochastic processes. Information theory is a recent addition.

The book "Probability-2" is basically concerned with stochastic processes with discrete time: random sequences. However, the material presented in the second chapter of "Probability-1" provides a solid foundation (particularly of a logical nature) for the study of the general theory of stochastic processes.

Although the present edition of "Probability-1" and "Probability-2" is devoted to *Probability Theory*, it will be appropriate now to say a few words about *Mathematical Statistics* and, more generally, about *Statistics* and relation of these disciplines to Probability Theory.

In many countries (e.g., in Great Britain) *Probability Theory* is regarded as "integral" part of *Statistics* handling its mathematical aspects. In this context *Statistics* is assumed to consist of *descriptive statistics* and *mathematical statistics*. (Many encyclopedias point out that the original meaning of the word *statistics* was the "study of the status of a state" (from Latin *status*). Formerly statistics was called "political arithmetics" and its aim was estimation of various numerical characteristics describing the status of the society, economics, etc., and recovery of various quantitative properties of mass phenomena from incomplete data.)

The *descriptive statistics* deals with representation of statistical data ("statistical raw material") in the form suitable for analysis. (The *key words* here are, e.g.: population, sample, frequency distributions and their histograms, relative frequencies and their histograms, frequency polygons, etc.)

Mathematical statistics is designed to produce mathematical processing of "statistical raw material" in order to estimate characteristics of the underlying distributions or underlying distributions themselves, or in general to make an appropriate statistical inference with indication of its accuracy. (Key words: point and interval estimation, testing statistical hypotheses, nonparametric tests, regression analysis, analysis of variance, statistics of random processes, etc.)

In Russian tradition *Mathematical Statistics* is regarded as a natural part of *Probability Theory* dealing with "inverse probabilistic problems," i.e., problems of finding the probabilistic model which most adequately fits the available statistical data.

This point of view, which regards mathematical statistics as part of probability theory, enables us to provide the rigorous mathematical background to statistical methods and conclusions and to present statistical inference in the form of rigorous probabilistic statements. (See, e.g., "Probability-1," Sect. 13, Chap. 3, "Fundamental theorems of mathematical statistics.") In this connection it might be appropriate to recall that the *first* limit theorem of probability theory—the Law of Large Numbers—arose in J. Bernoulli's "Ars Conjectandi" from his motivation to obtain the *mathematical* justification for using the "frequency" as an estimate of the "probability of success" in the scheme of "Bernoulli trials." (See in this regard "Probability-1," Sect. 7, Chap. 1.)

We conclude this Introduction with words of J. Bernoulli from "Ars Conjectandi" (Chap. 2 of Part 4)[2]:

> "We are said to *know* or to *understand* those things which are certain and beyond doubt; all other things we are said merely to *conjecture* or *guess about*.
>
> To *conjecture about* something is to measure its probability; and therefore, the *art of conjecturing* or the *stochastic art* is defined by us as the art of measuring as exactly as possible the probabilities of things with this end in mind: that in our decisions or actions we may be able always to choose or to follow what has been perceived as being superior, more advantageous, safer, or better considered; in this alone lies all the wisdom of the philosopher and all the discretion of the statesman."

To the Latin expression *ars conjectandi* (the art of conjectures) there corresponds the Greek expression $\sigma\tau o\chi\alpha\sigma\tau\iota\varkappa\acute{\eta}\ \tau\acute{\epsilon}\chi\nu\eta$ (with the second word often omitted). This expression derives from Greek $\sigma\tau\acute{o}\chi o\zeta$ meaning aim, conjecture, assumption.

Presently the word "stochastic" is widely used as a synonym of "random." For example, the expressions "stochastic processes" and "random processes" are regarded as equivalent. It is worth noting that *theory of random processes* and *statistics of random processes* are nowadays among basic and intensively developing areas of probability theory and mathematical statistics.

---

[2] Cited from: Translations from James Bernoulli, transl. by Bing Sung, Dept. Statist., Harvard Univ., Preprint No. 2 (1966); Chs. 1–4 also available on: http://cerebro.xu.edu/math/Sources/JakobBernoulli/ars_sung.pdf. (Transl. 2016 ed.).

Chapter 1
Elementary Probability Theory

> We call *elementary probability theory* that part of probability theory which deals with probabilities of only a finite number of events.
>
> A. N. Kolmogorov, "Foundations of the Theory of Probability" [51]

1 Probabilistic Model of an Experiment with a Finite Number of Outcomes

1. Let us consider an experiment of which all possible results are included in a finite number of outcomes $\omega_1, \ldots, \omega_N$. We do not need to know the nature of these outcomes, only that there are a finite number N of them.

We call $\omega_1, \ldots, \omega_N$ *elementary events*, or *sample points*, and the finite set

$$\Omega = \{\omega_1, \ldots, \omega_N\},$$

the (finite) *space of elementary events* or *the sample space*.

The choice of the space of elementary events is the *first step* in formulating a *probabilistic model* for an experiment. Let us consider some examples of sample spaces.

Example 1. For a single toss of a coin the sample space Ω consists of two points:

$$\Omega = \{H, T\},$$

where $H = $ "head" and $T = $ "tail."

Example 2. For n tosses of a coin the sample space is

$$\Omega = \{\omega: \omega = (a_1, \ldots, a_n), \ a_i = H \text{ or } T\}$$

and the total number $N(\Omega)$ of outcomes is 2^n.

© Springer Science+Business Media New York 2016
A.N. Shiryaev, *Probability-1*, Graduate Texts
in Mathematics 95, DOI 10.1007/978-0-387-72206-1_1

Example 3. First toss a coin. If it falls "head" then toss a die (with six faces numbered 1, 2, 3, 4, 5, 6); if it falls "tail," toss the coin again. The sample space for this experiment is

$$\Omega = \{H1, H2, H3, H4, H5, H6, TH, TT\}.$$

2. We now consider some more complicated examples involving the selection of n balls from an urn containing M distinguishable balls.

Example 4 (*Sampling with Replacement*). This is an experiment in which at each step one ball is drawn at random and returned again. The balls are numbered $1, \ldots, M$, so that each sample of n balls can be presented in the form $(a_1, \ldots, a_n)$, where a_i is the label of the ball drawn at the ith step. It is clear that in sampling with replacement each a_i can have any of the M values $1, 2, \ldots, M$. The description of the sample space depends in an essential way on whether we consider samples like, for example, $(4, 1, 2, 1)$ and $(1, 4, 2, 1)$ as different or the same. It is customary to distinguish two cases: *ordered* samples and *unordered* samples. In the first case samples containing the same elements, but arranged differently, are considered to be *different*. In the second case the order of the elements is disregarded and the two samples are considered to be *identical*. To emphasize which kind of sample we are considering, we use the notation $(a_1, \ldots, a_n)$ for *ordered* samples and $[a_1, \ldots, a_n]$ for *unordered* samples.

Thus for *ordered* samples *with replacement* the sample space has the form

$$\Omega = \{\omega : \omega = (a_1, \ldots, a_n),\ a_i = 1, \ldots, M\}$$

and the number of (different) outcomes, which in combinatorics are called *arrangements* of n out of M elements *with repetitions*, is

$$N(\Omega) = M^n. \tag{1}$$

If, however, we consider *unordered* samples *with replacement* (called in combinatorics *combinations* of n out of M elements *with repetitions*), then

$$\Omega = \{\omega : \omega = [a_1, \ldots, a_n],\ a_i = 1, \ldots, M\}.$$

Clearly the number $N(\Omega)$ of (different) unordered samples is smaller than the number of ordered samples. Let us show that in the present case

$$N(\Omega) = C^n_{M+n-1}, \tag{2}$$

where $C^l_k \equiv k!/[l!(k-l)!]$ is the number of combinations of k elements, taken l at a time.

We prove this by induction. Let $N(M, n)$ be the number of outcomes of interest. It is clear that when $k \leq M$ we have

$$N(k, 1) = k = C^1_k.$$

Now suppose that $N(k, n) = C_{k+n-1}^n$ for $k \leq M$; we will show that this formula continues to hold when n is replaced by $n+1$. For the unordered samples $[a_1, \ldots, a_{n+1}]$ that we are considering, we may suppose that the elements are arranged in nondecreasing order: $a_1 \leq a_2 \leq \cdots \leq a_{n+1}$. It is clear that the number of unordered samples of size $n + 1$ with $a_1 = 1$ is $N(M, n)$, the number with $a_1 = 2$ is $N(M - 1, n)$, etc. Consequently

$$
\begin{aligned}
N(M, n + 1) &= N(M, n) + N(M - 1, n) + \cdots + N(1, n) \\
&= C_{M+n-1}^n + C_{M-1+n-1}^n + \cdots + C_n^n \\
&= (C_{M+n}^{n+1} - C_{M+n-1}^{n+1}) + (C_{M-1+n}^{n+1} - C_{M-1+n-1}^{n+1}) \\
&\quad + \cdots + (C_{n+2}^{n+1} - C_{n+1}^{n+1}) + C_n^n = C_{M+n}^{n+1};
\end{aligned}
$$

here we have used the easily verified property

$$
C_k^{l-1} + C_k^l = C_{k+1}^l
$$

of the *binomial coefficients* C_k^l. (This is the property of the binomial coefficients which allows for counting them by means of "Pascal's triangle.")

Example 5 (*Sampling Without Replacement*). Suppose that $n \leq M$ and that the selected balls are not returned. In this case we again consider two possibilities, namely ordered and unordered samples.

For *ordered* samples *without replacement* (called in combinatorics *arrangements* of n out of M elements *without repetitions*) the sample space

$$
\Omega = \{\omega \colon \omega = (a_1, \ldots, a_n),\ a_k \neq a_l,\ k \neq l,\ a_i = 1, \ldots, M\},
$$

consists of $M(M - 1) \ldots (M - n + 1)$ elements. This number is denoted by $(M)_n$ or A_M^n and is called the number of *arrangements of n out of M elements.*

For *unordered* samples *without replacement* (called in combinatorics *combinations* of n out of M elements *without repetitions*) the sample space

$$
\Omega = \{\omega \colon \omega = [a_1, \ldots, a_n],\ a_k \neq a_l,\ k \neq l,\ a_i = 1, \ldots, M\}
$$

consists of

$$
N(\Omega) = C_M^n \tag{3}
$$

elements. In fact, from each unordered sample $[a_1, \ldots, a_n]$ consisting of distinct elements we can obtain $n!$ ordered samples. Consequently

$$
N(\Omega) \cdot n! = (M)_n
$$

and therefore

$$
N(\Omega) = \frac{(M)_n}{n!} = C_M^n.
$$

The results on the numbers of samples of size n from an urn with M balls are presented in Table 1.1.

Table 1.1

M^n	C_{M+n-1}^n	With replacement
$(M)_n$	C_M^n	Without replacement
Ordered	Unordered	Type / Sample

For the case $M = 3$ and $n = 2$, the corresponding sample spaces are displayed in Table 1.2.

Table 1.2

(1, 1) (1, 2) (1, 3) (2, 1) (2, 2) (2, 3) (3, 1) (3, 2) (3, 3)	[1, 1] [2, 2] [3, 3] [1, 2] [1, 3] [2, 3]	With replacement
(1, 2) (1, 3) (2, 1) (2, 3) (3, 1) (3, 2)	[1, 2] [1, 3] [2, 3]	Without replacement
Ordered	Unordered	Type / Sample

Example 6 (*Allocation of Objects Among Cells*). We consider the structure of the sample space in the problem of allocation of n objects (balls, etc.) among M cells (boxes, etc.). For example, such problems arise in statistical physics in studying the distribution of n objects (which might be protons, electrons, ...) among M states (which might be energy levels).

Let the cells be numbered $1, 2, \ldots, M$, and suppose first that the objects are distinguishable (numbered $1, 2, \ldots, n$). Then an allocation of the n objects among the M cells is completely described by an (ordered) collection $(a_1, \ldots, a_n)$, where a_i is the index of the cell containing the ith object. However, if the objects are indistinguishable their allocation among the M cells is completely determined by the unordered set $[a_1, \ldots, a_n]$, where a_i is the index of the cell into which an object is put at the ith step.

Comparing this situation with Examples 4 and 5, we have the following corre-spondences:

$$(\text{ordered samples}) \leftrightarrow (\text{distinguishable objects}),$$
$$(\text{unordered samples}) \leftrightarrow (\text{indistinguishable objects}),$$

by which we mean that to an instance of choosing an ordered (unordered) sample of n balls from an urn containing M balls there corresponds (one and only one) instance of distributing n distinguishable (indistinguishable) objects among M cells.

In a similar sense we have the following correspondences:

$$(\text{sampling with replacement}) \leftrightarrow \left(\begin{array}{l} \text{a cell may receive any number} \\ \text{of objects} \end{array} \right),$$

$$(\text{sampling without replacement}) \leftrightarrow \left(\begin{array}{l} \text{a cell may receive at most} \\ \text{one object} \end{array} \right).$$

These correspondences generate others of the same kind:

$$\left(\begin{array}{l} \text{an unordered sample in} \\ \text{sampling without} \\ \text{replacement} \end{array} \right) \leftrightarrow \left(\begin{array}{l} \text{indistinguishable objects in the} \\ \text{problem of allocation among cells} \\ \text{when each cell may receive at} \\ \text{most one object} \end{array} \right)$$

etc.; so that we can use Examples 4 and 5 to describe the sample space for the problem of allocation distinguishable or indistinguishable objects among cells either with exclusion (a cell may receive at most one object) or without exclusion (a cell may receive any number of objects).

Table 1.3 displays the allocation of two objects among three cells. For distin-guishable objects, we denote them by W (white) and B (black). For indistinguish-able objects, the presence of an object in a cell is indicated by a +.

The duality that we have observed between the two problems gives us an obvious way of finding the number of outcomes in the problem of placing objects in cells. The results, which include the results in Table 1.1, are given in Table 1.4.

In statistical physics one says that distinguishable (or indistinguishable, respec-tively) particles that are not subject to the Pauli exclusion principle* obey Maxwell–Boltzmann statistics (or, respectively, Bose–Einstein statistics). If, however, the par-ticles are indistinguishable and are subject to the exclusion principle, they obey Fermi–Dirac statistics (see Table 1.4). For example, electrons, protons and neu-trons obey Fermi–Dirac statistics. Photons and pions obey Bose–Einstein statistics. Distinguishable particles that are subject to the exclusion principle do not occur in physics.

3. In addition to the concept of *sample space* we now introduce the important con-cept of *event* playing a fundamental role in construction of any probabilistic model ("theory") of the experiment at hand.

---

* At most one particle in each cell. (Translator).

Table 1.3

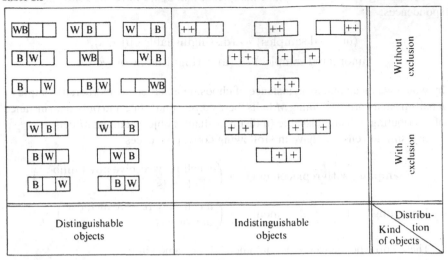

Distinguishable objects	Indistinguishable objects	Kind of objects / Distribution

Table 1.4

$N(\Omega)$ in the problem of placing n objects in M cells			
Kind of objects / Distribution	Distinguishable objects	Indistinguishable objects	
Without exclusion	M^n (Maxwell–Boltzmann statistics)	C^n_{M+n-1} (Bose–Einstein statistics)	With replacement
With exclusion	$(M)_n$	C^n_M (Fermi–Dirac statistics)	Without replacement
	Ordered samples	Unordered samples	Sample Type
$N(\Omega)$ in the problem of choosing n balls from an urn containing M balls			

Experimenters are ordinarily interested, not in what particular outcome occurs as the result of a trial, but in whether the outcome belongs to some subset of the set of all possible outcomes. We shall describe as *events* all subsets $A \subseteq \Omega$ for which, under the conditions of the experiment, it is possible to say either "the outcome $\omega \in A$" or "the outcome $\omega \notin A$."

For example, let a coin be tossed three times. The sample space Ω consists of the eight points

$$\Omega = \{HHH, HHT, \ldots, TTT\}$$

and if we are able to observe (determine, measure, etc.) the results of all three tosses, we say that the set

$$A = \{HHH, HHT, HTH, THH\}$$

is the event consisting of the appearance of at least two heads. If, however, we can determine only the result of the first toss, this set A cannot be considered to be an *event*, since there is no way to give either a positive or negative answer to the question of whether a specific outcome ω belongs to A.

Starting from a given collection of sets that are events, we can form new events by means of statements containing the logical connectives "or," "and" and "not," which correspond in the language of set theory to the operations "union," "intersection," and "complement."

If A and B are sets, their *union*, denoted by $A \cup B$, is the set of points that belong either to A or to B (or to both):

$$A \cup B = \{\omega \in \Omega : \omega \in A \text{ or } \omega \in B\}.$$

In the language of probability theory, $A \cup B$ is the event consisting of the realization of *at least one* of events A or B.

The *intersection* of A and B, denoted by $A \cap B$, or by AB, is the set of points that belong to both A and B:

$$A \cap B = \{\omega \in \Omega : \omega \in A \text{ and } \omega \in B\}.$$

The event $A \cap B$ consists of the simultaneous realization of both A and B.
For example, if $A = \{HH, HT, TH\}$ and $B = \{TT, TH, HT\}$ then

$$A \cup B = \{HH, HT, TH, TT\} \quad (= \Omega),$$
$$A \cap B = \{TH, HT\}.$$

If A is a subset of Ω, its *complement*, denoted by $\overline{A}$, is the set of points of Ω that do not belong to A.

If $B \backslash A$ denotes the *difference* of B and A (i.e., the set of points that belong to B but not to A) then $\overline{A} = \Omega \backslash A$. In the language of probability, $\overline{A}$ is the event consisting of the nonrealization of A. For example, if $A = \{HH, HT, TH\}$ then $\overline{A} = \{TT\}$, the event in which two successive tails occur.

The sets A and $\overline{A}$ have no points in common and consequently $A \cap \overline{A}$ is empty. We denote the empty set by $\varnothing$. In probability theory, $\varnothing$ is called an *impossible* event. The set Ω is naturally called the *certain* event.

When A and B are disjoint ($AB = \varnothing$), the union $A \cup B$ is called the *sum* of A and B and written $A + B$.

If we consider a collection $\mathscr{A}_0$ of sets $A \subseteq \Omega$ we may use the set-theoretic operators $\cup$, $\cap$ and $\backslash$ to form a new collection of sets from the elements of $\mathscr{A}_0$; these sets are again events. If we adjoin the certain and impossible events Ω and $\varnothing$ we obtain a collection $\mathscr{A}$ of sets which is an *algebra*, i.e. a collection of subsets of Ω for which

(1) $\Omega \in \mathscr{A}$,
(2) if $A \in \mathscr{A}$, $B \in \mathscr{A}$, the sets $A \cup B$, $A \cap B$, $A \backslash B$ also belong to $\mathscr{A}$.

It follows from what we have said that it will be advisable to consider collections of events that form algebras. In the future we shall consider only such collections.

Here are some examples of algebras of events:

(a) $\{\Omega, \varnothing\}$, the collection consisting of Ω and the empty set (we call this the *trivial* algebra);
(b) $\{A, \overline{A}, \Omega, \varnothing\}$, the collection generated by A;
(c) $\mathscr{A} = \{A : A \subseteq \Omega\}$, the collection consisting of *all* the subsets of Ω (including the empty set $\varnothing$).

It is easy to check that all these algebras of events can be obtained from the following principle.

We say that a collection

$$\mathscr{D} = \{D_1, \ldots, D_n\}$$

of sets is a *decomposition* of Ω, and call the D_i the *atoms* of the decomposition, if the D_i are not empty, are pairwise disjoint, and their sum is Ω:

$$D_1 + \cdots + D_n = \Omega.$$

For example, if Ω consists of three points, $\Omega = \{1, 2, 3\}$, there are five different decompositions:

$$
\begin{array}{lll}
\mathscr{D}_1 = \{D_1\} & \text{with} & D_1 = \{1, 2, 3\}; \\
\mathscr{D}_2 = \{D_1, D_2\} & \text{with} & D_1 = \{1, 2\}, D_2 = \{3\}; \\
\mathscr{D}_3 = \{D_1, D_2\} & \text{with} & D_1 = \{1, 3\}, D_2 = \{2\}; \\
\mathscr{D}_4 = \{D_1, D_2\} & \text{with} & D_1 = \{2, 3\}, D_2 = \{1\}; \\
\mathscr{D}_5 = \{D_1, D_2, D_3\} & \text{with} & D_1 = \{1\}, D_2 = \{2\}, D_3 = \{3\}.
\end{array}
$$

(For the general number of decompositions of a finite set see Problem 2.)

If we consider all unions of the sets in $\mathscr{D}$, the resulting collection of sets, together with the empty set, forms an algebra, called the *algebra induced by* $\mathscr{D}$, and denoted by $\alpha(\mathscr{D})$. Thus the elements of $\alpha(\mathscr{D})$ consist of the empty set together with the *sums* of sets which are atoms of $\mathscr{D}$.

Thus if $\mathscr{D}$ is a decomposition, there is associated with it a specific algebra $\mathscr{B} = \alpha(\mathscr{D})$.

The converse is also true. Let $\mathscr{B}$ be an algebra of subsets of a finite space Ω. Then there is a unique decomposition $\mathscr{D}$ whose atoms are the elements of $\mathscr{B}$, with $\mathscr{B} = \alpha(\mathscr{D})$. In fact, let $D \in \mathscr{B}$ and let D have the property that for every $B \in \mathscr{B}$ the set $D \cap B$ either coincides with D or is empty. Then this collection of sets D forms a decomposition $\mathscr{D}$ with the required property $\alpha(\mathscr{D}) = \mathscr{B}$. In Example (a), $\mathscr{D}$ is the trivial decomposition consisting of the single set $D_1 = \Omega$; in (b), $\mathscr{D} = \{A, \overline{A}\}$. The most fine-grained decomposition $\mathscr{D}$, which consists of the *singletons* $\{\omega_i\}$, $\omega_i \in \Omega$, induces the algebra in Example (c), i.e., the algebra of all subsets of Ω.

Let $\mathscr{D}_1$ and $\mathscr{D}_2$ be two decompositions. We say that $\mathscr{D}_2$ is finer than $\mathscr{D}_1$, and write $\mathscr{D}_1 \preccurlyeq \mathscr{D}_2$, if $\alpha(\mathscr{D}_1) \subseteq \alpha(\mathscr{D}_2)$.

Let us show that if Ω consists, as we assumed above, of a finite number of points $\omega_1, \ldots, \omega_N$, then the number $N(\mathscr{A})$ of sets in the collection $\mathscr{A}$ as in Example (c) is equal to 2^N. In fact, every nonempty set $A \in \mathscr{A}$ can be represented as $A = \{\omega_{i_1}, \ldots, \omega_{i_k}\}$, where $\omega_{i_j} \in \Omega$, $1 \le k \le N$. With this set we associate the sequence of zeros and ones

$$(0, \ldots, 0, 1, 0, \ldots, 0, 1, \ldots),$$

where there are ones in the positions $i_1, \ldots, i_k$ and zeros elsewhere. Then for a given k the number of different sets A of the form $\{\omega_{i_1}, \ldots, \omega_{i_k}\}$ is the same as the number of ways in which k ones (k indistinguishable objects) can be placed in N positions (N cells). According to Table 1.4 (see the lower right-hand square) we see that this number is C_N^k. Hence (counting the empty set) we find that

$$N(\mathscr{A}) = 1 + C_N^1 + \cdots + C_N^N = (1+1)^N = 2^N.$$

4. We have now taken the first two steps in defining a probabilistic model ("theory") of an experiment with a finite number of outcomes: we have selected a sample space and a collection $\mathscr{A}$ of its subsets, which form an algebra and are called events. (Sometimes the pair $\mathscr{E} = (\Omega, \mathscr{A})$ is regarded as an *experiment*.) We now take the next step, to assign to each sample point (outcome) $\omega_i \in \Omega$, $i = 1, \ldots, N$, a *weight*. This is denoted by $p(\omega_i)$ and called the *probability* of the outcome ω_i; we assume that it has the following properties:

(a) $0 \le p(\omega_i) \le 1$ (nonnegativity),
(b) $p(\omega_1) + \cdots + p(\omega_N) = 1$ (normalization).

Starting from the given probabilities $p(\omega_i)$ of the outcomes ω_i, we define the probability $\mathsf{P}(A)$ of any event $A \in \mathscr{A}$ by

$$\mathsf{P}(A) = \sum_{\{i:\omega_i \in A\}} p(\omega_i). \tag{4}$$

Definition. The "probability space"

$$(\Omega, \mathscr{A}, \mathsf{P}),$$

where $\Omega = \{\omega_1, \ldots, \omega_N\}$, $\mathscr{A}$ is an algebra of subsets of Ω, and

$$\mathsf{P} = \{\mathsf{P}(A); A \in \mathscr{A}\},$$

is said to specify the *probabilistic model* ("theory") of an experiment with a (finite) space Ω of outcomes (elementary events) and algebra $\mathscr{A}$ of events. (Clearly, $P(\{\omega_i\}) = p(\omega_i)$, $i = 1, \ldots, N$.) A probability space $(\Omega, \mathscr{A}, P)$ with a finite set Ω is called *discrete*.

The following properties of probability follow from (4):

$$P(\varnothing) = 0, \tag{5}$$

$$P(\Omega) = 1, \tag{6}$$

$$P(A \cup B) = P(A) + P(B) - P(A \cap B). \tag{7}$$

In particular, if $A \cap B = \varnothing$, then

$$P(A + B) = P(A) + P(B) \tag{8}$$

and

$$P(\overline{A}) = 1 - P(A). \tag{9}$$

5. In constructing a probabilistic model for a specific situation, the construction of the sample space Ω and the algebra $\mathscr{A}$ of events are ordinarily not difficult. In elementary probability theory one usually takes the algebra $\mathscr{A}$ to be the algebra of *all* subsets of Ω. Any difficulty that may arise is in assigning probabilities to the sample points. In principle, the solution to this problem lies outside the domain of probability theory, and we shall not consider it in detail. We consider that our fundamental problem is not the question of how to assign probabilities, but how to calculate the probabilities of complicated events (elements of $\mathscr{A}$) from the probabilities of the sample points.

It is clear from a mathematical point of view that for finite sample spaces we can obtain all conceivable (finite) probability spaces by assigning nonnegative numbers $p_1, \ldots, p_N$, satisfying the condition $p_1 + \cdots + p_N = 1$, to the outcomes $\omega_1, \ldots, \omega_N$.

The *validity* of the assignments of the numbers $p_1, \ldots, p_N$ can, in specific cases, be checked to a certain extent by using the *law of large numbers* (which will be discussed later on). It states that in a long series of "independent" experiments, carried out under identical conditions, the frequencies with which the elementary events appear are "close" to their probabilities.

In connection with the difficulty of assigning probabilities to outcomes, we note that there are many actual situations in which for reasons of *symmetry* or *homogeneity* it seems reasonable to consider all conceivable outcomes as *equally probable*. In such cases, if the sample space consists of points $\omega_1, \ldots, \omega_N$, with $N < \infty$, we put

$$p(\omega_1) = \cdots = p(\omega_N) = 1/N,$$

and consequently

$$P(A) = N(A)/N \tag{10}$$

for every event $A \in \mathscr{A}$, where $N(A)$ is the number of sample points in A.

This is called the *classical* method of assigning probabilities. It is clear that in this case the calculation of $P(A)$ reduces to calculating the *number* of outcomes belonging to A. This is usually done by combinatorial methods, so that *combinatorics*, dealing with finite sets, plays a significant role in the calculus of probabilities.

Example 7 (*Coincidence Problem*). Let an urn contain M balls numbered $1, 2, \ldots, M$. We draw an ordered sample of size n with replacement. It is clear that then

$$\Omega = \{\omega \colon \omega = (a_1, \ldots, a_n), \ a_i = 1, \ \ldots, \ M\}$$

and $N(\Omega) = M^n$. Using the classical assignment of probabilities, we consider the M^n outcomes equally probable and ask for the probability of the event

$$A = \{\omega \colon \omega = (a_1, \ldots, a_n), \ a_i \neq a_j, \ i \neq j\},$$

i.e., the event in which there is no repetition. Clearly $N(A) = M(M-1) \cdots (M - n + 1)$, and therefore

$$P(A) = \frac{(M)_n}{M^n} = \left(1 - \frac{1}{M}\right)\left(1 - \frac{2}{M}\right) \cdots \left(1 - \frac{n-1}{M}\right). \tag{11}$$

This problem has the following striking interpretation. Suppose that there are n students in a class and that each student's birthday is on one of 365 days with all days being equally probable. The question is, what is the probability P_n that there are at least two students in the class whose birthdays coincide? If we interpret selection of birthdays as selection of balls from an urn containing 365 balls, then by (11)

$$P_n = 1 - \frac{(365)_n}{365^n}.$$

The following table lists the values of P_n for some values of n:

n	4	16	22	23	40	64
P_n	0.016	0.284	0.476	0.507	0.891	0.997

For sufficiently large M

$$\log \frac{(M)_n}{M^n} = \sum_{k=1}^{n-1} \log\left(1 - \frac{k}{M}\right) \sim -\frac{1}{M}\sum_{k=1}^{n-1} k = -\frac{1}{M}\frac{n(n-1)}{2},$$

hence

$$P_M(n) \equiv 1 - \frac{(M)_n}{M^n} \sim 1 - e^{-\frac{n(n-1)}{2M}} \quad (\equiv \tilde{P}_M(n)), \quad M \to \infty.$$

The figures below present the graphs of $P_{365}(n)$ and $\tilde{P}_{365}(n)$ and the graph of their difference. The graphs of $P_{365}(n)$ and its approximation $\tilde{P}_{365}(n)$ shown in the left panel practically coincide. The maximal difference between them in the interval $[0, 60]$ equals approximately 0.01 (at about $n = 30$).

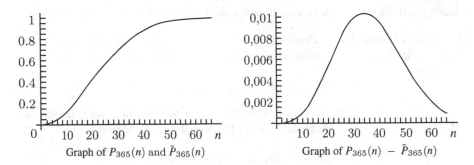

Graph of $P_{365}(n)$ and $\tilde{P}_{365}(n)$ Graph of $P_{365}(n) - \tilde{P}_{365}(n)$

It is interesting to note that (unexpectedly!) the size of class in which there is probability $\frac{1}{2}$ of finding at least two students with the same birthday is not very large: only 23.

Example 8 (*Prizes in a Lottery*). Consider a lottery that is run in the following way. There are M tickets numbered $1, 2, \ldots, M$, of which n, numbered $1, \ldots, n$, win prizes ($M \geq 2n$). You buy n tickets, and ask for the probability (P, say) of winning at least one prize.

Since the order in which the tickets are drawn plays no role in the presence or absence of winners in your purchase, we may suppose that the sample space has the form

$$\Omega = \{\omega: \omega = [a_1, \ldots, a_n], \ a_k \neq a_l, \ k \neq l, \ a_i = 1, \ldots, M\}.$$

By Table 1.1, $N(\Omega) = C_M^n$. Now let

$$A_0 = \{\omega: \omega = [a_1, \ldots, a_n], \ a_k \neq a_l, \ k \neq l, \ a_i = n+1, \ldots, M\}$$

be the event that there is *no winner* in the set of tickets you bought. Again by Table 1.1, $N(A_0) = C_{M-n}^n$. Therefore

$$P(A_0) = \frac{C_{M-n}^n}{C_M^n} = \frac{(M-n)_n}{(M)_n}$$

$$= \left(1 - \frac{n}{M}\right)\left(1 - \frac{n}{M-1}\right) \cdots \left(1 - \frac{n}{M-n+1}\right)$$

and consequently

$$P = 1 - P(A_0) = 1 - \left(1 - \frac{n}{M}\right)\left(1 - \frac{n}{M-1}\right) \cdots \left(1 - \frac{n}{M-n+1}\right).$$

If $M = n^2$ and $n \to \infty$, then $P(A_0) \to e^{-1}$ and

$$P \to 1 - e^{-1} \approx 0.632.$$

The convergence is quite fast: for $n = 10$ the probability is already $P = 0.670$.

6. PROBLEMS

1. Establish the following properties of the operators $\cap$ and $\cup$:

$$A \cup B = B \cup A, \quad AB = BA \quad \text{(commutativity)},$$
$$A \cup (B \cup C) = (A \cup B) \cup C, \quad A(BC) = (AB)C \quad \text{(associativity)},$$
$$A(B \cup C) = AB \cup AC, \quad A \cup (BC) = (A \cup B)(A \cup C) \quad \text{(distributivity)},$$
$$A \cup A = A, \quad AA = A \quad \text{(idempotency)}.$$

Show also the following *De Morgan's laws*:

$$\overline{A \cup B} = \overline{A} \cap \overline{B}, \quad \overline{AB} = \overline{A} \cup \overline{B}.$$

2. Let Ω contain N elements. Show that *Bell's number* B_N of different decompositions of Ω is given by the formula

$$B_N = e^{-1} \sum_{k=0}^{\infty} \frac{k^N}{k!}. \tag{12}$$

(Hint: Show that

$$B_N = \sum_{k=0}^{N-1} C_{N-1}^k B_k, \quad \text{where} \quad B_0 = 1,$$

and then verify that the series in (12) satisfy the same recurrence relations.)

3. For any finite collection of sets $A_1, \ldots, A_n$,

$$P(A_1 \cup \cdots \cup A_n) \leq P(A_1) + \cdots + P(A_n).$$

4. Let A and B be events. Show that $A\overline{B} \cup B\overline{A}$ is the event in which *exactly one* of A and B occurs. Moreover,

$$P(A\overline{B} \cup B\overline{A}) = P(A) + P(B) - 2P(AB).$$

5. Let $A_1, \ldots, A_n$ be events, and define $S_0, S_1, \ldots, S_n$ as follows: $S_0 - 1$,

$$S_r = \sum_{J_r} P(A_{k_1} \cap \cdots \cap A_{k_r}), \quad 1 \leq r \leq n,$$

where the sum is over the unordered subsets $J_r = [k_1, \ldots, k_r]$ of $\{1, \ldots, n\}$, $k_i \neq k_j, i \neq j$.

Let B_m be the event in which exactly m of the events $A_1, \ldots, A_n$ occur simultaneously. Show that

$$P(B_m) = \sum_{r=m}^{n} (-1)^{r-m} C_r^m S_r.$$

In particular, for $m = 0$

$$P(B_0) = 1 - S_1 + S_2 - \cdots \pm S_n.$$

Show also that the probability that *at least m* of the events $A_1, \ldots, A_n$ occur simultaneously is

$$P(B_m) + \cdots + P(B_n) = \sum_{r=m}^{n} (-1)^{r-m} C_{r-1}^{m-1} S_r.$$

In particular, the probability that at least one of the events $A_1, \ldots, A_n$ occurs is

$$P(B_1) + \cdots + P(B_n) = S_1 - S_2 + \cdots \mp S_n.$$

Prove the following properties:

(a) *Bonferroni's inequalities*: for any $k = 1, 2, \ldots$ such that $2k \leq n$,

$$S_1 - S_2 + \cdots - S_{2k} \leq P\left(\bigcup_{i=1}^{n} A_i\right) \leq S_1 - S_2 + \cdots + S_{2k-1};$$

(b) *Gumbel's inequalities*:

$$P\left(\bigcup_{r=1}^{n} A_i\right) \leq \frac{\tilde{S}_m}{C_{n-1}^{m-1}}, \quad m = 1, \ldots, n,$$

where

$$\tilde{S}_m = \sum_{1 \leq i_1 < \ldots < i_m \leq n} P(A_{i_1} \cup \ldots \cup A_{i_m});$$

(c) *Frechét's inequalities*:

$$P\left(\bigcup_{i=1}^{n} A_i\right) \leq \frac{\tilde{S}_{m+1}}{C_{n-1}^{m}} \leq \frac{\tilde{S}_m}{C_{n-1}^{m-1}}, \quad m = 1, \ldots, n-1.$$

6. Show that $P(A \cap B \cap C) \geq P(A) + P(B) + P(C) - 2$ and, by induction,

$$P\left(\bigcap_{i=1}^{n} A_i\right) \geq \sum_{i=1}^{n} P(A_i) - (n-1).$$

7. Explore the asymptotic behavior of the probabilities $P_M(n)$ in Example 7 under various assumptions about n and M (for example: $n = xM$, $M \to \infty$, or $n = x\sqrt{M}$, $M \to \infty$, where x is a fixed number). Compare the results with the local limit theorem in Sect. 6.

2 Some Classical Models and Distributions

1. Binomial distribution. Let a coin be tossed n times and record the results as an ordered set $(a_1, \ldots, a_n)$, where $a_i = 1$ for a head ("success") and $a_i = 0$ for a tail ("failure"). The sample space is

$$\Omega = \{\omega \colon \omega = (a_1, \ldots, a_n), \ a_i = 0, 1\}.$$

To each sample point $\omega = (a_1, \ldots, a_n)$ we assign the probability ("weight")

$$p(\omega) = p^{\Sigma a_i} q^{n - \Sigma a_i},$$

where the nonnegative numbers p and q satisfy $p + q = 1$. In the first place, we verify that this assignment of the weights $p(\omega)$ is consistent. It is enough to show that $\sum_{\omega \in \Omega} p(\omega) = 1$.

Consider all outcomes $\omega = (a_1, \ldots, a_n)$ for which $\sum_i a_i = k$, where $k = 0, 1, \ldots, n$. According to Table 1.4 (allocation of k indistinguishable objects over n places) the number of these outcomes is C_n^k. Therefore

$$\sum_{\omega \in \Omega} p(\omega) = \sum_{k=0}^{n} C_n^k p^k q^{n-k} = (p + q)^n = 1.$$

Thus the space Ω together with the collection $\mathscr{A}$ of all its subsets and the probabilities $P(A) = \sum_{\omega \in A} p(\omega)$, $A \in \mathscr{A}$ (in particular, $P(\{\omega\}) = p(\omega)$, $\omega \in \Omega$), defines a discrete probabilistic model. It is natural to call this the *probabilistic model for n tosses of a coin*. This model is also called the *Bernoulli scheme*.

In the case $n = 1$, when the sample space contains just the two points $\omega = 1$ ("success") and $\omega = 0$ ("failure"), it is natural to call $p(1) = p$ the probability of success. We shall see later that this model for n tosses of a coin can be thought of as the result of n "independent" experiments with probability p of success at each trial.

Let us consider the events

$$A_k = \{\omega : \omega = (a_1, \ldots, a_n), \ a_1 + \cdots + a_n = k\}, \quad k = 0, 1, \ldots, n,$$

containing exactly k successes. It follows from what we said above that

$$P(A_k) = C_n^k p^k q^{n-k}, \tag{1}$$

and $\sum_{k=0}^{n} P(A_k) = 1$.

The set of probabilities $(P(A_0), \ldots, P(A_n))$ is called the *binomial distribution* (the probability distribution of the number of successes in a sample of size n). This distribution plays an extremely important role in probability theory since it arises in the most diverse probabilistic models. We write $P_n(k) = P(A_k)$, $k = 0, 1, \ldots, n$. Figure 1 shows the binomial distribution in the case $p = \frac{1}{2}$ (symmetric coin) for $n = 5, 10, 20$.

We now present a different model (in essence, equivalent to the preceding one) which describes the random walk of a "particle."

Let the particle start at the origin, and after unit time let it take a unit step upward or downward (Fig. 2).

Consequently after n steps the particle can have moved at most n units up or n units down. It is clear that each path ω of the particle is completely specified by a vector $(a_1, \ldots, a_n)$, where $a_i = +1$ if the particle moves up at the ith step, and $a_i = -1$ if it moves down. Let us assign to each path ω the weight $p(\omega) = p^{\nu(\omega)} q^{n - \nu(\omega)}$, where $\nu(\omega)$ is the number of $+1$'s in the sequence $\omega = (a_1, \ldots, a_n)$,

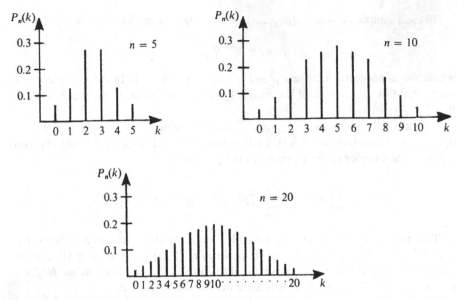

Fig. 1 Graph of the binomial probabilities $P_n(k)$ for $n = 5, 10, 20$

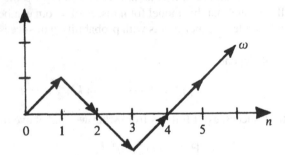

Fig. 2

i.e., $\nu(\omega) = [(a_1 + \cdots + a_n) + n]/2$, and the nonnegative numbers p and q satisfy $p + q = 1$.

Since $\sum_{\omega \in \Omega} p(\omega) = 1$, the set of probabilities $p(\omega)$ together with the space Ω of paths $\omega = (a_1, \ldots, a_n)$ and its subsets define an acceptable probabilistic model of the motion of the particle for n steps.

Let us ask the following question: What is the probability of the event A_k that after n steps the particle is at a point with ordinate k? This condition is satisfied by those paths ω for which $\nu(\omega) - (n - \nu(\omega)) = k$, i.e.,

$$\nu(\omega) = \frac{n+k}{2}, \quad k = -n, -n+2, \ldots, n.$$

The number of such paths (see Table 1.4) is $C_n^{(n+k)/2}$, and therefore

$$P(A_k) = C_n^{(n+k)/2} p^{(n+k)/2} q^{(n-k)/2}.$$

Consequently the binomial distribution $(P(A_{-n}), \ldots, P(A_0), \ldots, P(A_n))$ can be said to describe the probability distribution of the position of the particle after n steps.

Note that in the symmetric case $(p = q = \frac{1}{2})$ when the probabilities of the individual paths are equal to 2^{-n},

$$P(A_k) = C_n^{(n+k)/2} \cdot 2^{-n}.$$

Let us investigate the asymptotic behavior of these probabilities for large n.

If the number of steps is $2n$, it follows from the properties of the binomial coefficients that the largest of the probabilities $P(A_k)$, $|k| \leq 2n$ is

$$P(A_0) = C_{2n}^n \cdot 2^{-2n}.$$

From Stirling's formula (see formula (6) below)

$$n! \sim \sqrt{2\pi n} \, e^{-n} n^n.*$$

Consequently

$$C_{2n}^n = \frac{(2n)!}{(n!)^2} \sim 2^{2n} \cdot \frac{1}{\sqrt{\pi n}}$$

and therefore for large n

$$P(A_0) \sim \frac{1}{\sqrt{\pi n}}.$$

Figure 3 represents the beginning of the binomial distribution for $2n$ steps of a random walk (in contrast to Fig. 2, the time axis is now directed upward).

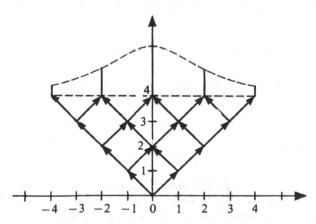

Fig. 3 Beginning of the binomial distribution

---

* The notation $f(n) \sim g(n)$ means that $f(n)/g(n) \to 1$ as $n \to \infty$.

2. Multinomial distribution. Generalizing the preceding model, we now suppose that the sample space is

$$\Omega = \{\omega \colon \omega = (a_1, \ldots, a_n), \ a_i = b_1, \ldots, b_r\},$$

where $b_1, \ldots, b_r$ are given numbers. Let $\nu_i(\omega)$ be the number of elements of $\omega = (a_1, \ldots, a_n)$ that are equal to b_i, $i = 1, \ldots, r$, and define the probability of ω by

$$p(\omega) = p_1^{\nu_1(\omega)} \cdots p_r^{\nu_r(\omega)},$$

where $p_i \geq 0$ and $p_1 + \cdots + p_r = 1$. Note that

$$\sum_{\omega \in \Omega} p(\omega) = \sum_{\left\{ \substack{n_1 \geq 0, \cdots, n_r \geq 0, \\ n_1 + \cdots + n_r = n} \right\}} C_n(n_1, \ldots, n_r) p_1^{n_1} \cdots p_r^{n_r},$$

where $C_n(n_1, \ldots, n_r)$ is the number of (ordered) sequences $(a_1, \ldots, a_n)$ in which b_1 occurs n_1 times, $\ldots$, b_r occurs n_r times. Since n_1 elements b_1 can be allocated among n positions in $C_n^{n_1}$ ways; n_2 elements b_2 among $n - n_1$ positions in $C_{n-n_1}^{n_2}$ ways, etc., we have

$$\begin{aligned}
C_n(n_1, \ldots, n_r) &= C_n^{n_1} \cdot C_{n-n_1}^{n_2} \cdots C_{n-(n_1+\cdots+n_{r-1})}^{n_r} \\
&= \frac{n!}{n_1!(n-n_1)!} \cdot \frac{(n-n_1)!}{n_2!(n-n_1-n_2)!} \cdots 1 \\
&= \frac{n!}{n_1! \cdots n_r!}.
\end{aligned}$$

Therefore

$$\sum_{\omega \in \Omega} P(\omega) = \sum_{\left\{ \substack{n_1 \geq 0, \ldots, n_r \geq 0. \\ n_1 + \cdots + n_r = n} \right\}} \frac{n_1!}{n! \cdots n_r!} p_1^{n_1} \cdots p_r^{n_r} = (p_1 + \cdots + p_r)^n = 1,$$

and consequently we have defined an acceptable method of assigning probabilities.
    Let

$$A_{n_1, \ldots, n_r} = \{\omega \colon \nu_1(\omega) = n_1, \ldots, \nu_r(\omega) = n_r\}.$$

Then

$$P(A_{n_1, \ldots, n_r}) = C_n(n_1, \ldots, n_r) \, p_1^{n_1} \cdots p_r^{n_r}. \tag{2}$$

The set of probabilities

$$\{P(A_{n_1, \ldots, n_r})\}$$

is called the *multinomial* (or polynomial) distribution.
    We emphasize that both this distribution and its special case, the binomial distribution, originate from problems about sampling *with replacement*.

3. The multidimensional hypergeometric distribution occurs in problems that involve sampling *without replacement*.

Consider, for example, an urn containing M balls numbered $1, 2, \ldots, M$, where M_1 balls have the color $b_1, \ldots, M_r$ balls have the color b_r, and $M_1 + \cdots + M_r = M$. Suppose that we draw a sample of size $n < M$ without replacement. The sample space is

$$\Omega = \{\omega : \omega = (a_1, \ldots, a_n),\ a_k \neq a_l,\ k \neq l,\ a_i = 1, \ldots, M\}$$

and $N(\Omega) = (M)_n$. Let us suppose that the sample points are equiprobable, and find the probability of the event $B_{n_1, \ldots, n_r}$ in which n_1 balls have color $b_1, \ldots, n_r$ balls have color b_r, where $n_1 + \cdots + n_r = n$. It is easy to show that

$$N(B_{n_1, \ldots, n_r}) = C_n(n_1, \ldots, n_r)(M_1)_{n_1} \cdots (M_r)_{n_r},$$

and therefore

$$P(B_{n_1, \ldots, n_r}) = \frac{N(B_{n_1, \ldots, n_r})}{N(\Omega)} = \frac{C_{M_1}^{n_1} \cdots C_{M_r}^{n_r}}{C_M^n}. \tag{3}$$

The set of probabilities $\{P(B_{n_1, \ldots, n_r})\}$ is called the *multidimensional hypergeometric distribution*. When $r = 2$ it is simply called the *hypergeometric* distribution because its "generating function" is a hypergeometric function.

The structure of the multidimensional hypergeometric distribution is rather complicated. For example, the probability

$$P(B_{n_1, n_2}) = \frac{C_{M_1}^{n_1} C_{M_2}^{n_2}}{C_M^n}, \quad n_1 + n_2 = n,\ M_1 + M_2 = M, \tag{4}$$

contains nine factorials. However, it is easily established that if $M, M_1 \to \infty$ in such a way that $M_1/M \to p$ (and therefore $M_2/M \to 1 - p$) then

$$P(B_{n_1, n_2}) \to C_{n_1 + n_2}^{n_1} p^{n_1} (1 - p)^{n_2}. \tag{5}$$

In other words, under the present hypotheses the hypergeometric distribution is approximated by the binomial; this is intuitively clear since when M and M_1 are large (but finite), sampling without replacement ought to give almost the same result as sampling with replacement.

Example. Let us use (4) to find the probability of picking six "lucky" numbers in a lottery of the following kind (this is an abstract formulation of the "Sportloto," which was well known in Russia in 1970s–80s):

There are 49 balls numbered from 1 to 49; six of them are lucky (colored red, say, whereas the rest are white). We draw a sample of six balls, without replacement. The question is, What is the probability that all six of these balls are lucky? Taking $M = 49$, $M_1 = 6$, $n_1 = 6$, $n_2 = 0$, we see that the event of interest, namely

$$B_{6,0} = \{6 \text{ balls, all lucky}\}$$

has, by (4), probability

$$P(B_{6,0}) = \frac{1}{C_{49}^6} \approx 7.2 \times 10^{-8}.$$

4. The numbers $n!$ increase extremely rapidly with n. For example,

$$10! = 3,628,800,$$
$$15! = 1,307,674,368,000,$$

and $100!$ has 158 digits. Hence from either the theoretical or the computational point of view, it is important to know *Stirling's formula,*

$$n! = \sqrt{2\pi n} \left(\frac{n}{e}\right)^n \exp\left(\frac{\theta_n}{12n}\right), \quad 0 < \theta_n < 1, \tag{6}$$

whose proof can be found in most textbooks on mathematical analysis.

5. PROBLEMS

1. Prove formula (5).
2. Show that for the multinomial distribution $\{P(A_{n_1,\ldots,n_r})\}$ the maximum probability is attained at a point $(k_1,\ldots,k_r)$ that satisfies the inequalities $np_i - 1 < k_i \leq (n + r - 1)p_i$, $i = 1,\ldots,r$.
3. *One-dimensional Ising model.* Consider n particles located at the points 1, 2, $\ldots$, n. Suppose that each particle is of one of two types, and that there are n_1 particles of the first type and n_2 of the second ($n_1 + n_2 = n$). We suppose that all $n!$ arrangements of the particles are equally probable.

   Construct a corresponding probabilistic model and find the probability of the event $A_n(m_{11}, m_{12}, m_{21}, m_{22}) = \{v_{11} = m_{11}, \ldots, v_{22} = m_{22}\}$, where v_{ij} is the number of particles of type i following particles of type j ($i, j = 1, 2$).
4. Prove the following equalities using probabilistic and combinatorial arguments:

$$\sum_{k=0}^{n} C_n^k = 2^n,$$

$$\sum_{k=0}^{n} (C_n^k)^2 = C_{2n}^n,$$

$$\sum_{k=0}^{n} (-1)^{n-k} C_m^k = C_{m-1}^n, \quad m \geq n + 1,$$

$$\sum_{k=0}^{m} k(k-1) C_m^k = m(m-1) 2^{m-2}, \quad m \geq 2,$$

$$k C_n^k = n C_{n-1}^{k-1},$$

$$C_n^m = \sum_{j=0}^{m} C_k^j C_{n-k}^{m-j},$$

where $0 \leq m \leq n$, $0 \leq k \leq n$ and we set $C_l^j = 0$ for $j < 0$ or $j > l$.

5. Suppose we want to estimate the size N of a population without total counting. Such a question may be of interest, for example, when we try to estimate the population of a country, or a town, etc.

In 1786 Laplace proposed the following method to estimate the number N of inhabitants of France.

Draw a sample of size M, say, from the population and mark its elements. Then return them into the initial population and assume that they become "well mixed" with unmarked elements. Then draw n elements from the "mixed" population. Suppose there are X marked elements among them.

Show that the corresponding probability $P_{N,M;n}\{X = m\}$ is given by the formula for the hypergeometric distribution (cf. (4)):

$$P_{N,M;n}\{X = m\} = \frac{C_M^n C_{N-M}^{n-m}}{C_N^n}.$$

For fixed M, n and m find N maximizing this probability, i.e., find the "most likely" size of the whole population (for fixed M and n) given that the number X of marked elements in the repeated sample is equal to m.

Show that the "most likely" value (to be denoted by $\hat{N}$) is given by the formula (with $[\cdot]$ denoting the integral part):

$$\hat{N} = [Mnm^{-1}].$$

The estimator $\hat{N}$ for N obtained in this way is called the *maximum likelihood estimator*.

(This problem is continued in Sect. 7 (Problem 4).)

6. (Compare with Problem 2 in Sect. 1.) Let Ω contain N elements and let $\tilde{d}(N)$ be the number of different decompositions of Ω with the property that each subset of the decomposition has odd number of elements. Show that

$$\tilde{d}(1) = 1, \quad \tilde{d}(2) = 1, \quad \tilde{d}(3) = 2,$$
$$\tilde{d}(4) = 5, \quad \tilde{d}(5) = 12, \quad \tilde{d}(6) = 37$$

and, in general,

$$\sum_{n=1}^{\infty} \frac{\tilde{d}(n)x^n}{n!} = e^{\sinh x} - 1, \quad |x| < 1.$$

3 Conditional Probability: Independence

1. The concept of probabilities of events lets us answer questions of the following kind: If there are M balls in an urn, M_1 white and M_2 black, what is the probability $P(A)$ of the event A that a selected ball is white? With the classical approach, $P(A) = M_1/M$.

The concept of *conditional probability*, which will be introduced below, lets us answer questions of the following kind: What is the probability that the second ball is white (event B) under the condition that the first ball was also white (event A)? (We are thinking of sampling without replacement.)

It is natural to reason as follows: if the first ball is white, then at the second step we have an urn containing $M - 1$ balls, of which $M_1 - 1$ are white and M_2 black; hence it seems reasonable to suppose that the (conditional) probability in question is $(M_1 - 1)/(M - 1)$.

We now give a definition of conditional probability that is consistent with our intuitive ideas.

Let $(\Omega, \mathscr{A}, \mathsf{P})$ be a (discrete) probability space and A an event (i.e. $A \in \mathscr{A}$).

Definition 1. The *conditional probability* of event B given that event A, $\mathsf{P}(A) > 0$, occurred (denoted by $\mathsf{P}(B\,|\,A)$) is

$$\frac{\mathsf{P}(AB)}{\mathsf{P}(A)}. \tag{1}$$

In the classical approach we have $\mathsf{P}(A) = N(A)/N(\Omega)$, $\mathsf{P}(AB) = N(AB)/N(\Omega)$, and therefore

$$\mathsf{P}(B\,|\,A) = \frac{N(AB)}{N(A)}. \tag{2}$$

From Definition 1 we immediately get the following properties of conditional probability:

$$\mathsf{P}(A\,|\,A) = 1,$$
$$\mathsf{P}(\varnothing\,|\,A) = 0,$$
$$\mathsf{P}(B\,|\,A) = 1, \quad B \supseteq A,$$
$$\mathsf{P}(B_1 + B_2\,|\,A) = \mathsf{P}(B_1\,|\,A) + \mathsf{P}(B_2\,|\,A).$$

It follows from these properties that for a given set A the conditional probability $\mathsf{P}(\cdot\,|\,A)$ has the same properties on the space $(\Omega \cap A, \mathscr{A} \cap A)$, where $\mathscr{A} \cap A = \{B \cap A \colon B \in \mathscr{A}\}$, that the original probability $\mathsf{P}(\cdot)$ has on $(\Omega, \mathscr{A})$.

Note that

$$\mathsf{P}(B\,|\,A) + \mathsf{P}(\overline{B}\,|\,A) = 1;$$

however in general

$$\mathsf{P}(B\,|\,A) + \mathsf{P}(B\,|\,\overline{A}) \neq 1,$$
$$\mathsf{P}(B\,|\,A) + \mathsf{P}(\overline{B}\,|\,\overline{A}) \neq 1.$$

Example 1. Consider a family with two children. We ask for the probability that both children are boys, assuming

(a) that the older child is a boy;
(b) that at least one of the children is a boy.

The sample space is

$$\Omega = \{BB, BG, GB, GG\},$$

where BG means that the older child is a boy and the younger is a girl, etc.
Let us suppose that all sample points are equally probable:

$$P(BB) = P(BG) = P(GB) = P(GG) = \tfrac{1}{4}.$$

Let A be the event that the older child is a boy, and B, that the younger child is a boy. Then $A \cup B$ is the event that at least one child is a boy, and AB is the event that both children are boys. In question (a) we want the conditional probability $P(AB \mid A)$, and in (b), the conditional probability $P(AB \mid A \cup B)$.

It is easy to see that

$$P(AB \mid A) = \frac{P(AB)}{P(A)} = \frac{\tfrac{1}{4}}{\tfrac{1}{2}} = \frac{1}{2},$$

$$P(AB \mid A \cup B) = \frac{P(AB)}{P(A \cup B)} = \frac{\tfrac{1}{4}}{\tfrac{3}{4}} = \frac{1}{3}.$$

2. The simple but important formula (3), below, is called the *formula for total probability*. It provides the basic means for calculating the probabilities of complicated events by using conditional probabilities.

Consider a decomposition $\mathscr{D} = \{A_1, \ldots, A_n\}$ with $P(A_i) > 0$, $i = 1, \ldots, n$ (such a decomposition is often called a *complete set of disjoint events*). It is clear that

$$B = BA_1 + \cdots + BA_n$$

and therefore

$$P(B) = \sum_{i=1}^{n} P(BA_i).$$

But

$$P(BA_i) = P(B \mid A_i) P(A_i).$$

Hence we have the *formula for total probability*:

$$P(B) = \sum_{i=1}^{n} P(B \mid A_i) P(A_i). \tag{3}$$

In particular, if $0 < P(A) < 1$, then

$$P(B) = P(B \mid A) P(A) + P(B \mid \overline{A}) P(\overline{A}). \tag{4}$$

Example 2. An urn contains M balls, m of which are "lucky." We ask for the probability that the second ball drawn is lucky (assuming that the result of the first draw is unknown, that a sample of size 2 is drawn without replacement, and that all outcomes are equally probable). Let A be the event that the first ball is lucky, and B the event that the second is lucky. Then

$$P(B \mid A) = \frac{P(BA)}{P(A)} = \frac{\frac{m(m-1)}{M(M-1)}}{\frac{m}{M}} = \frac{m-1}{M-1},$$

$$P(B \mid \bar{A}) = \frac{P(B\bar{A})}{P(\bar{A})} = \frac{\frac{m(M-m)}{M(M-1)}}{\frac{M-m}{M}} = \frac{m}{M-1}$$

and

$$P(B) = P(B \mid A) P(A) + P(B \mid \bar{A}) P(\bar{A})$$
$$= \frac{m-1}{M-1} \cdot \frac{m}{M} + \frac{m}{M-1} \cdot \frac{M-m}{M} = \frac{m}{M}.$$

It is interesting to observe that $P(A)$ is precisely m/M. Hence, when the nature of the first ball is unknown, it does not affect the probability that the second ball is lucky.

By the definition of conditional probability (with $P(A) > 0$),

$$P(AB) = P(B \mid A) P(A). \tag{5}$$

This formula, the *multiplication formula for probabilities*, can be generalized (by induction) as follows: If $A_1, \ldots, A_{n-1}$ are events with $P(A_1 \cdots A_{n-1}) > 0$, then

$$P(A_1 \cdots A_n) = P(A_1) P(A_2 \mid A_1) \cdots P(A_n \mid A_1 \cdots A_{n-1}) \tag{6}$$

(here $A_1 \cdots A_n = A_1 \cap A_2 \cap \cdots \cap A_n$).

3. Suppose that A and B are events with $P(A) > 0$ and $P(B) > 0$. Then along with (5) we have the parallel formula

$$P(AB) = P(A \mid B) P(B). \tag{7}$$

From (5) and (7) we obtain *Bayes's formula*

$$P(A \mid B) = \frac{P(A) P(B \mid A)}{P(B)}. \tag{8}$$

If the events $A_1, \ldots, A_n$ form a decomposition of Ω, (3) and (8) imply *Bayes's theorem*:

$$P(A_i \mid B) = \frac{P(A_i) P(B \mid A_i)}{\sum_{j=1}^{n} P(A_j) P(B \mid A_j)}. \tag{9}$$

In statistical applications, $A_1, \ldots, A_n$ $(A_1 + \cdots + A_n = \Omega)$ are often called hypotheses, and $P(A_i)$ is called the *prior* (or *a priori*)* probability of A_i. The condi-

---

* *Apriori*: before the experiment; *aposteriori*: after the experiment.

tional probability $P(A_i \mid B)$ is considered as the *posterior* (or the *a posteriori*) probability of A_i after the occurrence of event B.

Example 3. Let an urn contain two coins: A_1, a fair coin with probability $\frac{1}{2}$ of falling H; and A_2, a biased coin with probability $\frac{1}{3}$ of falling H. A coin is drawn at random and tossed. Suppose that it falls head. We ask for the probability that the fair coin was selected.

Let us construct the corresponding probabilistic model. Here it is natural to take the sample space to be the set $\Omega = \{A_1H, A_1T, A_2H, A_2T\}$, which describes all possible outcomes of a selection and a toss (A_1H means that coin A_1 was selected and fell heads, etc.) The probabilities $p(\omega)$ of the various outcomes have to be assigned so that, according to the statement of the problem,

$$P(A_1) = P(A_2) = \tfrac{1}{2}$$

and

$$P(H \mid A_1) = \tfrac{1}{2}, \qquad P(H \mid A_2) = \tfrac{1}{3}.$$

With these assignments, the probabilities of the sample points are uniquely determined:

$$P(A_1H) = \tfrac{1}{4}, \quad P(A_1T) = \tfrac{1}{4}, \quad P(A_2H) = \tfrac{1}{6}, \quad P(A_2T) = \tfrac{1}{3}.$$

Then by *Bayes's formula* the probability in question is

$$P(A_1 \mid H) = \frac{P(A_1)\,P(H \mid A_1)}{P(A_1)\,P(H \mid A_1) + P(A_2)\,P(H \mid A_2)} = \frac{3}{5},$$

and therefore

$$P(A_2 \mid H) - \tfrac{2}{5}.$$

4. In certain sense, the concept of *independence*, which we are now going to introduce, plays a central role in probability theory: it is precisely this concept that distinguishes probability theory from the general theory of measure spaces.

If A and B are two events, it is natural to say that B is *independent* of A if knowing that A has occurred has no effect on the probability of B. In other words, "B is independent of A" if

$$P(B \mid A) = P(B) \tag{10}$$

(we are supposing that $P(A) > 0$).

Since

$$P(B \mid A) = \frac{P(AB)}{P(A)},$$

it follows from (10) that

$$P(AB) = P(A)\,P(B). \tag{11}$$

In exactly the same way, if $P(B) > 0$ it is natural to say that "A is independent of B" if

$$P(A \mid B) = P(A).$$

Hence we again obtain (11), which is symmetric in A and B and still makes sense when the probabilities of these events are zero.

After these preliminaries, we introduce the following definition.

Definition 2. Events A and B are called *independent* or *statistically independent* (with respect to the probability P) if

$$P(AB) = P(A)\,P(B).$$

In probability theory we often need to consider not only independence of events (or sets) but also independence of *collections* of events (or sets).

Accordingly, we introduce the following definition.

Definition 3. Two algebras $\mathscr{A}_1$ and $\mathscr{A}_2$ of events (or sets) are called *independent* or *statistically independent* (with respect to the probability P) if all pairs of sets A_1 and A_2, belonging respectively to $\mathscr{A}_1$ and $\mathscr{A}_2$, are independent.

For example, let us consider the two algebras

$$\mathscr{A}_1 = \{A_1, \overline{A}_1, \varnothing, \Omega\} \quad \text{and} \quad \mathscr{A}_2 = \{A_2, \overline{A}_2, \varnothing, \Omega\},$$

where A_1 and A_2 are subsets of Ω. It is easy to verify that $\mathscr{A}_1$ and $\mathscr{A}_2$ are independent if and only if A_1 and A_2 are independent. In fact, the independence of $\mathscr{A}_1$ and $\mathscr{A}_2$ means the independence of the 16 pairs of events A_1 and A_2, A_1 and $\overline{A}_2, \ldots, \Omega$ and Ω. Consequently A_1 and A_2 are independent. Conversely, if A_1 and A_2 are independent, we have to show that the other 15 pairs of events are independent. Let us verify, for example, the independence of A_1 and $\overline{A}_2$. We have

$$P(A_1\overline{A}_2) = P(A_1) - P(A_1 A_2) = P(A_1) - P(A_1)\,P(A_2)$$
$$= P(A_1) \cdot (1 - P(A_2)) = P(A_1)\,P(\overline{A}_2).$$

The independence of the other pairs is verified similarly.

5. The concept of independence of two sets (events) or two algebras of sets can be extended to any finite number of sets or algebras of sets.

Definition 4. We say that the *sets (events)* $A_1, \ldots, A_n$ are mutually *independent* or *statistically independent* (with respect to the probability P) if for any $k = 1, \ldots, n$ and $1 \le i_1 < i_2 < \cdots < i_k \le n$

$$P(A_{i_1} \ldots A_{i_k}) = P(A_{i_1}) \ldots P(A_{i_k}). \tag{12}$$

Definition 5. The *algebras* $\mathscr{A}_1, \ldots, \mathscr{A}_n$ of sets (events) are called mutually *independent* or *statistically independent* (with respect to the probability P) if any sets $A_1, \ldots, A_n$ belonging respectively to $\mathscr{A}_1, \ldots, \mathscr{A}_n$ are independent.

Note that *pairwise independence* of events *does not imply* their independence. In fact if, for example, $\Omega = \{\omega_1, \omega_2, \omega_3, \omega_4\}$ and all outcomes are equiprobable, it is easily verified that the events

$$A = \{\omega_1, \omega_2\}, \quad B = \{\omega_1, \omega_3\}, \quad C = \{\omega_1, \omega_4\}$$

are pairwise independent, whereas

$$P(ABC) = \tfrac{1}{4} \neq (\tfrac{1}{2})^3 = P(A)\,P(B)\,P(C).$$

Also note that if

$$P(ABC) = P(A)\,P(B)\,P(C)$$

for events A, B and C, it by no means follows that these events are pairwise independent. In fact, let Ω consist of the 36 ordered pairs (i, j), where $i, j = 1, 2, \ldots, 6$ and all the pairs are equiprobable. Then if $A = \{(i, j) : j = 1,\ 2\ \text{or}\ 5\}$, $B = \{(i, j) : j = 4,\ 5\ \text{or}\ 6\}$, $C = \{(i, j) : i + j = 9\}$ we have

$$P(AB) = \tfrac{1}{6} \neq \tfrac{1}{4} = P(A)\,P(B),$$
$$P(AC) = \tfrac{1}{36} \neq \tfrac{1}{18} = P(A)\,P(C),$$
$$P(BC) = \tfrac{1}{12} \neq \tfrac{1}{18} = P(B)\,P(C),$$

but also

$$P(ABC) = \tfrac{1}{36} = P(A)\,P(B)\,P(C).$$

6. Let us consider in more detail, from the point of view of *independence*, the classical discrete model $(\Omega, \mathscr{A}, P)$ that was introduced in Sect. 2 and used as a basis for the binomial distribution.

In this model

$$\Omega = \{\omega : \omega = (a_1, \ldots, a_n),\ a_i = 0,\ 1\}, \quad \mathscr{A} = \{A : A \subseteq \Omega\}$$

and

$$p(\omega) = p^{\Sigma a_i} q^{n - \Sigma a_i}. \tag{13}$$

Consider an event $A \subseteq \Omega$. We say that this event depends on a trial at time k if it is determined by the value a_k alone. Examples of such events are

$$A_k = \{\omega : a_k = 1\}, \quad \overline{A}_k = \{\omega : a_k = 0\}.$$

Let us consider the sequence of algebras $\mathscr{A}_1, \mathscr{A}_2, \ldots, \mathscr{A}_n$, where $\mathscr{A}_k = \{A_k, \overline{A}_k, \varnothing, \Omega\}$ and show that under (13) these algebras are independent.

It is clear that

$$P(A_k) = \sum_{\{\omega\,:\,a_k=1\}} p(\omega) = \sum_{\{\omega\,:\,a_k=1\}} p^{\sum a_i} q^{n-\sum a_i}$$

$$= p \sum_{(a_1,\ldots,a_{k-1},a_{k+1},\ldots,a_n)} p^{a_1+\cdots+a_{k-1}+a_{k+1}+\cdots+a_n}$$

$$\times\, q^{(n-1)-(a_1+\cdots+a_{k-1}+a_{k+1}+\cdots+a_n)} = p \sum_{l=0}^{n-1} C_{n-1}^{l} p^l q^{(n-1)-l} = p,$$

and a similar calculation shows that $P(\overline{A}_k) = q$ and that, for $k \neq l$,

$$P(A_k A_l) = p^2, \quad P(A_k \overline{A}_l) = pq, \quad P(\overline{A}_k A_l) = pq, \quad P(\overline{A}_k \overline{A}_l) = q^2.$$

It is easy to deduce from this that $\mathscr{A}_k$ and $\mathscr{A}_l$ are independent for $k \neq l$.

It can be shown in the same way that $\mathscr{A}_1, \mathscr{A}_2, \ldots, \mathscr{A}_n$ are independent. This is the basis for saying that our model $(\Omega, \mathscr{A}, \mathsf{P})$ corresponds to "n independent trials with two outcomes and probability p of success." James Bernoulli was the first to study this model systematically, and established the law of large numbers (Sect. 5) for it. Accordingly, this model is also called the Bernoulli scheme with two outcomes (success and failure) and probability p of success.

A detailed study of the probability space for the Bernoulli scheme shows that it has the structure of a direct product of probability spaces, defined as follows.

Suppose that we are given a collection $(\Omega_1, \mathscr{B}_1, \mathsf{P}_1), \ldots, (\Omega_n, \mathscr{B}_n, \mathsf{P}_n)$ of discrete probability spaces. Form the space $\Omega = \Omega_1 \times \Omega_2 \times \cdots \times \Omega_n$ of points $\omega = (a_1, \ldots, a_n)$, where $a_i \in \Omega_i$. Let $\mathscr{A} = \mathscr{B}_1 \otimes \cdots \otimes \mathscr{B}_n$ be the algebra of the subsets of Ω that consists of sums of sets of the form

$$A = B_1 \times B_2 \times \cdots \times B_n$$

with $B_i \in \mathscr{B}_i$. Finally, for $\omega = (a_1, \ldots, a_n)$ take $p(\omega) = p_1(a_1) \cdots p_n(a_n)$ and define $\mathsf{P}(A)$ for the set $A = B_1 \times B_2 \times \cdots \times B_n$ by

$$\mathsf{P}(A) = \sum_{\{a_1 \in B_1,\ldots,a_n \in B_n\}} p_1(a_1) \ldots p_n(a_n).$$

It is easy to verify that $\mathsf{P}(\Omega) = 1$ and therefore the triple $(\Omega, \mathscr{A}, \mathsf{P})$ defines a probability space. This space is called the *direct product of the probability spaces* $(\Omega_1, \mathscr{B}_1, \mathsf{P}_1), \ldots, (\Omega_n, \mathscr{B}_n, \mathsf{P}_n)$.

We note an easily verified property of the direct product of probability spaces: with respect to P, the events

$$A_1 = \{\omega\,:\, a_1 \in B_1\}, \ldots, A_n = \{\omega\,:\, a_n \in B_n\},$$

where $B_i \in \mathscr{B}_i$, are independent. In the same way, the algebras of subsets of Ω,

$$\mathscr{A}_1 = \{A_1 : A_1 = \{\omega : a_1 \in B_1\}, \ B_1 \in \mathscr{B}_1\},$$

$$\cdots\cdots\cdots\cdots\cdots\cdots\cdots\cdots\cdots\cdots\cdots\cdots\cdots$$

$$\mathscr{A}_n = \{A_n : A_n = \{\omega : a_n \in B_n\}, \ B_n \in \mathscr{B}_n\}$$

are independent.

It is clear from our construction that the Bernoulli scheme

$$(\Omega, \mathscr{A}, P) \text{ with } \Omega = \{\omega : \omega = (a_1, \ldots, a_n), \ a_i = 0 \text{ or } 1\},$$
$$\mathscr{A} = \{A : A \subseteq \Omega\} \text{ and } p(\omega) = p^{\sum a_i} q^{n - \sum a_i}$$

can be thought of as the direct product of the probability spaces $(\Omega_i, \mathscr{B}_i, P_i)$, $i = 1$, $2, \ldots, n$, where

$$\Omega_i = \{0, 1\}, \quad \mathscr{B}_i = \{\{0\}, \{1\}, \varnothing, \Omega_i\},$$
$$P_i(\{1\}) = p, \qquad P_i(\{0\}) = q.$$

7. Problems

1. Give examples to show that in general the equations

$$P(B \mid A) + P(B \mid \overline{A}) = 1,$$
$$P(B \mid A) + P(\overline{B} \mid \overline{A}) = 1$$

are false.

2. An urn contains M balls, of which M_1 are white. Consider a sample of size n. Let B_j be the event that the ball selected at the jth step is white, and A_k the event that a sample of size n contains exactly k white balls. Show that

$$P(B_j \mid A_k) = k/n$$

both for sampling with replacement and for sampling without replacement.

3. Let $A_1, \ldots, A_n$ be independent events. Then

$$P\left(\bigcup_{i=1}^{n} A_i\right) = 1 - \prod_{i=1}^{n} P(\overline{A}_i).$$

4. Let $A_1, \ldots, A_n$ be independent events with $P(A_i) = p_i$. Then the probability P_0 that neither event occurs is

$$P_0 = \prod_{i=1}^{n} (1 - p_i).$$

5. Let A and B be independent events. In terms of $P(A)$ and $P(B)$, find the probabilities of the events that exactly k, at least k, and at most k of A and B occur $(k = 0, 1, 2)$.

6. Let event A be independent of itself, i.e., let A and A be independent. Show that $P(A)$ is either 0 or 1.

7. Let event A have $P(A) = 0$ or 1. Show that A and an arbitrary event B are independent.

8. Consider the electric circuit shown in Fig. 4. Each of the switches A, B, C, D, and E is independently open or closed with probabilities p and q, respectively. Find the probability that a signal fed in at "input" will be received at "output." If the signal is received, what is the conditional probability that E is open?

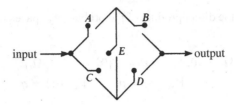

Fig. 4

9. Let $P(A + B) > 0$. Show that

$$P(A \mid A + B) = \frac{P(A)}{P(A) + P(B)}.$$

10. Let an event A be independent of events B_n, $n \geq 1$, such that $B_i \cap B_j = \varnothing$, $i \neq j$. Then A and $\bigcup_{n=1}^{\infty} B_n$ are independent.

11. Show that if $P(A \mid C) > P(B \mid C)$ and $P(A \mid \bar{C}) > P(B \mid \bar{C})$, then $P(A) > P(B)$.

12. Show that

$$P(A \mid B) = P(A \mid BC)\, P(C \mid B) + P(A \mid B\bar{C})\, P(\bar{C} \mid B).$$

13. Let X and Y be independent binomial random variables with parameters (n, p).* Show that

$$P(X = k \mid X + Y = m) = \frac{C_n^k C_n^{m-k}}{C_{2n}^m}, \quad k = 0, 1, \ldots, \min(m, n).$$

14. Let A, B, C be pairwise independent equiprobable events such that $A \cap B \cap C = \varnothing$. Find the largest possible value of the probability $P(A)$.

15. Into an urn containing one white ball another ball is added which is white or black with equal probabilities. Then one ball is drawn at random which occurred white. What is the conditional probability that the ball remaining in the urn is also white?

---

* See (2) in the next section. *Translator.*

4 Random Variables and Their Properties

1. Let $(\Omega, \mathscr{A}, \mathsf{P})$ be a discrete probabilistic model of an experiment with a *finite* number of outcomes, $N(\Omega) < \infty$, where $\mathscr{A}$ is the algebra of all subsets of Ω. We observe that in the examples above, where we calculated the probabilities of various events $A \in \mathscr{A}$, the specific nature of the sample space Ω was of no interest. We were interested only in numerical properties depending on the sample points. For example, we were interested in the probability of some number of successes in a series of n trials, in the probability distribution for the number of objects in cells, etc.

The concept "random variable," which we now introduce (later it will be given a more general form), serves to define quantities describing the results of "measurements" in random experiments.

Definition 1. Any numerical function $\xi = \xi(\omega)$ defined on a (finite) sample space Ω is called a (simple) *random variable*. (The reason for the term "simple" random variable will become clear after the introduction of the general concept of random variable in Sect. 4, Chap. 2).

Example 1. In the model of two tosses of a coin with sample space $\Omega = \{HH, HT, TH, TT\}$, define a random variable $\xi = \xi(\omega)$ by the table

ω	HH	HT	TH	TT
$\xi(\omega)$	2	1	1	0

Here, from its very definition, $\xi(\omega)$ is nothing but the number of heads in the outcome ω.

Another extremely simple example of a random variable is the *indicator* (or *characteristic function*) of a set $A \in \mathscr{A}$:

$$\xi = I_A(\omega),$$

where*

$$I_A(\omega) = \begin{cases} 1, & \omega \in A, \\ 0, & \omega \notin A. \end{cases}$$

When experimenters are concerned with random variables that describe observations, their main interest is in the probabilities with which the random variables take various values. From this point of view they are interested, not in the probability distribution P on $(\Omega, \mathscr{A})$, but in the probability distribution over the range of a random variable. Since we are considering the case when Ω contains only a finite number of points, the range X of the random variable ξ is also finite. Let $X = \{x_1, \ldots, x_m\}$, where the (different) numbers $x_1, \ldots, x_m$ exhaust the values of ξ.

---

* The notation $I(A)$ is also used. For frequently used properties of indicators see Problem 1.

Let $\mathcal{X}$ be the collection of all subsets of X, and let $B \in \mathcal{X}$. We can also interpret B as an event if the sample space is taken to be X, the set of values of ξ.

On $(X, \mathcal{X})$, consider the probability $P_\xi(\cdot)$ induced by ξ according to the formula

$$P_\xi(B) = \mathsf{P}\{\omega : \xi(\omega) \in B\}, \quad B \in \mathcal{X}.$$

It is clear that the values of this probability are completely determined by the probabilities

$$P_\xi(x_i) = \mathsf{P}\{\omega : \xi(\omega) = x_i\}, \quad x_i \in X.$$

The set of numbers $\{P_\xi(x_1), \ldots, P_\xi(x_m)\}$ is called the *probability distribution of the random variable* ξ.

Example 2. A random variable ξ that takes the two values 1 and 0 with probabilities p ("success") and q ("failure"), is called a Bernoulli* random variable. Clearly

$$P_\xi(x) = p^x q^{1-x}, \quad x = 0, 1. \tag{1}$$

A *binomial* (or binomially distributed) *random variable* ξ is a random variable that takes the $n + 1$ values $0, 1, \ldots, n$ with probabilities

$$P_\xi(x) = C_n^x p^x q^{n-x}, \quad x = 0, 1, \ldots, n. \tag{2}$$

Note that here and in many subsequent examples we do not specify the sample spaces $(\Omega, \mathcal{A}, \mathsf{P})$, but are interested only in the values of the random variables and their probability distributions.

The probabilistic structure of the random variables ξ is completely specified by the probability distributions $\{P_\xi(x_i), i = 1, \ldots, m\}$. The concept of distribution function, which we now introduce, yields an equivalent description of the probabilistic structure of the random variables.

Definition 2. Let $x \in R^1$. The function

$$F_\xi(x) = \mathsf{P}\{\omega : \xi(\omega) \leq x\}$$

is called the *distribution function* of the random variable ξ.

Clearly

$$F_\xi(x) = \sum_{\{i\,:\ x_i \leq x\}} P_\xi(x_i)$$

---

* We use the terms "Bernoulli, binomial, Poisson, Gaussian, …, random variables" for what are more usually called random variables with Bernoulli, binomial, Poisson, Gaussian, …, distributions.

and

$$P_\xi(x_i) = F_\xi(x_i) - F_\xi(x_i -),$$

where $F_\xi(x-) = \lim_{y \uparrow x} F_\xi(y)$.

If we suppose that $x_1 < x_2 < \cdots < x_m$ and put $F_\xi(x_0) = 0$, then

$$P_\xi(x_i) = F_\xi(x_i) - F_\xi(x_{i-1}), \quad i = 1, \ldots, m.$$

The following diagrams (Fig. 5) exhibit $P_\xi(x)$ and $F_\xi(x)$ for a binomial random variable.

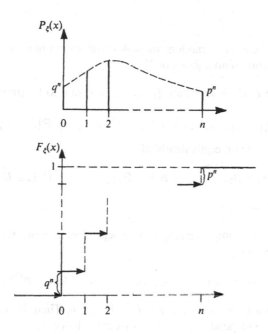

Fig. 5

It follows immediately from Definition 2 that the distribution function $F_\xi = F_\xi(x)$ has the following properties:

(1) $F_\xi(-\infty) = 0$, $F_\xi(+\infty) = 1$;
(2) $F_\xi(x)$ *is continuous on the right* $(F_\xi(x+) = F_\xi(x))$ *and piecewise constant.*

Along with random variables it is often necessary to consider *random vectors* $\xi = (\xi_1, \ldots, \xi_r)$ whose components are random variables. For example, when we considered the multinomial distribution we were dealing with a random vector $\nu = (\nu_1, \ldots, \nu_r)$, where $\nu_i = \nu_i(\omega)$ was the number of elements equal to b_i, $i = 1, \ldots, r$, in the sequence $\omega = (a_1, \ldots, a_n)$.

The set of probabilities

$$P_\xi(x_1, \ldots, x_r) = P\{\omega : \xi_1(\omega) = x_1, \ldots, \xi_r(\omega) = x_r\},$$

where $x_i \in X_i$, the range of ξ_i, is called the *probability distribution of the random vector* ξ, and the function

$$F_\xi(x_1, \ldots, x_r) = \mathsf{P}\{\omega: \xi_1(\omega) \le x_1, \ldots, \xi_r(\omega) \le x_r\},$$

where $x_i \in R^1$, is called the *distribution function of the random vector* $\xi = (\xi_1, \ldots, \xi_r)$.

For example, for the random vector $v = (v_1, \ldots, v_r)$ mentioned above,

$$P_v(n_1, \ldots, n_r) = C_n(n_1, \ldots, n_r)p_1^{n_1} \cdots p_r^{n_r}$$

(see (2), Sect. 2).

2. Let $\xi_1, \ldots, \xi_r$ be a set of random variables with values in a (finite) set $X \subseteq R^1$. Let $\mathscr{X}$ be the algebra of all subsets of X.

Definition 3. The random variables $\xi_1, \ldots, \xi_r$ are said to be (mutually) *independent* if

$$\mathsf{P}\{\xi_1 = x_1, \ldots, \xi_r = x_r\} = \mathsf{P}\{\xi_1 = x_1\} \cdots \mathsf{P}\{\xi_r = x_r\}$$

for all $x_1, \ldots, x_r \in X$; or, equivalently, if

$$\mathsf{P}\{\xi_1 \in B_1, \ldots, \xi_r \in B_r\} = \mathsf{P}\{\xi_1 \in B_1\} \cdots \mathsf{P}\{\xi_r \in B_r\}$$

for all $B_1, \ldots, B_r \in \mathscr{X}$.

We can get a very simple example of independent random variables from the Bernoulli scheme. Let

$$\Omega = \{\omega: \omega = (a_1, \ldots, a_n), a_i = 0,1\}, \quad p(\omega) = p^{\Sigma a_i} q^{n - \Sigma a_i}$$

and $\xi_i(\omega) = a_i$ for $\omega = (a_1, \ldots, a_n)$, $i = 1, \ldots, n$. Then the random variables $\xi_1, \xi_2, \ldots, \xi_n$ are independent, as follows from the independence of the events

$$A_1 = \{\omega: a_1 = 1\}, \ldots, A_n = \{\omega: a_n = 1\},$$

which was established in Sect. 3.

3. We shall frequently encounter the problem of finding the probability distributions of random variables that are *functions* $f(\xi_1, \ldots, \xi_r)$ of random variables $\xi_1, \ldots, \xi_r$. For the present we consider only the determination of the distribution of a *sum* $\zeta = \xi + \eta$ of random variables.

If ξ and η take values in the respective sets $X = \{x_1, \ldots, x_k\}$ and $Y = \{y_1, \ldots, y_l\}$, the random variable $\zeta = \xi + \eta$ takes values in the set $Z = \{z: z = x_i + y_j, i = 1, \ldots, k; j = 1, \ldots, l\}$. Then it is clear that

$$P_\zeta(z) = \mathsf{P}\{\zeta = z\} = \mathsf{P}\{\xi + \eta = z\} = \sum_{\{(i,j):\, x_i + y_j = z\}} \mathsf{P}\{\xi = x_i, \eta = y_j\}.$$

The case of *independent* random variables ξ and η is particularly important. In this case

$$P\{\xi = x_i, \eta = y_j\} = P\{\xi = x_i\} P\{\eta = y_j\},$$

and therefore

$$P_\zeta(z) = \sum_{\{(i,j):\ x_i + y_j = z\}} P_\xi(x_i)P_\eta(y_j) = \sum_{i=1}^{k} P_\xi(x_i)P_\eta(z - x_i) \tag{3}$$

for all $z \in Z$, where in the last sum $P_\eta(z - x_i)$ is taken to be zero if $z - x_i \notin Y$.

For example, if ξ and η are independent Bernoulli random variables, taking the values 1 and 0 with respective probabilities p and q, then $Z = \{0, 1, 2\}$ and

$$P_\zeta(0) = P_\xi(0)P_\eta(0) = q^2,$$
$$P_\zeta(1) = P_\xi(0)P_\eta(1) + P_\xi(1)P_\eta(0) = 2pq,$$
$$P_\zeta(2) = P_\xi(1)P_\eta(1) = p^2.$$

It is easy to show by induction that if $\xi_1, \xi_2, \ldots, \xi_n$ are independent Bernoulli random variables with $P\{\xi_i = 1\} = p$, $P\{\xi_i = 0\} = q$, then the random variable $\zeta = \xi_1 + \cdots + \xi_n$ has the *binomial* distribution

$$P_\zeta(k) = C_n^k p^k q^{n-k}, \quad k = 0, 1, \ldots, n. \tag{4}$$

4. We now turn to the important concept of the *expectation*, or *mean value*, of a random variable.

Let $(\Omega, \mathscr{A}, P)$ be a (discrete) probability space and $\xi = \xi(\omega)$ a random variable with values in the set $X = \{x_1, \ldots, x_k\}$. If we put $A_i = \{\omega: \xi = x_i\}$, $i = 1, \ldots, k$, then ξ can evidently be represented as

$$\xi(\omega) = \sum_{i=1}^{k} x_i I(A_i), \tag{5}$$

where the sets $A_1, \ldots, A_k$ form a decomposition of Ω (i.e., they are pairwise disjoint and their sum is Ω; see Subsection 3 of Sect. 1).

Let $p_i = P\{\xi = x_i\}$. It is intuitively plausible that if we observe the values of the random variable ξ in "n repetitions of identical experiments," the value x_i ought to be encountered about $p_i n$ times, $i = 1, \ldots, k$. Hence the mean value calculated from the results of n experiments is roughly

$$\frac{1}{n}[np_1x_1 + \cdots + np_kx_k] = \sum_{i=1}^{k} p_ix_i.$$

This discussion provides the motivation for the following definition.

Definition 4. The *expectation** or *mean value* of the random variable $\xi = \sum_{i=1}^{k} x_i$ $I(A_i)$ is the number

$$\mathsf{E}\,\xi = \sum_{i=1}^{k} x_i\,\mathsf{P}(A_i). \tag{6}$$

Since $A_i = \{\omega \colon \xi(\omega) = x_i\}$ and $P_\xi(x_i) = \mathsf{P}(A_i)$, we have

$$\mathsf{E}\,\xi = \sum_{i=1}^{k} x_i P_\xi(x_i). \tag{7}$$

Recalling the definition of $F_\xi = F_\xi(x)$ and writing

$$\Delta F_\xi(x) = F_\xi(x) - F_\xi(x-),$$

we obtain $P_\xi(x_i) = \Delta F_\xi(x_i)$ and consequently

$$\mathsf{E}\,\xi = \sum_{i=1}^{k} x_i \Delta F_\xi(x_i). \tag{8}$$

Before discussing the properties of the expectation, we remark that it is often convenient to use another representation of the random variable ξ, namely

$$\xi(\omega) = \sum_{j=1}^{l} x_j'\,I(B_j),$$

where $B_1 + \cdots + B_l = \Omega$, but some of the x_j' may be repeated. In this case $\mathsf{E}\,\xi$ can be calculated from the formula $\sum_{j=1}^{l} x_j'\,\mathsf{P}(B_j)$, which differs formally from (6) because in (6) the x_i are all different. In fact,

$$\sum_{\{j \colon x_j' = x_i\}} x_j'\,\mathsf{P}(B_j) = x_i \sum_{\{j \colon x_j' = x_i\}} \mathsf{P}(B_j) = x_i\,\mathsf{P}(A_i)$$

and therefore

$$\sum_{j=1}^{l} x_j'\,\mathsf{P}(B_j) = \sum_{i=1}^{k} x_i\,\mathsf{P}(A_i).$$

5. We list the basic properties of the expectation:

(1) *If* $\xi \geq 0$ *then* $\mathsf{E}\,\xi \geq 0$.
(2) $\mathsf{E}\,(a\xi + b\eta) = a\,\mathsf{E}\,\xi + b\,\mathsf{E}\,\eta$, *where a and b are constants.*
(3) *If* $\xi \geq \eta$ *then* $\mathsf{E}\,\xi \geq \mathsf{E}\,\eta$.
(4) $|\mathsf{E}\,\xi| \leq \mathsf{E}\,|\xi|$.

---

* Also known as mathematical expectation, or expected value, or (especially in physics) expectation value. (Translator of 1984 edition).

(5) If ξ and η are independent, then $\mathsf{E}\,\xi\eta = \mathsf{E}\,\xi \cdot \mathsf{E}\,\eta$.

(6) $(\mathsf{E}\,|\xi\eta|)^2 \le \mathsf{E}\,\xi^2 \cdot \mathsf{E}\,\eta^2$ (Cauchy–Bunyakovskii inequality).*

(7) If $\xi = I(A)$ then $\mathsf{E}\,\xi = \mathsf{P}(A)$.

Properties (1) and (7) are evident. To prove (2), let

$$\xi = \sum_i x_i\,I(A_i), \quad \eta = \sum_j y_j\,I(B_j).$$

Then

$$a\xi + b\eta = a\sum_{i,j} x_i\,I(A_i \cap B_j) + b\sum_{i,j} y_j\,I(A_i \cap B_j)$$

$$= \sum_{i,j}(ax_i + by_j)I(A_i \cap B_j)$$

and

$$\mathsf{E}(a\xi + b\eta) = \sum_{i,j}(ax_i + by_j)\,\mathsf{P}(A_i \cap B_j)$$

$$= \sum_i ax_i\,\mathsf{P}(A_i) + \sum_j by_j\,\mathsf{P}(B_j)$$

$$= a\sum_i x_i\,\mathsf{P}(A_i) + b\sum_j y_j\,\mathsf{P}(B_j) = a\,\mathsf{E}\,\xi + b\,\mathsf{E}\,\eta.$$

Property (3) follows from (1) and (2). Property (4) is evident, since

$$|\mathsf{E}\,\xi| = \left|\sum_i x_i\,\mathsf{P}(A_i)\right| \le \sum_i |x_i|\,\mathsf{P}(A_i) = \mathsf{E}\,|\xi|.$$

To prove (5) we note that

$$\mathsf{E}\,\xi\eta = \mathsf{E}\left(\sum_i x_i\,I(A_i)\right)\left(\sum_j y_j\,I(B_j)\right)$$

$$= \mathsf{E}\sum_{i,j} x_i\,y_j\,I(A_i \cap B_j) = \sum_{i,j}' x_i y_j\,\mathsf{P}(A_i \cap B_j)$$

$$= \sum_{i,j} x_i y_j\,\mathsf{P}(A_i)\,\mathsf{P}(B_j)$$

$$= \left(\sum_i x_i\,\mathsf{P}(A_i)\right) \cdot \left(\sum_j y_j\,\mathsf{P}(B_j)\right) = \mathsf{E}\,\xi \cdot \mathsf{E}\,\eta,$$

where we have used the property that for independent random variables the events

$$A_i = \{\omega : \xi(\omega) = x_i\} \quad \text{and} \quad B_j = \{\omega : \eta(\omega) = y_j\}$$

are independent: $\mathsf{P}(A_i \cap B_j) = \mathsf{P}(A_i)\,\mathsf{P}(B_j)$.

---

* Also known as the Cauchy–Schwarz or Schwarz inequality. (Translator of 1984 edition).

To prove property (6) we observe that

$$\xi^2 = \sum_i x_i^2 I(A_i), \quad \eta^2 = \sum_j y_j^2 I(B_j)$$

and

$$\mathsf{E}\,\xi^2 = \sum_i x_i^2\,\mathsf{P}(A_i), \quad \mathsf{E}\,\eta^2 = \sum_j y_j^2\,\mathsf{P}(B_j).$$

Let $\mathsf{E}\,\xi^2 > 0$, $\mathsf{E}\,\eta^2 > 0$. Put

$$\tilde{\xi} = \frac{\xi}{\sqrt{\mathsf{E}\,\xi^2}}, \quad \tilde{\eta} = \frac{\eta}{\sqrt{\mathsf{E}\,\eta^2}}.$$

Since $2|\tilde{\xi}\tilde{\eta}| \leq \tilde{\xi}^2 + \tilde{\eta}^2$, we have $2\mathsf{E}\,|\tilde{\xi}\tilde{\eta}| \leq \mathsf{E}\,\tilde{\xi}^2 + \mathsf{E}\,\tilde{\eta}^2 = 2$. Therefore $\mathsf{E}\,|\tilde{\xi}\tilde{\eta}| \leq 1$ and $(\mathsf{E}\,|\xi\eta|)^2 \leq \mathsf{E}\,\xi^2 \cdot \mathsf{E}\,\eta^2$.

However, if, say, $\mathsf{E}\,\xi^2 = 0$, this means that $\sum_i x_i^2\,\mathsf{P}(A_i) = 0$ and consequently the mean value of ξ is 0, and $\mathsf{P}\{\omega\colon \xi(\omega) = 0\} = 1$. Therefore if at least one of $\mathsf{E}\,\xi^2$ or $\mathsf{E}\,\eta^2$ is zero, it is evident that $\mathsf{E}\,|\xi\eta| = 0$ and consequently the Cauchy–Bunyakovskii inequality still holds.

Remark. Property (5) generalizes in an obvious way to any finite number of random variables: if $\xi_1, \ldots, \xi_r$ are *independent*, then

$$\mathsf{E}\,\xi_1 \cdots \xi_r = \mathsf{E}\,\xi_1 \cdots \mathsf{E}\,\xi_r.$$

The proof can be given in the same way as for the case $r = 2$, or by induction.

Example 3. Let ξ be a Bernoulli random variable, taking the values 1 and 0 with probabilities p and q. Then

$$\mathsf{E}\,\xi = 1 \cdot \mathsf{P}\{\xi = 1\} + 0 \cdot \mathsf{P}\{\xi = 0\} = p.$$

Example 4. Let $\xi_1, \ldots, \xi_n$ be n Bernoulli random variables with $\mathsf{P}\{\xi_i = 1\} = p$, $\mathsf{P}\{\xi_i = 0\} = q$, $p + q = 1$. Then if

$$S_n = \xi_1 + \cdots + \xi_n$$

we find that

$$\mathsf{E}\,S_n = np.$$

This result can also be obtained in a different way. It is easy to see that $\mathsf{E}\,S_n$ is not changed if we assume that the Bernoulli random variables $\xi_1, \ldots, \xi_n$ are independent. With this assumption, we have according to (4)

$$\mathsf{P}(S_n = k) = C_n^k p^k q^{n-k}, \quad k = 0, 1, \ldots, n.$$

Therefore

$$
\begin{aligned}
\mathsf{E}\, S_n &= \sum_{k=0}^{n} k\, \mathsf{P}(S_n = k) = \sum_{k=0}^{n} k C_n^k p^k q^{n-k} \\
&= \sum_{k=0}^{n} k \cdot \frac{n!}{k!(n-k)!} p^k q^{n-k} \\
&= np \sum_{k=1}^{n} \frac{(n-1)!}{(k-1)!((n-1)-(k-1))!} p^{k-1} q^{(n-1)-(k-1)} \\
&= np \sum_{l=0}^{n} \frac{(n-1)!}{l!((n-1)-l)!} p^l q^{(n-1)-l} = np.
\end{aligned}
$$

However, the first method is more direct.

6. Let $\xi = \sum_i x_i I(A_i)$, where $A_i = \{\omega : \xi(\omega) = x_i\}$, and let $\varphi = \varphi(\xi(\omega))$ be a function of $\xi(\omega)$. If $B_j = \{\omega : \varphi(\xi(\omega)) = y_j\}$, then

$$
\varphi(\xi(\omega)) = \sum_j y_j I(B_j),
$$

and consequently

$$
\mathsf{E}\,\varphi = \sum_j y_j \mathsf{P}(B_j) = \sum_j y_j P_\varphi(y_j). \tag{9}
$$

But it is also clear that

$$
\varphi(\xi(\omega)) = \sum_i \varphi(x_i) I(A_i).
$$

Hence, along with (9), the expectation of the random variable $\varphi = \varphi(\xi)$ can be calculated as

$$
\mathsf{E}\,\varphi(\xi) = \sum_i \varphi(x_i) P_\xi(x_i).
$$

7. The important notion of the *variance* of a random variable ξ indicates the amount of *scatter* of the values of ξ around $\mathsf{E}\,\xi$.

Definition 5. The *variance* of the random variable ξ (denoted by $\operatorname{Var}\xi$) is

$$
\operatorname{Var}\xi = \mathsf{E}(\xi - \mathsf{E}\,\xi)^2.
$$

The number $\sigma = +\sqrt{\operatorname{Var}\xi}$ is called the *standard deviation* (of ξ from the mean value $\mathsf{E}\,\xi$).

Since

$$
\mathsf{E}(\xi - \mathsf{E}\,\xi)^2 = \mathsf{E}(\xi^2 - 2\xi \cdot \mathsf{E}\,\xi + (\mathsf{E}\,\xi)^2) = \mathsf{E}\,\xi^2 - (\mathsf{E}\,\xi)^2,
$$

we have

$$\operatorname{Var} \xi = \mathsf{E}\, \xi^2 - (\mathsf{E}\, \xi)^2.$$

Clearly $\operatorname{Var} \xi \geq 0$. It follows from the definition that

$$\operatorname{Var}(a + b\xi) = b^2 \operatorname{Var} \xi, \quad \text{where } a \text{ and } b \text{ are constants.}$$

In particular, $\operatorname{Var} a = 0$, $\operatorname{Var}(b\xi) = b^2 \operatorname{Var} \xi$.

Let ξ and η be random variables. Then

$$\begin{aligned}
\operatorname{Var}(\xi + \eta) &= \mathsf{E}((\xi - \mathsf{E}\, \xi) + (\eta - \mathsf{E}\, \eta))^2 \\
&= \operatorname{Var} \xi + \operatorname{Var} \eta + 2\, \mathsf{E}(\xi - \mathsf{E}\, \xi)(\eta - \mathsf{E}\, \eta).
\end{aligned}$$

Write

$$\operatorname{Cov}(\xi, \eta) = \mathsf{E}(\xi - \mathsf{E}\, \xi)(\eta - \mathsf{E}\, \eta).$$

This number is called the *covariance* of ξ and η. If $\operatorname{Var} \xi > 0$ and $\operatorname{Var} \eta > 0$, then

$$\rho(\xi, \eta) = \frac{\operatorname{Cov}(\xi, \eta)}{\sqrt{\operatorname{Var} \xi \cdot \operatorname{Var} \eta}}$$

is called the *correlation coefficient* of ξ and η. It is easy to show (see Problem 7 below) that if $\rho(\xi, \eta) = \pm 1$, then ξ and η are linearly dependent:

$$\eta = a\xi + b,$$

with $a > 0$ if $\rho(\xi, \eta) = 1$ and $a < 0$ if $\rho(\xi, \eta) = -1$.

We observe immediately that if ξ and η are independent, so are $\xi - \mathsf{E}\, \xi$ and $\eta - \mathsf{E}\, \eta$. Consequently by Property (5) of expectations,

$$\operatorname{Cov}(\xi, \eta) = \mathsf{E}(\xi - \mathsf{E}\, \xi) \cdot \mathsf{E}(\eta - \mathsf{E}\, \eta) = 0.$$

Using the notation that we introduced for covariance, we have

$$\operatorname{Var}(\xi + \eta) = \operatorname{Var} \xi + \operatorname{Var} \eta + 2\operatorname{Cov}(\xi, \eta); \tag{10}$$

if ξ and η are *independent*, the *variance of the sum $\xi + \eta$ is equal to the sum of the variances*,

$$\operatorname{Var}(\xi + \eta) = \operatorname{Var} \xi + \operatorname{Var} \eta. \tag{11}$$

It follows from (10) that (11) is still valid under weaker hypotheses than the independence of ξ and η. In fact, it is enough to suppose that ξ and η are uncorrelated, i.e., $\operatorname{Cov}(\xi, \eta) = 0$.

Remark. If ξ and η are uncorrelated, it does not follow in general that they are independent. Here is a simple example. Let the random variable α take the values 0, $\pi/2$ and π with probability $\frac{1}{3}$. Then $\xi = \sin \alpha$ and $\eta = \cos \alpha$ are uncorrelated;

however, they are stochastically dependent (i.e., not independent with respect to the probability P):

$$P\{\xi = 1, \eta = 1\} = 0 \neq \tfrac{1}{9} = P\{\xi = 1\} \, P\{\eta = 1\}.$$

Properties (10) and (11) can be extended in the obvious way to any number of random variables:

$$\operatorname{Var}\left(\sum_{i=1}^{n} \xi_i\right) = \sum_{i=1}^{n} \operatorname{Var} \xi_i + 2 \sum_{i>j} \operatorname{Cov}(\xi_i, \xi_j). \tag{12}$$

In particular, if $\xi_1, \ldots, \xi_n$ are pairwise independent (pairwise uncorrelated is sufficient), then

$$\operatorname{Var}\left(\sum_{i=1}^{n} \xi_i\right) = \sum_{i=1}^{n} \operatorname{Var} \xi_i. \tag{13}$$

Example 5. If ξ is a Bernoulli random variable, taking the values 1 and 0 with probabilities p and q, then

$$\operatorname{Var} \xi = \mathsf{E}(\xi - \mathsf{E}\,\xi)^2 = \mathsf{E}(\xi - p)^2 = (1 - p)^2 p + p^2 q = pq.$$

It follows that if $\xi_1, \ldots, \xi_n$ are independent identically distributed Bernoulli random variables, and $S_n = \xi_1 + \cdots + \xi_n$, then

$$\operatorname{Var} S_n = npq. \tag{14}$$

8. Consider two random variables ξ and η. Suppose that only ξ can be observed. If ξ and η are correlated, we may expect that knowing the value of ξ allows us to make some inference about the values of the unobserved variable η.

Any function $f = f(\xi)$ of ξ is called an *estimator* for η. We say that an estimator $f^* = f^*(\xi)$ is *best (or optimal) in the mean-square sense* if

$$\mathsf{E}(\eta - f^*(\xi))^2 = \inf_f \mathsf{E}(\eta - f(\xi))^2.$$

Let us show how to find the best estimator in the class of *linear* estimators $\lambda(\xi) = a + b\xi$. Consider the function $g(a, b) = \mathsf{E}(\eta - (a + b\xi))^2$. Differentiating $g(a, b)$ with respect to a and b, we obtain

$$\frac{\partial g(a, b)}{\partial a} = -2\,\mathsf{E}[\eta - (a + b\xi)],$$

$$\frac{\partial g(a, b)}{\partial b} = -2\,\mathsf{E}[(\eta - (a + b\xi))\xi],$$

whence, setting the derivatives equal to zero, we find that the best mean-square linear estimator is $\lambda^*(\xi) = a^* + b^*\xi$, where

$$a^* = \mathsf{E}\,\eta - b^*\,\mathsf{E}\,\xi, \quad b^* = \frac{\operatorname{Cov}(\xi, \eta)}{\operatorname{Var} \xi}. \tag{15}$$

In other words,

$$\lambda^*(\xi) = \mathsf{E}\,\eta + \frac{\mathrm{Cov}(\xi, \eta)}{\mathrm{Var}\,\xi}(\xi - \mathsf{E}\,\xi). \tag{16}$$

The number $\mathsf{E}(\eta - \lambda^*(\xi))^2$ is called the *mean-square error of estimation*. An easy calculation shows that it is equal to

$$\Delta^* = \mathsf{E}(\eta - \lambda^*(\xi))^2 = \mathrm{Var}\,\eta - \frac{\mathrm{Cov}^2(\xi, \eta)}{\mathrm{Var}\,\xi} = \mathrm{Var}\,\eta \cdot [1 - \rho^2(\xi, \eta)]. \tag{17}$$

Consequently, the larger (in absolute value) the correlation coefficient $\rho(\xi, \eta)$ between ξ and η, the smaller the mean-square error of estimation Δ^*. In particular, if $|\rho(\xi, \eta)| = 1$ then $\Delta^* = 0$ (cf. Problem 7). On the other hand, if ξ and η are uncorrelated ($\rho(\xi, \eta) = 0$), then $\lambda^*(\xi) = \mathsf{E}\,\eta$, i.e., in the absence of correlation between ξ and η the best estimate of η in terms of ξ is simply $\mathsf{E}\,\eta$ (cf. Problem 4).

9. PROBLEMS.

1. Verify the following properties of indicators $I_A = I_A(\omega)$:

$$I_\varnothing = 0, \quad I_\Omega = 1, \quad I_A + I_{\bar{A}} = 1,$$

$$I_{AB} = I_A \cdot I_B,$$

$$I_{A \cup B} = I_A + I_B - I_{AB}.$$

The indicator of $\bigcup_{i=1}^n A_i$ is $1 - \prod_{i=1}^n (1 - I_{A_i})$, the indicator of $\overline{\bigcup_{i=1}^n A_i}$ is $\prod_{i=1}^n (1 - I_{A_i})$, the indicator of $\sum_{i=1}^n A_i$ is $\sum_{i=1}^n I_{A_i}$, and

$$I_{A \triangle B} = (I_A - I_B)^2 = I_A + I_B \pmod 2,$$

where $A \triangle B$ is the *symmetric difference* of A and B, i.e., the set $(A \backslash B) \cup (B \backslash A)$.

2. Let $\xi_1, \ldots, \xi_n$ be independent random variables and

$$\xi_{\min} = \min(\xi_1, \ldots, \xi_n), \quad \xi_{\max} = \max(\xi_1, \ldots, \xi_n).$$

Show that

$$\mathsf{P}\{\xi_{\min} \geq x\} = \prod_{i=1}^n \mathsf{P}\{\xi_i \geq x\}, \quad \mathsf{P}\{\xi_{\max} < x\} = \prod_{i=1}^n \mathsf{P}\{\xi_i < x\}.$$

3. Let $\xi_1, \ldots, \xi_n$ be independent Bernoulli random variables such that

$$\mathsf{P}\{\xi_i = 0\} = 1 - \lambda_i \Delta, \quad \mathsf{P}\{\xi_i = 1\} = \lambda_i \Delta,$$

where n and $\lambda_i > 0$, $i = 1, \ldots, n$, are fixed and $\Delta > 0$ is a small number. Show that

$$\mathsf{P}\{\xi_1 + \cdots + \xi_n = 1\} = \left(\sum_{i=1}^n \lambda_i\right)\Delta + O(\Delta^2),$$

$$\mathsf{P}\{\xi_1 + \cdots + \xi_n > 1\} = O(\Delta^2).$$

4. Show that $\inf_{-\infty<a<\infty} E(\xi - a)^2$ is attained for $a = E\xi$ and consequently

$$\inf_{-\infty<a<\infty} E(\xi - a)^2 = \operatorname{Var}\xi.$$

5. Let ξ be a random variable with distribution function $F_\xi(x)$ and let m_e be a median of $F_\xi(x)$, i.e., a point such that

$$F_\xi(m_e-) \le \tfrac{1}{2} \le F_\xi(m_e).$$

Show that

$$\inf_{-\infty<a<\infty} E|\xi - a| = E|\xi - m_e|.$$

6. Let $P_\xi(x) = P\{\xi = x\}$ and $F_\xi(x) = P(\xi \le x)$. Show that

$$P_{a\xi+b}(x) = P_\xi\left(\frac{x-b}{a}\right),$$

$$F_{a\xi+b}(x) = F_\xi\left(\frac{x-b}{a}\right)$$

for $a > 0$ and $-\infty < b < \infty$. If $y \ge 0$, then

$$F_{\xi^2}(y) = F_\xi(+\sqrt{y}) - F_\xi(-\sqrt{y}) + P_\xi(-\sqrt{y}).$$

Let $\xi^+ = \max(\xi, 0)$. Then

$$F_{\xi^+}(x) = \begin{cases} 0, & x < 0, \\ F_\xi(0), & x = 0, \\ F_\xi(x), & x > 0. \end{cases}$$

7. Let ξ and η be random variables with $\operatorname{Var}\xi > 0$, $\operatorname{Var}\eta > 0$, and let $\rho = \rho(\xi, \eta)$ be their correlation coefficient. Show that $|\rho| \le 1$. If $|\rho| = 1$, there are constants a and b such that $\eta = a\xi + b$. Moreover, if $\rho = 1$, then

$$\frac{\eta - E\eta}{\sqrt{\operatorname{Var}\eta}} = \frac{\xi - E\xi}{\sqrt{\operatorname{Var}\xi}}$$

(and therefore $a > 0$), whereas if $\rho = -1$, then

$$\frac{\eta - E\eta}{\sqrt{\operatorname{Var}\eta}} = -\frac{\xi - E\xi}{\sqrt{\operatorname{Var}\xi}}$$

(and therefore $a < 0$).

8. Let ξ and η be random variables with $E\xi = E\eta = 0$, $\operatorname{Var}\xi = \operatorname{Var}\eta = 1$ and correlation coefficient $\rho = \rho(\xi, \eta)$. Show that

$$E\max(\xi^2, \eta^2) \le 1 + \sqrt{1 - \rho^2}.$$

9. Use the equation

$$\left(\text{Indicator of } \overline{\bigcup_{i=1}^{n} A_i} \right) = \prod_{i=1}^{n} (1 - I_{A_i})$$

to prove the formula $P(B_0) = 1 - S_1 + S_2 + \cdots \pm S_n$ in Problem 4 of Sect. 1.

10. Let $\xi_1, \ldots, \xi_n$ be independent random variables, $\varphi_1 = \varphi_1(\xi_1, \ldots, \xi_k)$ and $\varphi_2 = \varphi_2(\xi_{k+1}, \ldots, \xi_n)$, functions respectively of $\xi_1, \ldots, \xi_k$ and $\xi_{k+1}, \ldots, \xi_n$. Show that the random variables φ_1 and φ_2 are independent.

11. Show that the random variables $\xi_1, \ldots, \xi_n$ are independent if and only if

$$F_{\xi_1, \ldots, \xi_n}(x_1, \ldots, x_n) = F_{\xi_1}(x_1) \cdots F_{\xi_n}(x_n)$$

for all $x_1, \ldots, x_n$, where $F_{\xi_1, \ldots, \xi_n}(x_1, \ldots, x_n) = P\{\xi_1 \leq x_1, \ldots, \xi_n \leq x_n\}$.

12. Show that the random variable ξ is independent of itself (i.e., ξ and ξ are independent) if and only if $\xi = \text{const}$.

13. Under what conditions on ξ are the random variables ξ and $\sin \xi$ independent?

14. Let ξ and η be independent random variables and $\eta \neq 0$. Express the probabilities of the events $P\{\xi\eta \leq z\}$ and $P\{\xi/\eta \leq z\}$ in terms of the probabilities $P_\xi(x)$ and $P_\eta(y)$.

15. Let ξ, η, ζ be random variables such that $|\xi| \leq 1, |\eta| \leq 1, |\zeta| \leq 1$. Prove the *Bell inequality*:

$$|\, \mathsf{E}\, \xi\zeta - \mathsf{E}\, \eta\zeta| \leq 1 - \mathsf{E}\, \xi\eta.$$

(See, e.g., [46].)

16. Let k balls be independently thrown into n urns. (Each ball falls into any specific urn with probability $1/n$.) Find the expectation of the number of nonempty urns.

5 The Bernoulli Scheme: I—The Law of Large Numbers

1. In accordance with the definitions given in Sect. 2, Subsection 1, for $n = 1, 2, \ldots$, a triple

$$(\Omega_n, \mathscr{A}_n, \mathsf{P}_n) \quad \text{with} \quad \Omega_n = \{\omega : \omega = (a_1, \ldots, a_n), \, a_i = 0, 1\},$$
$$\mathscr{A}_n = \{A : A \subseteq \Omega_n\}, \quad \mathsf{P}_n(\{\omega\}) = p^{\Sigma a_i} q^{n - \Sigma a_i} \quad (q = 1 - p) \tag{1}$$

is called a (discrete) probabilistic model of n independent experiments with two outcomes, or a *Bernoulli scheme*.

In this and the next section we study some limiting properties (in a sense described below) for Bernoulli schemes. These are best expressed in terms of random variables and of the probabilities of events connected with them.

We introduce random variables $\xi_{n1}, \ldots, \xi_{nn}$ by taking $\xi_{ni}(\omega) = a_i$, $i = 1, \ldots, n$, where $\omega = (a_1, \ldots, a_n)$. As we saw above, the Bernoulli variables $\xi_{ni}(\omega)$ are independent and identically distributed:

$$P_n\{\xi_{ni} = 1\} = p, \quad P_n\{\xi_{ni} = 0\} = q, \quad i = 1, \ldots, n.$$

It is natural to think of ξ_{ni} as describing the result of an experiment at the ith stage (or at time i).

Let us put $S_{n0}(\omega) \equiv 0$ and

$$S_{nk} = \xi_{n1} + \cdots + \xi_{nk}, \quad k = 1, \ldots, n.$$

For notational simplicity we will write S_n for S_{nn}. As we found above, $\mathsf{E}_n S_n = np$ and consequently

$$\mathsf{E}_n \frac{S_n}{n} = p. \tag{2}$$

In other words, the mean value of the frequency of "success," i.e., S_n/n, coincides with the probability p of success. Hence we are led to ask how much the frequency S_n/n of success differs from its probability p.

We first note that we cannot expect that, for a sufficiently small $\varepsilon > 0$ and for sufficiently large n, the deviation of S_n/n from p is less than ε for all ω, i.e., that

$$\left| \frac{S_n(\omega)}{n} - p \right| \leq \varepsilon, \quad \omega \in \Omega_n. \tag{3}$$

In fact, when $0 < p < 1$,

$$P_n\left\{ \frac{S_n}{n} = 1 \right\} = P_n\{\xi_{n1} = 1, \ldots, \xi_{nn} = 1\} = p^n,$$

$$P_n\left\{ \frac{S_n}{n} = 0 \right\} = P_n\{\xi_{n1} = 0, \ldots, \xi_{nn} = 0\} = q^n,$$

whence it follows that (3) is not satisfied for sufficiently small $\varepsilon > 0$.

We observe, however, that for large n the probabilities of the events $\{S_n/n = 1\}$ and $\{S_n/n = 0\}$ are small. It is therefore natural to expect that the total probability of the events for which $|[S_n(\omega)/n] - p| > \varepsilon$ will also be small when n is sufficiently large.

We shall accordingly try to estimate the probability of the event

$$\{\omega : |[S_n(\omega)/n] - p| > \varepsilon\}.$$

For $n \geq 1$ and $0 \leq k \leq n$, write

$$P_n(k) = C_n^k p^k q^{n-k}.$$

Then

$$P_n\left\{\left|\frac{S_n}{n} - p\right| \geq \varepsilon\right\} = \sum_{\{k:\ |(k/n)-p|\geq\varepsilon\}} P_n(k). \tag{4}$$

It was proved by J. Bernoulli that, as $n \to \infty$, the expression in the right-hand side and hence the probability in the left-hand side tend to 0. The latter statement is called the *law of large numbers*.

The analytic proof of this statement is rather involved, and we will prove that

$$P_n\left\{\left|\frac{S_n}{n} - p\right| \geq \varepsilon\right\} \to 0 \qquad \text{as} \quad n \to \infty \tag{5}$$

by probabilistic methods. For this purpose we will use the following inequality, which was established by Chebyshev.

Chebyshev's (Bienaymé–Chebyshev's) inequality. *Let $(\Omega, \mathscr{A}, P)$ be a (discrete) probability space and $\xi = \xi(\omega)$ a nonnegative random variable defined on $(\Omega, \mathscr{A})$. Then*

$$P\{\xi \geq \varepsilon\} \leq E\,\xi/\varepsilon \tag{6}$$

for all $\varepsilon > 0$.

PROOF. We notice that

$$\xi = \xi I(\xi \geq \varepsilon) + \xi I(\xi < \varepsilon) \geq \xi I(\xi \geq \varepsilon) \geq \varepsilon I(\xi \geq \varepsilon),$$

where $I(A)$ is the indicator of A. Then, by the properties of the expectation,

$$E\,\xi \geq \varepsilon\, E\, I(\xi \geq \varepsilon) = \varepsilon\, P(\xi \geq \varepsilon),$$

which establishes (6).

□

Corollary. *If ξ is any random variable defined on $(\Omega, \mathscr{A})$, we have for $\varepsilon > 0$,*

$$\begin{aligned}
P\{|\xi| \geq \varepsilon\} &\leq E\,|\xi|/\varepsilon, \\
P\{|\xi| \geq \varepsilon\} &= P\{\xi^2 \geq \varepsilon^2\} \leq E\,\xi^2/\varepsilon^2, \\
P\{|\xi - E\,\xi| \geq \varepsilon\} &\leq \operatorname{Var}\xi/\varepsilon^2, \\
P\left(|\xi - E\,\xi|/\sqrt{\operatorname{Var}\xi} \geq \varepsilon\right) &\leq 1/\varepsilon^2.
\end{aligned} \tag{7}$$

(The last inequality represents the form in which Chebyshev obtained the inequality in his paper [16].)

Now we turn again to the probability space (1). Take $\xi = S_n/n$ in the next-to-last of inequalities (7). Then using (14) of Sect. 4, we obtain

$$P_n\left\{\left|\frac{S_n}{n} - p\right| \geq \varepsilon\right\} \leq \frac{\operatorname{Var}_n(S_n/n)}{\varepsilon^2} = \frac{\operatorname{Var}_n S_n}{n^2\varepsilon^2} = \frac{npq}{n^2\varepsilon^2} = \frac{pq}{n\varepsilon^2}.$$

Therefore

$$P_n\left\{\left|\frac{S_n}{n} - p\right| \ge \varepsilon\right\} \le \frac{pq}{n\varepsilon^2} \le \frac{1}{4n\varepsilon^2}, \tag{8}$$

and, since $\varepsilon > 0$ is fixed, this implies the law of large numbers (5).

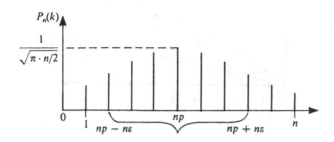

Fig. 6

It is clear from (4) and (5) that

$$\sum_{\{k:\ |(k/n)-p|\ge\varepsilon\}} P_n(k) \to 0, \quad n \to \infty. \tag{9}$$

We can clarify this graphically in the following way. Let us represent the binomial distribution $\{P_n(k),\ 0 \le k \le n\}$ as in Fig. 6.

Then as n increases the graph spreads out and becomes flatter. At the same time the sum of $P_n(k)$ over k, for which $np - n\varepsilon \le k < np + n\varepsilon$, tends to 1.

Let us think of the sequence of random variables $S_{n0}, S_{n1}, \ldots, S_{nn}$ as the *path* of a wandering particle. Then (9) has the following interpretation.

Let us draw lines from the origin of slopes kp, $k(p+\varepsilon)$, and $k(p-\varepsilon)$. Then on the average the path follows the kp line, and for every $\varepsilon > 0$ we can say that when n is sufficiently large there is a large probability that the point S_n specifying the position of the particle at time n lies in the interval $[n(p - \varepsilon),\ n(p + \varepsilon)]$; see Fig. 7.

The statement (5) goes by the name of **James Bernoulli's law of large numbers.** We may remark that to be precise, Bernoulli's proof consisted in establishing (9), which he did quite rigorously by using estimates for the "tails" of the binomial probabilities $P_n(k)$ (for the values of k for which $|(k/n) - p| \ge \varepsilon$). A direct calculation of the sum of the tail probabilities of the binomial distribution $\sum_{\{k:|(k/n)-p|\ge\varepsilon\}} P_n(k)$ is rather difficult problem for large n, and the resulting formulas are ill adapted for actual estimates of the probability with which the frequencies S_n/n differ from p by less than ε. Important progress resulted from the discovery by de Moivre (for $p = \frac{1}{2}$) and then by Laplace (for $0 < p < 1$) of simple asymptotic formulas for $P_n(k)$, which led not only to new proofs of the law of large numbers but also to more precise statements of both *local* and *integral limit theorems*, the essence of which is that for large n and at least for $k \sim np$,

$$P_n(k) \sim \frac{1}{\sqrt{2\pi npq}} e^{-(k-np)^2/(2npq)},$$

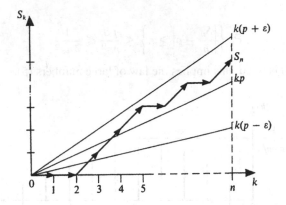

Fig. 7

and

$$\sum_{\{k:\ |(k/n)-p|\leq\varepsilon\}} P_n(k) \sim \frac{1}{\sqrt{2\pi}} \int_{-\varepsilon\sqrt{n/pq}}^{\varepsilon\sqrt{n/pq}} e^{-x^2/2} dx.$$

2. The next section will be devoted to precise statements and proofs of these results. For the present we consider the question of the real meaning of the law of large numbers, and of its empirical interpretation.

Let us carry out a large number, say N, of *series* of experiments, each of which consists of "n independent trials with probability p of the event C of interest." Let S_n^i/n be the frequency of event C in the ith series and N_ε the number of series in which the frequency deviates from p by less than ε:

$$N_\varepsilon \text{ is the number of } i\text{'s for which } |(S_n^i/n) - p| \leq \varepsilon.$$

Then by the law of large numbers

$$N_\varepsilon/N \sim P_\varepsilon \tag{10}$$

where $P_\varepsilon = \mathsf{P}_n\{|(S_n^1/n) - p| \leq \varepsilon\}$.

3. Let us apply the estimate obtained above,

$$\mathsf{P}\left\{\left|\frac{S_n}{n} - p\right| \geq \varepsilon\right\} = \sum_{\{k:\ |(k/n)-p|\geq\varepsilon\}} P_n(k) \leq \frac{1}{4n\varepsilon^2}, \tag{11}$$

to answer the following question that is typical of *mathematical statistics*: what is the least number n of observations which guarantees (for arbitrary $0 < p < 1$) that

$$\mathsf{P}\left\{\left|\frac{S_n}{n} - p\right| \leq \varepsilon\right\} \geq 1 - \alpha, \tag{12}$$

where α is a given number (usually small)? (Here and later we omit the index n of P and the like when the meaning of the notation is clear from the context.)

It follows from (11) that this number is the smallest integer n for which

$$n \geq \frac{1}{4\varepsilon^2 \alpha}. \tag{13}$$

For example, if $\alpha = 0.05$ and $\varepsilon = 0.02$, then 12500 observations guarantee that (12) will hold independently of the value of the unknown parameter p.

Later (Subsection 5, Sect. 6) we shall see that this number is much overstated; this came about because Chebyshev's inequality provides only a very crude upper bound for $P\{|(S_n/n) - p| \geq \varepsilon\}$.

4. Let us write

$$C(n, \varepsilon) = \left\{ \omega : \left| \frac{S_n(\omega)}{n} - p \right| \leq \varepsilon \right\}.$$

From the law of large numbers that we proved, it follows that for every $\varepsilon > 0$ and for sufficiently large n, the probability of the set $C(n, \varepsilon)$ is close to 1. In this sense it is natural to call paths (realizations) ω that are in $C(n, \varepsilon)$ *typical* (or (n, ε)-typical).

We ask the following question: How many typical realizations are there, and what is the weight $p(\omega)$ of a typical realization?

For this purpose we first notice that the total number $N(\Omega)$ of points is 2^n, and that if $p = 0$ or 1, the set of typical paths $C(n, \varepsilon)$ contains only the single path $(0, 0, \ldots, 0)$ or $(1, 1, \ldots, 1)$. However, if $p = \frac{1}{2}$, it is intuitively clear that "almost all" paths (all except those of the form $(0, 0, \ldots, 0)$ or $(1, 1, \ldots, 1)$) are typical and that consequently there should be about 2^n of them.

It turns out that we can give a definitive answer to the question when $0 < p < 1$; it will then appear that both the number of typical realizations and the weights $p(\omega)$ are determined by a function of p called the entropy.

In order to present the corresponding results in more depth, it will be helpful to consider the somewhat more general scheme of Subsection 2 of Sect. 2 instead of the Bernoulli scheme itself.

Let $(p_1, p_2, \ldots, p_r)$ be a finite probability distribution, i.e., a set of nonnegative numbers satisfying $p_1 + \cdots + p_r = 1$. The *entropy* of this distribution is

$$H = -\sum_{i=1}^{r} p_i \log p_i, \tag{14}$$

with $0 \cdot \log 0 = 0$. It is clear that $H \geq 0$, and $H = 0$ if and only if every p_i, with one exception, is zero. The function $f(x) = -x \log x$, $0 \leq x \leq 1$, is convex upward, so that, as we know from the theory of convex functions,

$$\frac{f(x_1) + \cdots + f(x_r)}{r} \leq f\left(\frac{x_1 + \cdots + x_r}{r} \right).$$

Consequently

$$H = -\sum_{i=1}^{r} p_i \log p_i \leq -r \cdot \frac{p_1 + \cdots + p_r}{r} \cdot \log\left(\frac{p_1 + \cdots + p_r}{r}\right) = \log r.$$

In other words, the entropy attains its largest value for $p_1 = \cdots = p_r = 1/r$ (see Fig. 8 for $H = H(p)$ in the case $r = 2$).

If we consider the probability distribution $(p_1, p_2, \ldots, p_r)$ as giving the probabilities for the occurrence of events $A_1, A_2, \ldots, A_r$, say, then it is quite clear that the "degree of indeterminacy" of an event will be different for different distributions. If, for example, $p_1 = 1$, $p_2 = \cdots = p_r = 0$, it is clear that this distribution does not admit any indeterminacy: we can say with complete certainty that the result of the experiment will be A_1. On the other hand, if $p_1 = \cdots = p_r = 1/r$, the distribution has maximal indeterminacy, in the sense that it is impossible to discover any preference for the occurrence of one event rather than another.

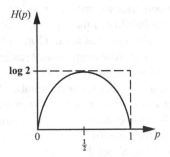

Fig. 8 The function $H(p) = -p\log p - (1-p)\log(1-p)$

Consequently it is important to have a *quantitative* measure of the *indeterminacy* of different probability distributions, so that we may compare them in this respect. As we will see, such a measure of indeterminacy is successfully provided by the entropy; it plays an important role in statistical mechanics and in many significant problems of coding and communication theory.

Suppose now that the sample space is

$$\Omega = \{\omega : \omega = (a_1, \ldots, a_n), \ a_i = 1, \ldots, r\}$$

and that $p(\omega) = p_1^{v_1(\omega)} \cdots p_r^{v_r(\omega)}$, where $v_i(\omega)$ is the number of occurrences of i in the sequence ω, and $(p_1, \ldots, p_r)$ is a probability distribution.

For $\varepsilon > 0$ and $n = 1, 2, \ldots$, let us put

$$C(n, \varepsilon) = \left\{\omega : \left|\frac{v_i(\omega)}{n} - p_i\right| < \varepsilon, \ i = 1, \ldots, r\right\}.$$

It is clear that

$$\mathsf{P}(C(n, \varepsilon)) \geq 1 - \sum_{i=1}^{r} \mathsf{P}\left\{\left|\frac{v_i(\omega)}{n} - p_i\right| \geq \varepsilon\right\},$$

and for sufficiently large n the probabilities $P\{|(v_i(\omega)/n) - p_i| \geq \varepsilon\}$ are arbitrarily small when n is sufficiently large, by the law of large numbers applied to the random variables

$$\xi_k(\omega) = \begin{cases} 1, & a_k = i, \\ 0, & a_k \neq i, \end{cases} \qquad k = 1, \ldots, n.$$

Hence for large n the probability of the event $C(n, \varepsilon)$ is close to 1. Thus, as in the case $n = 2$, a path in $C(n, \varepsilon)$ can be said to be *typical*.

If all $p_i > 0$, then for every $\omega \in \Omega$

$$p(\omega) = \exp\left\{-n \sum_{k=1}^{r} \left(-\frac{v_k(\omega)}{n} \log p_k\right)\right\}.$$

Consequently if ω is a typical path, we have

$$\left|\sum_{k=1}^{r} \left(-\frac{v_k(\omega)}{n} \log p_k\right) - H\right| \leq -\sum_{k=1}^{r} \left|\frac{v_k(\omega)}{n} - p_k\right| \log p_k \leq -\varepsilon \sum_{k=1}^{r} \log p_k.$$

It follows that for typical paths the probability $p(\omega)$ is close to e^{-nH} and—since, by the law of large numbers, the typical paths "almost" exhaust Ω when n is large— the number of such paths must be of order e^{nH}. These considerations lead us to the following proposition.

Theorem (Macmillan). *Let $p_i > 0$, $i = 1, \ldots, r$, and $0 < \varepsilon < 1$. Then there is an $n_0 = n_0(\varepsilon; p_1, \ldots, p_r)$ such that for all $n > n_0$*

(a) $e^{n(H-\varepsilon)} \leq N(C(n, \varepsilon_1)) \leq e^{n(H+\varepsilon)}$;
(b) $e^{-n(H+\varepsilon)} \leq p(\omega) \leq e^{-n(H-\varepsilon)}, \quad \omega \in C(n, \varepsilon_1)$;
(c) $P(C(n, \varepsilon_1)) = \sum_{\omega \in C(n, \varepsilon_1)} p(\omega) \to 1, \quad n \to \infty$,

where

$$\varepsilon_1 \text{ is the smaller of } \varepsilon \text{ and } \varepsilon \Big/ \left\{-2 \sum_{k=1}^{r} \log p_k\right\}.$$

PROOF. Conclusion (c) follows from the law of large numbers. To establish the other conclusions, we notice that if $\omega \in C(n, \varepsilon_1)$ then

$$np_k - \varepsilon_1 n < v_k(\omega) < np_k + \varepsilon_1 n, \quad k = 1, \ldots, r,$$

and therefore

$$p(\omega) = \exp\{-\sum v_k \log p_k\} < \exp\{-n \sum p_k \log p_k - \varepsilon_1 n \sum \log p_k\}$$
$$\leq \exp\{-n(H - \tfrac{1}{2}\varepsilon)\}.$$

Similarly

$$p(\omega) > \exp\{-n(H + \tfrac{1}{2}\varepsilon)\}.$$

Consequently (b) is now established.

Furthermore, since

$$P(C(n, \varepsilon_1)) \geq N(C(n, \varepsilon_1)) \cdot \min_{\omega \in C(n, \varepsilon_1)} p(\omega),$$

we have

$$N(C(n, \varepsilon_1)) \leq \frac{P(C(n, \varepsilon_1))}{\min\limits_{\omega \in C(n, \varepsilon_1)} p(\omega)} < \frac{1}{e^{-n(H+(1/2)\varepsilon)}} = e^{n(H+(1/2)\varepsilon)}$$

and similarly

$$N(C(n, \varepsilon_1)) \geq \frac{P(C(n, \varepsilon_1))}{\max\limits_{\omega \in C(n, \varepsilon_1)} p(\omega)} > P(C(n, \varepsilon_1))e^{n(H-(1/2)\varepsilon)}.$$

Since $P(C(n, \varepsilon_1)) \to 1$, $n \to \infty$, there is an n_1 such that $P(C(n, \varepsilon_1)) > 1 - \varepsilon$ for $n > n_1$, and therefore

$$N(C(n, \varepsilon_1)) \geq (1 - \varepsilon) \exp\{n(H - \tfrac{1}{2}\varepsilon)\}$$
$$= \exp\{n(H - \varepsilon) + (\tfrac{1}{2}n\varepsilon + \log(1 - \varepsilon))\}.$$

Let n_2 be such that

$$\tfrac{1}{2}n\varepsilon + \log(1 - \varepsilon) > 0$$

for $n > n_2$. Then when $n \geq n_0 = \max(n_1, n_2)$ we have

$$N(C(n, \varepsilon_1)) \geq e^{n(H-\varepsilon)}.$$

This completes the proof of the theorem.

□

5. The law of large numbers for Bernoulli schemes lets us give a simple and elegant proof of the Weierstrass theorem on the approximation of continuous functions by polynomials.

Let $f = f(p)$ be a continuous function on the interval $[0, 1]$. We introduce the polynomials

$$B_n(p) = \sum_{k=0}^{n} f\left(\frac{k}{n}\right) C_n^k p^k q^{n-k}, \quad q = 1 - p, \tag{15}$$

which are called Bernstein's polynomials after the inventor of this proof of Weierstrass's theorem.

If $\xi_1, \ldots, \xi_n$ is a sequence of independent Bernoulli random variables with $P\{\xi_i = 1\} = p$, $P\{\xi_i = 0\} = q$ and $S_n = \xi_1 + \cdots + \xi_n$, then

$$E f\left(\frac{S_n}{n}\right) = B_n(p).$$

Since the function $f = f(p)$, being continuous on $[0,1]$, is uniformly continuous, for every $\varepsilon > 0$ we can find $\delta > 0$ such that $|f(x) - f(y)| \leq \varepsilon$ whenever $|x - y| \leq \delta$. It is also clear that this function is bounded: $|f(x)| \leq M < \infty$.

Using this and (8), we obtain

$$|f(p) - B_n(p)| = \left| \sum_{k=0}^{n} \left[f(p) - f\left(\frac{k}{n}\right) \right] C_n^k p^k q^{n-k} \right|$$

$$\leq \sum_{\{k:|(k/n)-p|\leq\delta\}} \left| f(p) - f\left(\frac{k}{n}\right) \right| C_n^k p^k q^{n-k}$$

$$+ \sum_{\{k:|(k/n)-p|>\delta\}} \left| f(p) - f\left(\frac{k}{n}\right) \right| C_n^k p^k q^{n-k}$$

$$\leq \varepsilon + 2M \sum_{\{k:|(k/n)-p|>\delta\}} C_n^k p^k q^{n-k} \leq \varepsilon + \frac{2M}{4n\delta^2} = \varepsilon + \frac{M}{2n\delta^2}.$$

Hence for Bernstein's polynomials (15)

$$\lim_{n\to\infty} \max_{0\leq p\leq 1} |f(p) - B_n(p)| = 0,$$

which is the conclusion of the Weierstrass theorem.

6. Problems

1. Let ξ and η be random variables with correlation coefficient ρ. Establish the following *two-dimensional* analog of Chebyshev's inequality:

$$\mathsf{P}\{|\xi - \mathsf{E}\,\xi| \geq \varepsilon\sqrt{\operatorname{Var}\xi} \text{ or } |\eta - \mathsf{E}\,\eta| \geq \varepsilon\sqrt{\operatorname{Var}\eta}\} \leq \frac{1}{\varepsilon^2}(1 + \sqrt{1 - \rho^2}).$$

(Hint: Use the result of Problem 8 of Sect. 4.)

2. Let $f = f(x)$ be a nonnegative even function that is nondecreasing for positive x. Then for a random variable ξ with $|\xi(\omega)| \leq C$,

$$\mathsf{P}\{|\xi| \geq \varepsilon\} \geq \frac{\mathsf{E}f(\xi) - f(\varepsilon)}{f(C)}.$$

In particular, if $f(x) = x^2$,

$$\frac{\mathsf{E}\,\xi^2 - \varepsilon^2}{C^2} \leq \mathsf{P}\{|\xi - \mathsf{E}\,\xi| \geq \varepsilon\} \leq \frac{\operatorname{Var}\xi}{\varepsilon^2}.$$

3. Let $\xi_1, \ldots, \xi_n$ be a sequence of independent random variables with $\operatorname{Var}\xi_i \leq C$. Then

$$\mathsf{P}\left\{ \left| \frac{\xi_1 + \cdots + \xi_n}{n} - \frac{\mathsf{E}(\xi_1 + \cdots + \xi_n)}{n} \right| \geq \varepsilon \right\} \leq \frac{C}{n\varepsilon^2}. \qquad (16)$$

(Inequality (16) implies the validity of the law of large numbers in more general contexts than Bernoulli schemes.)

4. Let $\xi_1, \ldots, \xi_n$ be independent Bernoulli random variables with $P\{\xi_i = 1\} = p > 0$, $P\{\xi_i = -1\} = 1 - p$. Derive the following *Bernstein's inequality*: there is a number $a > 0$ such that

$$P\left\{\left|\frac{S_n}{n} - (2p - 1)\right| \geq \varepsilon\right\} \leq 2e^{-a\varepsilon^2 n},$$

where $S_n = \xi_1 + \cdots + \xi_n$ and $\varepsilon > 0$.

5. Let ξ be a nonnegative random variable and $a > 0$. Find $\sup P\{x \geq a\}$ over all distributions such that:
   (i) $E\xi = 20$;
   (ii) $E\xi = 20$, $\text{Var}\, \xi = 25$;
   (iii) $E\xi = 20$, $\text{Var}\, \xi = 25$ and ξ is symmetric about its mean value.

6 The Bernoulli Scheme: II—Limit Theorems (Local, de Moivre–Laplace, Poisson)

1. As in the preceding section, let

$$S_n = \xi_1 + \cdots + \xi_n.$$

Then

$$E\frac{S_n}{n} = p, \tag{1}$$

and by (14) of Sect. 4

$$E\left(\frac{S_n}{n} - p\right)^2 = \frac{pq}{n}. \tag{2}$$

Formula (1) implies that $\frac{S_n}{n} \sim p$, where the precise meaning of the equivalence sign $\sim$ has been provided by the law of large numbers in the form of bounds for probabilities $P\left\{\left|\frac{S_n}{n} - p\right| \geq \varepsilon\right\}$. We can naturally expect that the "relation"

$$\left|\frac{S_n}{n} - p\right| \sim \sqrt{\frac{pq}{n}}, \tag{3}$$

obtainable apparently as a consequence of (2), can also receive an exact probabilistic meaning by treating, for example, the probabilities of the form

$$P\left\{\left|\frac{S_n}{n} - p\right| \leq x\sqrt{\frac{pq}{n}}\right\}, \quad x \in R^1,$$

or equivalently

$$P\left\{\left|\frac{S_n - \mathsf{E}\,S_n}{\sqrt{\mathrm{Var}\,S_n}}\right| \leq x\right\}$$

(since $\mathsf{E}\,S_n = np$ and $\mathrm{Var}\,S_n = npq$).

If, as before, we write

$$P_n(k) = C_n^k p^k q^{n-k}, \quad 0 \leq k \leq n,$$

for $n \geq 1$, then

$$P\left\{\left|\frac{S_n - \mathsf{E}\,S_n}{\sqrt{\mathrm{Var}\,S_n}}\right| \leq x\right\} = \sum_{\{k:\ |(k-np)/\sqrt{npq}|\leq x\}} P_n(k). \tag{4}$$

We set the problem of finding convenient asymptotic formulas, as $n \to \infty$, for $P_n(k)$ and for their sum over the values of k that satisfy the condition on the right-hand side of (4).

The following result provides an answer not only for these values of k (that is, for those satisfying $|k - np| = O(\sqrt{npq})$) but also for those satisfying $|k - np| = o(npq)^{2/3}$.

Local Limit Theorem. *Let $0 < p < 1$; then*

$$P_n(k) \sim \frac{1}{\sqrt{2\pi npq}} e^{-(k-np)^2/(2npq)} \tag{5}$$

uniformly in k such that $|k - np| = o(npq)^{2/3}$, more precisely, as $n \to \infty$

$$\sup_{\{k:\ |k-np|\leq\varphi(n)\}} \left|\frac{P_n(k)}{\frac{1}{\sqrt{2\pi npq}} e^{-(k-np)^2/(2npq)}} - 1\right| \to 0, \tag{6}$$

where $\varphi(n) = o(npq)^{2/3}$.

THE PROOF depends on Stirling's formula (6) of Sect. 2

$$n! = \sqrt{2\pi n}\, e^{-n} n^n (1 + R(n)),$$

where $R(n) \to 0$ as $n \to \infty$.

Then if $n \to \infty$, $k \to \infty$, $n - k \to \infty$, we have

$$C_n^k = \frac{n!}{k!(n-k)!}$$

$$= \frac{\sqrt{2\pi n}\, e^{-n} n^n}{\sqrt{2\pi k \cdot 2\pi(n-k)}\, e^{-k} k^k \cdot e^{-(n-k)}(n-k)^{n-k}} \frac{1 + R(n)}{(1 + R(k))(1 + R(n-k))}$$

$$= \frac{1}{\sqrt{2\pi n \frac{k}{n}\left(1 - \frac{k}{n}\right)}} \cdot \frac{1 + \varepsilon(n,k,n-k)}{\left(\frac{k}{n}\right)^k \left(1 - \frac{k}{n}\right)^{n-k}},$$

where $\varepsilon = \varepsilon(n, k, n - k)$ is defined in an evident way and $\varepsilon \to 0$ as $n \to \infty$, $k \to \infty$, $n - k \to \infty$.

Therefore

$$P_n(k) = C_n^k p^k q^{n-k} = \frac{1}{\sqrt{2\pi n \frac{k}{n} \left(1 - \frac{k}{n}\right)}} \frac{p^k (1-p)^{n-k}}{\left(\frac{k}{n}\right)^k \left(1 - \frac{k}{n}\right)^{n-k}} (1 + \varepsilon).$$

Write $\hat{p} = k/n$. Then

$$P_n(k) = \frac{1}{\sqrt{2\pi n \hat{p}(1 - \hat{p})}} \left(\frac{p}{\hat{p}}\right)^k \left(\frac{1-p}{1-\hat{p}}\right)^{n-k} (1 + \varepsilon)$$

$$= \frac{1}{\sqrt{2\pi n \hat{p}(1 - \hat{p})}} \exp\left\{k \log \frac{p}{\hat{p}} + (n - k) \log \frac{1-p}{1-\hat{p}}\right\} \cdot (1 + \varepsilon)$$

$$= \frac{1}{\sqrt{2\pi n \hat{p}(1 - \hat{p})}} \exp\left\{n\left[\frac{k}{n} \log \frac{p}{\hat{p}} + \left(1 - \frac{k}{n}\right) \log \frac{1-p}{1-\hat{p}}\right]\right\} (1 + \varepsilon)$$

$$= \frac{1}{\sqrt{2\pi n \hat{p}(1 - \hat{p})}} \exp\{-nH(\hat{p})\}(1 + \varepsilon),$$

where

$$H(x) = x \log \frac{x}{p} + (1 - x) \log \frac{1-x}{1-p}.$$

We are considering values of k such that $|k - np| = o(npq)^{2/3}$, and consequently $p - \hat{p} \to 0$, $n \to \infty$.

Since, for $0 < x < 1$,

$$H'(x) = \log \frac{x}{p} - \log \frac{1-x}{1-p},$$

$$H''(x) = \frac{1}{x} + \frac{1}{1-x},$$

$$H'''(x) = -\frac{1}{x^2} + \frac{1}{(1-x)^2},$$

if we write $H(\hat{p})$ in the form $H(p + (\hat{p} - p))$ and use Taylor's formula, we find that as $n \to \infty$

$$H(\hat{p}) = H(p) + H'(p)(\hat{p} - p) + \tfrac{1}{2}H''(p)(\hat{p} - p)^2 + O(|\hat{p} - p|^3)$$

$$= \frac{1}{2}\left(\frac{1}{p} + \frac{1}{q}\right)(\hat{p} - p)^2 + O(|\hat{p} - p|^3).$$

Consequently

$$P_n(k) = \frac{1}{\sqrt{2\pi n \hat{p}(1 - \hat{p})}} \exp\left\{-\frac{n}{2pq}(\hat{p} - p)^2 + nO(|\hat{p} - p|^3)\right\}(1 + \varepsilon).$$

Notice that

$$\frac{n}{2pq}(\hat{p}-p)^2 = \frac{n}{2pq}\left(\frac{k}{n}-p\right)^2 = \frac{(k-np)^2}{2npq}.$$

Therefore

$$P_n(k) = \frac{1}{\sqrt{2\pi npq}}e^{-(k-np)^2/(2npq)}(1+\varepsilon'(n,k,n-k)),$$

where

$$1+\varepsilon'(n,k,n-k) = (1+\varepsilon(n,k,n-k))\exp\{n\,O(|p-\hat{p}|^3)\}\sqrt{\frac{p(1-p)}{\hat{p}(1-\hat{p})}}$$

and, as is easily seen,

$$\sup|\varepsilon'(n,k,n-k)| \to 0, \quad n \to \infty,$$

if the sup is taken over the values of k for which

$$|k-np| \le \varphi(n), \quad \varphi(n) = o(npq)^{2/3}.$$

This completes the proof. □

Corollary. *The conclusion of the local limit theorem can be put in the following equivalent form: For all $x \in R^1$ such that $x = o(npq)^{1/6}$, and for $np + x\sqrt{npq}$ an integer from the set $\{0,1,\ldots,n\}$,*

$$P_n(np + x\sqrt{npq}) \sim \frac{1}{\sqrt{2\pi npq}}e^{-x^2/2}, \tag{7}$$

i.e., as $n \to \infty$,

$$\sup_{\{x:\,|x|\le\psi(n)\}}\left|\frac{P_n(np+x\sqrt{npq})}{\frac{1}{\sqrt{2\pi npq}}e^{-x^2/2}}-1\right| \to 0, \tag{8}$$

where $\psi(n) = o(npq)^{1/6}$.

We can reformulate these results in probabilistic language in the following way:

$$P\{S_n = k\} \sim \frac{1}{\sqrt{2\pi npq}}e^{-(k-np)^2/(2npq)}, \quad |k-np| = o(npq)^{2/3}, \tag{9}$$

$$P\left\{\frac{S_n - np}{\sqrt{npq}} = x\right\} \sim \frac{1}{\sqrt{2\pi npq}}e^{-x^2/2}, \quad x = o(npq)^{1/6}. \tag{10}$$

(In the last formula $np + x\sqrt{npq}$ is assumed to have one of the values $0,1,\ldots,n$.)

If we put $t_k = (k-np)/\sqrt{npq}$ and $\Delta t_k = t_{k+1} - t_k = 1/\sqrt{npq}$, the preceding formula assumes the form

$$P\left\{\frac{S_n - np}{\sqrt{npq}} = t_k\right\} \sim \frac{\Delta t_k}{\sqrt{2\pi}}e^{-t_k^2/2}, \quad t_k = o(npq)^{1/6}. \tag{11}$$

It is clear that $\Delta t_k = 1/\sqrt{npq} \to 0$ and the set of points $\{t_k\}$ as it were "fills" the real line. It is natural to expect that (11) can be used to obtain the integral formula

$$\mathsf{P}\left\{a < \frac{S_n - np}{\sqrt{npq}} \le b\right\} \sim \frac{1}{\sqrt{2\pi}} \int_a^b e^{-x^2/2} dx, \quad -\infty < a \le b < \infty.$$

Let us now give a precise statement.

2. For $-\infty < a \le b < \infty$ let

$$P_n(a, b] = \sum_{a < x \le b} P_n(np + x\sqrt{npq}),$$

where the summation is over those x for which $np + x\sqrt{npq}$ is an integer.

It follows from the local theorem (see also (11)) that for all t_k defined by $k = np + t_k\sqrt{npq}$ and satisfying $|t_k| \le T < \infty$,

$$P_n(np + t_k\sqrt{npq}) = \frac{\Delta t_k}{\sqrt{2\pi}} e^{-t_k^2/2}[1 + \varepsilon(t_k, n)], \tag{12}$$

where

$$\sup_{|t_k| \le T} |\varepsilon(t_k, n)| \to 0, \quad n \to \infty. \tag{13}$$

Consequently, if a and b are given so that $-T \le a \le b \le T$, then

$$\sum_{a < t_k \le b} P_n(np + t_k\sqrt{npq}) = \sum_{a < t_k \le b} \frac{\Delta t_k}{\sqrt{2\pi}} e^{-t_k^2/2} + \sum_{a < t_k \le b} \varepsilon(t_k, n) \frac{\Delta t_k}{\sqrt{2\pi}} e^{-t_k^2/2}$$

$$= \frac{1}{\sqrt{2\pi}} \int_a^b e^{-x^2/2} dx + R_n^{(1)}(a, b) + R_n^{(2)}(a, b), \tag{14}$$

where

$$R_n^{(1)}(a, b) = \sum_{a < t_k \le b} \frac{\Delta t_k}{\sqrt{2\pi}} e^{-t_k^2/2} - \frac{1}{\sqrt{2\pi}} \int_a^b e^{-x^2/2} \, dx,$$

$$R_n^{(2)}(a, b) = \sum_{a < t_k \le b} \varepsilon(t_k, n) \frac{\Delta t_k}{\sqrt{2\pi}} e^{-t_k^2/2}.$$

From the standard properties of Riemann sums,

$$\sup_{-T \le a \le b \le T} |R_n^{(1)}(a, b)| \to 0, \quad n \to \infty. \tag{15}$$

It also clear that

$$
\sup_{-T \le a \le b \le T} |R_n^{(2)}(a, b)|
$$

$$
\le \sup_{|t_k| \le T} |\varepsilon(t_k, n)| \cdot \sum_{|t_k| \le T} \frac{\Delta t_k}{\sqrt{2\pi}} e^{-t_k^2/2}
$$

$$
\le \sup_{|t_k| \le T} |\varepsilon(t_k, n)| \tag{16}
$$

$$
\times \left[\frac{1}{\sqrt{2\pi}} \int_{-T}^{T} e^{-x^2/2}\, dx + \sup_{-T \le a \le b \le T} |R_n^{(1)}(a, b)| \right] \to 0,
$$

where the convergence of the right-hand side to zero follows from (15) and from

$$
\frac{1}{\sqrt{2\pi}} \int_{-T}^{T} e^{-x^2/2}\, dx \le \frac{1}{\sqrt{2\pi}} \int_{-\infty}^{\infty} e^{-x^2/2}\, dx = 1, \tag{17}
$$

the value of the last integral being well known. We write

$$
\Phi(x) = \frac{1}{\sqrt{2\pi}} \int_{-\infty}^{x} e^{-t^2/2}\, dt.
$$

Then it follows from (14)–(16) that

$$
\sup_{-T \le a \le b \le T} |P_n(a, b] - (\Phi(b) - \Phi(a))| \to 0, \quad n \to \infty. \tag{18}
$$

We now show that this result holds for $T = \infty$ as well as for finite T. By (17), corresponding to a given $\varepsilon > 0$ we can find a finite $T = T(\varepsilon)$ such that

$$
\frac{1}{\sqrt{2\pi}} \int_{-T}^{T} e^{-x^2/2}\, dx > 1 - \tfrac{1}{4}\varepsilon. \tag{19}
$$

According to (18), we can find an N such that for all $n > N$ and $T = T(\varepsilon)$ we have

$$
\sup_{-T \le a \le b \le T} |P_n(a, b] - (\Phi(b) - \Phi(a))| < \tfrac{1}{4}\varepsilon. \tag{20}
$$

It follows from this and (19) that

$$
P_n(-T, T] > 1 - \tfrac{1}{2}\varepsilon,
$$

and consequently

$$
P_n(-\infty, T] + P_n(T, \infty) \le \tfrac{1}{2}\varepsilon,
$$

where $P_n(-\infty, T] = \lim_{S \downarrow -\infty} P_n(S, T]$ and $P_n(T, \infty) = \lim_{S \uparrow \infty} P_n(T, S]$.

Therefore for $-\infty \le a \le -T < T \le b \le \infty$,

$$\left| P_n(a, b] - \frac{1}{\sqrt{2\pi}} \int_a^b e^{-x^2/2}\, dx \right|$$

$$\le \left| P_n(-T, T] - \frac{1}{\sqrt{2\pi}} \int_{-T}^T e^{-x^2/2}\, dx \right|$$

$$+ \left| P_n(a, -T] - \frac{1}{\sqrt{2\pi}} \int_a^{-T} e^{-x^2/2}\, dx \right| + \left| P_n(T, b] - \frac{1}{\sqrt{2\pi}} \int_T^b e^{-x^2/2}\, dx \right|$$

$$\le \frac{1}{4}\varepsilon + P_n(-\infty, -T] + \frac{1}{\sqrt{2\pi}} \int_{-\infty}^{-T} e^{-x^2/2}\, dx + P_n(T, \infty)$$

$$+ \frac{1}{\sqrt{2\pi}} \int_T^\infty e^{-x^2/2}\, dx \le \frac{1}{4}\varepsilon + \frac{1}{2}\varepsilon + \frac{1}{8}\varepsilon + \frac{1}{8}\varepsilon = \varepsilon.$$

By using (18) it is now easy to see that $P_n(a, b]$ tends to $\Phi(b) - \Phi(a)$ *uniformly* for $-\infty \le a < b \le \infty$.

Thus we have proved the following theorem.

De Moivre–Laplace Integral Theorem. *Let* $0 < p < 1$,

$$P_n(k) = C_n^k p^k q^{n-k}, \quad P_n(a, b] = \sum_{a < x \le b} P_n(np + x\sqrt{npq}).$$

Then

$$\sup_{-\infty \le a < b \le \infty} \left| P_n(a, b] - \frac{1}{\sqrt{2\pi}} \int_a^b e^{-x^2/2}\, dx \right| \to 0, \quad n \to \infty. \tag{21}$$

In probabilistic language (21) can be stated in the following way:

$$\sup_{-\infty \le a < b \le \infty} \left| P\left\{ a < \frac{S_n - \mathsf{E}\, S_n}{\sqrt{\mathrm{Var}\, S_n}} \le b \right\} - \frac{1}{\sqrt{2\pi}} \int_a^b e^{-x^2/2}]\, dx \right| \to 0, \quad n \to \infty.$$

It follows at once from this formula that

$$P\{A < S_n \le B\} - \left[\Phi\left(\frac{B - np}{\sqrt{npq}} \right) - \Phi\left(\frac{A - np}{\sqrt{npq}} \right) \right] \to 0, \tag{22}$$

as $n \to \infty$, whenever $-\infty \le A < B \le \infty$.

Example. A true die is tossed 12000 times. We ask for the probability P that the number of 6's lies in the interval (1800, 2100].

The required probability is

$$P = \sum_{1800 < k \le 2100} C_{12000}^k \left(\frac{1}{6} \right)^k \left(\frac{5}{6} \right)^{12000-k}.$$

An exact calculation of this sum would obviously be rather difficult. However, if we use the integral theorem we find that the probability P in question is approximately ($n = 12000$, $p = \frac{1}{6}$, $a = 1800$, $b = 2100$)

$$\Phi\left(\frac{2100 - 2000}{\sqrt{12000 \cdot \frac{1}{6} \cdot \frac{5}{6}}}\right) - \Phi\left(\frac{1800 - 2000}{\sqrt{12000 \cdot \frac{1}{6} \cdot \frac{5}{6}}}\right) = \Phi(\sqrt{6}) - \Phi(-2\sqrt{6})$$
$$\approx \Phi(2.449) - \Phi(-4.898) \approx 0.992,$$

where the values of $\Phi(2.449)$ and $\Phi(-4.898)$ were taken from tables of $\Phi(x)$ (this is the *normal distribution function*; see Subsection 6 below).

3. We have plotted a graph of $P_n(np+x\sqrt{npq})$ (with x assumed such that $np+x\sqrt{npq}$ is an integer) in Fig. 9.

Then the local theorem says that the curve $(1/\sqrt{2\pi npq})e^{-x^2/2}$ provides a close fit to $P_n(np + x\sqrt{npq})$ when $x = o(npq)^{1/6}$. On the other hand the integral theorem says that

$$P_n(a, b] = P\{a\sqrt{npq} < S_n - np \leq b\sqrt{npq}\}$$
$$= P\{np + a\sqrt{npq} < S_n \leq np + b\sqrt{npq}\}$$

is closely approximated by the integral $(1/\sqrt{2\pi}) \int_a^b e^{-x^2/2}\, dx$.

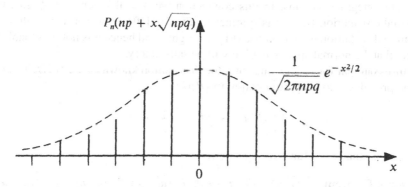

Fig. 9

We write

$$F_n(x) = P_n(-\infty, x] \quad \left(= P\left\{\frac{S_n - np}{\sqrt{npq}} \leq x\right\}\right).$$

Then it follows from (21) that

$$\sup_{-\infty \leq x \leq \infty} |F_n(x) - \Phi(x)| \to 0, \quad n \to \infty. \tag{23}$$

It is natural to ask how rapid the approach to zero is in (21) and (23), as $n \to \infty$. We quote a result in this direction (a special case of the Berry–Esseen theorem: see Sect. 11 in Chap. 3):

$$\sup_{-\infty \leq x \leq \infty} |F_n(x) - \Phi(x)| \leq \frac{p^2 + q^2}{\sqrt{npq}}. \tag{24}$$

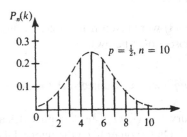 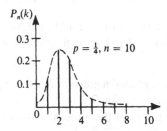

Fig. 10

It is important to recognize that the order of the estimate $(1/\sqrt{npq})$ cannot be improved; this means that the approximation of $F_n(x)$ by $\Phi(x)$ can be poor for values of p that are close to 0 or 1, even when n is large. This suggests the question of whether there is a better method of approximation for the probabilities of interest when p or q is small, something better than the normal approximation given by the local and integral theorems. In this connection we note that for $p = \frac{1}{2}$, say, the binomial distribution $\{P_n(k)\}$ is symmetric (Fig. 10, left). However, for small p the binomial distribution is *asymmetric* (Fig. 10, right), and hence it is not reasonable to expect that the normal approximation will be satisfactory.

4. It turns out that for small values of p the distribution known as the *Poisson distribution* provides a good approximation to $\{P_n(k)\}$.

Let

$$P_n(k) = \begin{cases} C_n^k p^k q^{n-k}, & k = 0, 1, \ldots, n, \\ 0, & k = n+1, n+2, \ldots, \end{cases}$$

and suppose that p is a function $p(n)$ of n.

Poisson's Theorem. *Let* $p(n) \to 0$, $n \to \infty$, *in such a way that* $np(n) \to \lambda$, *where* $\lambda > 0$. *Then for* $k = 1, 2, \ldots$,

$$P_n(k) \to \pi_k, \quad n \to \infty, \tag{25}$$

where

$$\pi_k = \frac{\lambda^k e^{-\lambda}}{k!}, \quad k = 0, 1, \ldots. \tag{26}$$

THE PROOF is extremely simple. Since $p(n) = (\lambda/n) + o(1/n)$ by hypothesis, for a given $k = 0, 1, \ldots$ and $n \to \infty$,

$$P_n(k) = C_n^k p^k q^{n-k}$$

$$= \frac{n(n-1)\cdots(n-k+1)}{k!} \left[\frac{\lambda}{n} + o\left(\frac{1}{n}\right) \right]^k \cdot \left[1 - \frac{\lambda}{n} + o\left(\frac{1}{n}\right) \right]^{n-k}.$$

But

$$n(n-1)\cdots(n-k+1) \left[\frac{\lambda}{n} + o\left(\frac{1}{n}\right) \right]^k$$

$$= \frac{n(n-1)\cdots(n-k+1)}{n^k} [\lambda + o(1)]^k \to \lambda^k, \quad n \to \infty,$$

and

$$\left[1 - \frac{\lambda}{n} + o\left(\frac{1}{n}\right) \right]^{n-k} \to e^{-\lambda}, \quad n \to \infty,$$

which establishes (25). $\square$

The set of numbers $\{\pi_k, \ k = 0, 1, \ldots\}$ defines the *Poisson probability distribution* $(\pi_k \geq 0, \ \sum_{k=0}^{\infty} \pi_k = 1)$. Notice that all the (discrete) distributions considered previously were concentrated at only a finite number of points. The Poisson distribution is the first example that we have encountered of a (discrete) distribution concentrated at a countable number of points.

The following result of Prokhorov exhibits the rate of convergence of $P_n(k)$ to π_k as $n \to \infty$: *if $np(n) = \lambda > 0$, then*

$$\sum_{k=0}^{\infty} |P_n(k) - \pi_k| \leq \frac{2\lambda}{n} \cdot \min(2, \lambda). \tag{27}$$

The proof of a somewhat weaker result is given in Sect. 12, Chap. 3.

5. Let us return to the de Moivre–Laplace limit theorem, and show how it implies the law of large numbers. Since

$$P\left\{ \left| \frac{S_n}{n} - p \right| \leq \varepsilon \right\} = P\left\{ \left| \frac{S_n - np}{\sqrt{npq}} \right| \leq \varepsilon \sqrt{\frac{n}{pq}} \right\},$$

it is clear from (21) that for $\varepsilon > 0$

$$P\left\{ \left| \frac{S_n}{n} - p \right| \leq \varepsilon \right\} - \frac{1}{\sqrt{2\pi}} \int_{-\varepsilon\sqrt{n/pq}}^{\varepsilon\sqrt{n/pq}} e^{-x^2/2} \, dx \to 0, \quad n \to \infty, \tag{28}$$

whence

$$P\left\{ \left| \frac{S_n}{n} - p \right| \leq \varepsilon \right\} \to 1, \quad n \to \infty,$$

which is the conclusion of the law of large numbers.

From (28)

$$P\left\{\left|\frac{S_n}{n} - p\right| \le \varepsilon\right\} \sim \frac{1}{\sqrt{2\pi}} \int_{-\varepsilon\sqrt{n/pq}}^{\varepsilon\sqrt{n/pq}} e^{-x^2/2} \, dx, \quad n \to \infty, \tag{29}$$

whereas Chebyshev's inequality yielded only

$$P\left\{\left|\frac{S_n}{n} - p\right| \le \varepsilon\right\} \ge 1 - \frac{pq}{n\varepsilon^2}.$$

It was shown in Subsection 3 of Sect. 5 that Chebyshev's inequality yielded the estimate

$$n \ge \frac{1}{4\varepsilon^2\alpha} \quad (= n_1(\alpha))$$

for the number of observations needed for the validity of the inequality

$$P\left\{\left|\frac{S_n}{n} - p\right| \le \varepsilon\right\} \ge 1 - \alpha.$$

Thus with $\varepsilon = 0.02$ and $\alpha = 0.05, 12500$ observations were needed. We can now solve the same problem by using the approximation (29).

We define the number $k(\alpha)$ by

$$\frac{1}{\sqrt{2\pi}} \int_{-k(\alpha)}^{k(\alpha)} e^{-x^2/2} \, dx = 1 - \alpha.$$

Since $\varepsilon\sqrt{(n/pq)} \ge 2\varepsilon\sqrt{n}$, if we define n as the smallest integer satisfying

$$2\varepsilon\sqrt{n} \ge k(\alpha) \tag{30}$$

we find that

$$P\left\{\left|\frac{S_n}{n} - p\right| \le \varepsilon\right\} \gtrsim 1 - \alpha. \tag{31}$$

We find from (30) that the smallest integer $n \ge n_2(\alpha)$ with

$$n_2(\alpha) = \left[\frac{k^2(\alpha)}{4\varepsilon^2}\right]$$

guarantees that (31) is satisfied, and the accuracy of the approximation can easily be established by using (24).

Taking $\varepsilon = 0.02$, $\alpha = 0.05$, we find that in fact 2500 observations suffice, rather than the 12500 found by using Chebyshev's inequality. The values of $k(\alpha)$ have been tabulated. We quote a number of values of $k(\alpha)$ for various values of α:

α	0,50	0,3173	0,10	0,05	0,0454	0,01	0,0027
$k(\alpha)$	0,675	1,000	1,645	1,960	2,000	2,576	3,000

6. The function

$$\Phi(x) = \frac{1}{\sqrt{2\pi}} \int_{-\infty}^{x} e^{-t^2/2} \, dt, \qquad (32)$$

which was introduced above and occurs in the de Moivre–Laplace integral theorem, plays an exceptionally important role in probability theory. It is known as the *normal* or *Gaussian distribution function* on the real line, with the (*normal or Gaussian*) density

$$\varphi(x) = \frac{1}{\sqrt{2\pi}} e^{-x^2/2}, \qquad x \in R^1.$$

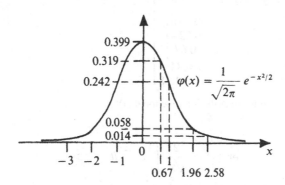

Fig. 11 Graph of the normal probability density $\varphi(x)$

We have already encountered (discrete) distributions concentrated on a finite or countable set of points. The normal distribution belongs to another important class of distributions that arise in probability theory. We have mentioned its exceptional role; this comes about, first of all, because under rather general hypotheses, sums of a large number of independent random variables (not necessarily Bernoulli variables) are closely approximated by the normal distribution (Sect. 4, Chap. 3). For the present we mention only some of the simplest properties of $\varphi(x)$ and $\Phi(x)$, whose graphs are shown in Figs. 11 and 12.

The function $\varphi(x)$ is a symmetric bell-shaped curve, decreasing very rapidly with increasing $|x|$: thus $\varphi(1) = 0.24197$, $\varphi(2) = 0.053991$, $\varphi(3) = 0.004432$, $\varphi(4) = 0.000134$, $\varphi(5) = 0.000016$. Its maximum is attained at $x = 0$ and is equal to $(2\pi)^{-1/2} \approx 0.399$.

The curve $\Phi(x) = (1/\sqrt{2\pi}) \int_{-\infty}^{x} e^{-t^2/2} dt$ approaches 1 very rapidly as x increases: $\Phi(1) = 0.841345$, $\Phi(2) = 0.977250$, $\Phi(3) = 0.998650$, $\Phi(4) = 0.999968$, $\Phi(4.5) = 0.999997$.

For tables of $\varphi(x)$ and $\Phi(x)$, as well as of other important functions that are used in probability theory and mathematical statistics, see [11].

It is worth to mention that for calculations, along with $\Phi(x)$, a closely related *error function*

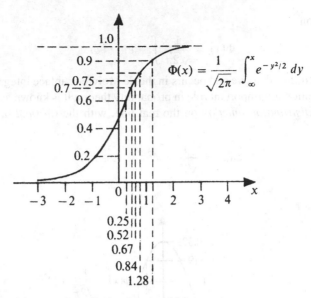

Fig. 12 Graph of the normal distribution function $\Phi(x)$

$$\mathrm{erf}(x) = \frac{2}{\sqrt{\pi}} \int_0^x e^{-t^2}\, dt, \quad x > 0,$$

is often used. Obviously, for $x > 0$,

$$\Phi(x) = \frac{1}{2}\left[1 + \mathrm{erf}\left(\frac{x}{\sqrt{2}}\right)\right], \quad \mathrm{erf}(x) = 2\Phi(\sqrt{2}x) - 1.$$

7. At the end of Subsection 3, Sect. 5, we noticed that the upper bound for the probability of the event $\{\omega\colon |(S_n/n) - p| \geq \varepsilon\}$, given by Chebyshev's inequality, was rather crude. That estimate was obtained from Chebyshev's inequality $\mathsf{P}\{X \geq \varepsilon\} \leq \mathsf{E}X^2/\varepsilon^2$ for nonnegative random variables $X \geq 0$. We may, however, use Chebyshev's inequality in the form

$$\mathsf{P}\{X \geq \varepsilon\} = \mathsf{P}\{X^{2k} \geq \varepsilon^{2k}\} \leq \frac{\mathsf{E}X^{2k}}{\varepsilon^{2k}}. \tag{33}$$

However, we can go further by using the "exponential form" of Chebyshev's inequality: if $X \geq 0$ and $\lambda > 0$, this states that

$$\mathsf{P}\{X \geq \varepsilon\} = \mathsf{P}\{e^{\lambda X} \geq e^{\lambda \varepsilon}\} \leq \mathsf{E}\,e^{\lambda(X-\varepsilon)}. \tag{34}$$

Since the positive number λ is arbitrary, it is clear that

$$\mathsf{P}\{X \geq \varepsilon\} \leq \inf_{\lambda > 0} \mathsf{E}\,e^{\lambda(X-\varepsilon)}. \tag{35}$$

Let us see what the consequences of this approach are in the case when $X = S_n/n$, $S_n = \xi_1 + \cdots + \xi_n$, $P(\xi_i = 1) = p$, $P(\xi_i = 0) = q$, $i \geq 1$.

Let us set $\varphi(\lambda) = E\, e^{\lambda \xi_1}$. Then

$$\varphi(\lambda) = 1 - p + p e^\lambda$$

and, under the hypothesis of the independence of $\xi_1, \xi_2, \ldots, \xi_n$,

$$E\, e^{\lambda S_n} = [\varphi(\lambda)]^n.$$

Therefore $(0 < a < 1)$

$$P\left\{\frac{S_n}{n} \geq a\right\} \leq \inf_{\lambda > 0} E\, e^{\lambda(S_n/n - a)} = \inf_{\lambda > 0} e^{-n[\lambda a/n - \log \varphi(\lambda/n)]}$$

$$= \inf_{s > 0} e^{-n[as - \log \varphi(s)]} = e^{-n \sup_{s>0}[as - \log \varphi(s)]}. \tag{36}$$

Similarly,

$$P\left\{\frac{S_n}{n} \leq a\right\} \leq e^{-n \sup_{s<0}[as - \log \varphi(s)]}. \tag{37}$$

The function $f(s) = as - \log[1 - p + p e^s]$ attains its maximum for $p \leq a \leq 1$ at the point s_0 $(f'(s_0) = 0)$ determined by the equation

$$e^{s_0} = \frac{a(1 - p)}{p(1 - a)}.$$

Consequently,

$$\sup_{s>0} f(s) = H(a),$$

where

$$H(a) = a \log \frac{a}{p} + (1 - a) \log \frac{1 - a}{1 - p}$$

is the function that was used in the proof of the local theorem (Subsection 1).

Thus, for $p \leq a \leq 1$

$$P\left\{\frac{S_n}{n} \geq a\right\} \leq e^{-nH(a)}, \tag{38}$$

and therefore, since $H(p + x) \geq 2x^2$ and $0 \leq p + x \leq 1$, we have, for $\varepsilon > 0$ and $0 \leq p \leq 1$,

$$P\left\{\frac{S_n}{n} - p \geq \varepsilon\right\} \leq e^{-2n\varepsilon^2}. \tag{39}$$

We can establish similarly that for $a \leq p \leq 1$

$$P\left\{\frac{S_n}{n} \leq a\right\} \leq e^{-nH(a)} \tag{40}$$

and consequently, for every $\varepsilon > 0$ and $0 \leq p \leq 1$,

$$P\left\{\frac{S_n}{n} - p \leq -\varepsilon\right\} \leq e^{-2n\varepsilon^2}. \tag{41}$$

Therefore,

$$P\left\{\left|\frac{S_n}{n} - p\right| \geq \varepsilon\right\} \leq 2e^{-2n\varepsilon^2}. \tag{42}$$

This implies that the number of observations $n_3(\alpha)$ which ensures the validity of the inequality

$$P\left\{\left|\frac{S_n}{n} - p\right| \leq \varepsilon\right\} \geq 1 - \alpha, \tag{43}$$

for any $0 \leq p \leq 1$ is given by the formula

$$n_3(\alpha) = \left[\frac{\log(2/\alpha)}{2\varepsilon^2}\right], \tag{44}$$

where $[x]$ is the integral part of x. If we neglect "integral parts" and compare $n_3(\alpha)$ with $n_1(\alpha) = [(4\alpha\varepsilon^2)^{-1}]$, we find that

$$\frac{n_1(\alpha)}{n_3(\alpha)} = \frac{1}{2\alpha \log \frac{2}{\alpha}} \uparrow \infty, \quad \alpha \downarrow 0.$$

It is clear from this that when $\alpha \downarrow 0$, an estimate of the *smallest* number of observations needed to ensure (43), which can be obtained from the exponential Chebyshev inequality, is more precise than the estimate obtained from the ordinary Chebyshev inequality, especially for small α.

Using the relation

$$\frac{1}{\sqrt{2\pi}} \int_x^\infty e^{-y^2/2}\, dy \sim \frac{1}{\sqrt{2\pi}x} e^{-x^2/2}, \quad x \to \infty,$$

which is easily established with the help of L'Hôpital's rule, one can show that $k^2(\alpha) \sim 2\log \frac{2}{\alpha}$, $\alpha \downarrow 0$. Therefore,

$$\frac{n_2(\alpha)}{n_3(\alpha)} \to 1, \quad \alpha \downarrow 0.$$

Inequalities like (38)–(42) are known as *inequalities for the probability of large deviations*. This terminology can be explained in the following way.

The de Moivre–Laplace integral theorem makes it possible to estimate in a simple way the probabilities of the events $\{|S_n - np| \leq x\sqrt{n}\}$ characterizing the "standard" deviation (up to order $\sqrt{n}$) of S_n from np, whereas the inequalities (39), (41), and (42) provide bounds for the probabilities of the events $\{\omega \colon |S_n - np| \leq xn\}$, describing deviations of order greater than $\sqrt{n}$, in fact of order n.

We shall continue the discussion of probabilities of large deviations in more general situations in Sect. 5, Chap. 4, Vol. 2.

8. PROBLEMS

1. Let $n = 100$, $p = 0.1, 0.2, 0.3, 0.4, 0.5$. Using tables (for example, those in [11]) of the binomial and Poisson distributions, compare the values of the probabilities

$$P\{10 < S_{100} \leq 12\}, \quad P\{20 < S_{100} \leq 22\},$$
$$P\{33 < S_{100} \leq 35\}, \quad P\{40 < S_{100} \leq 42\},$$
$$P\{50 < S_{100} \leq 52\}$$

with the corresponding values given by the normal and Poisson approximations.

2. Let $p = \frac{1}{2}$ and $Z_n = 2S_n - n$ (the *excess* of 1's over 0's in n trials). Show that

$$\sup_j \left| \sqrt{\pi n} P\{Z_{2n} = j\} - e^{-j^2/4n} \right| \to 0, \quad n \to \infty.$$

3. Show that the rate of convergence in Poisson's theorem (with $p = \lambda/n$) is given by

$$\sup_k \left| P_n(k) - \frac{\lambda^k e^{-\lambda}}{k!} \right| \leq \frac{2\lambda^2}{n}.$$

(It is advisable to read Sect. 12, Chap. 3.)

7 Estimating the Probability of Success in the Bernoulli Scheme

1. In the Bernoulli scheme $(\Omega, \mathscr{A}, P)$ with $\Omega = \{\omega : \omega = (x_1, \ldots, x_n), x_i = 0, 1)\}$, $\mathscr{A} = \{A : A \subseteq \Omega\}$, $P(\{\omega\}) = p(\omega)$, where

$$p(\omega) = p^{\Sigma x_i} q^{n - \Sigma x_i},$$

we supposed that p (the probability of "success") was *known*.

Let us now suppose that p is not known in advance and that we want to determine it by observing the outcomes of experiments; or, what amounts to the same thing, by observations of the random variables $\xi_1, \ldots, \xi_n$, where $\xi_i(\omega) = x_i$. This is a typical problem of mathematical statistics, which can be formulated in various ways. We shall consider two of the possible formulations: the problem of *point estimation* and the problem of *constructing confidence intervals*.

In the notation used in mathematical statistics, the unknown parameter is denoted by θ, assuming *a priori* that θ belongs to the set $\Theta = [0, 1]$. The set of objects

$$\mathscr{E} = (\Omega, \mathscr{A}, P_\theta; \theta \in \Theta) \quad \text{with} \quad P_\theta(\{\omega\}) = \theta^{\Sigma x_i}(1 - \theta)^{n - \Sigma x_i}$$

is often said to be the *probabilistic-statistical model* (corresponding to "n indepen-dent trials" with probability of "success" $\theta \in \Theta$), and any function $T_n = T_n(\omega)$ with values in Θ is called an *estimator*.

If $S_n = \xi_1 + \cdots + \xi_n$ and $T_n^* = S_n/n$, it follows from the law of large numbers that T_n^* is *consistent*, in the sense that ($\varepsilon > 0$)

$$P_\theta\{|T_n^* - \theta| \geq \varepsilon\} \to 0, \quad n \to \infty. \tag{1}$$

Moreover, this estimator is *unbiased*: for every θ

$$\mathsf{E}_\theta \, T_n^* = \theta, \tag{2}$$

where E_θ is the expectation corresponding to the probability P_θ.

The property of being unbiased is quite natural: it expresses the fact that any reasonable estimate ought, at least "on the average," to lead to the desired result. However, it is easy to see that T_n^* is not the only unbiased estimator. For example, the same property is possessed by every estimator

$$T_n = \frac{b_1 x_1 + \cdots + b_n x_n}{n},$$

where $b_1 + \cdots + b_n = n$. Moreover, the law of large numbers (1) is also satisfied by such estimators (at least if $|b_i| \leq K < \infty$); and so these estimators T_n are just as "good" as T_n^*.

By the very meaning of "estimator," it is natural to suppose that an estimator is the better, the smaller its deviation from the parameter that is being estimated. On this basis, we call an estimator $\tilde{T}_n$ *efficient* (in the class of unbiased estimators T_n) if

$$\mathrm{Var}_\theta \, \tilde{T}_n = \inf_{T_n} \mathrm{Var}_\theta \, T_n, \quad \theta \in \Theta, \tag{3}$$

where $\mathrm{Var}_\theta \, T_n$ is the variance of T_n, i.e., $\mathsf{E}_\theta (T_n - \theta)^2$.

Let us show that the estimator T_n^*, considered above, is efficient. We have

$$\mathrm{Var}_\theta \, T_n^* = \mathrm{Var}_\theta \left(\frac{S_n}{n} \right) = \frac{\mathrm{Var}_\theta \, S_n}{n^2} = \frac{n\theta(1-\theta)}{n^2} = \frac{\theta(1-\theta)}{n}. \tag{4}$$

Hence to establish that T_n^* is efficient, we have only to show that

$$\inf_{T_n} \mathrm{Var}_\theta \, T_n \geq \frac{\theta(1-\theta)}{n}. \tag{5}$$

This is obvious for $\theta = 0$ or 1. Let $\theta \in (0, 1)$ and

$$p_\theta(x_i) = \theta^{x_i}(1 - \theta)^{1-x_i}.$$

It is clear that $\mathsf{P}_\theta(\{\omega\}) = p_\theta(\omega)$, where

$$p_\theta(\omega) = \prod_{i=1}^{n} p_\theta(x_i).$$

Let us write

$$L_\theta(\omega) = \log p_\theta(\omega).$$

Then

$$L_\theta(\omega) = \log\theta \sum x_i + \log(1-\theta) \sum(1-x_i)$$

and

$$\frac{\partial L_\theta(\omega)}{\partial\theta} = \frac{\sum(x_i - \theta)}{\theta(1-\theta)}.$$

Since

$$1 = \mathsf{E}_\theta\, 1 = \sum_\omega p_\theta(\omega),$$

and since T_n is unbiased,

$$\theta \equiv \mathsf{E}_\theta\, T_n = \sum_\omega T_n(\omega)p_\theta(\omega),$$

after differentiating with respect to θ, we find that

$$0 = \sum_\omega \frac{\partial p_\theta(\omega)}{\partial\theta} = \sum_\omega \frac{\left(\frac{\partial p_\theta(\omega)}{\partial\theta}\right)}{p_\theta(\omega)} p_\theta(\omega) = \mathsf{E}_\theta\left[\frac{\partial L_\theta(\omega)}{\partial\theta}\right],$$

$$1 = \sum_\omega T_n \frac{\left(\frac{\partial p_\theta(\omega)}{\partial\theta}\right)}{p_\theta(\omega)} p_\theta(\omega) = \mathsf{E}_\theta\left[T_n \frac{\partial L_\theta(\omega)}{\partial\theta}\right].$$

Therefore

$$1 = \mathsf{E}_\theta\left[(T_n - \theta)\frac{\partial L_\theta(\omega)}{\partial\theta}\right]$$

and by the Cauchy–Bunyakovskii inequality,

$$1 < \mathsf{E}_0[T_n - \theta]^2 \cdot \mathsf{E}_\theta\left[\frac{\partial L_\theta(\omega)}{\partial\theta}\right]^2,$$

whence

$$\mathsf{E}_\theta[T_n - \theta]^2 \geq \frac{1}{I_n(\theta)}, \qquad (6)$$

where

$$I_n(\theta) = \left[\frac{\partial L_\theta(\omega)}{\partial\theta}\right]^2$$

is known as *Fisher's information*.

From (6) we can obtain a special case of the *Rao–Cramér inequality* for unbiased estimators T_n:

$$\inf_{T_n} \mathrm{Var}_\theta\, T_n \geq \frac{1}{I_n(\theta)}. \qquad (7)$$

In the present case

$$I_n(\theta) = \mathsf{E}_\theta \left[\frac{\partial L_\theta(\omega)}{\partial \theta} \right]^2 = \mathsf{E}_\theta \left[\frac{\sum (\xi_i - \theta)}{\theta(1-\theta)} \right]^2 = \frac{n\theta(1-\theta)}{[\theta(1-\theta)]^2} = \frac{n}{\theta(1-\theta)},$$

which also establishes (5), from which, as we already noticed, there follows the efficiency of the unbiased estimator $T_n^* = S_n/n$ for the unknown parameter θ.

2. It is evident that, in considering T_n^* as a point estimator for θ, we have introduced a certain amount of inaccuracy. It can even happen that the numerical value of T_n^* calculated from observations of $x_1, \ldots, x_n$ differs rather severely from the true value θ. Hence it would be advisable to determine the size of the error.

It would be too much to hope that $T_n^*(\omega)$ differs little from the true value θ for *all* sample points ω. However, we know from the law of large numbers that for every $\delta > 0$ the probability of the event $\{|\theta - T_n^*(\omega)| > \delta\}$ will be arbitrarily small for sufficiently large n.

By Chebyshev's inequality

$$\mathsf{P}_\theta \{|\theta - T_n^*| > \delta\} \leq \frac{\mathrm{Var}_\theta\, T_n^*}{\delta^2} = \frac{\theta(1-\theta)}{n\delta^2}$$

and therefore, for every $\lambda > 0$,

$$\mathsf{P}_\theta \left\{ |\theta - T_n^*| \leq \lambda\sqrt{\frac{\theta(1-\theta)}{n}} \right\} \geq 1 - \frac{1}{\lambda^2}.$$

If we take, for example, $\lambda = 3$, then with P_θ-probability greater than 0.8888 (since $1 - (1/3^2) = \frac{8}{9} \approx 0.8889$) the event

$$|\theta - T_n^*| \leq 3\sqrt{\frac{\theta(1-\theta)}{n}}$$

will be realized, and a fortiori the event

$$|\theta - T_n^*| \leq \frac{3}{2\sqrt{n}},$$

since $\theta(1-\theta) \leq \frac{1}{4}$.

Therefore

$$\mathsf{P}_\theta \left\{ |\theta - T_n^*| \leq \frac{3}{2\sqrt{n}} \right\} = \mathsf{P}_\theta \left\{ T_n^* - \frac{3}{2\sqrt{n}} \leq \theta \leq T_n^* + \frac{3}{2\sqrt{n}} \right\} \geq 0.8888.$$

In other words, we can say with probability greater than 0.8888 that the exact value of θ is in the interval $[T_n^* - (3/2\sqrt{n}), T_n^* + (3/2\sqrt{n})]$. This statement is sometimes written in the symbolic form

$$\theta \simeq T_n^* \pm \frac{3}{2\sqrt{n}} \quad (\geq 88\%),$$

where "$\geq 88\%$" means "in more than 88% of all cases."

The interval $[T_n^* - (3/2\sqrt{n}),\ T_n^* + (3/2\sqrt{n})]$ is an example of what are called confidence intervals for the unknown parameter.

Definition. An interval of the form

$$[\psi_1(\omega),\ \psi_2(\omega)]$$

where $\psi_1(\omega)$ and $\psi_2(\omega)$ are functions of sample points, is called a *confidence interval of reliability* $1 - \delta$ (or of *significance level* δ) if

$$\mathbf{P}_\theta\{\psi_1(\omega) \le \theta \le \psi_2(\omega)\} \ge 1 - \delta$$

for all $\theta \in \Theta$.

The preceding discussion shows that the interval

$$\left[T_n^* - \frac{\lambda}{2\sqrt{n}},\ T_n^* + \frac{\lambda}{2\sqrt{n}}\right]$$

has reliability $1 - (1/\lambda^2)$. In point of fact, the reliability of this confidence interval is considerably higher, since Chebyshev's inequality gives only crude estimates of the probabilities of events.

To obtain more precise results we notice that

$$\left\{\omega:\ |\theta - T_n^*| \le \lambda\sqrt{\frac{\theta(1 - \theta)}{n}}\right\} = \{\omega:\ \psi_1(T_n^*,\ n) \le \theta \le \psi_2(T_n^*,\ n)\},$$

where $\psi_1 = \psi_1(T_n^*,\ n)$ and $\psi_2 = \psi_2(T_n^*,\ n)$ are the roots of the quadratic equation

$$(\theta - T_n^*)^2 = \frac{\lambda^2}{n}\theta(1 - \theta),$$

which describes an ellipse situated as shown in Fig. 13.

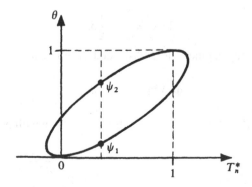

Fig. 13

Now let

$$F_\theta^n(x) = \mathsf{P}_\theta \left\{ \frac{S_n - n\theta}{\sqrt{n\theta(1-\theta)}} \le x \right\}.$$

Then by (24) of Sect. 6

$$\sup_x |F_\theta^n(x) - \Phi(x)| \le \frac{1}{\sqrt{n\theta(1-\theta)}}.$$

Therefore if we know *a priori* that

$$0 < \Delta \le \theta \le 1 - \Delta < 1,$$

where Δ is a constant, then

$$\sup_x |F_\theta^n(x) - \Phi(x)| \le \frac{1}{\Delta\sqrt{n}}$$

and consequently

$$\mathsf{P}_\theta\{\psi_1(T_n^*, n) \le \theta \le \psi_2(T_n^*, n)\} = \mathsf{P}_\theta \left\{ |\theta - T_n^*| \le \lambda\sqrt{\frac{\theta(1-\theta)}{n}} \right\}$$

$$= \mathsf{P}_\theta \left\{ \frac{|S_n - n\theta|}{\sqrt{n\theta(1-\theta)}} \le \lambda \right\}$$

$$\ge (2\Phi(\lambda) - 1) - \frac{2}{\Delta\sqrt{n}}.$$

Let λ^* be the smallest λ for which

$$(2\Phi(\lambda) - 1) - \frac{2}{\Delta\sqrt{n}} \ge 1 - \delta^*,$$

where δ^* is a given significance level. Putting $\delta = \delta^* - (2/\Delta\sqrt{n})$, we find that λ^* satisfies the equation

$$\Phi(\lambda) = 1 - \frac{1}{2}\delta.$$

For large n we may neglect the term $2/\Delta\sqrt{n}$ and assume that λ^* satisfies

$$\Phi(\lambda^*) = 1 - \frac{1}{2}\delta^*.$$

In particular, if $\lambda^* = 3$ then $1 - \delta^* = 0.9973\ldots$. Then with probability *approximately* 0.9973

$$T_n^* - 3\sqrt{\frac{\theta(1-\theta)}{n}} \le \theta \le T_n^* + 3\sqrt{\frac{\theta(1-\theta)}{n}} \tag{8}$$

or, after iterating and then suppressing terms of order $O(n^{-3/4})$, we obtain

$$T_n^* - 3\sqrt{\frac{T_n^*(1 - T_n^*)}{n}} \leq \theta \leq T_n^* + 3\sqrt{\frac{T_n^*(1 - T_n^*)}{n}}. \tag{9}$$

Hence it follows that the confidence interval

$$\left[T_n^* - \frac{3}{2\sqrt{n}}, \; T_n^* + \frac{3}{2\sqrt{n}} \right] \tag{10}$$

has (for large n) reliability 0.9973 (whereas Chebyshev's inequality only provided reliability approximately 0.8889).

To illustrate the practical meaning of this result, suppose that we carry out a large number N of series of experiments, in each of which we estimate the parameter θ after n observations. Then in about 99.73 % of the N cases the estimate will differ from the true value of the parameter by at most $\frac{3}{2\sqrt{n}}$. (On this topic see also the end of Sect. 5.)

One should remember that the confidence interval (10) is approximate and valid only for large n. For the construction of the exact confidence interval with appropriate tables and references see [11].

3. PROBLEMS

1. Let it be known *a priori* that θ takes values in the set $\Theta_0 \subseteq [0, 1]$. When does an unbiased estimator for θ exist, taking values only in Θ_0?
2. Under the conditions of the preceding problem, find an analog of the Rao–Cramér inequality and discuss the problem of efficient estimators.
3. Under the conditions of the first problem, discuss the construction of confidence intervals for θ.
4. In addition to Problem 5 in Sect. 2 discuss the problem of unbiasedness and efficiency of the estimator $\widehat{N}$ assuming that N is sufficiently large, $N \gg M, N \gg n$. By analogy with the confidence intervals for θ (see (8) and (9)), construct confidence intervals $[\widehat{N} - a(\widehat{N}), \widehat{N} + b(\widehat{N})]$ for N such that

$$P_{N, M; n}\{\widehat{N} - a(\widehat{N}) \leq N \leq \widehat{N} + b(\widehat{N})\} \approx 1 - \alpha,$$

where α is a small number.

8 Conditional Probabilities and Expectations with Respect to Decompositions

1. Let $(\Omega, \mathscr{A}, \mathsf{P})$ be a finite probability space and

$$\mathscr{D} = \{D_1, \ldots, D_k\}$$

a decomposition of Ω ($D_i \in \mathscr{A}$, $P(D_i) > 0$, $i = 1, \ldots, k$, and $D_1 + \cdots + D_k = \Omega$). Also let A be an event from $\mathscr{A}$ and $P(A \mid D_i)$ the conditional probability of A with respect to D_i.

With a set of conditional probabilities $\{P(A \mid D_i), i = 1, \ldots, k\}$ we may associate the random variable

$$\pi(\omega) = \sum_{i=1}^{k} P(A \mid D_i) I_{D_i}(\omega) \tag{1}$$

(cf. (5) of Sect. 4), that takes the values $P(A \mid D_i)$ on the atoms of D_i. To emphasize that this *random variable* is associated specifically with the decomposition $\mathscr{D}$, we denote it by

$$P(A \mid \mathscr{D}) \quad \text{or} \quad P(A \mid \mathscr{D})(\omega)$$

and call it the *conditional probability of the event A with respect to the decomposition $\mathscr{D}$*.

This concept, as well as the more general concept of *conditional probabilities with respect to a σ-algebra*, which will be introduced later, plays an important role in probability theory, a role that will be developed progressively as we proceed.

We mention some of the simplest properties of conditional probabilities:

$$P(A + B \mid \mathscr{D}) = P(A \mid \mathscr{D}) + P(B \mid \mathscr{D}); \tag{2}$$

if $\mathscr{D}$ is the trivial decomposition consisting of the single set Ω then

$$P(A \mid \Omega) = P(A). \tag{3}$$

The definition of $P(A \mid \mathscr{D})$ as a random variable lets us speak of its expectation; by using this, we can write the *formula for total probability* (see (3), Sect. 3) in the following compact form:

$$\mathsf{E}\, P(A \mid \mathscr{D}) = P(A). \tag{4}$$

In fact, since

$$P(A \mid \mathscr{D}) = \sum_{i=1}^{k} P(A \mid D_i) I_{D_i}(\omega),$$

then by the definition of expectation (see (5) and (6), Sect. 4)

$$\mathsf{E}\, P(A \mid \mathscr{D}) = \sum_{i=1}^{k} P(A \mid D_i)\, P(D_i) = \sum_{i=1}^{k} P(AD_i) = P(A).$$

Now let $\eta = \eta(\omega)$ be a random variable that takes the values $y_1, \ldots, y_k$ with positive probabilities:

$$\eta(\omega) = \sum_{j=1}^{k} y_j I_{D_j}(\omega),$$

where $D_j = \{\omega : \eta(\omega) = y_j\}$. The decomposition $\mathscr{D}_\eta = \{D_1, \ldots, D_k\}$ is called the decomposition induced by η. The conditional probability $P(A \mid \mathscr{D}_\eta)$ will be denoted by $P(A \mid \eta)$ or $P(A \mid \eta)(\omega)$, and called the *conditional probability of A with respect to the random variable* η. We also denote by $P(A \mid \eta = y_j)$ the conditional probability $P(A \mid D_j)$, where $D_j = \{\omega : \eta(\omega) = y_j\}$.

Similarly, if $\eta_1, \eta_2, \ldots, \eta_m$ are random variables and $\mathscr{D}_{\eta_1,\eta_2,\ldots,\eta_m}$ is the decomposition induced by $\eta_1, \eta_2, \ldots, \eta_m$ with atoms

$$D_{y_1,y_2,\ldots,y_m} = \{\omega : \eta_1(\omega) = y_1, \ldots, \eta_m(\omega) = y_m\},$$

then $P(A \mid D_{\eta_1,\eta_2,\ldots,\eta_m})$ will be denoted by $P(A \mid \eta_1, \eta_2, \ldots, \eta_m)$ and called the *conditional probability of A with respect to* $\eta_1, \eta_2, \ldots, \eta_m$.

Example 1. Let ξ and η be independent identically distributed random variables, each taking the values 1 and 0 with probabilities p and q. For $k = 0, 1, 2$, let us find the conditional probability $P(\xi + \eta = k \mid \eta)$ of the event $A = \{\omega : \xi + \eta = k\}$ with respect to η.

To do this, we first notice the following useful general fact: if ξ and η are independent random variables with respective values x and y, then

$$P(\xi + \eta = z \mid \eta = y) = P(\xi + y = z). \tag{5}$$

In fact,

$$
\begin{aligned}
P(\xi + \eta = z \mid \eta = y) &= \frac{P(\xi + \eta = z, \eta = y)}{P(\eta = y)} \\
&= \frac{P(\xi + y = z, \eta = y)}{P(\eta = y)} = \frac{P(\xi + y = z)\, P(\eta = y)}{P(\eta = y)} \\
&= P(\xi + y = z).
\end{aligned}
$$

Using this formula for the case at hand, we find that

$$
\begin{aligned}
P(\xi + \eta = k \mid \eta) &= P(\xi + \eta = k \mid \eta = 0) I_{\{\eta=0\}}(\omega) \\
&\quad + P(\xi + \eta = k \mid \eta = 1) I_{\{\eta=1\}}(\omega) \\
&= P(\xi = k) I_{\{\eta=0\}}(\omega) + P\{\xi = k - 1\} I_{\{\eta=1\}}(\omega).
\end{aligned}
$$

Thus

$$P(\xi + \eta = k \mid \eta) = \begin{cases} q I_{\{\eta=0\}}(\omega), & k = 0, \\ p I_{\{\eta=0\}}(\omega) + q I_{\{\eta=1\}}(\omega), & k = 1, \\ p I_{\{\eta=1\}}(\omega), & k = 2, \end{cases} \tag{6}$$

or equivalently

$$P(\xi + \eta = k \mid \eta) = \begin{cases} q(1 - \eta), & k = 0, \\ p(1 - \eta) + q\eta, & k = 1, \\ p\eta, & k = 2. \end{cases} \tag{7}$$

2. Let $\xi = \xi(\omega)$ be a random variable with values in the set $X = \{x_1, \ldots, x_n\}$:

$$\xi = \sum_{j=1}^{l} x_j I_{A_j}(\omega), \qquad A_j = \{\omega : \xi = x_j\},$$

and let $\mathscr{D} = \{D_1, \ldots, D_k\}$ be a decomposition. Just as we defined the expectation of ξ with respect to the probabilities $\mathsf{P}(A_j)$, $j = 1, \ldots, l$,

$$\mathsf{E}\xi = \sum_{j=1}^{l} x_j \mathsf{P}(A_j), \tag{8}$$

it is now natural to define the *conditional expectation of ξ with respect to $\mathscr{D}$* by using the conditional probabilities $\mathsf{P}(A_j \mid \mathscr{D})$, $j = 1, \ldots, l$. We denote this expectation by $\mathsf{E}(\xi \mid \mathscr{D})$ or $\mathsf{E}(\xi \mid \mathscr{D})(\omega)$, and define it by the formula

$$\mathsf{E}(\xi \mid \mathscr{D}) = \sum_{j=1}^{l} x_j \mathsf{P}(A_j \mid \mathscr{D}). \tag{9}$$

According to this definition the conditional expectation $\mathsf{E}(\xi \mid \mathscr{D})(\omega)$ is a *random variable* which, at all sample points ω belonging to the same atom D_i, takes the same value $\sum_{j=1}^{l} x_j \mathsf{P}(A_j \mid D_i)$. This observation shows that the definition of $\mathsf{E}(\xi \mid \mathscr{D})$ could have been expressed differently. In fact, we could first define $\mathsf{E}(\xi \mid D_i)$, the conditional expectation of ξ with respect to D_i, by

$$\mathsf{E}(\xi \mid D_i) = \sum_{j=1}^{l} x_j \mathsf{P}(A_j \mid D_i) \left(= \frac{\mathsf{E}[\xi I_{D_i}]}{\mathsf{P}(D_i)} \right), \tag{10}$$

and then define

$$\mathsf{E}(\xi \mid \mathscr{D})(\omega) = \sum_{i=1}^{k} \mathsf{E}(\xi \mid D_i) I_{D_i}(\omega) \tag{11}$$

(see the diagram in Fig. 14).

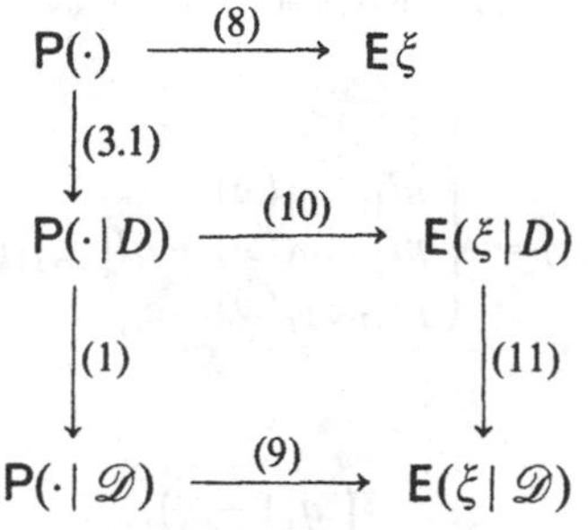

$$
\begin{array}{ccc}
\mathsf{P}(\cdot) & \xrightarrow{\;(8)\;} & \mathsf{E}\xi \\[4pt]
\Big\downarrow{\scriptstyle(3.1)} & & \\[6pt]
\mathsf{P}(\cdot \mid D) & \xrightarrow{\;(10)\;} & \mathsf{E}(\xi \mid D) \\[4pt]
\Big\downarrow{\scriptstyle(1)} & & \Big\downarrow{\scriptstyle(11)} \\[6pt]
\mathsf{P}(\cdot \mid \mathscr{D}) & \xrightarrow{\;(9)\;} & \mathsf{E}(\xi \mid \mathscr{D})
\end{array}
$$

Fig. 14

It is also useful to notice that $\mathsf{E}(\xi \mid D)$ and $\mathsf{E}(\xi \mid \mathscr{D})$ are independent of the representation of ξ.

The following properties of conditional expectations follow immediately from the definitions:

$$\mathsf{E}(a\xi + b\eta \mid \mathscr{D}) = a\,\mathsf{E}(\xi \mid \mathscr{D}) + b\,\mathsf{E}(\eta \mid \mathscr{D}), \quad a \text{ and } b \text{ constants;} \tag{12}$$

$$\mathsf{E}(\xi \mid \Omega) = \mathsf{E}\,\xi; \tag{13}$$

$$\mathsf{E}(C \mid \mathscr{D}) = C, \quad C \text{ constant;} \tag{14}$$

if $\xi = I_A(\omega)$ then

$$\mathsf{E}(\xi \mid \mathscr{D}) = \mathsf{P}(A \mid \mathscr{D}). \tag{15}$$

The last equation shows, in particular, that properties of conditional probabilities can be deduced directly from properties of conditional expectations.

The following important property extends the *formula for total probability* (4):

$$\mathsf{E}\,\mathsf{E}(\xi \mid \mathscr{D}) = \mathsf{E}\,\xi. \tag{16}$$

For the proof, it is enough to notice that by (4)

$$\mathsf{E}\,\mathsf{E}(\xi \mid \mathscr{D}) = \mathsf{E}\sum_{j=1}^{l} x_j\,\mathsf{P}(A_j \mid \mathscr{D}) = \sum_{j=1}^{l} x_j\,\mathsf{E}\,\mathsf{P}(A_j \mid \mathscr{D}) = \sum_{j=1}^{l} x_j\,\mathsf{P}(A_j) = \mathsf{E}\,\xi.$$

Let $\mathscr{D} = \{D_1, \ldots, D_k\}$ be a decomposition and $\eta = \eta(\omega)$ a random variable. We say that η is *measurable* with respect to this decomposition, or $\mathscr{D}$-measurable, if $\mathscr{D}_\eta \preccurlyeq \mathscr{D}$, i.e., $\eta = \eta(\omega)$ can be represented in the form

$$\eta(\omega) = \sum_{i=1}^{k} y_i I_{D_i}(\omega),$$

where some y_i might be equal. In other words, a random variable is $\mathscr{D}$-measurable if and only if it takes constant values on the atoms of $\mathscr{D}$.

Example 2. If $\mathscr{D}$ is the trivial decomposition, $\mathscr{D} = \{\Omega\}$, then η is $\mathscr{D}$-measurable if and only if $\eta \equiv C$, where C is a constant. Every random variable η is measurable with respect to $\mathscr{D}_\eta$.

Suppose that the random variable η is $\mathscr{D}$-measurable. Then

$$\mathsf{E}(\xi\eta \mid \mathscr{D}) = \eta\,\mathsf{E}(\xi \mid \mathscr{D}) \tag{17}$$

and in particular

$$\mathsf{E}(\eta \mid \mathscr{D}) = \eta \quad (\mathsf{E}(\eta \mid \mathscr{D}_\eta) = \eta). \tag{18}$$

To establish (17) we observe that if $\xi = \sum_{j=1}^{l} x_j I_{A_j}$, then

$$\xi\eta = \sum_{j=1}^{l}\sum_{i=1}^{k} x_j y_i I_{A_j D_i}$$

and therefore

$$
\begin{aligned}
\mathsf{E}(\xi\eta\mid\mathscr{D}) &= \sum_{j=1}^{l}\sum_{i=1}^{k} x_j y_i\, \mathsf{P}(A_j D_i\mid\mathscr{D}) \\
&= \sum_{j=1}^{l}\sum_{i=1}^{k} x_j y_i \sum_{m=1}^{k} \mathsf{P}(A_j D_i\mid D_m) I_{D_m}(\omega) \\
&= \sum_{j=1}^{l}\sum_{i=1}^{k} x_j y_i\, \mathsf{P}(A_j D_i\mid D_i) I_{D_i}(\omega) \\
&= \sum_{j=1}^{l}\sum_{i=1}^{k} x_j y_i\, \mathsf{P}(A_j\mid D_i) I_{D_i}(\omega).
\end{aligned} \tag{19}
$$

On the other hand, since $I_{D_i}^2 = I_{D_i}$ and $I_{D_i}\cdot I_{D_m} = 0$, $i\neq m$, we obtain

$$
\begin{aligned}
\eta\, \mathsf{E}(\xi\mid\mathscr{D}) &= \left[\sum_{i=1}^{k} y_i I_{D_i}(\omega)\right]\cdot\left[\sum_{j=1}^{l} x_j\, \mathsf{P}(A_j\mid\mathscr{D})\right] \\
&= \left[\sum_{i=1}^{k} y_i I_{D_i}(\omega)\right]\cdot\sum_{m=1}^{k}\left[\sum_{j=1}^{l} x_j\, \mathsf{P}(A_j\mid D_m)\right]\cdot I_{D_m}(\omega) \\
&= \sum_{i=1}^{k}\sum_{j=1}^{l} y_i x_j\, \mathsf{P}(A_j\mid D_i)\cdot I_{D_i}(\omega),
\end{aligned}
$$

which, with (19), establishes (17).

We shall establish another important property of conditional expectations. Let $\mathscr{D}_1$ and $\mathscr{D}_2$ be two decompositions, with $\mathscr{D}_1 \preccurlyeq \mathscr{D}_2$ ($\mathscr{D}_2$ is "finer" than $\mathscr{D}_1$). Then the following "telescopic" property holds:

$$
\mathsf{E}[\mathsf{E}(\xi\mid\mathscr{D}_2)\mid\mathscr{D}_1] = \mathsf{E}(\xi\mid\mathscr{D}_1). \tag{20}
$$

For the proof, suppose that

$$
\mathscr{D}_1 = \{D_{11},\dots,D_{1m}\}, \qquad \mathscr{D}_2 = \{D_{21},\dots,D_{2n}\}.
$$

Then if $\xi = \sum_{j=1}^{l} x_j I_{A_j}$, we have

$$
\mathsf{E}(\xi\mid\mathscr{D}_2) = \sum_{j=1}^{l} x_j\, \mathsf{P}(A_j\mid\mathscr{D}_2),
$$

and it is sufficient to establish that

$$
\mathsf{E}[\mathsf{P}(A_j\mid\mathscr{D}_2)\mid\mathscr{D}_1] = \mathsf{P}(A_j\mid\mathscr{D}_1). \tag{21}
$$

Since

$$P(A_j \mid \mathscr{D}_2) = \sum_{q=1}^{n} P(A_j \mid D_{2q}) I_{D_{2q}},$$

we have

$$
\begin{aligned}
\mathsf{E}[P(A_j \mid \mathscr{D}_2) \mid \mathscr{D}_1] &= \sum_{q=1}^{n} P(A_j \mid D_{2q}) \, P(D_{2q} \mid \mathscr{D}_1) \\
&= \sum_{q=1}^{n} P(A_j \mid D_{2q}) \left[\sum_{p=1}^{m} P(D_{2q} \mid D_{1p}) I_{D_{1p}} \right] \\
&= \sum_{p=1}^{m} I_{D_{1p}} \cdot \sum_{q=1}^{n} P(A_j \mid D_{2q}) \, P(D_{2q} \mid D_{1p}) \\
&= \sum_{p=1}^{m} I_{D_{1p}} \cdot \sum_{\{q:\, D_{2q} \subseteq D_{1p}\}} P(A_j \mid D_{2q}) \, P(D_{2q} \mid D_{1p}) \\
&= \sum_{p=1}^{m} I_{D_{1p}} \cdot \sum_{\{q:\, D_{2q} \subseteq D_{1p}\}} \frac{P(A_j D_{2q})}{P(D_{2q})} \cdot \frac{P(D_{2q})}{P(D_{1p})} \\
&= \sum_{p=1}^{m} I_{D_{1p}} \cdot P(A_j \mid D_{1p}) = P(A_j \mid \mathscr{D}_1),
\end{aligned}
$$

which establishes (21).

When $\mathscr{D}$ is induced by the random variables $\eta_1, \ldots, \eta_k$ (i.e., $\mathscr{D} = \mathscr{D}_{\eta_1,\ldots,\eta_k}$), the conditional expectation $\mathsf{E}(\xi \mid \mathscr{D}_{\eta_1,\ldots,\eta_k})$ will be denoted by $\mathsf{E}(\xi \mid \eta_1, \ldots, \eta_k)$, or $\mathsf{E}(\xi \mid \eta_1, \ldots, \eta_k)(\omega)$, and called the *conditional expectation of ξ with respect to* $\eta_1, \ldots, \eta_k$.

It follows immediately from the definition of $\mathsf{E}(\xi \mid \eta)$ that if ξ and η are *independent*, then

$$\mathsf{E}(\xi \mid \eta) = \mathsf{E}\,\xi. \tag{22}$$

From (18) it also follows that

$$\mathsf{E}(\eta \mid \eta) = \eta. \tag{23}$$

Using the notation $\mathsf{E}(\xi \mid \eta)$ for $\mathsf{E}(\xi \mid \mathscr{D}_\eta)$, formula (16), which restates the formula for total probability (4), can be written in the following widely used form:

$$\mathsf{E}\,\mathsf{E}(\xi \mid \eta) = \mathsf{E}\,\xi. \tag{24}$$

(See also property (27) in Problem 3.)

Property (22) admits the following generalization. Let ξ be independent of $\mathscr{D}$ (i.e., for each $D_i \in \mathscr{D}$ the random variables ξ and I_{D_i} are independent). Then

$$\mathsf{E}(\xi \mid \mathscr{D}) = \mathsf{E}\,\xi.$$

As a special case of (20) we obtain the following useful formula:

$$E[E(\xi \mid \eta_1, \eta_2) \mid \eta_1] = E(\xi \mid \eta_1). \tag{25}$$

Example 3. Let us find $E(\xi + \eta \mid \eta)$ for the random variables ξ and η considered in Example 1. By (22) and (23),

$$E(\xi + \eta \mid \eta) = E\xi + \eta = p + \eta.$$

This result can also be obtained by starting from (8):

$$E(\xi + \eta \mid \eta) = \sum_{k=0}^{2} k\, P(\xi + \eta = k \mid \eta) = p(1 - \eta) + q\eta + 2p\eta = p + \eta.$$

Example 4. Let ξ and η be independent and identically distributed random variables. Then
$$E(\xi \mid \xi + \eta) = E(\eta \mid \xi + \eta) = \frac{\xi + \eta}{2}. \tag{26}$$

In fact, if we assume for simplicity that ξ and η take the values $1, 2, \ldots, m$, we find ($1 \le k \le m,\ 2 \le l \le 2m$)

$$
\begin{aligned}
P(\xi = k \mid \xi + \eta = l) &= \frac{P(\xi = k, \xi + \eta = l)}{P(\xi + \eta = l)} = \frac{P(\xi = k, \eta = l - k)}{P(\xi + \eta = l)} \\
&= \frac{P(\xi = k)\, P(\eta = l - k)}{P(\xi + \eta = l)} = \frac{P(\eta = k)\, P(\xi = l - k)}{P(\xi + \eta = l)} \\
&= P(\eta = k \mid \xi + \eta = l).
\end{aligned}
$$

This establishes the first equation in (26). To prove the second, it is enough to notice that

$$2\,E(\xi \mid \xi + \eta) = E(\xi \mid \xi + \eta) + E(\eta \mid \xi + \eta) = E(\xi + \eta \mid \xi + \eta) = \xi + \eta.$$

3. We have already noticed in Sect. 1 that to each decomposition $\mathscr{D} = \{D_1, \ldots, D_k\}$ of the finite set Ω there corresponds an algebra $\alpha(\mathscr{D})$ of subsets of Ω. The converse is also true: every algebra $\mathscr{B}$ of subsets of the finite space Ω generates a decomposition $\mathscr{D}(\mathscr{B} = \alpha(\mathscr{D}))$. Consequently there is a one-to-one correspondence between algebras and decompositions of a finite space Ω. This should be kept in mind in connection with the concept, which will be introduced later, of conditional expectation with respect to the special systems of sets called σ-algebras.

For *finite* spaces, the concepts of algebra and σ-algebra coincide. It will turn out that if $\mathscr{B}$ is an algebra, the conditional expectation $E(\xi \mid \mathscr{B})$ of a random variable ξ with respect to $\mathscr{B}$ (to be introduced in Sect. 7, Chap. 2) simply coincides with $E(\xi \mid \mathscr{D})$, the expectation of ξ with respect to the decomposition $\mathscr{D}$ such that $\mathscr{B} = \alpha(\mathscr{D})$. In this sense we can, in dealing with finite spaces in the future, not distinguish between $E(\xi \mid \mathscr{B})$ and $E(\xi \mid \mathscr{D})$, understanding in each case that $E(\xi \mid \mathscr{B})$ is simply defined to be $E(\xi \mid \mathscr{D})$.

4. PROBLEMS.

1. Give an example of random variables ξ and η which are not independent but for which

$$\mathsf{E}(\xi \mid \eta) = \mathsf{E}\,\xi.$$

   (Cf. (22).)

2. The conditional variance of ξ with respect to $\mathscr{D}$ is the random variable

$$\mathrm{Var}(\xi \mid \mathscr{D}) = \mathsf{E}[(\xi - \mathsf{E}(\xi \mid \mathscr{D}))^2 \mid \mathscr{D}].$$

   Show that

$$\mathrm{Var}\,\xi = \mathsf{E}\,\mathrm{Var}(\xi \mid \mathscr{D}) + \mathrm{Var}\,\mathsf{E}(\xi \mid \mathscr{D}).$$

3. Starting from (17), show that for every function $f = f(\eta)$ the conditional expectation $\mathsf{E}(\xi \mid \eta)$ has the property

$$\mathsf{E}[f(\eta)\,\mathsf{E}(\xi \mid \eta)] = \mathsf{E}[\xi f(\eta)]. \tag{27}$$

4. Let ξ and η be random variables. Show that $\inf_f \mathsf{E}(\eta - f(\xi))^2$ is attained for $f^*(\xi) = \mathsf{E}(\eta \mid \xi)$. (Consequently, the *best estimator* for η in terms of ξ, in the mean-square sense, is the conditional expectation $\mathsf{E}(\eta \mid \xi)$).

5. Let $\xi_1, \ldots, \xi_n$, τ be *independent* random variables, where $\xi_1, \ldots, \xi_n$ are identically distributed and τ takes the values $1, 2, \ldots, n$. Show that if $S_\tau = \xi_1 + \cdots + \xi_\tau$ is the *sum of a random number of the random variables*, then

$$\mathsf{E}(S_\tau \mid \tau) = \tau\,\mathsf{E}\,\xi_1, \qquad \mathrm{Var}(S_\tau \mid \tau) = \tau\,\mathrm{Var}\,\xi_1$$

   and

$$\mathsf{E}\,S_\tau = \mathsf{E}\,\tau \cdot \mathsf{E}\,\xi_1, \qquad \mathrm{Var}\,S_\tau = \mathsf{E}\,\tau \cdot \mathrm{Var}\,\xi_1 + \mathrm{Var}\,\tau \cdot (\mathsf{E}\,\xi_1)^2.$$

6. Establish equation (24).

9 Random Walk: I—Probabilities of Ruin and Mean Duration in Coin Tossing

1. The value of the limit theorems of Sect. 6 for Bernoulli schemes is not just that they provide convenient formulas for calculating probabilities $P(S_n = k)$ and $P(A < S_n \leq B)$. They have the additional significance of being of a *universal* nature, i.e., they remain useful not only for independent Bernoulli random variables that have only two values, but also for variables of much more general character. In this sense the Bernoulli scheme appears as the simplest model, on the basis of which we can recognize many probabilistic regularities which are inherent also in much more general models.

In this and the next section we shall discuss a number of new probabilistic regularities, some of which are quite surprising. The ones that we discuss are again based on the Bernoulli scheme, although many results on the nature of random oscillations remain valid for random walks of a more general kind.

2. Consider the Bernoulli scheme $(\Omega, \mathscr{A}, \mathsf{P})$, where $\Omega = \{\omega : \omega = (x_1, \ldots, x_n), x_i = \pm 1\}$, $\mathscr{A}$ consists of all subsets of Ω, and $\mathsf{P}(\{\omega\}) = p^{\nu(\omega)} q^{n-\nu(\omega)}$ with $\nu(\omega) = (\sum x_i + n)/2$. Let $\xi_i(\omega) = x_i$, $i = 1, \ldots, n$. Then, as we know, the sequence $\xi_1, \ldots, \xi_n$ is a sequence of independent Bernoulli random variables,

$$\mathsf{P}(\xi_i = 1) = p, \quad \mathsf{P}(\xi_i = -1) = q, \quad p + q = 1.$$

Let us put $S_0 = 0$, $S_k = \xi_1 + \cdots + \xi_k$, $1 \leq k \leq n$. The sequence $S_0, S_1, \ldots, S_n$ can be considered as the path of the random motion of a particle starting at zero. Here $S_{k+1} = S_k + \xi_k$, i.e., if the particle has reached the point S_k at time k, then at time $k + 1$ it is displaced either one unit up (with probability p) or one unit down (with probability q).

Let A and B be integers, $A \leq 0 \leq B$. An interesting problem about this random walk is *to find the probability that after n steps the moving particle has left the interval (A, B)*. It is also of interest to ask with what probability the particle leaves (A, B) at A or at B.

That these are natural questions to ask becomes particularly clear if we interpret them in terms of a gambling game. Consider two players (first and second) who start with respective bankrolls $(-A)$ and B. If $\xi_i = +1$, we suppose that the second player pays one unit to the first; if $\xi_i = -1$, the first pays the second. Then $S_k = \xi_1 + \cdots + \xi_k$ can be interpreted as the amount won by the first player from the second (if $S_k < 0$, this is actually the amount lost by the first player to the second) after k turns.

At the instant $k \leq n$ at which for the first time $S_k = B$ ($S_k = A$) the bank-roll of the second (first) player is reduced to zero; in other words, that player is ruined. (If $k < n$, we suppose that the game ends at time k, although the random walk itself is well defined up to time n, inclusive.)

Before we turn to a precise formulation, let us introduce some notation.

Let x be an integer in the interval $[A, B]$ and for $0 \leq k \leq n$ let $S_k^x = x + S_k$,

$$\tau_k^x = \min\{0 \leq l \leq k : S_l^x = A \text{ or } B\}, \tag{1}$$

where we agree to take $\tau_k^x = k$ if $A < S_l^x < B$ for all $0 \leq l \leq k$.

For each k in $0 \leq k \leq n$ and $x \in [A, B]$, the instant τ_k^x, called a *stopping time* (see Sect. 11), is an integer-valued random variable defined on the sample space Ω (the dependence of τ_k^x on ω is not explicitly indicated).

It is clear that for all $l < k$ the set $\{\omega : \tau_k^x = l\}$ is the event that the random walk $\{S_i^x, 0 \leq i \leq k\}$, starting at time zero at the point x, leaves the interval (A, B) at time l. It is also clear that when $l \leq k$ the sets $\{\omega : \tau_k^x = l, S_l^x = A\}$ and $\{\omega : \tau_k^x = l, S_l^x = B\}$ represent the events that the wandering particle leaves the interval (A, B) at time l through A or B respectively.

For $0 \leq k \leq n$, we write

$$
\begin{aligned}
\mathscr{A}_k^x &= \sum_{0 \leq l \leq k} \{\omega : \tau_k^x = l, \, S_l^x = A\}, \\
\mathscr{B}_k^x &= \sum_{0 \leq l \leq k} \{\omega : \tau_k^x = l, \, S_l^x = B\},
\end{aligned}
\tag{2}
$$

and let

$$
\alpha_k(x) = \mathsf{P}(\mathscr{A}_k^x), \qquad \beta_k(x) = \mathsf{P}(\mathscr{B}_k^x)
$$

be the probabilities that the particle leaves (A, B) through A or B respectively, during the time interval $[0, k]$. For these probabilities we can find recurrent relations from which we can successively determine $\alpha_1(x), \ldots, \alpha_n(x)$ and $\beta_1(x), \ldots, \beta_n(x)$.

Let, then, $A < x < B$. It is clear that $\alpha_0(x) = \beta_0(x) = 0$. Now suppose $1 \leq k \leq n$. Then by (3) of Sect. 3

$$
\begin{aligned}
\beta_k(x) &= \mathsf{P}(\mathscr{B}_k^x) = \mathsf{P}(\mathscr{B}_k^x \mid S_1^x = x + 1) \, \mathsf{P}(\xi_1 = 1) \\
&\quad + \mathsf{P}(\mathscr{B}_k^x \mid S_1^x = x - 1) \, \mathsf{P}(\xi_1 = -1) \\
&= p\mathsf{P}(\mathscr{B}_k^x \mid S_1^x = x + 1) + q\mathsf{P}(\mathscr{B}_k^x \mid S_1^x = x - 1).
\end{aligned}
\tag{3}
$$

We now show that

$$
\mathsf{P}(\mathscr{B}_k^x \mid S_1^x = x + 1) = \mathsf{P}(\mathscr{B}_{k-1}^{x+1}), \quad \mathsf{P}(\mathscr{B}_k^x \mid S_1^x = x - 1) = \mathsf{P}(\mathscr{B}_{k-1}^{x-1}).
$$

To do this, we notice that $\mathscr{B}_k^x$ can be represented in the form

$$
\mathscr{B}_k^x = \{\omega : (x, x + \xi_1, \ldots, x + \xi_1 + \cdots + \xi_k) \in B_k^x\},
$$

where B_k^x is the set of paths of the form

$$
(x, x + x_1, \ldots, x + x_1 + \cdots + x_k)
$$

with $x_1 = \pm 1$, which during the time $[0, k]$ first leave (A, B) at B (Fig. 15).

We represent B_k^x in the form $B_k^{x,x+1} + B_k^{x,x-1}$, where $B_k^{x,x+1}$ and $B_k^{x,x-1}$ are the paths in B_k^x for which $x_1 = +1$ or $x_1 = -1$, respectively.

Notice that the paths $(x, x + 1, x + 1 + x_2, \ldots, x + 1 + x_2 + \cdots + x_k)$ in $B_k^{x,x+1}$ are in one-to-one correspondence with the paths

$$
(x + 1, x + 1 + x_2, \ldots, x + 1 + x_2, \ldots, x + 1 + x_2 + \cdots + x_k)
$$

in B_{k-1}^{x+1}. The same is true for the paths in $B_k^{x,x-1}$. Using these facts, together with independence, the identical distribution of $\xi_1, \ldots, \xi_k$, and (6) of Sect. 8, we obtain

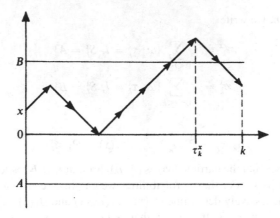

Fig. 15 Example of a path from the set B_k^x

$$
\begin{aligned}
P&(\mathscr{B}_k^x \mid S_1^x = x+1)\\
&= P(\mathscr{B}_k^x \mid \xi_1 = 1)\\
&= P\{(x, x+\xi_1, \ldots, x+\xi_1+\cdots+\xi_k) \in B_k^x \mid \xi_1 = 1\}\\
&= P\{(x+1, x+1+\xi_2, \ldots, x+1+\xi_2+\cdots+\xi_k) \in B_{k-1}^{x+1}\}\\
&= P\{(x+1, x+1+\xi_1, \ldots, x+1+\xi_1+\cdots+\xi_{k-1}) \in B_{k-1}^{x+1}\}\\
&= P(\mathscr{B}_{k-1}^{x+1}).
\end{aligned}
$$

In the same way,

$$
P(\mathscr{B}_k^x \mid S_1^x = x-1) = P(\mathscr{B}_{k-1}^{x-1}).
$$

Consequently, by (3), for $x \in (A, B)$ and $k \leq n$,

$$
\beta_k(x) = p\beta_{k-1}(x+1) + q\beta_{k-1}(x-1), \tag{4}
$$

where

$$
\beta_l(B) = 1, \quad \beta_l(A) = 0, \quad 0 \leq l \leq n. \tag{5}
$$

Similarly

$$
\alpha_k(x) = p\alpha_{k-1}(x+1) + q\alpha_{k-1}(x-1) \tag{6}
$$

with

$$
\alpha_l(A) = 1, \quad \alpha_l(B) = 0, \quad 0 \leq l \leq n.
$$

Since $\alpha_0(x) = \beta_0(x) = 0$, $x \in (A, B)$, these recurrent relations can (at least in principle) be solved for the probabilities

$$
\alpha_1(x), \ldots, \alpha_n(x) \quad \text{and} \quad \beta_1(x), \ldots, \beta_n(x).
$$

Putting aside any explicit calculation of the probabilities, we ask for their values for large n.

For this purpose we notice that since $\mathscr{B}_{k-1}^x \subset \mathscr{B}_k^x$, $k \leq n$, we have $\beta_{k-1}(x) \leq \beta_k(x) \leq 1$. It is therefore natural to expect (and this is actually the case; see Subsection 3) that for sufficiently large n the probability $\beta_n(x)$ will be close to the solution $\beta(x)$ of the equation

$$\beta(x) = p\beta(x+1) + q\beta(x-1) \tag{7}$$

with the boundary conditions

$$\beta(B) = 1, \qquad \beta(A) = 0 \tag{8}$$

that result from a formal approach to the limit in (4) and (5).

To solve the problem in (7) and (8), we first suppose that $p \neq q$. We see easily that the equation has the *two particular solutions* a and $b(q/p)^x$, where a and b are constants. Hence we look for a general solution of the form

$$\beta(x) = a + b(q/p)^x. \tag{9}$$

Taking account of (8), we find that for $A \leq x \leq B$

$$\beta(x) = \frac{(q/p)^x - (q/p)^A}{(q/p)^B - (q/p)^A}. \tag{10}$$

Let us show that this is the *only* solution of our problem. It is enough to show that all solutions of the problem in (7) and (8) admit the representation (9).

Let $\tilde{\beta}(x)$ be a solution with $\tilde{\beta}(A) = 0$, $\tilde{\beta}(B) = 1$. We can always find constants $\tilde{a}$ and $\tilde{b}$ such that

$$\tilde{a} + \tilde{b}(q/p)^A = \tilde{\beta}(A), \qquad \tilde{a} + \tilde{b}(q/p)^{A+1} = \tilde{\beta}(A+1).$$

Then it follows from (7) that

$$\tilde{\beta}(A+2) = \tilde{a} + \tilde{b}(q/p)^{A+2}$$

and generally

$$\tilde{\beta}(x) = \tilde{a} + \tilde{b}(q/p)^x.$$

Consequently the solution (10) is the only solution of our problem.

A similar discussion shows that the only solution of

$$\alpha(x) = p\alpha(x+1) + q\alpha(x-1), \quad x \in (A, B), \tag{11}$$

with the boundary conditions

$$\alpha(A) = 1, \qquad \alpha(B) = 0 \tag{12}$$

is given by the formula

$$\alpha(x) = \frac{(q/p)^B - (q/p)^x}{(q/p)^B - (q/p)^A}, \qquad A \leq x \leq B. \tag{13}$$

If $p = q = \frac{1}{2}$, the only solutions $\beta(x)$ and $\alpha(x)$ of (7), (8) and (11), (12) are respectively

$$\beta(x) = \frac{x - A}{B - A} \tag{14}$$

and

$$\alpha(x) = \frac{B - x}{B - A}. \tag{15}$$

We note that

$$\alpha(x) + \beta(x) = 1 \tag{16}$$

for $0 \le p \le 1$.

We call $\alpha(x)$ and $\beta(x)$ the *probabilities of ruin for the first and second players*, respectively (when the first player's bankroll is $x - A$, and the second player's is $B - x$) under the assumption of infinitely many turns, which of course presupposes an infinite sequence of independent Bernoulli random variables $\xi_1, \xi_2, \ldots$, where $\xi_i = +1$ is treated as a gain for the first player, and $\xi_i = -1$ as a loss. The probability space $(\Omega, \mathscr{A}, \mathsf{P})$ considered at the beginning of this section turns out to be too small to allow such an *infinite* sequence of independent variables. We shall see later (Sect. 9, Chap. 2) that such a sequence can actually be constructed and that $\beta(x)$ and $\alpha(x)$ are in fact the probabilities of ruin in an unbounded number of steps.

We now take up some corollaries of the preceding formulas.

If we take $A = 0$, $0 \le x \le B$, then the definition of $\beta(x)$ implies that this is the probability that a particle starting at x arrives at B before it reaches 0. It follows from (10) and (14) (Fig. 16) that

$$\beta(x) = \begin{cases} x/B, & p = q = 1/2, \\ \frac{(q/p)^x - 1}{(q/p)^B - 1}, & p \ne q. \end{cases} \tag{17}$$

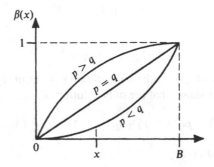

$\beta(x)$

$p > q$

$p = q$

$p < q$

Fig. 16 Graph of $\beta(x)$, the probability that a particle starting from x reaches B before reaching 0

Now let $q > p$, which means that the game is *unfavorable* for the first player, whose limiting probability of being ruined, namely $\alpha = \alpha(0)$, is given by

$$\alpha = \frac{(q/p)^B - 1}{(q/p)^B - (q/p)^A}.$$

Next suppose that the rules of the game are changed: the original bankrolls of the players are still $(-A)$ and B, but the payoff for each player is now $\frac{1}{2}$, rather than 1 as before. In other words, now let $\mathsf{P}(\xi_i = \frac{1}{2}) = p$, $\mathsf{P}(\xi_i = -\frac{1}{2}) = q$. In this case let us denote the limiting probability of ruin for the first player by $\alpha_{1/2}$. Then

$$\alpha_{1/2} = \frac{(q/p)^{2B} - 1}{(q/p)^{2B} - (q/p)^{2A}},$$

and therefore

$$\alpha_{1/2} = \alpha \cdot \frac{(q/p)^B + 1}{(q/p)^B + (q/p)^A} > \alpha,$$

if $q > p$.

Hence we can draw the following conclusion: *if the game is unfavorable to the first player (i.e., $q > p$) then doubling the stake decreases the probability of ruin.*

3. We now turn to the question of how fast $\alpha_n(x)$ and $\beta_n(x)$ approach their limiting values $\alpha(x)$ and $\beta(x)$.

Let us suppose for simplicity that $x = 0$ and put

$$\alpha_n = \alpha_n(0), \qquad \beta_n = \beta_n(0), \qquad \gamma_n = 1 - (\alpha_n + \beta_n).$$

It is clear that

$$\gamma_n = \mathsf{P}\{A < S_k < B, \ 0 \le k \le n\},$$

where $\{A < S_k < B, \ 0 \le k \le n\}$ denotes the event

$$\bigcap_{0 \le k \le n} \{A < S_k < B\}.$$

Let $n = rm$, where r and m are integers and

$$\zeta_1 = \xi_1 + \cdots + \xi_m,$$
$$\zeta_2 = \xi_{m+1} + \cdots + \xi_{2m},$$
$$\cdots\cdots\cdots\cdots\cdots\cdots\cdots\cdots$$
$$\zeta_r = \xi_{m(r-1)+1} + \cdots + \xi_{rm}.$$

Then if $C = |A| + B$, it is easy to see that

$$\{A < S_k < B, \ 1 \le k \le rm\} \subseteq \{|\zeta_1| < C, \ldots, |\zeta_r| < C\},$$

and therefore, since $\zeta_1, \ldots, \zeta_r$ are independent and identically distributed,

$$\gamma_n \le \mathsf{P}\{|\zeta_1| < C, \ldots, |\zeta_r| < C\} = \prod_{i=1}^{r} \mathsf{P}\{|\zeta_i| < C\} = (\mathsf{P}\{|\zeta_1| < C\})^r. \tag{18}$$

We notice that $\operatorname{Var} \zeta_1 = m[1 - (p - q)^2]$. Hence, for $0 < p < 1$ and sufficiently large m,

$$P\{|\zeta_1| < C\} \leq \varepsilon_1, \tag{19}$$

where $\varepsilon_1 < 1$, since $\operatorname{Var} \zeta_1 \leq C^2$ if $P\{|\zeta_1| \leq C\} = 1$.

If $p = 0$ or $p = 1$, then $P\{|\zeta_1| < C\} = 0$ for sufficiently large m, and consequently (19) is satisfied for $0 \leq p \leq 1$.

It follows from (18) and (19) that for sufficiently large n

$$\gamma_n \leq \varepsilon^n, \tag{20}$$

where $\varepsilon = \varepsilon_1^{1/m} < 1$.

According to (16), $\alpha + \beta = 1$. Therefore

$$(\alpha - \alpha_n) + (\beta - \beta_n) = \gamma_n,$$

and since $\alpha \geq \alpha_n$, $\beta \geq \beta_n$, we have

$$0 \leq \alpha - \alpha_n \leq \gamma_n \leq \varepsilon^n,$$
$$0 \leq \beta - \beta_n \leq \gamma_n \leq \varepsilon^n, \qquad \varepsilon < 1.$$

There are similar inequalities for the differences $\alpha(x) - \alpha_n(x)$ and $\beta(x) - \beta_n(x)$.

4. We now consider the question of the *mean duration* of the random walk.

Let $m_k(x) = \mathsf{E}\,\tau_k^x$ be the expectation of the stopping time τ_k^x, $k \leq n$. Proceeding as in the derivation of the recurrent relations for $\beta_k(x)$, we find that, for $x \in (A, B)$,

$$
\begin{aligned}
m_k(x) = \mathsf{E}\,\tau_k^x &= \sum_{1 \leq l \leq k} l\,\mathsf{P}(\tau_k^x = l) \\
&= \sum_{1 \leq l \leq k} l \cdot [p\,\mathsf{P}(\tau_k^x = l \mid \xi_1 = 1) + q\,\mathsf{P}(\tau_k^x = l \mid \xi_1 = -1)] \\
&= \sum_{1 \leq l \leq k} l \cdot [p\,\mathsf{P}(\tau_{k-1}^{x+1} = l - 1) + q\,\mathsf{P}(\tau_{k-1}^{x-1} = l - 1)] \\
&= \sum_{0 \leq l \leq k-1} (l + 1)[p\,\mathsf{P}(\tau_{k-1}^{x+1} = l) + q\,\mathsf{P}(\tau_{k-1}^{x-1} = l)] \\
&= pm_{k-1}(x + 1) + qm_{k-1}(x - 1) \\
&\quad + \sum_{0 \leq l \leq k-1} [p\,\mathsf{P}(\tau_{k-1}^{x+1} = l) + q\,\mathsf{P}(\tau_{k-1}^{x-1} = l)] \\
&= pm_{k-1}(x + 1) + qm_{k-1}(x - 1) + 1.
\end{aligned}
$$

Thus, for $x \in (A, B)$ and $0 \leq k \leq n$, the functions $m_k(x)$ satisfy the recurrent relations

$$m_k(x) = 1 + pm_{k-1}(x + 1) + qm_{k-1}(x - 1), \tag{21}$$

with $m_0(x) = 0$. From these equations together with the boundary conditions

$$m_k(A) = m_k(B) = 0, \tag{22}$$

we can successively find $m_1(x), \ldots, m_n(x)$.

Since $m_k(x) \leq m_{k+1}(x)$, the limit

$$m(x) = \lim_{n \to \infty} m_n(x)$$

exists, and by (21) it satisfies the equation

$$m(x) = 1 + pm(x+1) + qm(x-1) \tag{23}$$

with the boundary conditions

$$m(A) = m(B) = 0. \tag{24}$$

To solve this equation, we first suppose that

$$m(x) < \infty, \quad x \in (A, B). \tag{25}$$

Then if $p \neq q$ there is a particular solution of the form $x/(q-p)$ and the general solution (see (9)) can be written in the form

$$m(x) = \frac{-x}{p-q} + a + b\left(\frac{q}{p}\right)^x.$$

Then by using the boundary conditions $m(A) = m(B) = 0$ we find that

$$m(x) = \frac{1}{p-q}(B\beta(x) + A\alpha(x) - x], \tag{26}$$

where $\beta(x)$ and $\alpha(x)$ are defined by (10) and (13). If $p = q = \frac{1}{2}$, the general solution of (23) has the form

$$m(x) = a + bx - x^2,$$

and since $m(A) = m(B) = 0$ we have

$$m(x) = (B - x)(x - A). \tag{27}$$

It follows, in particular, that if the players start with equal bankrolls $(B = -A)$, then

$$m(0) = B^2.$$

If we take $B = 10$, and suppose that each turn takes a second, then the (limiting) time to the ruin of one player is rather long: 100 seconds.

We obtained (26) and (27) under the assumption that $m(x) < \infty$, $x \in (A, B)$. Let us now show that in fact $m(x)$ is finite for all $x \in (A, B)$. We consider only the case $x = 0$; the general case can be analyzed similarly.

Let $p = q = \frac{1}{2}$. We introduce the random variable $S_{\tau_n} = S_{\tau_n}(\omega)$ defined in terms of the sequence $S_0, S_1, \ldots, S_n$ and the stopping time $\tau_n = \tau_n^0$ by the equation

$$S_{\tau_n} = \sum_{k=0}^{n} S_k(\omega) I_{\{\tau_n=k\}}(\omega). \tag{28}$$

The descriptive meaning of S_{τ_n} is clear: it is the position reached by the random walk at the stopping time τ_n. Thus, if $\tau_n < n$, then $S_{\tau_n} = A$ or B; if $\tau_n = n$, then $A \leq S_{\tau_n} \leq B$.

Let us show that when $p = q = \frac{1}{2}$,

$$\mathsf{E}\, S_{\tau_n} = 0, \tag{29}$$

$$\mathsf{E}\, S_{\tau_n}^2 = \mathsf{E}\, \tau_n. \tag{30}$$

To establish the first equation we notice that

$$\mathsf{E}\, S_{\tau_n} = \sum_{k=0}^{n} \mathsf{E}[S_k I_{\{\tau_n=k\}}(\omega)]$$

$$= \sum_{k=0}^{n} \mathsf{E}[S_n I_{\{\tau_n=k\}}(\omega)] + \sum_{k=0}^{n} \mathsf{E}[(S_k - S_n) I_{\{\tau_n=k\}}(\omega)]$$

$$= \mathsf{E}\, S_n + \sum_{k=0}^{n} \mathsf{E}[(S_k - S_n) I_{\{\tau_n=k\}}(\omega)], \tag{31}$$

where we evidently have $\mathsf{E}\, S_n = 0$. Let us show that

$$\sum_{k=0}^{n} \mathsf{E}[(S_k - S_n) I_{\{\tau_n=k\}}(\omega)] = 0.$$

To do this, we notice that $\{\tau_n > k\} = \{A < S_1 < B, \ldots, A < S_k < B\}$ when $0 \leq k < n$. The event $\{A < S_1 < B, \ldots, A < S_k < B\}$ can evidently be written in the form

$$\{\omega : (\xi_1, \ldots, \xi_k) \in A_k\}, \tag{32}$$

where A_k is a subset of $\{-1, +1\}^k$. In other words, this set is determined by just the values of $\xi_1, \ldots, \xi_k$ and does not depend on $\xi_{k+1}, \ldots, \xi_n$. Since

$$\{\tau_n = k\} = \{\tau_n > k - 1\} \backslash \{\tau_n > k\},$$

this is also a set of the form (32). It then follows from the independence of $\xi_1, \ldots, \xi_n$ and from Problem 10 of Sect. 4 that for any $0 \leq k < n$ the random variables $S_n - S_k$ and $I_{\{\tau_n=k\}}$ are independent, and therefore

$$\mathsf{E}[(S_n - S_k) I_{\{\tau_n=k\}}] = \mathsf{E}[S_n - S_k] \cdot \mathsf{E}\, I_{\{\tau_n=k\}} = 0.$$

Hence we have established (29).

We can prove (30) by the same method:

$$
E S_{\tau_n}^2 = \sum_{k=0}^{n} E S_k^2 I_{\{\tau_n=k\}} = \sum_{k=0}^{n} E([S_n + (S_k - S_n)]^2 I_{\{\tau_n=k\}})
$$

$$
= \sum_{k=0}^{n} [E S_n^2 I_{\{\tau_n=k\}} + 2 E S_n(S_k - S_n) I_{\{\tau_n=k\}}
$$

$$
+ E(S_n - S_k)^2 I_{\{\tau_n=k\}}] = E S_n^2 - \sum_{k=0}^{n} E(S_n - S_k)^2 I_{\{\tau_n=k\}}
$$

$$
= n - \sum_{k=0}^{n} (n - k) P(\tau_n = k) = \sum_{k=0}^{n} k P(\tau_n = k) = E \tau_n.
$$

Thus we have (29) and (30) when $p = q = \frac{1}{2}$. For general p and q $(p + q = 1)$ it can be shown similarly that

$$
E S_{\tau_n} = (p - q) \cdot E \tau_n, \tag{33}
$$

$$
E[S_{\tau_n} - \tau_n \cdot E \xi_1]^2 = \operatorname{Var} \xi_1 \cdot E \tau_n, \tag{34}
$$

where $E \xi_1 = p - q$, $\operatorname{Var} \xi_1 = 1 - (p - q)^2$.

With the aid of the results obtained so far we can now show that $\lim_{n \to \infty} m_n(0) = m(0) < \infty$.

If $p = q = \frac{1}{2}$, then by (30)

$$
E \tau_n \le \max(A^2, B^2). \tag{35}
$$

If $p \ne q$, then by (33),

$$
E \tau_n \le \frac{\max(|A|, B)}{|p - q|}, \tag{36}
$$

from which it is clear that $m(0) < \infty$.

We also notice that when $p = q = \frac{1}{2}$

$$
E \tau_n = E S_{\tau_n}^2 = A^2 \alpha_n + B^2 \beta_n + E[S_n^2 I_{\{A < S_n < B\}} I_{\{\tau_n = n\}}]
$$

and therefore

$$
A^2 \alpha_n + B^2 \beta_n \le E \tau_n \le A^2 \alpha_n + B^2 \beta_n + \max(A^2, B^2) \gamma_n.
$$

It follows from this and (20) that as $n \to \infty$, $E \tau_n$ converges with *exponential* rate to

$$
m(0) = A^2 \alpha + B^2 \beta = A^2 \cdot \frac{B}{B - A} - B^2 \cdot \frac{A}{B - A} = |AB|.
$$

There is a similar result when $p \neq q$:

$$\mathsf{E}\,\tau_n \to m(0) = \frac{\alpha A + \beta B}{p - q} \quad \textit{exponentially} \text{ fast.}$$

5. PROBLEMS

1. Establish the following generalizations of (33) and (34):

$$\mathsf{E}\,S^x_{\tau^x_n} = x + (p - q)\,\mathsf{E}\,\tau^x_n,$$
$$\mathsf{E}[S^x_{\tau^x_n} - \tau^x_n \cdot \mathsf{E}\,\xi_1]^2 = \operatorname{Var}\xi_1 \cdot \mathsf{E}\,\tau^x_n + x^2.$$

2. Investigate the limits of $\alpha(x)$, $\beta(x)$, and $m(x)$ when the level $A \downarrow -\infty$.
3. Let $p = q = \frac{1}{2}$ in the Bernoulli scheme. What is the order of $\mathsf{E}\,|S_n|$ for large n?
4. Two players toss their own symmetric coins, independently. Show that the probability that each has the same number of heads after n tosses is $2^{-2n}\sum_{k=0}^{n}(C_n^k)^2$. Hence deduce the equation $\sum_{k=0}^{n}(C_n^k)^2 = C_{2n}^n$ (see also Problem 4 in Sect. 2).

   Let σ_n be the first time when the number of heads for the first player coincides with the number of heads for the second player (if this happens within n tosses; $\sigma_n = n + 1$ if there is no such time). Find $\mathsf{E}\min(\sigma_n, n)$.

10 Random Walk: II—Reflection Principle—Arcsine Law

1. As in the preceding section, we suppose that $\xi_1, \xi_2, \ldots, \xi_{2n}$ is a sequence of independent identically distributed Bernoulli random variables with

$$\mathsf{P}(\xi_i = 1) = p, \quad \mathsf{P}(\xi_i = -1) = q,$$
$$S_k = \xi_1 + \cdots + \xi_k, \quad 1 \leq k \leq 2n; \quad S_0 = 0.$$

We define

$$\sigma_{2n} = \min\{1 \leq k \leq 2n \colon S_k = 0\},$$

putting $\sigma_{2n} = \infty$ if $S_k \neq 0$ for $1 \leq k \leq 2n$.

The descriptive meaning of σ_{2n} is clear: it is the time of *first return* to zero. Properties of this time are studied in the present section, where we assume that the random walk is *symmetric*, i.e., $p = q = \frac{1}{2}$.

For $0 \leq k \leq n$ we write

$$u_{2k} = \mathsf{P}(S_{2k} = 0), \quad f_{2k} = \mathsf{P}(\sigma_{2n} = 2k). \tag{1}$$

It is clear that $u_0 = 1$ and

$$u_{2k} = C_{2k}^k \cdot 2^{-2k}.$$

Our immediate aim is to show that for $1 \le k \le n$ the probability f_{2k} is given by

$$f_{2k} = \frac{1}{2k} u_{2(k-1)}. \tag{2}$$

It is clear that

$$\{\sigma_{2n} = 2k\} = \{S_1 \ne 0, S_2 \ne 0, \ldots, S_{2k-1} \ne 0, S_{2k} = 0\}$$

for $1 \le k \le n$, and by symmetry

$$\begin{aligned} f_{2k} &= \mathsf{P}\{S_1 \ne 0, \ldots, S_{2k-1} \ne 0, S_{2k} = 0\} \\ &= 2\,\mathsf{P}\{S_1 > 0, \ldots, S_{2k-1} > 0, S_{2k} = 0\}. \end{aligned} \tag{3}$$

A sequence $(S_0, \ldots, S_k)$ is called a *path* of length k; we denote by $L_k(A)$ the number of paths of length k having some specified property A. Then

$$\begin{aligned} f_{2k} = 2 \sum_{(a_{2k+1}, \ldots, a_n)} & L_{2n}(S_1 > 0, \ldots, S_{2k-1} > 0, S_{2k} = 0, \\ & S_{2k+1} = a_{2k+1}, \ldots, S_{2n} = a_{2k+1} + \cdots + a_{2n}) \cdot 2^{-2n} \\ = 2L_{2k}(S_1 > 0, & \ldots, S_{2k-1} > 0, S_{2k} = 0) \cdot 2^{-2k}, \end{aligned} \tag{4}$$

where the summation is over all sets $(a_{2k+1}, \ldots, a_{2n})$ with $a_i = \pm 1$.

Consequently the determination of the probability f_{2k} reduces to calculating the number of paths $L_{2k}(S_1 > 0, \ldots, S_{2k-1} > 0, S_{2k} = 0)$.

Lemma 1. *Let a and b be nonnegative integers, $a - b > 0$ and $k = a + b$. Then*

$$L_k(S_1 > 0, \ldots, S_{k-1} > 0, S_k = a - b) = \frac{a-b}{k} C_k^a. \tag{5}$$

PROOF. In fact,

$$\begin{aligned} L_k(S_1 > 0, & \ldots, S_{k-1} > 0, S_k = a - b) \\ &= L_k(S_1 = 1, S_2 > 0, \ldots, S_{k-1} > 0, S_k = a - b) \\ &= L_k(S_1 = 1, S_k = a - b) - L_k(S_1 = 1, S_k = a - b; \\ &\quad \text{and } \exists\, i, 2 \le i \le k - 1, \text{ such that } S_i \le 0). \end{aligned} \tag{6}$$

In other words, the number of *positive* paths $(S_1, S_2, \ldots, S_k)$ that originate at $(1, 1)$ and terminate at $(k, a - b)$ is the same as the *total* number of paths from $(1, 1)$ to $(k, a - b)$ after excluding the paths that touch or intersect the time axis.*

---

* A path $(S_1, \ldots, S_k)$ is called *positive* (or nonnegative) if all $S_i > 0$ ($S_i \ge 0$); a path is said to *touch* the time axis if $S_j \ge 0$ or else $S_j \le 0$, for $1 \le j \le k$, and there is an i, $1 \le i \le k$, such that $S_i = 0$; and a path is said to *intersect* the time axis if there are two times i and j such that $S_i > 0$ and $S_j < 0$.

We now notice that

$$L_k(S_1 = 1, S_k = a - b; \exists i, 2 \leq i \leq k-1, \text{ such that } S_i \leq 0)$$
$$= L_k(S_1 = -1, S_k = a - b), \tag{7}$$

i.e., *the number of paths from* $\alpha = (1,1)$ *to* $\beta = (k, a-b)$, *which touch or intersect the time axis, is equal to the total number of paths that connect* $\alpha^* = (1, -1)$ *with* β. The proof of this statement, known as the *reflection principle*, follows from the easily established one-to-one correspondence between the paths $A = (S_1, \ldots, S_a, S_{a+1}, \ldots, S_k)$ joining α and β, and paths $B = (-S_1, \ldots, -S_a, S_{a+1}, \ldots, S_k)$ joining α^* and β (Fig. 17); a is the first point where A and B reach zero.

From (6) and (7) we find

$$L_k(S_1 > 0, \ldots, S_{k-1} > 0, S_k = a - b)$$
$$= L_k(S_1 = 1, S_k = a - b) - L_k(S_1 = -1, S_k = a - b)$$
$$= C_{k-1}^{a-1} - C_{k-1}^a = \frac{a-b}{k} C_k^a,$$

which establishes (5).

$\square$

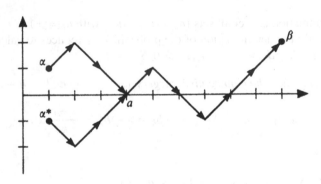

Fig. 17 The reflection principle

Turning to the calculation of f_{2k}, we find that by (4) and (5) (with $a = k$, $b = k - 1$),

$$f_{2k} = 2L_{2k}(S_1 > 0, \ldots, S_{2k-1} > 0, S_{2k} = 0) \cdot 2^{-2k}$$
$$= 2L_{2k-1}(S_1 > 0, \ldots, S_{2k-1} = 1) \cdot 2^{-2k}$$
$$= 2 \cdot 2^{-2k} \cdot \frac{1}{2k-1} C_{2k-1}^k = \frac{1}{2k} u_{2(k-1)}.$$

Hence (2) is established.

We present an alternative proof of this formula, based on the following observation. A straightforward verification shows that

$$\frac{1}{2k}u_{2(k-1)} = u_{2(k-1)} - u_{2k}. \tag{8}$$

At the same time, it is clear that

$$\{\sigma_{2n} = 2k\} = \{\sigma_{2n} > 2(k-1)\}\backslash\{\sigma_{2n} > 2k\},$$
$$\{\sigma_{2n} > 2l\} = \{S_1 \neq 0, \ldots, S_{2l} \neq 0\}$$

and therefore

$$\{\sigma_{2n} = 2k\} = \{S_1 \neq 0, \ldots, S_{2(k-1)} \neq 0\}\backslash\{S_1 \neq 0, \ldots, S_{2k} \neq 0\}.$$

Hence

$$f_{2k} = P\{S_1 \neq 0, \ldots, S_{2(k-1)} \neq 0\} - P\{S_1 \neq 0, \ldots, S_{2k} \neq 0\},$$

and consequently, because of (8), in order to show that $f_{2k} = (1/2k)u_{2(k-1)}$ it is enough to show only that

$$L_{2k}(S_1 \neq 0, \ldots, S_{2k} \neq 0) = L_{2k}(S_{2k} = 0). \tag{9}$$

For this purpose we notice that evidently

$$L_{2k}(S_1 \neq 0, \ldots, S_{2k} \neq 0) = 2L_{2k}(S_1 > 0, \ldots, S_{2k} > 0).$$

Hence to verify (9) we need only to establish that

$$2L_{2k}(S_1 > 0, \ldots, S_{2k} > 0) = L_{2k}(S_1 \geq 0, \ldots, S_{2k} \geq 0) \tag{10}$$

and

$$L_{2k}(S_1 \geq 0, \ldots, S_{2k} \geq 0) = L_{2k}(S_{2k} = 0). \tag{11}$$

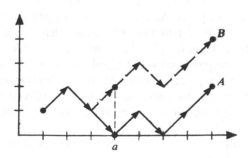

Fig. 18

Now (10) will be established if we show that we can establish a one-to-one correspondence between the paths $A = (S_1, \ldots, S_{2k})$ for which at least one $S_i = 0$, and the positive paths $B = (S_1, \ldots, S_{2k})$.

Let $A = (S_1, \ldots, S_{2k})$ be a nonnegative path for which the first zero occurs at the point a (i.e., $S_a = 0$). Let us construct the path, starting at $(a, 2)$, $(S_a + 2, S_{a+1} + 2, \ldots, S_{2k} + 2)$ (indicated by the broken lines in Fig. 18). Then the path $B = (S_1, \ldots, S_{a-1}, S_a + 2, \ldots, S_{2k} + 2)$ is positive.

Conversely, let $B = (S_1, \ldots, S_{2k})$ be a positive path and b the last instant at which $S_b = 1$ (Fig. 19). Then the path

$$A = (S_1, \ldots, S_b, S_{b+1} - 2, \ldots, S_k - 2)$$

is nonnegative. It follows from these constructions that there is a one-to-one correspondence between the positive paths and the nonnegative paths with at least one $S_i = 0$. Therefore formula (10) is established.

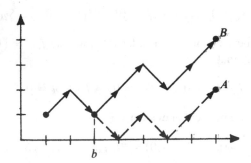

Fig. 19

We now establish (11). From symmetry and (10) it is enough to show that

$$L_{2k}(S_1 > 0, \ldots, S_{2k} > 0) + L_{2k}(S_1 \geq 0, \ldots, S_{2k} \geq 0 \text{ and } \exists\, i,\, 1 \leq i \leq 2k, \text{ such that } S_i = 0) = L_{2k}(S_{2k} = 0).$$

The set of paths $(S_{2k} = 0)$ can be represented as the sum of the two sets $\mathscr{C}_1$ and $\mathscr{C}_2$, where $\mathscr{C}_1$ contains the paths $(S_0, \ldots, S_{2k})$ that have just one minimum, and $\mathscr{C}_2$ contains those for which the minimum is attained at at least two points.

Let $C_1 \in \mathscr{C}_1$ (Fig. 20) and let γ be the minimum point. We put the path $C_1 = (S_0, S_1, \ldots, S_{2k})$ in correspondence with the path C_1^* obtained in the following way (Fig. 21). We reflect $(S_0, S_1, \ldots, S_l)$ around the vertical line through the point l, and displace the resulting path to the right and upward, thus releasing it from the point $(2k, 0)$. Then we move the origin to the point $(l, -m)$. The resulting path C_1^* will be positive.

In the same way, if $C_2 \in \mathscr{C}_2$ we can use the same device to put it into correspondence with a nonnegative path C_2^*.

Conversely, let $C_1^* = (S_1 > 0, \ldots, S_{2k} > 0)$ be a positive path with $S_{2k} = 2m$ (see Fig. 21). We make it correspond to the path C_1 that is obtained in the following

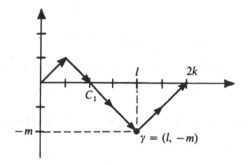

Fig. 20

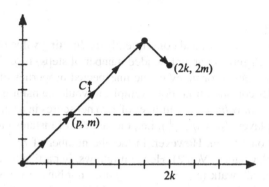

Fig. 21

way. Let p be the last point at which $S_p = m$. Reflect $(S_p, \ldots, S_{2m})$ with respect to the vertical line $x = p$ and displace the resulting path downward and to the left until its right-hand end coincides with the point $(0, 0)$. Then we move the origin to the left-hand end of the resulting path (this is just the path drawn in Fig. 20). The resulting path $C_1 = (S_0, \ldots, S_{2k})$ has a unique minimum and $S_{2k} = 0$. A similar construction applied to paths $(S_1 \geq 0, \ldots, S_{2k} \geq 0$ and $\exists\, i,\ 1 \leq i \leq 2k$, with $S_i = 0)$ leads to paths for which there are at least two minima and $S_{2k} = 0$. Hence we have established a one-to-one correspondence, which establishes (11).

Therefore we have established (9) and consequently also the formula $f_{2k} = u_{2(k-1)} - u_{2k} = (1/2k)u_{2(k-1)}$.

By Stirling's formula

$$u_{2k} = C_{2k}^k \cdot 2^{-2k} \sim \frac{1}{\sqrt{\pi k}}, \quad k \to \infty.$$

Therefore

$$f_{2k} \sim \frac{1}{2\sqrt{\pi}k^{3/2}}, \quad k \to \infty.$$

Hence it follows that the expectation of the first time when zero is reached, namely

$$\mathsf{E}\min(\sigma_{2n}, 2n) = \sum_{k=1}^{n} 2k\,\mathsf{P}(\sigma_{2n} = 2k) + 2nu_{2n}$$

$$= \sum_{k=1}^{n} u_{2(k-1)} + 2nu_{2n},$$

can be arbitrarily large.

In fact, $\sum_{k=1}^{\infty} u_{2(k-1)} = \infty$, and consequently the limiting value of the mean time for the walk to reach zero (in an unbounded number of steps) is ∞.

This property accounts for many of the unexpected properties of the symmetric random walk under consideration. For example, it would be natural to suppose that after time $2n$ the *mean value* of the number of zero net scores in a game between two equally matched players ($p = q = \frac{1}{2}$), i.e., the number of instants i at which $S_i = 0$, would be proportional to $2n$. However, in fact the number of zeros has order $\sqrt{2n}$ (see (17) in Sect. 9, Chap. 7, Vol. 2). Hence it follows, in particular, that, contrary to intuition, the "typical" walk ($S_0, S_1, \ldots, S_n$) does not have a *sinusoidal* character (so that roughly half the time the particle would be on the positive side and half the time on the negative side), but instead must resemble a *stretched-out wave*. The precise formulation of this statement is given by the *arcsine law*, which we proceed to investigate.

2. Let $P_{2k, 2n}$ be the probability that during the interval $[0, 2n]$ the particle spends $2k$ units of time on the positive side.*

Lemma 2. *Let $u_0 = 1$ and $0 \le k \le n$. Then*

$$P_{2k, 2n} = u_{2k} \cdot u_{2n-2k}. \tag{12}$$

PROOF. It was shown above that $f_{2k} = u_{2(k-1)} - u_{2k}$. Let us show that

$$u_{2k} = \sum_{r=1}^{k} f_{2r} \cdot u_{2(k-r)}. \tag{13}$$

---

* We say that the particle is on the positive side in the interval $[m - 1, m]$ if one, at least, of the values S_{m-1} and S_m is positive.

Since $\{S_{2k} = 0\} \subseteq \{\sigma_{2n} \leq 2k\}$, we have

$$\{S_{2k} = 0\} = \{S_{2k} = 0\} \cap \{\sigma_{2n} \leq 2k\} = \sum_{1 \leq l \leq k} \{S_{2k} = 0\} \cap \{\sigma_{2n} = 2l\}.$$

Consequently

$$u_{2k} = P(S_{2k} = 0) = \sum_{1 \leq l \leq k} P(S_{2k} = 0, \sigma_{2n} = 2l)$$

$$= \sum_{1 \leq l \leq k} P(S_{2k} = 0 \mid \sigma_{2k} = 2l) P(\sigma_{2n} = 2l).$$

But

$$P(S_{2k} = 0 \mid \sigma_{2n} = 2l) = P(S_{2k} = 0 \mid S_1 \neq 0, \ldots, S_{2l-1} \neq 0, S_{2l} = 0)$$

$$= P(S_{2l} + (\xi_{2l+1} + \cdots + \xi_{2k}) = 0 \mid S_1 \neq 0, \ldots, S_{2l-1} \neq 0, S_{2l} = 0)$$

$$= P(S_{2l} + (\xi_{2l+1} + \cdots + \xi_{2k}) = 0 \mid S_{2l} = 0)$$

$$= P(\xi_{2l+1} + \cdots + \xi_{2k} = 0) = P(S_{2(k-l)} = 0).$$

Therefore

$$u_{2k} = \sum_{1 \leq l \leq k} P\{S_{2(k-l)} = 0\} \, P\{\sigma_{2n} = 2l\},$$

which proves (13).

We turn now to the proof of (12). It is obviously true for $k = 0$ and $k = n$. Now let $1 \leq k \leq n - 1$. If the particle is on the positive side for exactly $2k < 2n$ instants, it must pass through zero. Let $2r$ be the time of first return to zero. There are two possibilities: either $S_l > 0$ for all $0 < l < 2r$, or $S_l < 0$ for all $0 < l < 2r$.

The number of paths of the first kind is easily seen to be

$$\left(\tfrac{1}{2} 2^{2r} f_{2r}\right) 2^{2(n-r)} P_{2(k-r),2(n-r)} = \tfrac{1}{2} \cdot 2^{2n} f_{2r} P_{2(k-r),2(n-r)}.$$

The corresponding number of paths of the second kind is

$$\tfrac{1}{2} \cdot 2^{2n} f_{2r} P_{2k,2(n-r)}.$$

Consequently, for $1 \leq k \leq n - 1$,

$$P_{2k,2n} = \frac{1}{2} \sum_{r=1}^{k} f_{2r} P_{2(k-r),2(n-r)} + \frac{1}{2} \sum_{r=1}^{k} f_{2r} P_{2k,2(n-r)}. \tag{14}$$

Let us suppose that $P_{2k, 2m} = u_{2k} \cdot u_{2m-2k}$ holds for $m = 1, \ldots, n - 1$. Then we find from (13) and (14) that

$$P_{2k,2n} = \tfrac{1}{2} u_{2n-2k} \sum_{r=1}^{k} f_{2r} u_{2k-2r} + \tfrac{1}{2} u_{2k} \sum_{r=1}^{k} f_{2r} u_{2n-2r-2k}$$

$$= \tfrac{1}{2} u_{2n-2k} u_{2k} + \tfrac{1}{2} u_{2k} u_{2n-2k} = u_{2k} u_{2n-2k}.$$

This completes the proof of the lemma.

□

Now let $\gamma(2n)$ be the number of time units that the particle spends on the *positive* axis in the interval $[0, 2n]$. Then, when $x < 1$,

$$\mathsf{P}\left\{ \frac{1}{2} < \frac{\gamma(2n)}{2n} \leq x \right\} = \sum_{\{k:\ 1/2 < (2k/2n) \leq x\}} P_{2k,2n}.$$

Since

$$u_{2k} \sim \frac{1}{\sqrt{\pi k}}$$

as $k \to \infty$, we have

$$P_{2k,2n} = u_{2k} u_{2(n-k)} \sim \frac{1}{\pi \sqrt{k(n-k)}},$$

as $k \to \infty$ and $n - k \to \infty$.

Therefore

$$\sum_{\{k:\ 1/2 < (2k/2n) \leq x\}} P_{2k,2n} - \sum_{\{k:\ 1/2 < (2k/2n) \leq x\}} \frac{1}{\pi n} \left[\frac{k}{n} \left(1 - \frac{k}{n} \right) \right]^{-1/2} \to 0, \quad n \to \infty,$$

whence

$$\sum_{\{k:\ 1/2 < (2k/2n) \leq x\}} P_{2k,2n} - \frac{1}{\pi} \int_{1/2}^{x} \frac{dt}{\sqrt{t(1-t)}} \to 0, \quad n \to \infty.$$

But, by symmetry,

$$\sum_{\{k:\ k/n \leq 1/2\}} P_{2k,2n} \to \frac{1}{2}$$

and

$$\frac{1}{\pi} \int_{1/2}^{x} \frac{dt}{\sqrt{t(1-t)}} = \frac{2}{\pi} \arcsin \sqrt{x} - \frac{1}{2}.$$

Consequently we have proved the following theorem.

Theorem (Arcsine Law). *The probability that the fraction of the time spent by the particle on the positive side is at most x tends to* $2\pi^{-1} \arcsin \sqrt{x}$:

$$\sum_{\{k:\, k/n \le x\}} P_{2k,2n} \to 2\pi^{-1} \arcsin \sqrt{x}. \tag{15}$$

We remark that the integrand $u(t) = (t(1-t))^{-1/2}$ in the integral

$$\frac{1}{\pi} \int_0^x \frac{dt}{\sqrt{t(1-t)}}$$

represents a U-shaped curve that tends to infinity as $t \to 0$ or 1.

Hence it follows that, for large n,

$$P\left\{0 < \frac{\gamma(2n)}{2n} \le \Delta\right\} > P\left\{\frac{1}{2} < \frac{\gamma(2n)}{2n} \le \frac{1}{2} + \Delta\right\},$$

i.e., it is more likely that the fraction of the time spent by the particle on the positive side is close to zero or one, than to the intuitive value $\frac{1}{2}$.

Using a table of arcsines and noting that the convergence in (15) is indeed quite rapid, we find that

$$P\left\{\frac{\gamma(2n)}{2n} \le 0.024\right\} \approx 0.1,$$

$$P\left\{\frac{\gamma(2n)}{2n} \le 0.1\right\} \approx 0.2,$$

$$P\left\{\frac{\gamma(2n)}{2n} \le 0.2\right\} \approx 0.3,$$

$$P\left\{\frac{\gamma(2n)}{2n} \le 0.65\right\} \approx 0.6.$$

Hence if, say, $n = 1000$, then in about one case in ten, the particle spends only 24 units of time on the positive axis and therefore spends the greatest amount of time, 976 units, on the negative axis.

3. PROBLEMS

1. How fast does $E \min(\sigma_{2n}, 2n) \to \infty$ as $n \to \infty$?
2. Let $\tau_n = \min\{1 \le k \le n : S_k = 1\}$, where we take $\tau_n = \infty$ if $S_k < 1$ for $1 \le k \le n$. What is the limit of $E \min(\tau_n, n)$ as $n \to \infty$ for symmetric $(p = q = \frac{1}{2})$ and for asymmetric $(p \ne q)$ walks?

3. Using the ideas and methods of Sect. 10, show that the symmetric ($p = q = 1/2$) Bernoulli random walk $\{S_k, k \leq n\}$ with $S_0 = 0$, $S_k = \xi_1 + \cdots + \xi_k$ fulfills the following equations (N is a positive integer):

$$P\left\{\max_{1 \leq k \leq n} S_k \geq N, S_n < N\right\} = P\{S_n > N\},$$

$$P\left\{\max_{1 \leq k \leq n} S_k \geq N\right\} = 2\,P\{S_n \geq N\} - P\{S_n = N\},$$

$$P\left\{\max_{1 \leq k \leq n} S_k = N\right\} = P\{S_n = N\} + P\{S_n = N + 1\}.$$

11 Martingales: Some Applications to the Random Walk

1. The Bernoulli random walk discussed above was generated by a sequence $\xi_1, \ldots, \xi_n$ of *independent* random variables. In this and the next section we introduce two important classes of *dependent* random variables, those that constitute *martingales* and *Markov chains*.

The theory of martingales will be developed in detail in Chapter 7, Vol. 2. Here we shall present only the essential definitions, prove a theorem on the preservation of the martingale property for stopping times, and apply this to deduce the " ballot theorem." In turn, the latter theorem will be used for another proof of the statement (5), Sect. 10, which was obtained above by applying the reflection principle.

2. Let $(\Omega, \mathscr{A}, \mathsf{P})$ be a finite probability space and $\mathscr{D}_1 \preccurlyeq \mathscr{D}_2 \preccurlyeq \cdots \preccurlyeq \mathscr{D}_n$ a sequence of decompositions.

Definition 1. A sequence of random variables $\xi_1, \ldots, \xi_n$ is called a *martingale* (with respect to the decompositions $\mathscr{D}_1 \preccurlyeq \mathscr{D}_2 \preccurlyeq \cdots \preccurlyeq \mathscr{D}_n$) if

(1) ξ_k is $\mathscr{D}_k$-measurable,
(2) $\mathsf{E}(\xi_{k+1} \mid \mathscr{D}_k) = \xi_k$, $1 \leq k \leq n - 1$.

In order to emphasize the *system of decompositions* with respect to which the random variables $\xi = (\xi_1, \ldots, \xi_n)$ form a martingale, we shall use the notation

$$\xi = (\xi_k, \mathscr{D}_k)_{1 \leq k \leq n}, \tag{1}$$

where for the sake of simplicity we often do not mention explicitly that $1 \leq k \leq n$.
When $\mathscr{D}_k$ is induced by $\xi_1, \ldots, \xi_n$, i.e.,

$$\mathscr{D}_k = \mathscr{D}_{\xi_1, \ldots, \xi_k},$$

instead of saying that $\xi = (\xi_k, \mathscr{D}_k)$ is a martingale, we simply say that the sequence $\xi = (\xi_k)$ is a martingale.

Here are some examples of martingales.

Example 1. Let $\eta_1, \ldots, \eta_n$ be independent Bernoulli random variables with

$$P(\eta_k = 1) = P(\eta_k = -1) = \tfrac{1}{2},$$

$$S_k = \eta_1 + \cdots + \eta_k \quad \text{and} \quad \mathscr{D}_k = \mathscr{D}_{\eta_1,\ldots,\eta_k}.$$

We observe that the decompositions $\mathscr{D}_k$ have a simple structure:

$$\mathscr{D}_1 = \{D^+, D^-\},$$

where $D^+ = \{\omega: \eta_1 = +1\}, D^- = \{\omega: \eta_1 = -1\};$

$$\mathscr{D}_2 = \{D^{++}, D^{+-}, D^{-+}, D^{--}\},$$

where $D^{++} = \{\omega: \eta_1 = +1, \eta_2 = +1\}, \ldots, D^{--} = \{\omega: \eta_1 = -1, \eta_2 = -1\}$, etc.

It is also easy to see that $\mathscr{D}_{\eta_1,\ldots,\eta_k} = \mathscr{D}_{S_1,\ldots,S_k}.$

Let us show that $(S_k, \mathscr{D}_k)_{1 \leq k \leq n}$ form a martingale. In fact, S_k is $\mathscr{D}_k$-measurable, and by (12) and (18) of Sect. 8

$$E(S_{k+1} \mid \mathscr{D}_k) = E(S_k + \eta_{k+1} \mid \mathscr{D}_k)$$
$$= E(S_k \mid \mathscr{D}_k) + E(\eta_{k+1} \mid \mathscr{D}_k) = S_k + E\,\eta_{k+1} = S_k.$$

If we put $S_0 = 0$ and take $D_0 = \{\Omega\}$, the trivial decomposition, then the sequence $(S_k, \mathscr{D}_k)_{0 \leq k \leq n}$ also forms a martingale.

Example 2. Let $\eta_1, \ldots, \eta_n$ be independent Bernoulli random variables with $P(\eta_i = 1) = p$, $P(\eta_i = -1) = q$. If $p \neq q$, each of the sequences $\xi = (\xi_k)$ with

$$\xi_k = \left(\frac{q}{p}\right)^{S_k}, \qquad \xi_k = S_k - k(p - q), \quad \text{where} \quad S_k = \eta_1 + \cdots + \eta_k,$$

is a martingale.

Example 3. Let η be a random variable, $\mathscr{D}_1 \leqslant \cdots \leqslant \mathscr{D}_n$, and

$$\xi_k = E(\eta \mid \mathscr{D}_k). \tag{2}$$

Then the sequence $\xi = (\xi_k, \mathscr{D}_k)_{1 \leq k \leq n}$ is a martingale. In fact, it is evident that $E(\eta \mid \mathscr{D}_k)$ is $\mathscr{D}_k$-measurable, and by (20) of Sect. 8

$$E(\xi_{k+1} \mid \mathscr{D}_k) = E[E(\eta \mid \mathscr{D}_{k+1}) \mid \mathscr{D}_k] = E(\eta \mid \mathscr{D}_k) = \xi_k.$$

In this connection we notice that if $\xi = (\xi_k, \mathscr{D}_k)$ is any martingale, then by (20) of Sect. 8

$$\xi_k = \mathsf{E}(\xi_{k+1} \mid \mathscr{D}_k) = \mathsf{E}[\mathsf{E}(\xi_{k+2} \mid \mathscr{D}_{k+1}) \mid \mathscr{D}_k]$$
$$= \mathsf{E}(\xi_{k+2} \mid \mathscr{D}_k) = \cdots = \mathsf{E}(\xi_n \mid \mathscr{D}_k). \tag{3}$$

Consequently the set of martingales $\xi = (\xi_k, \mathscr{D}_k)$ is exhausted by the martingales of the form (2). (We note that for infinite sequences $\xi = (\xi_k, \mathscr{D}_k)_{k \geq 1}$ this is, in general, no longer the case; see Problem 6 in Sect. 1 of Chap. 7, Vol. 2.)

Example 4. Let $\eta_1, \ldots, \eta_n$ be a sequence of independent identically distributed random variables, $S_k = \eta_1 + \cdots + \eta_k$, and $\mathscr{D}_1 = \mathscr{D}_{S_n}$, $\mathscr{D}_2 = \mathscr{D}_{S_n, S_{n-1}}, \ldots, \mathscr{D}_n = \mathscr{D}_{S_n, S_{n-1}, \ldots, S_1}$. Let us show that the sequence $\xi = (\xi_k, \mathscr{D}_k)$ with

$$\xi_1 = \frac{S_n}{n}, \ \xi_2 = \frac{S_{n-1}}{n-1}, \ \ldots, \ \xi_k = \frac{S_{n+1-k}}{n+1-k}, \ \ldots, \ \xi_n = S_1$$

is a martingale. In the first place, it is clear that $\mathscr{D}_k \leqslant \mathscr{D}_{k+1}$ and ξ_k is $\mathscr{D}_k$-measurable. Moreover, we have by symmetry, for $j \leq n - k + 1$,

$$\mathsf{E}(\eta_j \mid \mathscr{D}_k) = \mathsf{E}(\eta_1 \mid \mathscr{D}_k) \tag{4}$$

(compare (26), Sect. 8). Therefore

$$(n - k + 1)\,\mathsf{E}(\eta_1 \mid \mathscr{D}_k) = \sum_{j=1}^{n-k+1} \mathsf{E}(\eta_j \mid \mathscr{D}_k) = \mathsf{E}(S_{n-k+1} \mid \mathscr{D}_k) = S_{n-k+1},$$

and consequently

$$\xi_k = \frac{S_{n-k+1}}{n-k+1} = \mathsf{E}(\eta_1 \mid \mathscr{D}_k),$$

and it follows from Example 3 that $\xi = (\xi_k, \mathscr{D}_k)$ is a martingale.

Remark. From this martingale property of the sequence $\xi = (\xi_k, \mathscr{D}_k)_{1 \leq k \leq n}$, it is clear why we will sometimes say that the sequence $(S_k/k)_{1 \leq k \leq n}$ forms a *reversed martingale*. (Compare Problem 5 in Sect. 1, Chap. 7, Vol. 2).

Example 5. Let $\eta_1, \ldots, \eta_n$ be independent Bernoulli random variables with

$$\mathsf{P}(\eta_i = +1) = \mathsf{P}(\eta_i = -1) = \tfrac{1}{2},$$

$S_k = \eta_1 + \cdots + \eta_k$. Let A and B be integers, $A < 0 < B$. Then with $0 < \lambda < \pi/2$, the sequence $\xi = (\xi_k, \mathscr{D}_k)$ with $\mathscr{D}_k = \mathscr{D}_{S_1, \ldots, S_k}$ and

$$\xi_k = (\cos \lambda)^{-k} \exp\left\{ i\lambda \left(S_k - \frac{B+A}{2} \right) \right\}, \quad 1 \leq k \leq n, \tag{5}$$

is a complex martingale (i.e., the real and imaginary parts of ξ_k, $1 \leq k \leq n$, form martingales).

3. It follows from the definition of a martingale that the expectation $\mathsf{E}\,\xi_k$ is the same for every k:

$$\mathsf{E}\,\xi_k = \mathsf{E}\,\xi_1.$$

It turns out that this property persists if time k is replaced by a *stopping time*. In order to formulate this property we introduce the following definition.

Definition 2. A random variable $\tau = \tau(\omega)$ that takes the values $1, 2, \ldots, n$ is called a *stopping time* (with respect to decompositions $(\mathscr{D}_k)_{1 \le k \le n}$, $\mathscr{D}_1 \preceq \mathscr{D}_2 \preceq \cdots \preceq \mathscr{D}_n$) if, for any $k = 1, \ldots, n$, the random variable $I_{\{\tau=k\}}(\omega)$ is $\mathscr{D}_k$-measurable.

If we consider $\mathscr{D}_k$ as the decomposition induced by observations for k steps (for example, $\mathscr{D}_k = \mathscr{D}_{\eta_1, \ldots, \eta_k}$, the decomposition induced by the variables $\eta_1, \ldots, \eta_k$), then the $\mathscr{D}_k$-measurability of $I_{\{\tau=k\}}(\omega)$ means that the realization or nonrealization of the event $\{\tau = k\}$ is determined only by observations for k steps (and is independent of the "future").

If $\mathscr{B}_k = \alpha(\mathscr{D}_k)$, then the $\mathscr{D}_k$-measurability of $I_{\{\tau=k\}}(\omega)$ is equivalent to the assumption that

$$\{\tau = k\} \in \mathscr{B}_k. \tag{6}$$

We have already encountered specific examples of stopping times: the times τ_k^x and σ_{2n} introduced in Sects. 9 and 10. Those times are special cases of stopping times of the form

$$
\begin{aligned}
\tau^A &= \min\{0 < k \le n \colon \xi_k \in A\}, \\
\sigma^A &= \min\{0 \le k \le n \colon \xi_k \in A\},
\end{aligned}
\tag{7}
$$

which are the times (respectively the *first time after zero* and the *first time*) for a sequence $\xi_0, \xi_1, \ldots, \xi_n$ to attain a point of the set A.

4. Theorem 1. *Let $\xi = (\xi_k, \mathscr{D}_k)_{1 \le k \le n}$ be a martingale and τ a stopping time with respect to the decompositions $(\mathscr{D}_k)_{1 \le k \le n}$. Then*

$$\mathsf{E}(\xi_\tau \mid \mathscr{D}_1) = \xi_1, \tag{8}$$

where

$$\xi_\tau = \sum_{k=1}^{n} \xi_k I_{\{\tau=k\}}(\omega) \tag{9}$$

and

$$\mathsf{E}\,\xi_\tau = \mathsf{E}\,\xi_1. \tag{10}$$

PROOF (compare the proof of (29) in Sect. 9). Let $D \in \mathscr{D}_1$. Using (3) and the properties of conditional expectations, we find that

$$E(\xi_\tau \mid D) = \frac{E(\xi_\tau I_D)}{P(D)}$$

$$= \frac{1}{P(D)} \sum_{l=1}^{n} E(\xi_l \cdot I_{\{\tau=l\}} \cdot I_D)$$

$$= \frac{1}{P(D)} \sum_{l=1}^{n} E[E(\xi_n \mid \mathscr{D}_l) \cdot I_{\{\tau=l\}} \cdot I_D]$$

$$= \frac{1}{P(D)} \sum_{l=1}^{n} E[E(\xi_n I_{\{\tau=l\}} \cdot I_D \mid \mathscr{D}_l)]$$

$$= \frac{1}{P(D)} \sum_{l=1}^{n} E[\xi_n I_{\{\tau=l\}} \cdot I_D]$$

$$= \frac{1}{P(D)} E(\xi_n I_D) = E(\xi_n \mid D),$$

and consequently

$$E(\xi_\tau \mid \mathscr{D}_1) = E(\xi_n \mid \mathscr{D}_1) = \xi_1.$$

The equation $E\,\xi_\tau = E\,\xi_1$ then follows in an obvious way.

This completes the proof of the theorem.

□

Corollary. *For the martingale* $(S_k, \mathscr{D}_k)_{1\leq k\leq n}$ *of Example* 1, *and any stopping time* τ *(with respect to* $(\mathscr{D}_k)$*) we have the formulas*

$$E\,S_\tau = 0, \qquad E\,S_\tau^2 = E\,\tau, \tag{11}$$

known as Wald's identities (cf. (29) *and* (30) *in* Sect. 9; *see also Problem* 1 *and Theorem* 3 *in* Sect. 2, *Chap.* 7, *Vol.* 2).

Let us use Theorem 1 to establish the following proposition.

Theorem 1 (Ballot Theorem). *Let* $\eta_1, \ldots, \eta_n$ *be a sequence of independent identically distributed random variables taking finitely many values from the set* $(0, 1, \ldots)$ *and*

$$S_k = \eta_1 + \cdots + \eta_k, \quad 1 \leq k \leq n.$$

Then (P-a.s.)

$$P\{S_k < k \text{ for all } k, 1 \leq k \leq n \mid S_n\} = \left(1 - \frac{S_n}{n}\right)^+, \tag{12}$$

where $a^+ = \max(a, 0)$.

PROOF. On the set $\{\omega : S_n \geq n\}$ the formula is evident. We therefore prove (12) for the sample points at which $S_n < n$.

Let us consider the martingale $\xi = (\xi_k, \mathscr{D}_k)_{1 \leq k \leq n}$ introduced in Example 4, with $\xi_k = S_{n+1-k}/(n+1-k)$ and $\mathscr{D}_k = \mathscr{D}_{S_{n+1-k},\ldots,S_n}$.

We define

$$\tau = \min\{1 \leq k \leq n : \xi_k \geq 1\},$$

taking $\tau = n$ on the set $\{\xi_k < 1 \text{ for all } k, 1 \leq k \leq n\} = \{\max_{1 \leq l \leq n}(S_l/l) < 1\}$. It is clear that $\xi_\tau = \xi_n = S_1 = 0$ on this set, and therefore

$$\left\{\max_{1 \leq l \leq n} \frac{S_l}{l} < 1\right\} = \left\{\max_{1 \leq l \leq n} \frac{S_l}{l} < 1, \, S_n < n\right\} \subseteq \{\xi_\tau = 0\}. \tag{13}$$

Now let us consider the outcomes for which simultaneously $\max_{1 \leq l \leq n}(S_l/l) \geq 1$ and $S_n < n$. Write $\sigma = n + 1 - \tau$. It is easy to see that

$$\sigma = \max\{1 \leq k \leq n : S_k \geq k\}$$

and therefore (since $S_n < n$) we have $\sigma < n$, $S_\sigma \geq \sigma$, and $S_{\sigma+1} < \sigma + 1$. Consequently $\eta_{\sigma+1} = S_{\sigma+1} - S_\sigma < (\sigma + 1) - \sigma = 1$, i.e., $\eta_{\sigma+1} = 0$. Therefore $\sigma \leq S_\sigma = S_{\sigma+1} < \sigma + 1$, and consequently $S_\sigma = \sigma$ and

$$\xi_\tau = \frac{S_{n+1-\tau}}{n+1-\tau} = \frac{S_\sigma}{\sigma} = 1.$$

Therefore

$$\left\{\max_{1 \leq l \leq n} \frac{S_l}{l} \geq 1, \, S_n < n\right\} \subseteq \{\xi_\tau = 1\}. \tag{14}$$

From (13) and (14) we find that

$$\left\{\max_{1 \leq l \leq n} \frac{S_l}{l} \geq 1, \, S_n < n\right\} = \{\xi_\tau = 1\} \cap \{S_n < n\}.$$

Therefore, on the set $\{S_n < n\}$, we have

$$\mathsf{P}\left\{\max_{1 \leq l \leq n} \frac{S_l}{l} \geq 1 \,\Big|\, S_n\right\} - \mathsf{P}\{\xi_\tau - 1 \,|\, S_n\} = \mathsf{E}(\xi_\tau \,|\, S_n),$$

where the last equation follows because ξ_τ takes only the two values 0 and 1.

Let us notice now that $\mathsf{E}(\xi_\tau \,|\, S_n) = \mathsf{E}(\xi_\tau \,|\, \mathscr{D}_1)$, and (by Theorem 1) $\mathsf{E}(\xi_\tau \,|\, \mathscr{D}_1) = \xi_1 = S_n/n$. Consequently, on the set $\{S_n < n\}$ we have

$$\mathsf{P}\{S_k < k \text{ for all } k \text{ such that } 1 \leq k \leq n \,|\, S_n\} = 1 - (S_n/n).$$

This completes the proof of the theorem.

$\square$

We now apply this theorem to obtain a different proof of Lemma 1 of Sect. 10, and explain why it is called the *ballot theorem*.

Let $\xi_1, \ldots, \xi_n$ be independent Bernoulli random variables with

$$P(\xi_i = 1) = P(\xi_i = -1) = \tfrac{1}{2},$$

$S_k = \xi_1 + \cdots + \xi_k$ and a, b nonnegative integers such that $a - b > 0$, $a + b = n$. We are going to show that

$$P\{S_1 > 0, \ldots, S_n > 0 \mid S_n = a - b\} = \frac{a - b}{a + b}. \qquad (15)$$

In fact, by symmetry,

$$
\begin{aligned}
P\{S_1 & > 0, \ldots, S_n > 0 \mid S_n = a - b\} \\
&= P\{S_1 < 0, \ldots, S_n < 0 \mid S_n = -(a - b)\} \\
&= P\{S_1 + 1 < 1, \ldots, S_n + n < n \mid S_n + n = n - (a - b)\} \\
&= P\{\eta_1 < 1, \ldots, \eta_1 + \cdots + \eta_n < n \mid \eta_1 + \cdots + \eta_n = n - (a - b)\} \\
&= \left[1 - \frac{n - (a - b)}{n}\right]^{+} = \frac{a - b}{n} = \frac{a - b}{a + b},
\end{aligned}
$$

where we have put $\eta_k = \xi_k + 1$ and applied (12).

Now formula (5) of Sect. 10 established in Lemma 1 of Sect. 10 by using the reflection principle follows from (15) in an evident way.

Let us interpret $\xi_i = +1$ as a vote for candidate A and $\xi_i = -1$ as a vote for B. Then S_k is the difference between the numbers of votes cast for A and B at the time when k votes have been recorded, and

$$P\{S_1 > 0, \ldots, S_n > 0 \mid S_n = a - b\}$$

is the probability that A was *always ahead* of B given that A received a votes in all, B received b votes, and $a - b > 0$, $a + b = n$. According to (15) this probability is $(a - b)/n$.

5. PROBLEMS

1. Let $\mathscr{D}_0 \leqslant \mathscr{D}_1 \leqslant \cdots \leqslant \mathscr{D}_n$ be a sequence of decompositions with $\mathscr{D}_0 = \{\Omega\}$, and let η_k be $\mathscr{D}_k$-measurable variables, $1 \leq k \leq n$. Show that the sequence $\xi = (\xi_k, \mathscr{D}_k)$ with

$$\xi_k = \sum_{l=1}^{k} [\eta_l - \mathsf{E}(\eta_l \mid \mathscr{D}_{l-1})]$$

is a martingale.

2. Let the random variables $\eta_1, \ldots, \eta_k$ satisfy $\mathsf{E}(\eta_k \mid \eta_1, \ldots, \eta_{k-1}) = 0, 2 \leq k \leq n$. Show that the sequence $\xi = (\xi_k)_{1 \leq k \leq n}$ with $\xi_1 = \eta_1$ and

$$\xi_{k+1} = \sum_{i=1}^{k} \eta_{i+1} f_i(\eta_1, \ldots, \eta_i), \quad 1 \leq k \leq n - 1,$$

where f_i are given functions, is a martingale.

3. Show that every martingale $\xi = (\xi_i, \mathcal{D}_k)$ has *uncorrelated increments*: if $a < b < c < d$ then

$$\operatorname{Cov}(\xi_d - \xi_c, \xi_b - \xi_a) = 0.$$

4. Let $\xi = (\xi_1, \ldots, \xi_n)$ be a random sequence such that ξ_k is $\mathcal{D}_k$-measurable $(\mathcal{D}_1 \leqslant \mathcal{D}_2 \leqslant \cdots \leqslant \mathcal{D}_n)$. Show that a necessary and sufficient condition for this sequence to be a martingale (with respect to the system $(\mathcal{D}_k)$) is that $\mathsf{E}\,\xi_\tau = \mathsf{E}\,\xi_1$ for every stopping time τ (with respect to $(\mathcal{D}_k)$). (The phrase "for every stopping time" can be replaced by "for every stopping time that assumes two values.")

5. Show that if $\xi = (\xi_k, \mathcal{D}_k)_{1 \leq k \leq n}$ is a martingale and τ is a stopping time, then

$$\mathsf{E}[\xi_n I_{\{\tau=k\}}] = \mathsf{E}[\xi_k I_{\{\tau=k\}}]$$

for every k.

6. Let $\xi = (\xi_k, \mathcal{D}_k)$ and $\eta = (\eta_k, \mathcal{D}_k)$ be two martingales, $\xi_1 = \eta_1 = 0$. Show that

$$\mathsf{E}\,\xi_n \eta_n = \sum_{k=2}^{n} \mathsf{E}(\xi_k - \xi_{k-1})(\eta_k - \eta_{k-1})$$

and in particular that

$$\mathsf{E}\,\xi_n^2 = \sum_{k=2}^{n} \mathsf{E}(\xi_k - \xi_{k-1})^2.$$

7. Let $\eta_1, \ldots, \eta_n$ be a sequence of independent identically distributed random variables with $\mathsf{E}\,\eta_i = 0$. Show that the sequences $\xi = (\xi_k)$ with

$$\xi_k = \left(\sum_{i=1}^{k} \eta_i\right)^2 - k\,\mathsf{E}\,\eta_1^2,$$

$$\xi_k = \frac{\exp\{\lambda(\eta_1 + \cdots + \eta_k)\}}{(\mathsf{E}\exp\lambda\eta_1)^k}$$

are martingales.

8. Let $\eta_1, \ldots, \eta_n$ be a sequence of independent identically distributed random variables taking values in a finite set Y. Let $f_0(y) = \mathsf{P}(\eta_1 = y) > 0$, $y \in Y$, and let $f_1(y)$ be a nonnegative function with $\sum_{y \in Y} f_1(y) = 1$. Show that the sequence $\xi = (\xi_k, \mathcal{D}_k^\eta)$ with $\mathcal{D}_k^\eta = D_{\eta_1, \ldots, \eta_k}$,

$$\xi_k = \frac{f_1(\eta_1) \cdots f_1(\eta_k)}{f_0(\eta_1) \cdots f_0(\eta_k)},$$

is a martingale. (The variables ξ_k, known as *likelihood ratios*, are extremely important in mathematical statistics.)

12 Markov Chains: Ergodic Theorem, Strong Markov Property

1. We have discussed the Bernoulli scheme with

$$\Omega = \{\omega: \omega = (x_1, \ldots, x_n), \ x_i = 0, 1\},$$

where the probability $P(\{\omega\}) = p(\omega)$ of each outcome is given by

$$p(\omega) = p(x_1) \cdots p(x_n), \tag{1}$$

with $p(x) = p^x q^{1-x}$. With these hypotheses, the variables $\xi_1, \ldots, \xi_n$ with $\xi_i(\omega) = x_i$ are *independent* and *identically distributed* with

$$P(\xi_1 = x) = \cdots = P(\xi_n = x) = p(x), \quad x = 0, 1.$$

If we replace (1) by
$$p(\omega) = p_1(x_1) \cdots p_n(x_n),$$

where $p_i(x) = p_i^x(1 - p_i)^{1-x}$, $0 \le p_i \le 1$, the random variables $\xi_1, \ldots, \xi_n$ are still *independent*, but in general are *differently distributed*:

$$P(\xi_1 = x) = p_1(x), \ldots, P(\xi_n = x) = p_n(x).$$

We now consider a generalization that leads to *dependent* random variables that form what is known as a *Markov chain*.

Let us suppose that

$$\Omega = \{\omega: \omega = (x_0, x_1, \ldots, x_n), \ x_i \in X\},$$

where X is a finite set. Let there be given nonnegative functions $p_0(x)$, $p_1(x, y), \ldots,$ $p_n(x, y)$ such that

$$\sum_{x \in X} p_0(x) = 1,$$

$$\sum_{y \in X} p_k(x, y) = 1, \quad k = 1, \ldots, n; \ x \in X. \tag{2}$$

For each $\omega = (x_0, x_1, \ldots, x_n)$, put $P(\{\omega\}) = p(\omega)$, where

$$p(\omega) = p_0(x_0) p_1(x_0, x_1) \cdots p_n(x_{n-1}, x_n). \tag{3}$$

It is easily verified that $\sum_{\omega \in \Omega} p(\omega) = 1$, and consequently the set of numbers $p(\omega)$ together with the space Ω and the collection of its subsets defines a probabilistic model $(\Omega, \mathscr{A}, P)$, which it is usual to call a *model of experiments that form a Markov chain*.

Let us introduce the random variables $\xi_0, \xi_1, \ldots, \xi_n$ with $\xi_i(\omega) = x_i$ for $\omega = (x_1, \ldots, x_n)$. A simple calculation shows that

$$P(\xi_0 = a) = p_0(a),$$
$$P(\xi_0 = a_0, \ldots, \xi_k = a_k) = p_0(a_0)p_1(a_0, a_1) \cdots p_k(a_{k-1}, a_k). \quad (4)$$

We now establish the validity of the following fundamental property of conditional probabilities in the probability model $(\Omega, \mathscr{A}, P)$ at hand:

$$P\{\xi_{k+1} = a_{k+1} \mid \xi_k = a_k, \ldots, \xi_0 = a_0\} = P\{\xi_{k+1} = a_{k+1} \mid \xi_k = a_k\} \quad (5)$$

(under the assumption that $P(\xi_k = a_k, \ldots, \xi_0 = a_0) > 0$).

By (4) and the definition of conditional probabilities (Sect. 3)

$$P\{\xi_{k+1} = a_{k+1} \mid \xi_k = a_k, \ldots, \xi_0 = a_0\}$$
$$= \frac{P\{\xi_{k+1} = a_{k+1}, \ldots, \xi_0 = a_0\}}{P\{\xi_k = a_k, \ldots, \xi_0 = a_0\}}$$
$$= \frac{p_0(a_0)p_1(a_0, a_1) \cdots p_{k+1}(a_k, a_{k+1})}{p_0(a_0) \cdots p_k(a_{k-1}, a_k)} = p_{k+1}(a_k, a_{k+1}).$$

In a similar way we verify

$$P\{\xi_{k+1} = a_{k+1} \mid \xi_k = a_k\} = p_{k+1}(a_k, a_{k+1}), \quad (6)$$

which establishes (5).

Let $\mathscr{D}_k^\xi = \mathscr{D}_{\xi_0, \ldots, \xi_k}$ be the decomposition induced by $\xi_0, \ldots, \xi_k$, and $\mathscr{B}_k^\xi = \alpha(\mathscr{D}_k^\xi)$.

Then, in the notation introduced in Sect. 8, it follows from (5) that

$$P\{\xi_{k+1} = a_{k+1} \mid \mathscr{B}_k^\xi\} = P\{\xi_{k+1} = a_{k+1} \mid \xi_k\} \quad (7)$$

or

$$P\{\xi_{k+1} = a_{k+1} \mid \xi_0, \ldots, \xi_k\} = P\{\xi_{k+1} = a_{k+1} \mid \xi_k\}.$$

Remark 1. We interrupt here our exposition in order to make an important comment regarding the formulas (5) and (7) and events of *zero probability*.

Formula (5) was established assuming that $P\{\xi_k = a_k, \ldots, \xi_0 = a_0\} > 0$ (hence also $P\{\xi_k = a_k\} > 0$). In essence, this was needed only because conditional probabilities $P(A \mid B)$ have been defined (so far!) only under the assumption $P(B) > 0$.

Let us notice, however, that if $B = \{\xi_k = a_k, \ldots, \xi_0 = a_0\}$ and $P(B) = 0$ (and therefore also $P(C) = 0$ for $C = \{\xi_k = a_k\}$), then the "path" $\{\xi_0 = a_0, \ldots, \xi_k = a_k\}$ has to be viewed as *unrealizable* one, and then the question about the conditional probability of the event $\{\xi_{k+1} = a_k\}$ given that this *unrealizable* "path" occurs is of no practical interest.

In this connection we will for definiteness *define* the conditional probability $P(A \mid B)$ by the formula

$$P(A \mid B) = \begin{cases} \frac{P(AB)}{P(B)}, & \text{if } P(B) > 0, \\ 0, & \text{if } P(B) = 0. \end{cases}$$

With this definition the formulas (5) and (7) hold without any additional assumptions like $P\{\xi_k = a_k, \ldots, \xi_0 = a_0\} > 0$.

Let us emphasize that the difficulty related to the events of zero probability is very common for probability theory. We will give in Sect. 7, Chap. 2, a general definition of conditional probabilities (with respect to *arbitrary* decompositions, σ-algebras, etc.), which is both very natural and "works" in "zero probability" setups.

Now, if we use the evident equation

$$P(AB \mid C) = P(A \mid BC) P(B \mid C),$$

we find from (7) that

$$P\{\xi_n = a_n, \ldots, \xi_{k+1} = a_{k+1} \mid \mathscr{B}_k^\xi\} = P\{\xi_n = a_n, \ldots, \xi_{k+1} = a_{k+1} \mid \xi_k\} \qquad (8)$$

or

$$P\{\xi_n = a_n, \ldots, \xi_{k+1} = a_{k+1} \mid \xi_0, \ldots, \xi_k\}$$
$$= P\{\xi_n = a_n, \ldots, \xi_{k+1} = a_{k+1} \mid \xi_k\}. \qquad (9)$$

This equation admits the following intuitive interpretation. Let us think of ξ_k as the position of a particle "at present," with $(\xi_0, \ldots, \xi_{k-1})$ being the "past," and $(\xi_{k+1}, \ldots, \xi_n)$ the "future." Then (9) says that if the past and the present are given, the future depends only on the present and is independent of how the particle arrived at ξ_k, i.e., is independent of the past $(\xi_0, \ldots, \xi_{k-1})$.

Let *

$$P = \{\xi_{k-1} = a_{k-1}, \ldots, \xi_0 = a_0\},$$
$$N = \{\xi_k = a_k\},$$
$$F = \{\xi_n = a_n, \ldots, \xi_{k+1} = a_{k+1}\}.$$

Then it follows from (9) that

$$P(F \mid NP) = P(F \mid N),$$

from which we easily find that

$$P(FP \mid N) = P(F \mid N) P(P \mid N). \qquad (10)$$

In other words, it follows from (7) that for a given present N, the future F and the past P are independent. It is easily shown that the converse also holds: if (10) holds for all $k = 0, 1, \ldots, n - 1$, then (7) holds for every $k = 0, 1, \ldots, n - 1$.

---

* "Present" is denoted by N ("Now") to distinguish from P = "Past".—*Translator.*

The property of the independence of future and past, or, what is the same thing, the lack of dependence of the future on the past when the present is given, is called the *Markov property*, and the corresponding sequence of random variables $\xi_0, \ldots, \xi_n$ is a *Markov chain*.

Consequently if the probabilities $p(\omega)$ of the sample points are given by (3), the sequence $(\xi_0, \ldots, \xi_n)$ with $\xi_i(\omega) = x_i$ forms a Markov chain.

We give the following formal definition.

Definition. Let $(\Omega, \mathscr{A}, \mathsf{P})$ be a (finite) probability space and let $\xi = (\xi_0, \ldots, \xi_n)$ be a sequence of random variables with values in a (finite) set X. If (7) is satisfied, the sequence $\xi = (\xi_0, \ldots, \xi_n)$ is called a (finite) *Markov chain*.

The set X is called the *phase space* or *state space* of the chain. The set of probabilities $(p_0(x))$, $x \in X$, with $p_0(x) = \mathsf{P}(\xi_0 = x)$ is the *initial distribution*, and the matrix $\|p_k(x,y)\|$, $x, y \in X$, with $p_k(x,y) = \mathsf{P}\{\xi_k = y \mid \xi_{k-1} = x\}$ is the *matrix of transition probabilities* (from state x to state y) at time $k = 1, \ldots, n$.

When the transition probabilities $p_k(x,y)$ *do not depend* on k, that is, $p_k(x,y) = p(x,y)$, the sequence $\xi = (\xi_0, \ldots, \xi_n)$ is called a *homogeneous* Markov chain with transition matrix $\|p(x,y)\|$.

Notice that the matrix $\|p(x,y)\|$ is *stochastic*: its elements are nonnegative and the sum of the elements in each row is 1: $\sum_y p(x,y) = 1$, $x \in X$.

We shall suppose that the phase space X is a finite set of integers ($X = \{0, 1, \ldots, N\}$, $X = \{0, \pm 1, \ldots, \pm N\}$, etc.), and use the traditional notation $p_i = p_0(i)$ and $p_{ij} = p(i,j)$.

It is clear that the properties of homogeneous Markov chains are completely determined by the initial probabilities p_i and the transition probabilities p_{ij}. In specific cases we describe the evolution of the chain, not by writing out the matrix $\|p_{ij}\|$ explicitly, but by a (directed) graph whose vertices are the states in X, and an arrow from state i to state j with the number p_{ij} over it indicates that it is possible to pass from point i to point j with probability p_{ij}. When $p_{ij} = 0$, the corresponding arrow is omitted.

Example 1. Let $X = \{0, 1, 2\}$ and

$$\|p_{ij}\| = \begin{pmatrix} 1 & 0 & 0 \\ \frac{1}{2} & 0 & \frac{1}{2} \\ \frac{2}{3} & 0 & \frac{1}{3} \end{pmatrix}.$$

The following graph corresponds to this matrix:

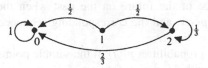

Here state 0 is said to be absorbing: if the particle gets into this state it remains there, since $p_{00} = 1$. From state 1 the particle goes to the adjacent states 0 or 2 with equal probabilities; state 2 has the property that the particle remains there with probability $\frac{1}{3}$ and goes to state 0 with probability $\frac{2}{3}$.

Example 2. Let $X = \{0, \pm 1, \ldots, \pm N\}$, $p_0 = 1$, $p_{NN} = p_{-N,-N} = 1$, and, for $|i| < N$,

$$p_{ij} = \begin{cases} p, & j = i+1, \\ q, & j = i-1, \\ 0 & \text{otherwise.} \end{cases} \tag{11}$$

The transitions corresponding to this chain can be presented graphically in the following way $(N = 3)$:

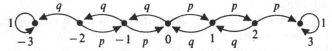

This chain corresponds to the two-player game discussed earlier, when each player has a bankroll N and at each turn the first player wins $+1$ from the second with probability p, and loses (wins -1) with probability q. If we think of state i as the amount won by the first player from the second, then reaching state N or $-N$ means the ruin of the second or first player, respectively.

In fact, if $\eta_1, \eta_2, \ldots, \eta_n$ are independent Bernoulli random variables with $P(\eta_i = +1) = p$, $P(\eta_i = -1) = q$, $S_0 = 0$ and $S_k = \eta_1 + \cdots + \eta_k$ the amounts won by the first player from the second, then the sequence $S_0, S_1, \ldots, S_n$ is a Markov chain with $p_0 = 1$ and transition matrix (11), since

$$\begin{aligned} P\{S_{k+1} = j \,|\, S_k = i_k, \, S_{k-1} = i_{k-1}, \ldots, S_1 = i_1\} \\ = P\{S_k + \eta_{k+1} = j \,|\, S_k = i_k, \, S_{k-1} = i_{k-1}, \ldots, S_1 = i_1\} \\ = P\{S_k + \eta_{k+1} = j \,|\, S_k = i_k\} = P\{\eta_{k+1} = j - i_k\}. \end{aligned}$$

This Markov chain has a very simple structure:

$$S_{k+1} = S_k + \eta_{k+1}, \quad 0 \le k \le n - 1,$$

where $\eta_1, \eta_2, \ldots, \eta_n$ is a sequence of independent random variables.

The same considerations show that if $\xi_0, \eta_1, \ldots, \eta_n$ are independent random variables then the sequence $\xi_0, \xi_1, \ldots, \xi_n$ with

$$\xi_{k+1} = f_k(\xi_k, \eta_{k+1}), \quad 0 \le k \le n - 1, \tag{12}$$

is also a Markov chain.

It is worth noting in this connection that a Markov chain constructed in this way can be considered as a natural probabilistic analog of a (deterministic) sequence $x = (x_0, \ldots, x_n)$ generated by the recurrent equations

$$x_{k+1} = f_k(x_k).$$

We now give another example of a Markov chain of the form (12); this example arises in queueing theory.

Example 3. At a taxi stand let taxis arrive at unit intervals of time (one at a time). If no one is waiting at the stand, the taxi leaves immediately. Let η_k be the number of passengers who arrive at the stand at time k, and suppose that $\eta_1, \ldots, \eta_n$ are independent random variables. Let ξ_k be the length of the waiting line at time k, $\xi_0 = 0$. Then if $\xi_k = i$, at the next time $k + 1$ the length ξ_{k+1} of the waiting line is equal to

$$j = \begin{cases} \eta_{k+1} & \text{if } i = 0, \\ i - 1 + \eta_{k+1} & \text{if } i \geq 1. \end{cases}$$

In other words,

$$\xi_{k+1} = (\xi_k - 1)^+ + \eta_{k+1}, \quad 0 \leq k \leq n - 1,$$

where $a^+ = \max(a, 0)$, and therefore the sequence $\xi = (\xi_0, \ldots, \xi_n)$ is a Markov chain.

Example 4. This example comes from the theory of *branching processes*. A branching process with discrete time is a sequence of random variables $\xi_0, \xi_1, \ldots, \xi_n$, where ξ_k is interpreted as the number of particles in existence at time k, and the process of creation and annihilation of particles is as follows: each particle, independently of the other particles and of the "prehistory" of the process, is transformed into j particles with probability p_j, $j = 0, 1, \ldots, M$. (This model of the *process of creation and annihilation* is called the *Galton–Watson model*, see [6] and Problem 18 in Sect. 5, Chap. VIII of [90]).

We suppose that at the initial time there is just one particle, $\xi_0 = 1$. If at time k there are ξ_k particles (numbered $1, 2, \ldots, \xi_k$), then by assumption ξ_{k+1} is given as a *random sum of random variables*,

$$\xi_{k+1} = \eta_1^{(k)} + \cdots + \eta_{\xi_k}^{(k)},$$

where $\eta_i^{(k)}$ is the number of particles produced by particle number i. It is clear that if $\xi_k = 0$ then $\xi_{k+1} = 0$. If we suppose that all the random variables $\eta_j^{(k)}$, $k \geq 0$, are independent of each other, we obtain

$$\begin{aligned} \mathsf{P}\{\xi_{k+1} = i_{k+1} \mid \xi_k = i_k, \, \xi_{k-1} = i_{k-1}, \ldots\} &= \mathsf{P}\{\xi_{k+1} = i_{k+1} \mid \xi_k = i_k\} \\ &= \mathsf{P}\{\eta_1^{(k)} + \cdots + \eta_{i_k}^{(k)} = i_{k+1}\}. \end{aligned}$$

It is evident from this that the sequence $\xi_0, \xi_1, \ldots, \xi_n$ is a Markov chain.

A particularly interesting case is that in which each particle either vanishes with probability q or divides in two with probability p, $p + q = 1$. In this case it is easy to calculate that

$$p_{ij} = \mathsf{P}\{\xi_{k+1} = j \mid \xi_k = i\}$$

is given by the formula

$$p_{ij} = \begin{cases} C_i^{j/2} p^{j/2} q^{i-j/2}, & j = 0, 2, \ldots, 2i, \\ 0 & \text{in all other cases.} \end{cases}$$

2. Let $\xi = (\xi_k, \mathbf{p}, \mathbb{P})$ be a homogeneous Markov chain with initial vector (row) $\mathbf{p} = \|p_i\|$ and transition matrix $\mathbb{P} = \|p_{ij}\|$. It is clear that

$$p_{ij} = \mathsf{P}\{\xi_1 = j \mid \xi_0 = i\} = \cdots = \mathsf{P}\{\xi_n = j \mid \xi_{n-1} = i\}.$$

We shall use the notation

$$p_{ij}^{(k)} = \mathsf{P}\{\xi_k = j \mid \xi_0 = i\} \quad (= \mathsf{P}\{\xi_{k+l} = j \mid \xi_l = i\})$$

for the probability of a transition from state i to state j in k steps, and

$$p_j^{(k)} = \mathsf{P}\{\xi_k = j\}$$

for the probability of the particle to be at point j at time k. Also let

$$\mathbf{p}^{(k)} = \|p_i^k\|, \quad \mathbb{P}^{(k)} = \|p_{ij}^{(k)}\|.$$

Let us show that the transition probabilities $p_{ij}^{(k)}$ satisfy the *Kolmogorov–Chapman equation*

$$p_{ij}^{(k+l)} = \sum_{\alpha} p_{i\alpha}^{(k)} p_{\alpha j}^{(l)}, \tag{13}$$

or, in matrix form,

$$\mathbb{P}^{(k+l)} = \mathbb{P}^{(k)} \cdot \mathbb{P}^{(l)}. \tag{14}$$

The proof is extremely simple: using the formula for total probability and the Markov property, we obtain

$$p_{ij}^{(k+l)} = \mathsf{P}(\xi_{k+l} = j \mid \xi_0 = i) = \sum_{\alpha} \mathsf{P}(\xi_{k+l} = j, \ \xi_k = \alpha \mid \xi_0 = i)$$

$$= \sum_{\alpha} \mathsf{P}(\xi_{k+l} = j \mid \xi_k = \alpha) \, \mathsf{P}(\xi_k = \alpha \mid \xi_0 = i) = \sum_{\alpha} p_{\alpha j}^{(l)} p_{i\alpha}^{(k)}.$$

The following two cases of (13) are particularly important: the *backward equation*

$$p_{ij}^{(l+1)} = \sum_{\alpha} p_{i\alpha} p_{\alpha j}^{(l)} \tag{15}$$

and the *forward equation*

$$p_{ij}^{(k+1)} = \sum_{\alpha} p_{i\alpha}^{(k)} p_{\alpha j} \qquad (16)$$

(see Figs. 22 and 23). The forward and backward equations can be written in the following matrix forms

$$\mathbb{P}^{(k+1)} = \mathbb{P}^{(k)} \cdot \mathbb{P}, \qquad (17)$$

$$\mathbb{P}^{(k+1)} = \mathbb{P} \cdot \mathbb{P}^{(k)}. \qquad (18)$$

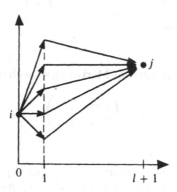

Fig. 22 For the backward equation

Similarly, we find for the (unconditional) probabilities $p_j^{(k)}$ that

$$p_j^{(k+l)} = \sum_{\alpha} p_{\alpha}^{(k)} p_{\alpha j}^{(l)}, \qquad (19)$$

or in matrix form

$$\mathbf{p}^{(k+l)} = \mathbf{p}^{(k)} \cdot \mathbb{P}^{(l)}.$$

In particular,

$$\mathbf{p}^{(k+1)} = \mathbf{p}^{(k)} \cdot \mathbb{P} \quad \textit{(forward equation)}$$

and

$$\mathbf{p}^{(k+1)} = \mathbf{p}^{(1)} \cdot \mathbb{P}^{(k)} \quad \textit{(backward equation)}.$$

Since $\mathbb{P}^{(1)} = \mathbb{P}, \mathbf{p}^{(0)} = \mathbf{p}$, it follows from these equations that

$$\mathbb{P}^{(k)} = \mathbb{P}^k, \quad \mathbf{p}^{(k)} = \mathbf{p} \cdot \mathbb{P}^k.$$

Consequently for homogeneous Markov chains the k-step transition probabilities $p_{ij}^{(k)}$ are the elements of the kth powers of the matrix $\mathbb{P}$, so that many properties of such chains can be investigated by the methods of matrix analysis.

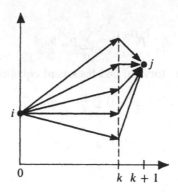

Fig. 23 For the forward equation

Example 5. Consider a homogeneous Markov chain with the two states 0 and 1 and the transition matrix

$$\mathbb{P} = \begin{pmatrix} p_{00} & p_{01} \\ p_{10} & p_{11} \end{pmatrix}.$$

It is easy to calculate that

$$\mathbb{P}^2 = \begin{pmatrix} p_{00}^2 + p_{01}p_{10} & p_{01}(p_{00} + p_{11}) \\ p_{10}(p_{00} + p_{11}) & p_{11}^2 + p_{01}p_{10} \end{pmatrix}$$

and (by induction)

$$\mathbb{P}^n = \frac{1}{2 - p_{00} - p_{11}} \begin{pmatrix} 1 - p_{11} & 1 - p_{00} \\ 1 - p_{11} & 1 - p_{00} \end{pmatrix}$$

$$+ \frac{(p_{00} + p_{11} - 1)^n}{2 - p_{00} - p_{11}} \begin{pmatrix} 1 - p_{00} & -(1 - p_{00}) \\ -(1 - p_{11}) & 1 - p_{11} \end{pmatrix}$$

(under the assumption that $|p_{00} + p_{11} - 1| < 1$).

Hence it is clear that if the elements of $\mathbb{P}$ satisfy $|p_{00} + p_{11} - 1| < 1$ (in particular, if all the transition probabilities p_{ij} are positive), then as $n \to \infty$

$$\mathbb{P}^n \to \frac{1}{2 - p_{00} - p_{11}} \begin{pmatrix} 1 - p_{11} & 1 - p_{00} \\ 1 - p_{11} & 1 - p_{00} \end{pmatrix}, \tag{20}$$

and therefore

$$\lim_n p_{i0}^{(n)} = \frac{1 - p_{11}}{2 - p_{00} - p_{11}}, \qquad \lim_n p_{i1}^{(n)} = \frac{1 - p_{00}}{2 - p_{00} - p_{11}}, \qquad i = 0, 1.$$

Consequently if $|p_{00} + p_{11} - 1| < 1$, such a Markov chain exhibits regular behavior of the following kind: the influence of the initial state on the probability of finding the particle in one state or another eventually becomes *negligible* ($p_{ij}^{(n)}$

approach limits π_j, *independent of i* and forming a probability distribution: $\pi_0 \geq 0$, $\pi_1 \geq 0$, $\pi_0 + \pi_1 = 1$); if also all $p_{ij} > 0$ then $\pi_0 > 0$ and $\pi_1 > 0$. (Compare with Theorem 1 below.)

3. The following theorem describes a wide class of Markov chains that have the property called *ergodicity*: the limits $\pi_j = \lim_n p_{ij}$ not only exist, are independent of i, and form a probability distribution ($\pi_j \geq 0$, $\sum_j \pi_j = 1$), but also $\pi_j > 0$ for all j (such a distribution π_j is said to be *ergodic*, see Sect. 3, Chap. 8, Vol. 2 for more detail).

Theorem 1 (Ergodic Theorem). *Let $\mathbb{P} = \|p_{ij}\|$ be the transition matrix of a Markov chain with a finite state space $X = \{1, 2, \ldots, N\}$.*

(a) *If there is an n_0 such that*

$$\min_{i,j} p_{ij}^{(n_0)} > 0, \tag{21}$$

then there are numbers $\pi_1, \ldots, \pi_N$ such that

$$\pi_j > 0, \qquad \sum_j \pi_j = 1 \tag{22}$$

and

$$p_{ij}^{(n)} \to \pi_j, \qquad n \to \infty \tag{23}$$

for every $j \in X$ and $i \in X$.

(b) *Conversely, if there are numbers $\pi_1, \ldots, \pi_N$ satisfying (22) and (23), there is an n_0 such that (21) holds.*

(c) *The numbers $(\pi_1, \ldots, \pi_N)$ satisfy the equations*

$$\pi_j = \sum_\alpha \pi_\alpha p_{\alpha j}, \qquad j = 1, \ldots, N. \tag{24}$$

PROOF. (a) Let

$$m_j^{(n)} = \min_i p_{ij}^{(n)}, \qquad M_j^{(n)} = \max_i p_{ij}^{(n)}.$$

Since

$$p_{ij}^{(n+1)} = \sum_\alpha p_{i\alpha} p_{\alpha j}^{(n)}, \tag{25}$$

we have

$$m_j^{(n+1)} = \min_i p_{ij}^{(n+1)} = \min_i \sum_\alpha p_{i\alpha} p_{\alpha j}^{(n)} \geq \min_i \sum_\alpha p_{i\alpha} \min_\alpha p_{\alpha j}^{(n)} = m_j^{(n)},$$

whence $m_j^{(n)} \leq m_j^{(n+1)}$ and similarly $M_j^{(n)} \geq M_j^{(n+1)}$. Consequently, to establish (23) it will be enough to prove that

$$M_j^{(n)} - m_j^{(n)} \to 0, \qquad n \to \infty, \quad j = 1, \ldots, N.$$

Let $\varepsilon = \min_{i,j} p_{ij}^{(n_0)} > 0$. Then

$$p_{ij}^{(n_0+n)} = \sum_{\alpha} p_{i\alpha}^{(n_0)} p_{\alpha j}^{(n)} = \sum_{\alpha} [p_{i\alpha}^{(n_0)} - \varepsilon p_{j\alpha}^{(n)}] p_{\alpha j}^{(n)} + \varepsilon \sum_{\alpha} p_{j\alpha}^{(n)} p_{\alpha j}^{(n)}$$

$$= \sum_{\alpha} [p_{i\alpha}^{(n_0)} - \varepsilon p_{j\alpha}^{(n)}] p_{\alpha j}^{(n)} + \varepsilon p_{jj}^{(2n)}.$$

But $p_{i\alpha}^{(n_0)} - \varepsilon p_{j\alpha}^{(n)} \geq 0$; therefore

$$p_{ij}^{(n_0+n)} \geq m_j^{(n)} \cdot \sum_{\alpha} [p_{i\alpha}^{(n_0)} - \varepsilon p_{j\alpha}^{(n)}] + \varepsilon p_{jj}^{(2n)} = m_j^{(n)}(1-\varepsilon) + \varepsilon p_{jj}^{(2n)},$$

and consequently

$$m_j^{(n_0+n)} \geq m_j^{(n)}(1-\varepsilon) + \varepsilon p_{jj}^{(2n)}.$$

In a similar way

$$M_j^{(n_0+n)} \leq M_j^{(n)}(1-\varepsilon) + \varepsilon p_{jj}^{(2n)}.$$

Combining these inequalities, we obtain

$$M_j^{(n_0+n)} - m_j^{(n_0+n)} \leq (M_j^{(n)} - m_j^{(n)}) \cdot (1-\varepsilon)$$

and consequently

$$M_j^{(kn_0+n)} - m_j^{(kn_0+n)} \leq (M_j^{(n)} - m_j^{(n)})(1-\varepsilon)^k \downarrow 0, \qquad k \to \infty.$$

Thus $M_j^{(n_\beta)} - m_j^{(n_\beta)} \to 0$ for some subsequence n_β, $n_\beta \to \infty$. But the difference $M_j^{(n)} - m_j^{(n)}$ is monotonic in n, and therefore $M_j^{(n)} - m_j^{(n)} \to 0$, $n \to \infty$.

If we put $\pi_j = \lim_n m_j^{(n)}$, it follows from the preceding inequalities that

$$|p_{ij}^{(n)} - \pi_j| \leq M_j^{(n)} - m_j^{(n)} \leq (1-\varepsilon)^{[n/n_0]-1}$$

for $n \geq n_0$, that is, $p_{ij}^{(n)}$ converges to its limit π_j *geometrically* (i.e., as fast as a geometric progression).

It is also clear that $m_j^{(n)} \geq m_j^{(n_0)} \geq \varepsilon > 0$ for $n \geq n_0$, and therefore $\pi_j > 0$.

(b) Inequality (21) follows from (23) since the number of states is finite and $\pi_j > 0$.

(c) Equations (24) follow from (23) and (25).

This completes the proof of the theorem. $\square$

4. The system of equations (compare with (24))

$$x_j = \sum_{\alpha} x_\alpha p_{\alpha j}, \quad j = 1, \ldots, N, \tag{24*}$$

plays a major role in the theory of Markov chains. Any *nonnegative* solution
$\mathbf{q} = (q_1, \ldots, q_N)$ of this system satisfying $\sum_\alpha q_\alpha = 1$ is said to be a *stationary* or
invariant probability distribution for the Markov chain with transition matrix $\|p_{ij}\|$.
The reason for this terminology is as follows.

Take the distribution $\mathbf{q} = (q_1, \ldots, q_N)$ for the initial one, i.e., let $p_j = q_j, j = 1, \ldots, N$. Then

$$p_j^{(1)} = \sum_\alpha q_\alpha p_{\alpha j} = q_j$$

and in general $p_j^{(n)} = q_j$. In other words, if we take $\mathbf{q} = (q_1, \ldots, q_N)$ as the initial
distribution, this distribution is unchanged as time goes on, i.e., for any k

$$\mathsf{P}(\xi_k = j) = \mathsf{P}(\xi_0 = j), \quad j = 1, \ldots, N.$$

Moreover, with initial distribution $\mathbf{q} = (q_1, \ldots, q_N)$ the Markov chain $\xi = (\xi, \mathbf{q}, \mathbb{P})$
is *stationary*: the joint distribution of the vector $(\xi_k, \xi_{k+1}, \ldots, \xi_{k+l})$ is independent
of k for all l (assuming that $k + l \leq n$).

Property (21) guarantees both the existence of limits $\pi_j = \lim p_{ij}^{(n)}$, which are
independent of i, and the existence of an ergodic distribution, i.e., of a distribution
$(\pi_1, \ldots, \pi_N)$ with $\pi_j > 0$. The distribution $(\pi_1, \ldots, \pi_N)$ is also a *stationary* distribution. Let us now show that the set $(\pi_1, \ldots, \pi_N)$ is the *only* stationary distribution.
In fact, let $(\tilde{\pi}_1, \ldots, \tilde{\pi}_N)$ be another stationary distribution. Then

$$\tilde{\pi}_j = \sum_\alpha \tilde{\pi}_\alpha p_{\alpha j} = \cdots = \sum_\alpha \tilde{\pi}_\alpha p_{\alpha j}^{(n)},$$

and since $p_{\alpha j}^{(n)} \to \pi_j$ we have

$$\tilde{\pi}_j = \sum_\alpha (\tilde{\pi}_\alpha \cdot \pi_j) = \pi_j.$$

We note that a stationary probability distribution (even unique) *may exist for a
nonergodic chain*. In fact, if

$$\mathbb{P} = \begin{pmatrix} 0 & 1 \\ 1 & 0 \end{pmatrix},$$

then

$$\mathbb{P}^{2n} = \begin{pmatrix} 0 & 1 \\ 1 & 0 \end{pmatrix}, \qquad \mathbb{P}^{2n+1} = \begin{pmatrix} 1 & 0 \\ 0 & 1 \end{pmatrix},$$

and consequently the limits $\lim p_{ij}^{(n)}$ do not exist. At the same time, the system

$$q_j = \sum_\alpha q_\alpha p_{\alpha j}, \qquad j = 1, 2,$$

reduces to

$$q_1 = q_2,$$
$$q_2 = q_1,$$

of which the unique solution (q_1, q_2) satisfying $q_1 + q_2 = 1$ is $(\frac{1}{2}, \frac{1}{2})$.

We also notice that for this example the system (24*) with $x_j = q_j$ has the form

$$q_0 = q_0 p_{00} + q_1 p_{10},$$
$$q_1 = q_0 p_{01} + q_1 p_{11},$$

from which, by the condition $q_0 + q_1 = 1$, we find that the unique stationary distribution (q_0, q_1) coincides with the one obtained above:

$$q_0 = \frac{1 - p_{11}}{2 - p_{00} - p_{11}}, \qquad q_1 = \frac{1 - p_{00}}{2 - p_{00} - p_{11}}.$$

We now consider some corollaries of the ergodic theorem.

Let $A \subseteq X$ be a set of states, and

$$I_A(x) = \begin{cases} 1, & x \in A, \\ 0, & x \notin A. \end{cases}$$

Consider

$$\nu_A(n) = \frac{I_A(\xi_0) + \cdots + I_A(\xi_n)}{n + 1}$$

which is the *fraction of the time* spent by the particle in the set A. Since

$$\mathsf{E}[I_A(\xi_k) \mid \xi_0 = i] = \mathsf{P}(\xi_k \in A \mid \xi_0 = i) = \sum_{j \in A} p_{ij}^{(k)} \quad (= p_i^{(k)}(A)),$$

we have

$$\mathsf{E}[\nu_A(n) \mid \xi_0 = i] = \frac{1}{n + 1} \sum_{k=0}^{n} p_i^{(k)}(A)$$

and in particular

$$\mathsf{E}[\nu_{\{j\}}(n) \mid \xi_0 = i] = \frac{1}{n + 1} \sum_{k=0}^{n} p_{ij}^{(k)}.$$

It is known from analysis (see also Lemma 1 in Sect. 3, Chap. 4, Vol. 2) that if $a_n \to a$ then $(a_0 + \cdots + a_n)/(n + 1) \to a$, $n \to \infty$. Hence if $p_{ij}^{(k)} \to \pi_j$, $k \to \infty$, then

$$\mathsf{E}\,\nu_{\{j\}}(n) \to \pi_j, \qquad \mathsf{E}\,\nu_A(n) \to \pi_A, \qquad \text{where} \quad \pi_A = \sum_{j \in A} \pi_j.$$

For ergodic chains one can in fact prove more, namely that the following result holds for $I_A(\xi_0), \ldots, I_A(\xi_n), \ldots$.

Law of Large Numbers. *If $\xi_0, \xi_1, \ldots$ form an ergodic Markov chain with a finite state space, then*

$$P\{|v_A(n) - \pi_A| > \varepsilon\} \to 0, \qquad n \to \infty, \tag{26}$$

for every $\varepsilon > 0$, every set $A \subseteq X$ and every initial distribution.

Before we undertake the proof, let us notice that we cannot apply the results of Sect. 5 directly to $I_A(\xi_0), \dots, I_A(\xi_n), \dots$, since these variables are, in general, dependent. However, the proof can be carried through along the same lines as for independent variables if we again use Chebyshev's inequality, and apply the fact that for an ergodic chain with finitely many states there is a number ρ, $0 < \rho < 1$, such that

$$|p_{ij}^{(n)} - \pi_j| \leq C \cdot \rho^n. \tag{27}$$

Let us consider states i and j (which might be the same) and show that, for $\varepsilon > 0$,

$$P\{|v_{\{j\}}(n) - \pi_j| > \varepsilon \mid \xi_0 = i\} \to 0, \qquad n \to \infty. \tag{28}$$

By Chebyshev's inequality,

$$P\{|v_{\{j\}}(n) - \pi_j| > \varepsilon \mid \xi_0 = i\} < \frac{E\{|v_{\{j\}}(n) - \pi_j|^2 \mid \xi_0 = i\}}{\varepsilon^2}.$$

Hence we have only to show that

$$E\{|v_{\{j\}}(n) - \pi_j|^2 \mid \xi_0 = i\} \to 0, \qquad n \to \infty.$$

A simple calculation shows that

$$E\{|v_{\{j\}}(n) - \pi_j|^2 \mid \xi_0 = i\} = \frac{1}{(n+1)^2} \cdot E\left\{\left[\sum_{k=0}^{n}(I_{\{j\}}(\xi_k) - \pi_j)\right]^2 \middle| \xi_0 = i\right\}$$

$$= \frac{1}{(n+1)^2} \sum_{k=0}^{n}\sum_{l=0}^{n} m_{ij}^{(k,l)},$$

where

$$m_{ij}^{(k,l)} = E\{[I_{\{j\}}(\xi_k)I_{\{j\}}(\xi_l)] \mid \xi_0 = i\}$$
$$- \pi_j \cdot E[I_{\{j\}}(\xi_k) \mid \xi_0 = i] - \pi_j \cdot E[I_{\{j\}}(\xi_l) \mid \xi_0 = i] + \pi_j^2$$
$$= p_{ij}^{(s)} \cdot p_{jj}^{(t)} - \pi_j \cdot p_{ij}^{(k)} - \pi_j \cdot p_{ij}^{(l)} + \pi_j^2,$$
$$s = \min(k, l) \quad \text{and} \quad t = |k - l|.$$

By (27),

$$p_{ij}^{(n)} = \pi_j + \varepsilon_{ij}^{(n)}, \qquad |\varepsilon_{ij}^{(n)}| \leq C\rho^n.$$

Therefore

$$|m_{ij}^{(k,l)}| \leq C_1[\rho^s + \rho^t + \rho^k + \rho^l],$$

where C_1 is a constant. Consequently

$$\frac{1}{(n+1)^2} \sum_{k=0}^{n} \sum_{l=0}^{n} m_{ij}^{(k,l)} \leq \frac{C_1}{(n+1)^2} \sum_{k=0}^{n} \sum_{l=0}^{n} [\rho^s + \rho^t + \rho^k + \rho^l]$$

$$\leq \frac{4C_1}{(n+1)^2} \cdot \frac{2(n+1)}{1-\rho} = \frac{8C_1}{(n+1)(1-\rho)} \to 0, \quad n \to \infty.$$

Then (28) follows from this, and we obtain (26) in an obvious way.

5. In Sect. 9 we gave, for a random walk S_0, S_1, ... generated by a Bernoulli scheme, recurrent equations for the probability and the expectation of the exit time at either boundary. We now derive similar equations for Markov chains.

Let $\xi = (\xi_0, \ldots, \xi_N)$ be a Markov chain with transition matrix $\|p_{ij}\|$ and phase space $X = \{0, \pm 1, \ldots, \pm N\}$. Let A and B be two integers, $-N \leq A \leq 0 \leq B \leq N$, and $x \in X$. Let $\mathscr{B}_{k+1}$ be the set of paths $(x_0, x_1, \ldots, x_k)$, $x_i \in X$, that leave the interval (A, B) for the first time at the upper end, i.e., leave (A, B) by going into the set $(B, B+1, \ldots, N)$.

For $A \leq x \leq B$, put

$$\beta_k(x) = \mathsf{P}\{(\xi_0, \ldots, \xi_k) \in \mathscr{B}_{k+1} \mid \xi_0 = x\}.$$

In order to find these probabilities (for the first exit of the Markov chain from (A, B) through the upper boundary) we use the method that was applied in the deduction of the backward equations.

We have

$$\beta_k(x) = \mathsf{P}\{(\xi_0, \ldots, \xi_k) \in \mathscr{B}_{k+1} \mid \xi_0 = x\}$$
$$= \sum_y p_{xy} \cdot \mathsf{P}\{(\xi_0, \ldots, \xi_k) \in \mathscr{B}_{k+1} \mid \xi_0 = x, \xi_1 = y\},$$

where, as is easily seen by using the Markov property and the homogeneity of the chain,

$$\mathsf{P}\{(\xi_0, \ldots, \xi_k) \in \mathscr{B}_{k+1} \mid \xi_0 = x, \xi_1 = y\}$$
$$= \mathsf{P}\{(x, y, \xi_2, \ldots, \xi_k) \in \mathscr{B}_{k+1} \mid \xi_0 = x, \xi_1 = y\}$$
$$= \mathsf{P}\{(y, \xi_2, \ldots, \xi_k) \in \mathscr{B}_k \mid \xi_1 = y\}$$
$$= \mathsf{P}\{(y, \xi_1, \ldots, \xi_{k-1}) \in \mathscr{B}_k \mid \xi_0 = y\} = \beta_{k-1}(y).$$

Therefore

$$\beta_k(x) = \sum_y p_{xy} \beta_{k-1}(y)$$

for $A < x < B$ and $1 \leq k \leq n$. Moreover, it is clear that

$$\beta_k(x) = 1, \qquad x = B, B+1, \ldots, N,$$

and
$$\beta_k(x) = 0, \qquad x = -N, \dots, A.$$

In a similar way we can find equations for $\alpha_k(x)$, the probabilities for first exit from (A, B) through the lower boundary.

Let $\tau_k = \min\{0 \leq l \leq k : \xi_l \notin (A, B)\}$, where $\tau_k = k$ if the set $\{\cdot\} = \varnothing$. Then the same method, applied to $m_k(x) = E(\tau_k \mid \xi_0 = x)$, leads to the following recurrent equations:

$$m_k(x) = 1 + \sum_y m_{k-1}(y) p_{xy}$$

(here $1 \leq k \leq n$, $A < x < B$). We define

$$m_k(x) = 0, \qquad x \notin (A, B).$$

It is clear that if the transition matrix is given by (11) the equations for $\alpha_k(x)$, $\beta_k(x)$ and $m_k(x)$ become the corresponding equations from Sect. 9, where they were obtained by essentially the same method that was used here.

These equations have the most interesting applications in the limiting case when the walk continues for an *unbounded length of time*. Just as in Sect. 9, the corresponding equations can be obtained by a formal limiting process $(k \to \infty)$.

By way of example, we consider the Markov chain with states $\{0, 1, \dots, B\}$ and transition probabilities

$$p_{00} = 1, \qquad p_{BB} = 1,$$

and

$$p_{ij} = \begin{cases} p_i > 0, & j = i+1, \\ r_i, & j = i, \\ q_i > 0, & j = i-1, \end{cases}$$

for $1 \leq i \leq B - 1$, where $p_l + q_l + r_l = 1$.

For this chain, the corresponding graph is

It is clear that states 0 and B are absorbing, whereas for every other state i the particle stays there with probability r_i, moves one step to the right with probability p_i, and to the left with probability q_i.

Let us find $\alpha(x) = \lim_{k \to \infty} \alpha_k(x)$, the limit of the probability that a particle starting at the point x arrives at state zero before reaching state B. Taking limits as $k \to \infty$ in the equations for $\alpha_k(x)$, we find that

$$\alpha(j) = q_j \alpha(j-1) + r_j \alpha(j) + p_j \alpha(j+1)$$

when $0 < j < B$, with the boundary conditions

$$\alpha(0) = 1, \quad \alpha(B) = 0.$$

Since $r_j = 1 - q_j - p_j$, we have

$$p_j(\alpha(j+1) - \alpha(j)) = q_j(\alpha(j) - \alpha(j-1))$$

and consequently

$$\alpha(j+1) - \alpha(j) = \rho_j(\alpha(1) - 1),$$

where

$$\rho_j = \frac{q_1 \cdots q_j}{p_1 \cdots p_j}, \quad \rho_0 = 1.$$

But

$$\alpha(j+1) - 1 = \sum_{i=0}^{j} (\alpha(i+1) - \alpha(i)).$$

Therefore

$$\alpha(j+1) - 1 = (\alpha(1) - 1) \cdot \sum_{i=0}^{j} \rho_i.$$

If $j = B - 1$, we have $\alpha(j+1) = \alpha(B) = 0$, and therefore

$$\alpha(1) - 1 = -\frac{1}{\sum_{i=0}^{B-1} \rho_i},$$

whence

$$\alpha(1) = \frac{\sum_{i=1}^{B-1} \rho_i}{\sum_{i=0}^{B-1} \rho_i} \quad \text{and} \quad \alpha(j) = \frac{\sum_{i=j}^{B-1} \rho_i}{\sum_{i=0}^{B-1} \rho_i}, \quad j = 1, \ldots, B.$$

(This should be compared with the results of Sect. 9.)

Now let $m(x) = \lim_k m_k(x)$, the limiting value of the average time taken to arrive at one of the states 0 or B. Then $m(0) = m(B) = 0$,

$$m(x) = 1 + \sum_y m(y)p_{xy}$$

and consequently for the example that we are considering,

$$m(j) = 1 + q_j m(j-1) + r_j m(j) + p_j m(j+1)$$

for all $j = 1, 2, \ldots, B - 1$. To find $m(j)$ we put

$$M(j) = m(j) - m(j-1), \quad j = 1, \ldots, B.$$

Then
$$p_j M(j+1) = q_j M(j) - 1, \qquad j = 1, \ldots, B-1,$$

and consequently we find that

$$M(j+1) = \rho_j M(1) - R_j,$$

where

$$\rho_j = \frac{q_1 \cdots q_j}{p_1 \cdots p_j}, \qquad R_j = \frac{1}{p_j} \left[1 + \frac{q_j}{p_{j-1}} + \cdots + \frac{q_j \cdots q_2}{p_{j-1} \cdots p_1} \right].$$

Therefore

$$m(j) = m(j) - m(0) = \sum_{i=0}^{j-1} M(i+1)$$

$$= \sum_{i=0}^{j-1} (\rho_i m(1) - R_i) = m(1) \sum_{i=0}^{j-1} \rho_i - \sum_{i=0}^{j-1} R_i.$$

It remains only to determine $m(1)$. But $m(B) = 0$, and therefore

$$m(1) = \frac{\sum_{i=0}^{B-1} R_i}{\sum_{i=0}^{B-1} \rho_i},$$

and for $1 < j \le B$,

$$m(j) = \sum_{i=0}^{j-1} \rho_i \cdot \frac{\sum_{i=0}^{B-1} R_i}{\sum_{i=0}^{B-1} \rho_i} - \sum_{i=0}^{j-1} R_i.$$

(This should be compared with the results in Sect. 9 for the case $r_i = 0$, $p_i = p$, $q_i = q$.)

6. In this subsection we consider a stronger version of the Markov property (8), namely that it remains valid if time k is replaced by a random time (see Theorem 2 below). The significance of this, the *strong Markov property*, will be illustrated in particular by the example of the derivation of the recurrent relations (38), which play an important role in the classification of the states of Markov chains (Chapter 8, Vol. 2).

Let $\xi = (\xi_1, \ldots, \xi_n)$ be a homogeneous Markov chain with transition matrix $\|p_{ij}\|$; let $\mathscr{D}^{\xi} = (\mathscr{D}_k^{\xi})_{0 \le k \le n}$ be a system of decompositions, $\mathscr{D}_k^{\xi} = \mathscr{D}_{\xi_0, \ldots, \xi_k}$. Let $\mathscr{B}_k^{\xi}$ denote the algebra $\alpha(\mathscr{D}_k^{\xi})$ generated by the decomposition $\mathscr{D}_k^{\xi}$.

We first put the Markov property (8) into a somewhat different form. Let $B \in \mathscr{B}_k^{\xi}$. Let us show that then

$$P\{\xi_n = a_n, \ldots, \xi_{k+1} = a_{k+1} \mid B \cap (\xi_k = a_k)\}$$
$$= P\{\xi_n = a_n, \ldots, \xi_{k+1} = a_{k+1} \mid \xi_k = a_k\} \qquad (29)$$

(assuming that $P\{B \cap (\xi_k = a_k)\} > 0$). In fact, B can be represented in the form

$$B = \sum\nolimits^* \{\xi_0 = a_0^*, \ldots, \xi_k = a_k^*\},$$

where $\sum^*$ extends over some set of collections $(a_0^*, \ldots, a_k^*)$. Consequently

$$
\begin{aligned}
& P\{\xi_n = a_n, \ldots, \xi_{k+1} = a_{k+1} \mid B \cap (\xi_k = a_k)\} \\
&= \frac{P\{(\xi_n = a_n, \ldots, \xi_k = a_k) \cap B\}}{P\{(\xi_k = a_k) \cap B\}} \\
&= \frac{\sum^* P\{(\xi_n = a_n, \ldots, \xi_k = a_k) \cap (\xi_0 = a_0^*, \ldots, \xi_k = a_k^*)\}}{P\{(\xi_k = a_k) \cap B\}}.
\end{aligned}
\tag{30}
$$

But, by the Markov property,

$$
\begin{aligned}
& P\{(\xi_n = a_n, \ldots, \xi_k = a_k) \cap (\xi_0 = a_0^*, \ldots, \xi_k = a_k^*)\} \\
&= \begin{cases} P\{\xi_n = a_n, \ldots, \xi_{k+1} = a_{k+1} \mid \xi_0 = a_0^*, \ldots, \xi_k = a_k^*\} \\ \quad \times P\{\xi_0 = a_0^*, \ldots, \xi_k = a_k^*\} & \text{if } a_k = a_k^*, \\ 0 & \text{if } a_k \neq a_k^*, \end{cases} \\
&= \begin{cases} P\{\xi_n = a_n, \ldots, \xi_{k+1} = a_{k+1} \mid \xi_k = a_k\} P\{\xi_0 = a_0^*, \ldots, \xi_k = a_k^*\} \\ \quad \text{if } a_k = a_k^*, \\ 0 \quad \text{if } a_k \neq a_k^*, \end{cases} \\
&= \begin{cases} P\{\xi_n = a_n, \ldots, \xi_{k+1} = a_{k+1} \mid \xi_k = a_k\} P\{(\xi_k = a_k) \cap B\} \\ \quad \text{if } a_k = a_k^*, \\ 0 \quad \text{if } a_k \neq a_k^*. \end{cases}
\end{aligned}
$$

Therefore the sum $\sum^*$ in (30) is equal to

$$P\{\xi_n = a_n, \ldots, \xi_{k+1} = a_{k+1} \mid \xi_k = a_k\} P\{(\xi_k = a_k) \cap B\}.$$

This establishes (29).

Let τ be a stopping time (with respect to the system $\mathcal{D}^\xi = (\mathcal{D}_k^\xi)_{0 \le k \le n}$; see Definition 2 in Sect. 11).

Definition. We say that a set B in the algebra $\mathscr{B}_n^\xi$ belongs to the system of sets $\mathscr{B}_\tau^\xi$ if, for each k, $0 \le k \le n$,

$$B \cap \{\tau = k\} \in \mathscr{B}_k^\xi. \tag{31}$$

It is easily verified that the collection of such sets B forms an algebra (called the algebra of events observed at time τ).

Theorem 2. *Let* $\xi = (\xi_0, \ldots, \xi_n)$ *be a homogeneous Markov chain with transition matrix* $\|p_{ij}\|$, τ *a stopping time (with respect to* $\mathcal{D}^\xi$*),* $B \in \mathscr{B}_\tau^\xi$ *and* $A = \{\omega : \tau + l \le n\}$. *Then if* $P\{A \cap B \cap (\xi_\tau = a_0)\} > 0$, *the following* strong Markov properties *hold*:

$$
\begin{aligned}
& P\{\xi_{\tau+l} = a_l, \ldots, \xi_{\tau+1} = a_1 \mid A \cap B \cap (\xi_\tau = a_0)\} \\
&\quad = P\{\xi_{\tau+l} = a_l, \ldots, \xi_{\tau+1} = a_1 \mid A \cap (\xi_\tau = a_0)\},
\end{aligned}
\tag{32}
$$

and if $P\{A \cap (\xi_\tau = a_0)\} > 0$ *then*

$$P\{\xi_{\tau+l} = a_l, \ldots, \xi_{\tau+1} = a_1 \mid A \cap (\xi_\tau = a_0)\} = p_{a_0 a_1} \cdots p_{a_{l-1} a_l}. \tag{33}$$

PROOF. For the sake of simplicity, we give the proof only for the case $l = 1$. Since $B \cap (\tau = k) \in \mathscr{B}_k^\xi$, we have, according to (29),

$$
\begin{aligned}
P\{\xi_{\tau+1} &= a_1, A \cap B \cap (\xi_\tau = a_0)\} \\
&= \sum_{k \leq n-1} P\{\xi_{k+1} = a_1, \xi_k = a_0, \tau = k, B\} \\
&= \sum_{k \leq n-1} P\{\xi_{k+1} = a_1 \mid \xi_k = a_0, \tau = k, B\} P\{\xi_k = a_0, \tau = k, B\} \\
&= \sum_{k \leq n-1} P\{\xi_{k+1} = a_1 \mid \xi_k = a_0\} P\{\xi_k = a_0, \tau = k, B\} \\
&= p_{a_0 a_1} \sum_{k \leq n-1} P\{\xi_k = a_0, \tau = k, B\} = p_{a_0 a_1} P\{A \cap B \cap (\xi_\tau = a_0)\},
\end{aligned}
$$

which simultaneously establishes (32) and (33) (for (33) we have to take $B = \Omega$).
□

Remark 2. When $l = 1$, the strong Markov property (32), (33) is evidently equivalent to the property that

$$P\{\xi_{\tau+1} \in C \mid A \cap B \cap (\xi_\tau = a_0)\} = P_{a_0}(C), \tag{34}$$

for every $C \subseteq X$, where

$$P_{a_0}(C) = \sum_{a_1 \in C} p_{a_0 a_1}.$$

In turn, (34) can be restated as follows: on the set $A = \{\tau \leq n - 1\}$,

$$P\{\xi_{\tau+1} \in C \mid \mathscr{B}_\tau^\xi\} = P_{\xi_\tau}(C), \tag{35}$$

which is a form of the strong Markov property that is commonly used in the general theory of homogeneous Markov processes.

Remark 3. If we use the conventions described in Remark 1, the properties (32) and (33) remain valid without assuming that the probabilities of the events $A \cap \{\xi_\tau = a_0\}$ and $A \cap B \cap \{\xi_\tau = a_0\}$ are positive.

7. Let $\xi = (\xi_0, \ldots, \xi_n)$ be a homogeneous Markov chain with transition matrix $\|p_{ij}\|$, and let

$$f_{ii}^{(k)} = \mathsf{P}\{\xi_k = i, \, \xi_l \neq i, \, 1 \leq l \leq k - 1 \,|\, \xi_0 = i\} \tag{36}$$

and

$$f_{ij}^{(k)} = \mathsf{P}\{\xi_k = j, \, \xi_l \neq j, \, 1 \leq l \leq k - 1 \,|\, \xi_0 = i\} \tag{37}$$

for $i \neq j$ be respectively the probability of *first return* to state i at time k and the probability of *first arrival* at state j at time k.

Let us show that

$$p_{ij}^{(n)} = \sum_{k=1}^{n} f_{ij}^{(k)} p_{jj}^{(n-k)}, \qquad \text{where} \quad p_{jj}^{(0)} = 1. \tag{38}$$

The intuitive meaning of the formula is clear: to go from state i to state j in n steps, it is necessary to reach state j for the first time in k steps ($1 \leq k \leq n$) and then to go from state j to state j in $n - k$ steps. We now give a rigorous derivation.

Let j be given and

$$\tau = \min\{1 \leq k \leq n : \xi_k = j\},$$

assuming that $\tau = n + 1$ if $\{\cdot\} = \varnothing$. Then $f_{ij}^{(k)} = \mathsf{P}\{\tau = k \,|\, \xi_0 = i\}$ and

$$\begin{aligned}
p_{ij}^{(n)} &= \mathsf{P}\{\xi_n = j \,|\, \xi_0 = i\} \\
&= \sum_{1 \leq k \leq n} \mathsf{P}\{\xi_n = j, \, \tau = k \,|\, \xi_0 = i\} \\
&= \sum_{1 \leq k \leq n} \mathsf{P}\{\xi_{\tau+n-k} = j, \, \tau = k \,|\, \xi_0 = i\},
\end{aligned} \tag{39}$$

where the last equation follows because $\xi_{\tau+n-k} = \xi_n$ on the set $\{\tau = k\}$. Moreover, the set $\{\tau = k\} = \{\tau = k, \, \xi_\tau = j\}$ for every k, $1 \leq k \leq n$. Therefore if $\mathsf{P}\{\xi_0 = i, \, \tau = k\} > 0$, it follows from Theorem 2 that

$$\begin{aligned}
\mathsf{P}\{\xi_{\tau+n-k} = j \,|\, \xi_0 = i, \, \tau = k\} &= \mathsf{P}\{\xi_{\tau+n-k} = j \,|\, \xi_0 = i, \, \tau = k, \, \xi_\tau = j\} \\
&= \mathsf{P}\{\xi_{\tau+n-k} = j \,|\, \xi_\tau = j\} = p_{jj}^{(n-k)}
\end{aligned}$$

and by (37)

$$\begin{aligned}
p_{ij}^{(n)} &= \sum_{k=1}^{n} \mathsf{P}\{\xi_{\tau+n-k} = j \,|\, \xi_0 = i, \, \tau = k\} \, \mathsf{P}\{\tau = k \,|\, \xi_0 = i\} \\
&= \sum_{k=1}^{n} p_{jj}^{(n-k)} f_{ij}^{(k)},
\end{aligned}$$

which establishes (38).

8. Problems

1. Let $\xi = (\xi_0, \ldots, \xi_n)$ be a Markov chain with values in X and $f = f(x)$ $(x \in X)$ a function. Will the sequence $(f(\xi_0), \ldots, f(\xi_n))$ form a Markov chain? Will the "reversed" sequence

$$(\xi_n, \xi_{n-1}, \ldots, \xi_0)$$

form a Markov chain?

2. Let $\mathbb{P} = \|p_{ij}\|$, $1 \leq i, j \leq r$, be a stochastic matrix and λ an eigenvalue of the matrix, i.e., a root of the characteristic equation $\det \|\mathbb{P} - \lambda E\| = 0$. Show that $\lambda_1 = 1$ is an eigenvalue and that all the other eigenvalues have moduli not exceeding 1. If all the eigenvalues $\lambda_1, \ldots, \lambda_r$ are distinct, then $p_{ij}^{(k)}$ admits the representation

$$p_{ij}^{(k)} = \pi_j + a_{ij}(2)\lambda_2^k + \cdots + a_{ij}(r)\lambda_r^k,$$

where $\pi_j, a_{ij}(2), \ldots, a_{ij}(r)$ can be expressed in terms of the elements of $\mathbb{P}$. (It follows from this algebraic approach to the study of Markov chains that, in particular, when $|\lambda_2| < 1, \ldots, |\lambda_r| < 1$, the limit $\lim_k p_{ij}^{(k)}$ exists for every j and is independent of i.)

3. Let $\xi = (\xi_0, \ldots, \xi_n)$ be a homogeneous Markov chain with state space X and transition matrix $\mathbb{P} = \|p_{xy}\|$. Let us denote by

$$T\varphi(x) = \mathsf{E}[\varphi(\xi_1) \mid \xi_0 = x] \quad \left(= \sum_y \varphi(y) p_{xy} \right)$$

the *operator of transition* for one step. Let the nonnegative function φ satisfy

$$T\varphi(x) = \varphi(x), \quad x \in X.$$

Show that the sequence of random variables

$$\zeta = (\zeta_k, \mathscr{D}_k^\xi) \quad \text{with} \quad \zeta_k = \varphi(\xi_k)$$

is a martingale.

4. Let $\xi = (\xi_n, \mathbf{p}, \mathbb{P})$ and $\tilde{\xi} = (\tilde{\xi}_n, \tilde{\mathbf{p}}, \mathbb{P})$ be two Markov chains with different initial distributions $\mathbf{p} = (p_1, \ldots, p_r)$ and $\tilde{\mathbf{p}} = (\tilde{p}_1, \ldots, \tilde{p}_r)$. Let $\mathbf{p}^{(n)} = (p_1^{(n)}, \ldots, p_r^{(n)})$ and $\tilde{\mathbf{p}}^{(n)} = (\tilde{p}_1^{(n)}, \ldots, \tilde{p}_r^{(n)})$. Show that if $\min_{i,j} p_{ij} \geq \varepsilon > 0$ then

$$\sum_{i=1}^r |\tilde{p}_i^{(n)} - p_i^{(n)}| \leq 2(1 - r\varepsilon)^n.$$

5. Let P and Q be stochastic matrices. Show that PQ and $\alpha P + (1-\alpha)Q$ with $0 \leq \alpha \leq 1$ are also stochastic matrices.

6. Consider a homogeneous Markov chain $(\xi_0, \ldots, \xi_n)$ with values in $X = \{0, 1\}$ and transition matrix

$$\begin{pmatrix} 1-p & p \\ q & 1-q \end{pmatrix},$$

where $0 < p < 1, 0 < q < 1$. Let $S_n = \xi_0 + \cdots + \xi_n$. As a generalization of the de Moivre–Laplace theorem (Sect. 6) show that

$$P\left\{ \frac{S_n - \frac{p}{p+q}n}{\sqrt{\frac{npq(2-p-q)}{(p+q)^3}}} \leq x \right\} \to \Phi(x), \quad n \to \infty.$$

Check that when $p + q = 1$, the variables $\xi_0, \ldots, \xi_n$ are independent and the above statement reduces to

$$P\left\{ \frac{S_n - pn}{\sqrt{npq}} \leq x \right\} \to \Phi(x), \quad n \to \infty.$$

13 Generating Functions

1. In discrete probability theory, which deals with a finite or countable set of outcomes, and more generally, in discrete mathematics, the method of generating functions, going back to L. Euler (eighteenth century), is one of the most powerful algebraic tools for solving combinatorial problems, arising, in particular, in probability theory.

Before giving formal definitions related to generating functions, we will formulate two probabilistic problems, for which the generating functions provide a very useful method of solving them.

2. Galileo's problem. Three true dice with faces marked as $1, 2, \ldots, 6$ are thrown simultaneously and independently. Find the probability P that the sum of the scores equals 10. (It will be shown that $P = \frac{1}{8}$.)

3. Lucky tickets problem. We buy a ticket chosen at random from the set of tickets bearing six-digits numbers from 000 000 to 999 999 (totally 10^6 tickets). What is the probability P that the number of the ticket we bought is such that the sum of the first three digits equals the sum of the last three? (The random choice of a ticket means that each of them can be bought with equal probability 10^{-6}.)

We will see in what follows that the method of generating functions is useful not only for solving probabilistic problems. This method is applicable to obtaining formulas for elements of sequences satisfying certain *recurrence relations*. For example, the Fibonacci numbers F_n, $n \geq 0$, satisfy the recurrence $F_n = F_{n-1} + F_{n-2}$ for $n \geq 2$ with $F_0 = F_1 = 1$. We will obtain by the method of generating functions (Subsection 6) that

$$F_n = \frac{1}{\sqrt{5}}\left[\left(\frac{1+\sqrt{5}}{2} \right)^{n+1} - \left(\frac{1-\sqrt{5}}{2} \right)^{n+1} \right].$$

It will also be shown how this method can be used for finding *integer-valued* solutions of equations

$$X_1 + \cdots + X_n = r$$

under various restrictions on X_i, $i = 1, \ldots, n$, and a given r from the set $\{0, 1, 2, \ldots\}$.

4. Now we will give the formal definitions.

Let $A = A(x)$ be a real-valued function, $x \in R$, which is representable, for $|x| < \lambda$, $\lambda > 0$, by a series

$$A(x) = a_0 + a_1 x + a_2 x^2 + \cdots \tag{1}$$

with coefficients

$$a = (a_0, a_1, a_2, \ldots).$$

It is clear that the knowledge of $A(x)$, $|x| < \lambda$, enables us to uniquely recover the coefficients $a = (a_0, a_1, a_2, \ldots)$. This explains why $A = A(x)$ is called the *generating function of the sequence* $a = (a_0, a_1, a_2, \ldots)$.

Along with functions $A = A(x)$ determined by series (1), it is often expedient to use the *exponential generating functions*

$$E(x) = a_0 + a_1 \frac{x}{1!} + a_2 \frac{x^2}{2!} + \cdots, \tag{2}$$

which are named so for the obvious reason that the sequence $(a_0, a_1, a_2, \ldots) \equiv (1, 1, 1, \ldots)$ generates the exponential function $\exp(x)$ (see Example 3 in Sect. 14).

In many problems it is useful to employ generating functions of two-sided infinite sequences

$$a = (\ldots, a_{-2}, a_{-1}, a_0, a_1, a_2, \ldots)$$

(see the example in Subsection 7).

If $\xi = \xi(\omega)$ is a random variable taking the values $0, 1, 2, \ldots$ with probabilities $p_0, p_1, p_2, \ldots$ (i.e., $p_i = \mathsf{P}(\xi = i)$, $\sum_{i=0}^{\infty} p_i = 1$), then the function

$$G(x) = \sum_{i=0}^{\infty} p_i x^i \tag{3}$$

is certainly defined for $|x| \leq 1$.

This function, which is the generating function of the sequence $(p_0, p_1, p_2, \ldots)$, is nothing but the expectation $\mathsf{E} x^\xi$, which was defined in Sect. 4 for finite sequences $(p_0, p_1, p_2, \ldots)$, and will be defined in the general case in Sect. 6, Chap. 2.

In the probability theory the function

$$G(x) = \mathsf{E} x^\xi \left(= \sum_{i=0}^{\infty} p_i x^i \right), \tag{4}$$

for obvious reasons, is called the *generating function of the random variable* $\xi = \xi(\omega)$.

5. Let us indicate some useful properties, which follow from the bijective (one-to-one) correspondence

$$(a_n)_{n\geq 0} \;\leftrightarrow\; A(x), \tag{5}$$

where $A(x)$ is determined by the series (1).

If along with (5) we have

$$(b_n)_{n\geq 0} \;\leftrightarrow\; B(x), \tag{6}$$

then for any constants c and d

$$(ca_n + db_n)_{n\geq 0} \;\leftrightarrow\; cA(x) + dB(x); \tag{7}$$

moreover, $A(x)$ and $B(x)$ fulfill the *convolution* property

$$\left(\sum_{i=0}^{n} a_i b_{n-i}\right)_{n\geq 0} \;\leftrightarrow\; A(x)B(x). \tag{8}$$

Besides the convolution operation we point out the following ones:

– the *composition* (or substitution) operation

$$(A \circ B)(x) = A(B(x)), \tag{9}$$

which means that if A and B satisfy (5) and (6), then

$$(A \circ B)(x) = \sum_{n\geq 0} a_n \left(\sum_{i\geq 0} b_i x^i\right)^n; \tag{10}$$

– the (formal) *differentiation operation* acting on $A(x)$ by the formula

$$D(A(x)) = \sum_{n=0}^{\infty} (n+1)a_{n+1}x^n. \tag{11}$$

The operator D has the following properties, which are well known for the ordinary differentiation:

$$D(AB) = D(A)B + A D(B), \quad D(A \circ B) = (D(A) \circ B) D(B). \tag{12}$$

6. To illustrate the method of generating functions we start with the example of finding the number $N(r; n)$ of solutions $(X_1, \ldots, X_n)$ of the equation

$$X_1 + \cdots + X_n = r \tag{13}$$

with nonnegative integer-valued X_j's subject to the constraints

$$X_1 \in \{k_1^{(1)}, k_1^{(2)}, \ldots\},$$

$$\ldots\ldots\ldots\ldots\ldots$$ (14)

$$X_n \in \{k_n^{(1)}, k_n^{(2)}, \ldots\},$$

where $0 \leq k_j^{(1)} < k_j^{(2)} < \ldots \leq r, j = 1, \ldots, n$, and $r \in \{0, 1, 2, \ldots\}$.

For example, if we ask for integer-valued solutions X_1, X_2, X_3 of the equation

$$X_1 + X_2 + X_3 = 3$$

subject to the constraints $0 \leq X_i \leq 3$, $i = 1, 2, 3$, then we can find by exhaustive search all the 10 solutions: $(1, 1, 1)$, $(0, 1, 2)$, $(0, 2, 1)$, $(1, 0, 2)$, $(1, 2, 0)$, $(2, 0, 1)$, $(2, 1, 0)$, $(0, 0, 3)$, $(0, 3, 0)$, $(3, 0, 0)$. But of course such a search becomes very time consuming for large n and r.

Let us introduce the following generating functions:

$$A_1(x) = x^{k_1^{(1)}} + x^{k_1^{(2)}} + \cdots,$$

$$\ldots\ldots\ldots\ldots\ldots\ldots\ldots$$ (15)

$$A_n(x) = x^{k_n^{(1)}} + x^{k_n^{(2)}} + \cdots,$$

constructed in accordance with conditions (14), and consider their product $A(x) = A_1(x) \cdots A_n(x)$:

$$A(x) = \left(x^{k_1^{(1)}} + x^{k_1^{(2)}} + \cdots\right) \cdots \left(x^{k_n^{(1)}} + x^{k_n^{(2)}} + \cdots\right).$$ (16)

When we multiply out the expressions in parentheses, there appear terms of the form x^k with some coefficients. Take $k = r$. Then we see that x^r is the sum of the products of the terms $x^{k_1^{(i_1)}}, \cdots, x^{k_n^{(i_n)}}$ with

$$k_1^{(i_1)} + \cdots + k_n^{(i_n)} = r.$$ (17)

But the number of different possibilities of obtaining (17) is exactly the number $N(r; n)$ of solutions to the system (13)–(14). Thus we have established the following simple, but important lemma.

Lemma. *The number $N(r; n)$ of integer-valued solutions to the system (13)–(14) is the coefficient of x^r in the product $A_1(x) \cdots A_n(x)$ of generating functions (15).*

7. Examples. (a) Let all the sets $\{k_i^{(1)}, k_i^{(2)}, \ldots\}$ have the form $\{0, 1, 2, \ldots\}$. Then

$$A_1(x) \cdots A_n(x) = (1 + x + x^2 + \cdots)^n$$

and the coefficient $N(r; n)$ of x^r in the expansion of $(1 + x + x^2 + \cdots)^n$ equals the *number of integer-valued solutions* of (13) subject to the condition $X_i \geq 0$.

Since

$$(1 + x + x^2 + \cdots)^n = \sum_{r \geq 0} C^r_{n+r-1} x^r, \qquad (18)$$

in this case the number of solutions is

$$N(r; n) = C^r_{n+r-1} \qquad (19)$$

(see Problem 7).

(b) Now we turn to *Galileo's problem* formulated in Subsection 1.

If the ith die falls on the number X_i $(X_i = 1, \ldots, 6)$, $i = 1, 2, 3$, then the number of all possibilities of summing them to 10 is the number $N(10; 3)$ of integer-valued solutions of the equation

$$X_1 + X_2 + X_3 = 10$$

such that $1 \leq X_i \leq 6$, $i = 1, 2, 3$.

The total number of possibilities (X_1, X_2, X_3) when throwing three dice is $6^3 = 216$. Therefore the probability of interest equals

$$P = \frac{N(10; 3)}{216}.$$

By the Lemma, $N(10; 3)$ equals the coefficient of x^{10} in the expansion of $(x + x^2 + \cdots + x^6)^3$ in powers of x. In order to find this coefficient, notice that

$$x + x^2 + \cdots + x^6 = x(1 + x + \cdots + x^5) = x(1 - x^6)(1 + x + x^2 + \cdots + x^5 + \cdots).$$

Consequently

$$(x + x^2 + \cdots + x^6)^3 = x^3(1 - x^6)^3(1 + x + x^2 + \cdots)^3. \qquad (20)$$

As we have seen in (18),

$$(1 + x + x^2 + \cdots)^3 = C^0_2 + C^1_3 x + C^2_4 x^2 + \cdots \qquad (21)$$

By the binomial formula

$$(a + b)^n = \sum_{k=0}^{n} C^k_n a^k b^{n-k} \qquad (22)$$

we find that

$$(1 - x^6)^3 = C^0_3 - C^1_3 x^6 + C^2_3 x^{12} - C^3_3 x^{18}.$$

Thus using (20) and (21) we obtain

$$(x + x^2 + \cdots + x^6)^3 = x^3(1 + 3x + 6x^2 + \cdots + 36x^7)(1 - 3x^6 + 3x^{12} - x^{18}).$$

Hence we see that the coefficient of x^{10} is equal to $36 - 9 = 27$. Therefore the required probability is

$$P = \frac{27}{216} = \frac{1}{8}.$$

(c) The problem on *lucky tickets* is solved in a similar manner.

Indeed, let the vector $(X_1, \ldots, X_6)$ consist of independent identically distributed random variables such that

$$p_k \equiv \mathsf{P}(X_i = k) = \frac{1}{10}$$

for all $k = 0, 1, \ldots, 9$ and $i = 1, \ldots, 6$.

The generating function of X_i is

$$G_{X_i}(x) = \mathsf{E}\, x^{X_i} = \sum_{k=0}^{9} p_k x^k = \frac{1}{10}(1 + x + \cdots + x^9) = \frac{1}{10} \frac{1 - x^{10}}{1 - x}.$$

Since $X_1, \ldots, X_6$ are assumed independent, we have

$$G_{X_1 + X_2 + X_3}(x) = G_{X_4 + X_5 + X_6}(x) = \frac{1}{10^3} \frac{(1 - x^{10})^3}{(1 - x)^3},$$

and the generating function $G_Y(x)$ of $Y = (X_1 + X_2 + X_3) - (X_4 + X_5 + X_6)$ is given by the formula

$$G_Y(x) = G_{X_1 + X_2 + X_3}(x)\, G_{X_4 + X_5 + X_6}\left(\frac{1}{x}\right) = \frac{1}{10^6} \frac{1}{x^{27}} \left(\frac{1 - x^{10}}{1 - x}\right)^6. \tag{23}$$

Writing $G_Y(x)$ as

$$G_Y(x) = \sum_{k=-\infty}^{\infty}{}' q_k x^k, \tag{24}$$

we see that the required probability

$$\mathsf{P}(X_1 + X_2 + X_3 = X_4 + X_5 + X_6),$$

i.e., $\mathsf{P}(Y = 0)$, equals the coefficient q_0 (of x^0) in the representation (24) for $G_Y(x)$.

We have

$$\frac{1}{x^{27}} \left(\frac{1 - x^{10}}{1 - x}\right)^6 = \frac{1}{x^{27}} (1 - x^{10})^6 (1 + x + x^2 + \cdots)^6.$$

It is seen from (18) that

$$(1 + x + x^2 + \cdots)^6 = \sum_{r \geq 0} C^r_{r+5} x^r. \tag{25}$$

The binomial formula (22) yields

$$(1 - x^{10})^6 = \sum_{k=0}^{6} (-1)^k C^k_6 x^{10k}. \tag{26}$$

Thus

$$G_Y(x) = \frac{1}{10^6} \frac{1}{x^{27}} \left(\sum_{r \geq 0} C^r_{r+5} x^r \right) \left(\sum_{k=0}^{6} (-1)^k C^k_6 x^{10k} \right) \quad \left(= \sum_{k=0}^{\infty} q_k x^k \right). \tag{27}$$

Multiplying the sums involved we can find (after simple, but rather tedious calculations) that the coefficient asked for is equal to

$$q_0 = \frac{55\,252}{10^6} = 0.055252, \tag{28}$$

which is the required probability $P(Y = 0)$ that $X_1 + X_2 + X_3 = X_4 + X_5 + X_6$ (i.e., that the sum of the first three digits equals the sum of the last three).

(d) Now we apply the method of generating functions to obtaining the solutions of *difference equations*, which appear in the problems of combinatorial probability.

In Subsection 1 we mentioned the Fibonacci numbers F_n, which first appeared (in connection with counting the number of rabbits in the nth generation) in the book *Liber abacci* published in 1220 by Leonardo Pisano known also as Fibonacci.

In terms of the number theory F_n is the number of representations of n as an *ordered sum* of 1s and 2s. Clearly, $F_1 = 1$, $F_2 = 2$ (since $2 = 1 + 1 = 2$), $F_3 = 3$ (since $3 = 1 + 1 + 1 = 1 + 2 = 2 + 1$).

Using this interpretation one can show that F_n's satisfy the recurrence relations

$$F_n = F_{n-1} + F_{n-2}, \quad n \geq 2 \tag{29}$$

(with $F_0 = F_1 = 1$). Let

$$G(x) = \sum_{n=0}^{\infty} F_n x^n \tag{30}$$

be the generating function of the sequence $\{F_n, n \geq 0\}$. Clearly,

$$G(x) = 1 + x + \sum_{n \geq 0} F_{n+2} x^{n+2}. \tag{31}$$

By (29) we have

$$F_{n+2}x^{n+2} = x(F_{n+1}x^{n+1}) + x^2(F_nx^n),$$

and therefore

$$\sum_{n\geq 0} F_{n+2}x^{n+2} = x\sum_{n\geq 0} F_{n+1}x^{n+1} + x^2\sum_{n\geq 0} F_nx^n.$$

This equality and (31) yield

$$G(x) - 1 - x = x[G(x) - 1] + x^2G(x).$$

Hence

$$G(x) = \frac{1}{1 - x - x^2}. \tag{32}$$

Note that

$$1 - x - x^2 = -(x - a)(x - b), \tag{33}$$

where

$$a = \frac{1}{2}(-1 - \sqrt{5}), \quad b = \frac{1}{2}(-1 + \sqrt{5}). \tag{34}$$

Now (32) and (33) imiply

$$G(x) = \frac{1}{a - b}\left[\frac{1}{a - x} - \frac{1}{b - x}\right] = \frac{1}{a - b}\left[\frac{1}{a}\sum_{n\geq 0}\left(\frac{x}{a}\right)^n - \frac{1}{b}\sum_{n\geq 0}\left(\frac{x}{b}\right)^n\right]$$

$$= \sum_{n\geq 0} x^n\left[\frac{1}{a - b}\left(\frac{1}{a^{n+1}} - \frac{1}{b^{n+1}}\right)\right]. \tag{35}$$

Since $G(x) = \sum_{n\geq 0} x^n F_n$, we obtain from (35) that

$$F_n = \frac{1}{a - b}\left[\frac{1}{a^{n+1}} - \frac{1}{b^{n+1}}\right].$$

Substituting for a and b their values (34) we find that

$$F_n = \frac{1}{\sqrt{5}}\left[\left(\frac{1 + \sqrt{5}}{2}\right)^{n+1} - \left(\frac{1 - \sqrt{5}}{2}\right)^{n+1}\right], \tag{36}$$

where

$$\frac{1 + \sqrt{5}}{2} = 1,6180\ldots, \quad \frac{1 - \sqrt{5}}{2} = -0,6180\ldots$$

(e) The method of generating functions is very efficient for finding various probabilities related to *random walks*. We will illustrate it by two models of a simple random walk considered in Sects. 2 and 9.

Let $S_0 = 0$ and $S_k = \xi_1 + \cdots + \xi_k$ be a sum of independent identically distributed random variables, $k \geq 1$, where each ξ_i takes two values $+1$ and -1 with probabilities p and q. Let $P_n(i) = \mathsf{P}(S_n = i)$ be the probability that the random walk $\{S_k, 0 \leq k \leq n\}$ occurs at the point i at time n. (Clearly, $i \in \{0, \pm 1, \ldots, \pm n\}$.)

In Sect. 2 (Subsection 1) we obtained by means of combinatorial arguments that

$$P_n(i) = C_n^{\frac{n+i}{2}} p^{\frac{n+i}{2}} q^{\frac{n-i}{2}} \tag{37}$$

(for integer $\frac{n+i}{2}$). Let us show how we could arrive at this solution using the method of generating functions.

By the formula for total probability (see Sects. 3 and 8)

$$
\begin{aligned}
\mathsf{P}(S_n = i) &= \mathsf{P}(S_n = i \mid S_{n-1} = i - 1)\, \mathsf{P}(S_{n-1} = i - 1) \\
&\quad + \mathsf{P}(S_n = i \mid S_{n-1} = i + 1)\, \mathsf{P}(S_{n-1} = i + 1) \\
&= \mathsf{P}(\xi_1 = 1)\, \mathsf{P}(S_{n-1} = i - 1) + \mathsf{P}(\xi_1 = -1)\, \mathsf{P}(S_{n-1} = i + 1) \\
&= p\, \mathsf{P}(S_{n-1} = i - 1) + q\, \mathsf{P}(S_{n-1} = i + 1).
\end{aligned}
$$

Consequently, for $i = 0, \pm 1, \pm 2, \ldots$, we obtain the recurrence relations

$$P_n(i) = p\, P_{n-1}(i - 1) + q\, P_{n-1}(i + 1), \tag{38}$$

which hold for any $n \geq 1$ with $P_0(0) = 1$, $P_0(i) = 0$ for $i \neq 0$.

Let us introduce the generating functions

$$G_k(x) = \sum_{i=-\infty}^{\infty} P_k(i) x^i \quad \left(= \sum_{i=-k}^{k} P_k(i) x^i \right). \tag{39}$$

By (38)

$$
\begin{aligned}
G_n(x) &= \sum_{i=-\infty}^{\infty} p P_{n-1}(i - 1) x^i + \sum_{i=-\infty}^{\infty} q P_{n-1}(i + 1) x^i \\
&= px \sum_{i=-\infty}^{\infty} P_{n-1}(i - 1) x^{i-1} + qx^{-1} \sum_{i=-\infty}^{\infty} P_{n-1}(i + 1) x^{i+1} \\
&= (px + qx^{-1}) G_{n-1}(x) = \ldots = (px + qx^{-1})^n G_0(x) \\
&= (px + qx^{-1})^n,
\end{aligned}
$$

since $G_0(x) = 1$.

The Lemma implies that $P_n(i)$ is the coefficient of x^i in the expansion of the generating function $G_n(x)$ in powers of x. Using the binomial formula (22) we obtain

$$(px + qx^{-1})^n = \sum_{k=0}^{n} C_n^k (px)^k (qx^{-1})^{n-k}$$

$$= \sum_{k=0}^{n} C_n^k p^k q^{n-k} x^{2k-n} = \sum_{i=-n}^{n} C_n^{\frac{n+i}{2}} p^{\frac{n+i}{2}} q^{\frac{n-i}{2}} x^i.$$

Hence the probability $P_n(i)$, which is the coefficient of x^i, is given by (37).

Let now $S_0 = 0$, $S_k = \xi_1 + \ldots + \xi_k$ be again the sum of independent identically distributed random variables ξ_i taking in this case the values 1 and 0 with probabilities p and q.

According to formula (1) of Sect. 2,

$$P_n(i) = C_n^i p^i q^{n-i}. \tag{40}$$

This formula can be obtained by the method of generating functions as follows. Since for $i \geq 1$

$$P_n(i) = p P_{n-1}(i-1) + q P_{n-1}(i)$$

and $P_n(0) = q^n$, we find for the generating functions

$$G_k(x) = \sum_{i=0}^{\infty} P_k(i) x^i$$

that

$$G_n(x) = (px + q) G_{n-1}(x) = \ldots = (px + q)^n G_0(x) = (px + q)^n,$$

because $G_0(x) = 1$.

The formula $G_n(x) = (px + q)^n$ just obtained and the binomial formula (22) imply that the coefficient of x^i in the expansion of $G_n(x)$ is $C_n^i p^i q^{n-i}$, which by the Lemma yields (40).

(f) The following example shows how we can prove various combinatorial identities using the properties of generating functions (and, in particular, the convolution property (9)).

For example, there is a well-known identity:

$$(C_n^0)^2 + (C_n^1)^2 + \cdots + (C_n^n)^2 = C_{2n}^n. \tag{41}$$

Let us prove it using the above Lemma.

By the binomial formula (22),

$$(1 + x)^{2n} = C_{2n}^0 + C_{2n}^1 x + C_{2n}^2 x^2 + \cdots + C_{2n}^n x^n + \cdots + C_{2n}^{2n}. \tag{42}$$

Hence

$$G_{2n}(x) = (1 + x)^{2n} \tag{43}$$

is the generating function of the sequence $\{C_{2n}^k, 0 \leq k \leq 2n\}$ with C_{2n}^n being the coefficient of x^n in (42).

Rewrite $(1 + x)^{2n}$ as

$$(1 + x)^{2n} = (1 + x)^n (1 + x)^n. \tag{44}$$

Then

$$G_{2n}(x) = G_n^{(a)}(x) G_n^{(b)}(x)$$

with

$$G_n^{(a)}(x) = (1 + x)^n = \sum_{k=0}^{n} a_k x^k \quad \text{and} \quad G_n^{(b)}(x) = (1 + x)^n = \sum_{k=0}^{n} b_k x^k,$$

where, obviously, using again (22)

$$a_k = b_k = C_n^k.$$

We employ now the convolution formula (9) to find that the coefficient of x^n in the product $G_n^{(a)}(x) G_n^{(b)}(x)$ equals

$$a_0 b_n + a_1 b_{n-1} + \cdots + a_n b_0 = C_n^0 C_n^n + C_n^1 C_n^{n-1} + \cdots + C_n^n C_n^0$$
$$= (C_n^0)^2 + (C_n^1)^2 + \cdots + (C_n^n)^2, \tag{45}$$

because $C_n^{n-k} = C_n^k$.

Since at the same time the coefficient of x^n in the expansion of $G_{2n}(x)$ is C_{2n}^n and $G_{2n}(x) = G_n^{(a)}(x) G_n^{(b)}(x)$, the required formula (41) follows from (45).

If we examine the above proof of (41), which relies on the equality (44), we can easily see that making use of the equality

$$(1 + x)^n (1 + x)^m = (1 + x)^{n+m}$$

instead of (44) and applying again the convolution formula (9), we obtain the following identity:

$$\sum_{j=1}^{n} C_n^j C_m^{k-j} = C_{n+m}^k, \tag{46}$$

which is known as "Wandermonde's convolution."

(g) Finally, we consider a classical example of application of the method of generating functions for finding the extinction probability of a branching process.

We continue to consider Example 4 of Sect. 12 assuming that the time parameter k is not limited by n, but may take any value $0, 1, 2, \ldots$.

Let ξ_k, $k \geq 0$, be the number of particles (individuums) at time k, $\xi_0 = 1$. According to Example 4 of Sect. 12 we assume that

$$\xi_{k+1} = \eta_1^{(k)} + \cdots + \eta_{\xi_k}^{(k)}, \qquad k \geq 0 \tag{47}$$

(the *Galton–Watson model*; see [6] and Problem 18 in Sect. 5, Chap. VIII of [90]), where $\{\eta_i^{(k)}, i \geq 1, k \geq 0\}$ is a sequence of independent identically distributed random variables having the same distribution as a random variable η, with probabilities

$$p_k = \mathsf{P}(\eta = k), \quad k \geq 0, \qquad \sum_{k=0}^{\infty} p_k = 1.$$

It is also assumed that for each k the random variables $\eta_i^{(k)}$ are independent of $\xi_1, \ldots, \xi_k$. (Thus the process of "creation–annihilation" evolves in such a way that each particle independently of others and of the "prehistory" turns into j particles with probabilities $p_j, j \geq 0$.)

Let $\tau = \inf\{k \geq 0 : \xi_k = 0\}$ be the extinction time (of the family). It is customary to let $\tau = \infty$ if $\xi_k > 0$ for all $k \geq 0$.

In the theory of branching processes the variable ξ_{k+1} is interpreted as the number of "parents" in the $(k+1)$th generation and $\eta_i^{(k)}$ as the number of "children" produced by the ith parent of the kth generation.

Let

$$G(x) = \sum_{k=0}^{\infty} p_k x^k, \qquad |x| \leq 1,$$

be the generating function of the random variable η (i.e., $G(x) = \mathsf{E}\,x^\eta$) and $F_k(x) = \mathsf{E}\,x^{\xi_k}$ the generating function of ξ_k.

The recurrence formula (47) and the property (16) (Sect. 8) of conditional expectations imply that

$$F_{k+1}(x) = \mathsf{E}\,x^{\xi_{k+1}} = \mathsf{E}\,\mathsf{E}(x^{\xi_{k+1}} \mid \xi_k).$$

We have by the independence assumption

$$\mathsf{E}(x^{\xi_{k+1}} \mid \xi_k = i) = \mathsf{E}(x^{\eta_1^{(k)} + \cdots + \eta_i^{(k)}}) = [G(x)]^i,$$

and therefore

$$F_{k+1}(x) = \mathsf{E}[G(x)]^{\xi_k} = F_k(G(x)).$$

According to (9) we may write

$$F_{k+1}(x) = (F_k \circ G)(x),$$

i.e., the generating function F_{k+1} is a *composition* of generating functions F_k and G.

One of the central problems of the theory of branching processes is obtaining the probability

$$q = \mathsf{P}(\tau < \infty),$$

i. e. the *probability of extinction* for a finite time.

Note that since $\{\xi_k = 0\} \subseteq \{\xi_{k+1} = 0\}$, we have

$$\mathsf{P}(\tau < \infty) = \mathsf{P}\left(\bigcup_{k=1}^{\infty} \{\xi_k = 0\}\right) = \lim_{N \to \infty} \mathsf{P}(\xi_N = 0).$$

But $P(\xi_N = 0) = F_N(0)$. Hence

$$q = \lim_{N \to \infty} F_N(0).$$

Note also that if $\xi_0 = 1$, then

$$F_0(x) = x, \quad F_1(x) = (F_0 \circ G)(x) = (G \circ F_0)(x),$$
$$F_2(x) = (F_1 \circ G)(x) = (G \circ G)(x) = (G \circ F_1)(x)$$

and, in general,

$$F_N(x) = (F_{N-1} \circ G)(x) = \underbrace{[G \circ G \circ \cdots \circ G]}_{N}(x) = G \circ \underbrace{[G \circ \cdots \circ G]}_{N-1}(x),$$

whence

$$F_N(x) = (G \circ F_{N-1})(x),$$

so that $q = \lim_{N \to \infty} F_N(0)$ satisfies the equation

$$q = G(q), \qquad 0 \le q \le 1.$$

The examination of the solutions to this equation shows that

(a) if $E\eta > 1$, then the extinction probability $0 < q < 1$;
(b) if $E\eta \le 1$, then the extinction probability $q = 1$.
   (For details and an outline of the proof see [10], Sect. 36, Chap. 8, and [90], Sect. 5, Chap. VIII, Problems 18–21.)

8. The above problems show that we often need to determine the coefficients a_i in the expansion of a generating function

$$G(x) = \sum_{i \ge 0} a_i x^i.$$

Here we state some standard generating functions whose series expansions are well known from calculus:

$$\frac{1}{1-x} = 1 + x + x^2 + \cdots$$

and therefore $(1-x)^{-1}$ is the generating function of the sequence $(1, 1, \ldots)$;

$$(1+x)^n = 1 + C_n^1 x^1 + C_n^2 x^2 + \cdots + C_n^n x^n;$$
$$\frac{1-x^{m+1}}{1-x} = 1 + x + x^2 + \cdots + x^m;$$
$$\frac{1}{(1-x)^n} = 1 + C_n^1 x^1 + C_{n+1}^2 x^2 + \cdots + C_{n+k-1}^k x^k + \cdots$$

Many sequences well known in calculus (the Bernoulli and Euler numbers, and so on) are defined by means of exponential generating functions:

Bernoulli numbers $(b_0, b_1, \ldots)$:

$$\frac{x}{e^x - 1} = \sum_{n=0}^{\infty} b_n \frac{x^n}{n!}$$

$$\left(b_0 = 1, \ \ b_1 = -\frac{1}{2}, \ \ b_2 = \frac{1}{6}, \ \ b_3 = 0, \ \ b_4 = -\frac{1}{30}, \ \ b_5 = 0, \ \ \ldots \right);$$

Euler numbers $(e_0, e_1, \ldots)$:

$$\frac{2e^x}{e^{2x} + 1} = \sum_{n=0}^{\infty} e_n \frac{x^n}{n!}$$

$$(e_0 = 1, \ \ e_1 = 0, \ \ e_2 = -1, \ \ e_3 = 0, \ \ e_4 = -5, \ \ e_5 = 0,$$
$$e_6 = -61, \ \ e_7 = 0, \ \ e_8 = 1385, \ \ \ldots).$$

9. Let ξ_1 and ξ_2 be two independent random variables having the Poisson distributions with parameters $\lambda_1 > 0$ and $\lambda_2 > 0$ respectively (see Sect. 6 and Table 1.2 in Sect. 2, Chap. 2):

$$P(\xi_i = k) = \frac{\lambda_i^k e^{-\lambda_i}}{k!}, \qquad k = 0, 1, 2, \ldots, \quad i = 1, 2. \tag{48}$$

It is not hard to calculate the generating functions $G_{\xi_i}(x)$ ($|x| \le 1$):

$$G_{\xi_i}(x) = E\, x^{\xi_i} = \sum_{i=0}^{\infty} P(\xi_i = k) x^k = e^{-\lambda_i (1-x)}. \tag{49}$$

Hence, by independence of ξ_1 and ξ_2, we find that the generating function $G_{\xi_1 + \xi_2}(x)$ of the sum $\xi_1 + \xi_2$ is given by

$$G_{\xi_1 + \xi_2}(x) = E\, x^{\xi_1 + \xi_2} = E(x^{\xi_1} x^{\xi_2}) = E\, x^{\xi_1} \cdot E\, x^{\xi_2}$$
$$= G_{\xi_1}(x) \cdot G_{\xi_2}(x) = e^{-\lambda_1(1-x)} \cdot e^{-\lambda_2(1-x)} = e^{-(\lambda_1 + \lambda_2)(1-x)}. \tag{50}$$

(The equality $E(x^{\xi_1} x^{\xi_2}) = E\, x^{\xi_1} \cdot E\, x^{\xi_2}$, which was used here, follows from independence of x^{ξ_1} and x^{ξ_2} due to the properties of expectation, see property (5) in Subsection 5 of Sect. 4 and Theorem 6 in Sect. 6, Chap. 2.)

We see from (48)–(50) that the sum of independent random variables ξ_1 and ξ_2 having the Poisson distributions with parameters λ_1 and λ_2 has again the Poisson distribution with parameter $\lambda_1 + \lambda_2$.

Obtaining the distribution of the difference $\xi_1 - \xi_2$ of the random variables ξ_1 and ξ_2 at hand is a more difficult problem. Using again their independence we find that

$$G_{\xi_1-\xi_2}(x) = G_{\xi_1}(x)\, G_{\xi_2}\left(\tfrac{1}{x}\right) = e^{-\lambda_1(1-x)}\, e^{-\lambda_2(1-1/x)}$$
$$= e^{-(\lambda_1+\lambda_2)+\lambda_1 x+\lambda_2(1/x)} = e^{-(\lambda_1+\lambda_2)}\, e^{\sqrt{\lambda_1\lambda_2}(t+1/t)},$$

where $t = x\sqrt{\lambda_1/\lambda_2}$.

It is known from calculus that for $\lambda \in R$

$$e^{\lambda(t+1/t)} = \sum_{k=-\infty}^{\infty} t^k I_k(2\lambda),$$

where $I_k(2\lambda)$ is the modified Bessel function of the first kind of order k (see, e.g. [40], vol. 2, pp. 504–507]):

$$I_k(2\lambda) = \lambda^k \sum_{r=0}^{\infty} \frac{\lambda^{2r}}{r!\,\Gamma(k+r+1)}, \qquad k = 0, \pm 1, \pm 2, \dots$$

Thus

$$P(\xi_1 - \xi_2 = k) = e^{-(\lambda_1+\lambda_2)}\left(\frac{\lambda_1}{\lambda_2}\right)^{k/2} I_k\left(2\sqrt{\lambda_1\lambda_2}\right)$$

for $k = 0, \pm 1, \pm 2, \dots$.

10. Problems

1. Find the generating functions of the sequences $\{a_r, r \geq 0\}$ with

$$(a)\ \ a_r = r!, \qquad (b)\ \ a_r = 2r^2, \qquad (c)\ \ a_r = \frac{1}{r}.$$

2. Find the generating functions for the number of integer-valued solutions of the systems

$$X_1 + X_2 + \cdots + X_n \leq r \quad \text{with } 1 \leq X_i \leq 4,$$
$$X_1 + 2X_2 + \cdots + nX_n = r \quad \text{with } X_i \geq 0.$$

3. Using the method of generating functions compute

$$\sum_{k=0}^{n} k(C_n^k)^2, \qquad \sum_{k=0}^{n} k^2 C_n^k.$$

4. Let $a_1, \ldots, a_k$ be different positive numbers. Show that the number $N(k; n)$ of partitions of the number n into a sum of the numbers from $\{a_1, \ldots, a_k\}$ is the coefficient of x^n in the series expansion of the product

$$(1 + x^{a_1} + x^{2a_1} + \cdots)(1 + x^{a_2} + x^{2a_2} + \cdots) \cdots (1 + x^{a_k} + x^{2a_k} + \cdots).$$

(The partitions allow for repeated a_i's, e.g., $8 = 2+2+2+2 = 3+3+2$, but the order of summation is immaterial, i.e., the representations $3 + 2$ and $2 + 3$ of the number 5 are counted as the same.)

5. Let a and b be two different positive numbers. Show that the number of non-negative solutions of the system

$$aX_1 + bX_2 = n$$

is the coefficient of x^n in the expansion of

$$(1 + x^a + x^{2a} + \cdots)(1 + x^b + x^{2b} + \cdots).$$

6. Show that:
   (a) The number of possible allocations of n indistinguishable balls over m different boxes equals C_{n+m-1}^n;
   (b) The number of vectors $(X_1, \ldots, X_m)$ with nonnegative integer components satisfying the equation $X_1 + \cdots + X_m = n$, is also equal to C_{n+m-1}^n;
   (c) The number of possible choices of n balls when sampling with replacement from an urn with m different balls is again equal to C_{n+m-1}^n.
   Hint. Establish that in each of the cases (a), (b) and (c) the required number is the coefficient of x^n in the expansion of $(1 + x + x^2 + \cdots)^m$.

7. Prove the formula (18).
8. Establish the formula (28) for q_0.
9. (*Euler's problem.*) There are loads of integer weight (in grams). The question is which loads can be weighed using weights of $1, 2, 2^2, 2^3, \ldots, 2^m, \ldots$ grams and in how many ways it can be done.
10. Let $G_\xi(x) = \mathsf{E} x^\xi \left(= \sum_{k \geq 0} x^k p_k \right)$ be the generating function of a random variable ξ. Show that

$$\mathsf{E}\,\xi = G_\xi'(1),$$
$$\mathsf{E}\,\xi^2 = G_\xi''(1) + G_\xi'(1),$$
$$\operatorname{Var}\xi = G_\xi''(1) + G_\xi'(1) - (G_\xi'(1))^2,$$

where $G_\xi'(x)$ and $G_\xi''(x)$ are the first and second derivatives (with respect to x) of $G_\xi(x)$.

11. Let ξ be an integer-valued random variable taking values $0, 1, \ldots$. Define $m_{(r)} = \mathsf{E}\,\xi(\xi - 1)\cdots(\xi - r + 1)$, $r = 1, 2, \ldots$. Show that the quantities $m_{(r)}$, which are called *factorial moments* (of order r), can be obtained from the generating function $G_\xi(x)$ by the formulas:

$$m_{(r)} = G_\xi^{(r)}(1),$$

where $G_\xi^{(r)}$ is the rth derivative of $G_\xi(x)$.

14 Inclusion–Exclusion Principle

1. When dealing with subsets $A, B, C, \ldots$ of a finite set Ω it is very suitable to use the so-called *Venn diagrams*, which provide an intuitively conceivable way of counting the number of outcomes $\omega \in \Omega$ contained in combinations of these sets such as, e.g., $A \cup B \cup C, A \cup B \cap \overline{C}$ etc.

If we denote by $N(D)$ the number of elements of a set D, then we see from the above diagrams that

$$N(A \cup B) = N(A) + N(B) - N(AB), \tag{1}$$

where the term $N(AB)$, i.e., the number of elements in the intersection $AB = A \cap B$ is subtracted ("exclusion") because in the sum $N(A) + N(B)$ ("inclusion) the elements of the intersection of A and B are counted twice.

In a similar way, for three sets A, B, and C we find that

$$N(A \cup B \cup C) = \big[N(A) + N(B) + N(C)\big]$$
$$- \big[N(AB) + N(AC) + N(BC)\big] + N(ABC). \tag{2}$$

The three terms appearing here correspond to "inclusion," "exclusion" and "inclusion."

When using the classical method of assigning probabilities $\mathsf{P}(A)$ by the formula

$$\mathsf{P}(A) = \frac{N(A)}{N(\Omega)}$$

(formula (10) in Sect. 1, Chap. 1), and in the general case (Sect. 2, Chap. 2), where we employ the finite additivity property of probability, we obtain the following formulas similar to (1) and (2):

$$P(A \cup B) = P(A) + P(B) - P(AB), \tag{3}$$

$$P(A \cup B \cup C) = [P(A) + P(B) + P(C)]$$
$$- [P(AB) + P(AC) + P(BC)]$$
$$+ P(ABC). \tag{4}$$

If we use the (easily verifiable) *De Morgan's laws* (see Problem 1 in Sect. 1)

$$\overline{A \cup B} = \overline{A} \cap \overline{B}, \qquad \overline{A \cap B} = \overline{A} \cup \overline{B}, \tag{5}$$

which establish relations between the three basic set operations (union, intersection, and taking the complement), then we find from (1) that

$$N(\overline{A}\,\overline{B}) = N(\Omega) - [N(A) + N(B)] + N(AB), \tag{6}$$

and similarly

$$N(\overline{A}\,\overline{B}\,\overline{C}) = N(\Omega) - [N(A) + N(B) + N(C)]$$
$$+ [N(AB) + N(AC) + N(BC)]$$
$$- N(ABC). \tag{7}$$

The event $\overline{A}\,\overline{B}\,(= \overline{A} \cap \overline{B})$ consists of outcomes ω which belong both to $\overline{A}$ and $\overline{B}$, i.e., of those ω which belong neither to A, nor to B. Now we give examples where this interpretation allows us to reduce the problem of counting the number of outcomes under consideration to counting the numbers of the type $N(\overline{A}\,\overline{B})$, $N(\overline{A}\,\overline{B}\,\overline{C})$, which may be counted using formulas (6) and (7). (An extension of these formulas to an arbitrary number of sets is given in the theorem below.)

Example 1. Consider a group of 30 students ($N(\Omega) = 30$). In this group 10 students study the foreign language A ($N(A) = 10$) and 15 students study the language B ($N(B) = 15$), while 5 of them study both languages ($N(AB) = 5$). The question is, how many students study neither of these languages. Clearly, this number is $N(\overline{A}\,\overline{B})$. According to (6)
$$N(\overline{A}\,\overline{B}) = 30 - [10 + 15 - 5] = 10.$$

Thus there are 10 students who do not study either of these languages.

Example 2 (From Number Theory). How many integers between 1 and 300
    (A) are not divisible by 3?
    (B) are not divisible by 3 or 5?
    (C) are not divisible by 3 or 5 or 7?
Here $N(\Omega) = 300$.

(A) Let $N(A)$ be the number of integers (in the interval $[1,\ldots,300]$) divisible by 3. Clearly, $N(A) = \frac{1}{3} \cdot 300 = 100$. Therefore the number of integers which are not divisible by 3 is $N(\overline{A}) = N(\Omega) - N(A) = 300 - 100 = 200$.

(B) Let $N(B)$ be the number of integers divisible by 5. Then $N(B) = \frac{1}{5} \cdot 300 = 60$. Further, $N(AB)$ is the number of integers divisible both by 3 and 5. Clearly, $N(AB) = \frac{1}{15} \cdot 300 = 20$.

The required number of integers divisible neither by 3, nor by 5 is

$$N(\overline{A}\,\overline{B}) = N(\Omega) - N(A) - N(B) + N(AB) = 300 - 100 - 60 + 20 = 160.$$

(C) Let $N(C)$ be the number of integers divisible by 7. Then $N(C) = \left\lfloor \frac{300}{7} \right\rfloor = 42$ and $N(AC) = \left\lfloor \frac{300}{21} \right\rfloor = 14$, $N(BC) = \left\lfloor \frac{300}{35} \right\rfloor = 8$, $N(AB) = 20$, $N(ABC) = 2$. Consequently by formula (7)

$$N(\overline{A}\,\overline{B}\,\overline{C}) = 300 - [100 + 60 + 42] + [20 + 14 + 8] - 2 = 138.$$

Thus the number of integers between 1 and 300 not divisible by any of 3, 5 or 7 is 138.

2. The formulas (1), (2), (6), and (7) can be extended in a natural way to an arbitrary number of subsets $A_1, \ldots, A_n$, $n \geq 2$, of Ω.

Theorem. *The following formulas of inclusion–exclusion hold*:

(a)

$$N(A_1 \cup \ldots \cup A_n) = \sum_{1 \leq i \leq n} N(A_i) - \sum_{1 \leq i_1 < i_2 \leq n} N(A_{i_1} \cap A_{i_1}) + \cdots$$
$$+ (-1)^{m+1} \sum_{1 \leq i_1 < \ldots < i_m \leq n} N(A_{i_1} \cap \cdots \cap A_{i_m}) + \cdots$$
$$+ (-1)^{n+1} N(A_1 \cap \cdots \cap A_n), \qquad (8)$$

or, in a more concise form,

$$N\left(\bigcup_{i=1}^{n} A_i\right) = \sum_{\varnothing \neq S \subseteq T} (-1)^{N(S)+1} N\left(\bigcap_{i \in S} A_i\right), \qquad (9)$$

where $T = \{1, \ldots, n\}$;

(b)

$$N(A_1 \cap \ldots \cap A_n) = \sum_{1 \leq i \leq n} N(A_i) - \sum_{1 \leq i_1 < i_2 \leq n} N(A_{i_1} \cup A_{i_2}) + \cdots$$
$$+ (-1)^{m+1} \sum_{1 \leq i_1 < \ldots < i_m \leq n} N(A_{i_1} \cup \cdots \cup A_{i_m}) + \cdots$$
$$+ (-1)^{n+1} N(A_1 \cup \cdots \cup A_n), \qquad (10)$$

or, in a more concise form,

$$N\left(\bigcap_{i=1}^{n} A_i\right) = \sum_{\emptyset \neq S \subseteq T} (-1)^{N(S)+1} N\left(\bigcup_{i \in S} A_i\right); \tag{11}$$

(c)

$$N(\overline{A}_1 \cup \cdots \cup \overline{A}_n) = N(\Omega) - N(A_1 \cap \cdots \cap A_n), \tag{12}$$

$$N(\overline{A}_1 \cap \cdots \cap \overline{A}_n) = N(\Omega) - N(A_1 \cup \cdots \cup A_n), \tag{13}$$

or, taking into account (11) *and* (8),

$$N\left(\bigcup_{i=1}^{n} \overline{A}_i\right) = \sum_{S \subseteq T} (-1)^{N(S)} N\left(\bigcup_{i \in S} A_i\right), \tag{14}$$

$$N\left(\bigcap_{i=1}^{n} \overline{A}_i\right) = \sum_{S \subseteq T} (-1)^{N(S)} N\left(\bigcap_{i \in S} A_i\right). \tag{15}$$

PROOF. It suffices to prove only formula (8) because all others can be derived from it substituting the events by their complements (by De Morgan's laws (5)).

Formula (8) could be proved by induction (Problem 1). But the proof relying on the properties of indicators of sets is more elegant.

Let

$$I_A(\omega) = \begin{cases} 1, & \omega \in A, \\ 0, & \omega \notin A, \end{cases}$$

be the indicator of the set A. If $B = A_1 \cup \cdots \cup A_n$, then by (5)

$$\overline{B} = \overline{A}_1 \cap \cdots \cap \overline{A}_n$$

and

$$I_B = 1 - I_{\overline{B}} = 1 - I_{\overline{A}_1} \cdots I_{\overline{A}_n} = 1 - (1 - I_{A_1}) \cdots (1 - I_{A_n}) =$$

$$= \sum_{i=1}^{n} I_{A_i} - \sum_{1 \leq i_1 < i_2 \leq n} I_{A_{i_1}} I_{A_{i_2}} + \cdots$$

$$+ (-1)^{m+1} \sum_{1 \leq i_1 < \ldots < i_m \leq n} I_{A_{i_1}} \ldots I_{A_{i_m}} + \cdots + (-1)^{n} I_{A_1} \ldots I_{A_n}. \tag{16}$$

All the indicators involved here are functions of ω: $I_B = I_B(\omega)$, $I_{A_i} = I_{A_i}(\omega)$, $I_{\overline{A}_i} = I_{\overline{A}_i}(\omega)$. Summing in (16) over all $\omega \in \Omega$ and taking into account that

$$\sum_{\omega \in \Omega} I_B(\omega) = N(B), \qquad \sum_{\omega \in \Omega} \left(\sum_{i=1}^{n} I_{A_i}(\omega) \right) = \sum_{i=1}^{n} N(A_i),$$

$$\sum_{\omega \in \Omega} \left(\sum_{1 \le i_1 < i_2 \le n} I_{A_{i_1}}(\omega) I_{A_{i_1}}(\omega) \right) \tag{17}$$

$$= \sum_{\omega \in \Omega} \left(\sum_{1 \le i_1 < i_2 \le n} I_{A_{i_1} \cap A_{i_1}}(\omega) \right) = \sum_{1 \le i_1 < i_2 \le n} N(A_{i_1} \cap A_{i_1})$$

etc., we arrive at (8). □

Remark. It is worth to note that the summation $\sum_{1 \le i_1 < \cdots < i_m \le n}$ in the above formulas is extended over all unordered subsets of cardinality m of the set $\{1, 2, \ldots, n\}$. The number of such subsets is C_n^m (see Table 1.1 in Sect. 1.)

3. The above formulas were derived under the assumption that Ω consists of finitely many outcomes ($N(\Omega) < \infty$). Under this assumption the classical definition of the probability of an event A as

$$P(A) = \frac{N(A)}{N(\Omega)}, \qquad A \subseteq \Omega, \tag{18}$$

immediately shows that the theorem remains valid if we replace throughout $N(A_i)$ by $P(A_i)$.

For example, formula (9) for the probability of the event $\bigcup_{i=1}^{n} A_i$ becomes

$$P\left(\bigcup_{i=1}^{n} A_i \right) = \sum_{\varnothing \ne S \subseteq T} (-1)^{N(S)+1} P\left(\bigcap_{i \in S} A_i \right), \tag{19}$$

where $T = \{1, \ldots, n\}$.

In fact, all these formulas for the probabilities of events $\bigcup_{i=1}^{n} A_i$, $\bigcap_{i=1}^{n} A_i$, $\bigcup_{i=1}^{n} \overline{A}_i$, $\bigcap_{i=1}^{n} \overline{A}_i$ remain valid not only in the case of the classical definition of probabilities by (18).

Indeed, the proof for $N(\Omega) < \infty$ was based on relation (16) for the indicators of events. If we take expectations of the left- and right-hand sides of this relation, then in the left-hand side we obtain $E I_B = P(B)$, while in the right-hand side we obtain combinations of terms like $E I_{A_{i_1}} \cdots I_{A_{i_m}}$. Since $I_{A_{i_1}} \cdots I_{A_{i_m}} = I_{A_{i_1} \cap \cdots \cap A_{i_m}}$, we have

$$E I_{A_{i_1}} \cdots I_{A_{i_m}} = E I_{A_{i_1} \cap \cdots \cap A_{i_m}} = P(A_{i_1} \cap \cdots \cap A_{i_m}).$$

From this formula and (16) we obtain the required inclusion–exclusion formula for $P(B) = P(A_1 \cup \ldots \cup A_n)$, which implies all other formulas for the events $\bigcap_{i=1}^{n} A_i$, $\bigcup_{i=1}^{n} \overline{A}_i$, $\bigcap_{i=1}^{n} \overline{A}_i$.

It is noticeable that in the proof just performed the assumption $N(\Omega) < \infty$ was not actually used, and this proof remains valid also for general probability spaces $(\Omega, \mathscr{F}, P)$ to be treated in the next chapter.

4. In addition to the two examples illustrating the use of the inclusion–exclusion formula consider the *derangements problem*.

Example 3. Suppose that Ω consists of $n!$ permutations of $(1, 2, \ldots, n)$. Consider the permutations

$$\begin{pmatrix} 1, & 2, & \ldots, & n \\ a_1, & a_2, & \ldots, & a_n \end{pmatrix}$$

with the property that there is no number i such that $a_i = i$.

We will show that the number D_n, $n \geq 1$, of such permutations ("the number of derangements") is given by the formula

$$D_n = n! \sum_{k=0}^{n} \frac{(-1)^k}{k!} \sim \frac{n!}{e} \quad (n \to \infty). \tag{20}$$

(For $n = 0$ this formula yields $D_0 = 1$.)

Let A_i be the event that $a_i = i$. It is clear that

$$N(A_i) = (n-1)!.$$

Similarly, for $i \neq j$

$$N(A_i A_j) = (n-2)!$$

and, in general,

$$N(A_{i_1} A_{i_2} \cdots A_{i_k}) = (n-k)!. \tag{21}$$

Using (13) and (8), the number of interest,

$$D_n = N(\overline{A}_1 \cap \overline{A}_2 \cap \cdots \cap \overline{A}_n),$$

can be represented as

$$D_n = N(\Omega) - S_1 + S_2 + \cdots + (-1)^m S_m + \cdots + (-1)^n S_n, \tag{22}$$

where

$$S_m = \sum_{1 \leq i_1 < i_2 < \ldots < i_m \leq n} N(A_{i_1} A_{i_2} \cdots A_{i_m}). \tag{23}$$

As was pointed out in the remark at the end of Subsection 2, the number of terms in the sum (23) equals C_n^m. Hence, taking into account (21), we find that

$$S_m = (n-m)! \, C_n^m, \tag{24}$$

and therefore (22) implies

$$D_n = n! - C_n^1 \cdot (n-1)! + C_n^2 \cdot (n-2)! + \cdots$$
$$+ (-1)^m \cdot C_n^m (n-m)! + \cdots + (-1)^n C_n^n \cdot 0!$$
$$= \sum_{m=0}^{n} (-1)^m C_n^m (n-m)!.$$

We have here $C_n^m (n-m)! = \frac{n!}{m!}$. Consequently,

$$D_n = n! \sum_{m=0}^{n} \frac{(-1)^m}{m!}, \qquad (25)$$

so that the probability of the "complete derangement" is

$$P(\bar{A}_1 \cdots \bar{A}_n) = \frac{N(\bar{A}_1 \cdots \bar{A}_n)}{n!} = \sum_{m=0}^{n} \frac{(-1)^m}{m!} = 1 - 1 + \frac{1}{2!} - \frac{1}{3!} + \cdots + \frac{(-1)^n}{n!}.$$

Since $\sum_{m=0}^{\infty} \frac{(-1)^m}{m!} = \frac{1}{e}$, we have

$$P(\bar{A}_1 \cdots \bar{A}_n) = \frac{1}{e} + O\left(\frac{1}{(n+1)!}\right)$$

and

$$D_n \sim \frac{n!}{e}.$$

The formula (25) could also be obtained by the method of (exponential) generating functions.

Namely, let $D_0 = 1$ (according to (20)) and note that $D_1 = 0$ and D_n for any $n \geq 2$ satisfies the following recurrence relations:

$$D_n = (n-1)\left[D_{n-1} + D_{n-2}\right] \qquad (26)$$

(the proof of this is left to the reader as Problem 3), which imply that

$$D_n - nD_{n-1} = -\left[D_{n-1} - (n-1)D_{n-2}\right].$$

Hence by the downward induction we find that

$$D_n = nD_{n-1} + (-1)^n,$$

whence we obtain for the exponential generating function $E(x)$ of the sequence $\{D_n, n \geq 0\}$

$$E(x) = \sum_{n=0}^{\infty} D_n \frac{x^n}{n!} = 1 + \sum_{n\geq 2} D_n \frac{x^n}{n!} = 1 + \sum_{n\geq 2} \{nD_n + (-1)^n\} \frac{x^n}{n!}$$

$$= 1 + \sum_{n\geq 0} \frac{D_n}{n!} x^n + \left[e^{-x} - (1-x) \right] = xE(x) + \left[e^{-x} - (1-x) \right].$$

Thus

$$E(x) = \frac{e^{-x}}{1-x} \tag{27}$$

and therefore

$$E(x) = \left(1 - x + \frac{x^2}{2} - \frac{x^3}{3!} + \cdots \right)(1 + x + x^2 + \cdots). \tag{28}$$

Writing the right-hand side of this formula as $\sum_{n\geq 0} D_n \frac{x^n}{n!}$ one can derive that the coefficients D_n have the form (25). However, we see that this result obtains by the method of inclusion–exclusion easier than by the method of generating functions.

8. Problems

1. Prove the formulas (8)–(9) and (10)–(11) by induction.
2. Let B_1 be the event that exactly one of the events $A_1, \ldots, A_n$ occurs. Show that

$$P(B_1) = \sum_{i=1}^{n} P(A_i) - 2 \sum_{1\leq i_1 < i_2 \leq n} P(A_{i_1} A_{i_2}) + \cdots + (-1)^{n+1} n \, P(A_1 \cdots A_n).$$

Hint. Use the formula

$$I_{B_1} = \sum_{i=1}^{n} I_{A_i} \prod_{j\neq i} (1 - I_{A_j}).$$

3. Prove the following generalization of (15): for any set I such that $\varnothing \subseteq I \subseteq T = \{1, \ldots, n\}$ it holds

$$N\left(\left(\bigcap_{i\in T\setminus I} \bar{A}_i \right) \cap \left(\bigcap_{i\in I} A_i \right) \right) = \sum_{I \subseteq S \subseteq T} (-1)^{N(S\setminus I)} N\left(\bigcap_{i\in S} A_i \right).$$

4. Using the method of inclusion–exclusion find the number of all integer-valued solutions of the equation $X_1 + X_2 + X_3 + X_4 = 25$ satisfying the constraints

$$-10 \leq X_i \leq 10.$$

Solve this problem also by the method of generating functions. Which method gives the solution of this problem faster?
5. Prove the formula (26).
6. Using the formula (8) find the number of allocations of n different balls over m different boxes subject to the constraint that at least one box remains empty.

7. Find the number of allocations of n different balls over m indistinguishable boxes subject to the condition that neither box remains empty.

8. Let $A = A(n)$ and $B = B(m)$ be two sets consisting of n and m elements respectively.

   A mapping $\mathbb{F}: A \to B$ is said to be a *function* if to each $a \in A$ it makes correspond some $b \in B$.

   A mapping $\mathbb{I}: A \to B$ is said to be an *injection* if to different elements of A it makes correspond different elements of B. (In this case $n \leq m$.)

   A mapping $\mathbb{S}: A \to B$ is said to be a *surjection* (or onto function) if for any $b \in B$ there is an $a \in A$ such that $S(a) = b$. (In this case $n \geq m$.)

   A mapping $\mathbb{B}: A \to B$ is said to be a *bijection* if it is both injection and surjection. (In this case $n = m$.)

   Using the inclusion–exclusion principle show that $N(\mathbb{F})$, $N(\mathbb{I})$, $N(\mathbb{S})$, and $N(\mathbb{B})$ (i.e., the number of functions, injections, surjections, and bijections) are given by the following formulas:

   $$N(\mathbb{F}) = m^n,$$
   $$N(\mathbb{I}) = (m)_n \ (= m(m-1) \cdots (m-n+1)),$$
   $$N(\mathbb{S}) = \sum_{i=0}^{m} (-1)^i C_m^i (m-i)^n,$$
   $$N(\mathbb{B}) = n! \ \left(= \sum_{i=0}^{n} (-1)^i C_n^i (n-i)^n \right).$$

Chapter 2
Mathematical Foundations of Probability Theory

The theory of probability, as a mathematical discipline, can and should be developed from axioms in exactly the same way as Geometry and Algebra. This means that after we have defined the elements to be studied and their basic relations, and have stated the axioms by which these relations are to be governed, all further exposition must be based exclusively on these axioms, independent of the usual concrete meaning of these elements and their relations.

A. N. Kolmogorov, "Foundations of the Theory of Probability" [51].

1 Probabilistic Model for an Experiment with Infinitely Many Outcomes: Kolmogorov's Axioms

1. The models introduced in the preceding chapter enabled us to give a probabilistic–statistical description of experiments with a finite number of outcomes. For example, the triple $(\Omega, \mathscr{A}, \mathsf{P})$ with

$$\Omega = \{\omega \colon \omega = (a_1, \ldots, a_n), \ a_i = 0, \ 1\}, \quad \mathscr{A} = \{A \colon A \subseteq \Omega\}$$

and $p(\omega) = p^{\Sigma a_i} q^{n - \Sigma a_i}$ is a model for the experiment in which a coin is tossed n times "independently" with probability p of falling head. In this model the number $N(\Omega)$ of outcomes, i.e., the number of points in Ω, is the finite number 2^n.

We now consider the problem of constructing a probabilistic model for the experiment consisting of an infinite number of independent tosses of a coin when at each step the probability of falling head is p.

It is natural to take the set of outcomes to be the set

$$\Omega = \{\omega \colon \omega = (a_1, a_2, \ldots), a_i = 0, 1\},$$

i.e., the space of sequences $\omega = (a_1, a_2, \ldots)$ whose elements are 0 or 1.

© Springer Science+Business Media New York 2016
A.N. Shiryaev, *Probability-1*, Graduate Texts
in Mathematics 95, DOI 10.1007/978-0-387-72206-1_2

What is the cardinality $N(\Omega)$ of Ω? It is well known that every number $a \in [0, 1)$ has a unique binary expansion (containing an infinite number of zeros)

$$a = \frac{a_1}{2} + \frac{a_2}{2^2} + \cdots \qquad (a_i = 0, 1).$$

Hence it is clear that there is a one-to-one correspondence between the points ω of Ω and the points a of the set $[0, 1)$, and therefore Ω has the cardinality of the *continuum*.

Consequently if we wish to construct a probabilistic model to describe experiments like tossing a coin *infinitely* many times, we must consider spaces Ω of a rather complicated nature.

We shall now try to see what probabilities ought reasonably to be assigned (or assumed) in a model of infinitely many independent tosses of a fair coin $\left(p + q = \frac{1}{2}\right)$.

Since we may take Ω to be the set $[0, 1)$, our problem can be considered as the problem of choosing points at random from this set. For reasons of symmetry, it is clear that all outcomes ought to be equiprobable. But the set $[0, 1)$ is uncountable, and if we suppose that its probability is 1, then it follows that the probability $p(\omega)$ of each outcome certainly must equal zero. However, this assignment of probabilities $(p(\omega) = 0, \ \omega \in [0, 1))$ does not lead very far. The fact is that we are ordinarily not interested in the probability of one outcome or another, but in the probability that the result of the experiment is in one or another specified set A of outcomes (an event). In elementary probability theory we use the probabilities $p(\omega)$ to find the probability $\mathsf{P}(A)$ of the event A: $\mathsf{P}(A) = \sum_{\omega \in A} p(\omega)$. In the present case, with $p(\omega) = 0$, $\omega \in [0, 1)$, we cannot define, for example, the probability that a point chosen at random from $[0, 1)$ belongs to the set $[0, \frac{1}{2})$. At the same time, it is intuitively clear that this probability should be $\frac{1}{2}$.

These remarks should suggest that in constructing probabilistic models for uncountable spaces Ω we must assign probabilities not to individual outcomes but to subsets of Ω. The same reasoning as in the first chapter shows that the collection of sets to which probabilities are assigned must be closed with respect to unions, intersections, and complements. Here the following definition is useful.

Definition 1. Let Ω be a set of points ω. A system $\mathscr{A}$ of subsets of Ω is called an *algebra* if

(a) $\Omega \in \mathscr{A}$,
(b) $A, B \in \mathscr{A} \Rightarrow A \cup B \in \mathscr{A}, \quad A \cap B \in \mathscr{A}$,
(c) $A \in \mathscr{A} \Rightarrow \overline{A} \in \mathscr{A}$.

(Notice that in condition (b) it is sufficient to require only that either $A \cup B \in \mathscr{A}$ or that $A \cap B \in \mathscr{A}$, since by De Morgan's laws (see (5) in Sect. 14, Chap. 1) $A \cup B = \overline{\overline{A} \cap \overline{B}}$ and $A \cap B = \overline{\overline{A} \cup \overline{B}}$.)

The next definition is needed in formulating the concept of a probabilistic model.

Definition 2. Let $\mathscr{A}$ be an algebra of subsets of Ω. A *set function* $\mu = \mu(A)$, $A \in \mathscr{A}$, taking values in $[0, \infty]$, is called a *finitely additive measure* defined on $\mathscr{A}$ if

$$\mu(A + B) = \mu(A) + \mu(B) \tag{1}$$

for every pair of disjoint sets A and B in $\mathscr{A}$.

A finitely additive measure μ with $\mu(\Omega) < \infty$ is called *finite*, and when $\mu(\Omega) = 1$ it is called a *finitely additive probability measure*, or a *finitely additive probability*.

2. We now define a probabilistic model (in the extended sense) of an experiment with outcomes in the set Ω.

Definition 3. An ordered triple $(\Omega, \mathscr{A}, \mathsf{P})$, where

(a) Ω is a *set* of points ω;
(b) $\mathscr{A}$ is an *algebra* of subsets of Ω;
(c) P is a *finitely additive probability* on A,

is a *probabilistic model*, or a *probabilistic "theory"* (of an experiment) *in the extended sense*.

It turns out, however, that this model is too broad to lead to a fruitful mathematical theory. Consequently we must restrict both the class of subsets of Ω that we consider, and the class of admissible probability measures.

Definition 4. A system $\mathscr{F}$ of subsets of Ω is a *σ-algebra* if it is an algebra and satisfies the following additional condition (stronger than (b) of Definition 1):

(b*) if $A_n \in \mathscr{F}$, $n = 1, 2, \ldots$, then

$$\bigcup A_n \in \mathscr{F}, \qquad \bigcap A_n \in \mathscr{F}$$

(it is sufficient to require either that $\bigcup A_n \in \mathscr{F}$ or that $\bigcap A_n \in \mathscr{F}$).

Definition 5. The space Ω together with a σ-algebra $\mathscr{F}$ of its subsets is a *measurable space*, and is denoted by $(\Omega, \mathscr{F})$.

Definition 6. A finitely additive measure μ defined on an algebra $\mathscr{A}$ of subsets of Ω is *countably additive* (or *σ-additive*), or simply a *measure*, if, for any pairwise disjoint subsets $A_1, A_2, \ldots$ of Ω with $\sum A_n \in \mathscr{A}$

$$\mu\left(\sum_{n=1}^{\infty} A_n\right) = \sum_{n=1}^{\infty} \mu(A_n).$$

A measure μ is said to be *σ-finite* if Ω can be represented in the form

$$\Omega = \sum_{n=1}^{\infty} \Omega_n, \qquad \Omega_n \in \mathscr{A},$$

with $\mu(\Omega_n) < \infty$, $n = 1, 2, \ldots$.

If a *measure* (let us stress that we mean a *countably additive measure*) P on the σ-algebra A satisfies $P(\Omega) = 1$, it is called a *probability measure* or a *probability* (defined on the sets that belong to the σ-algebra $\mathscr{A}$).

Probability measures have the following properties.

If $\varnothing$ is the empty set then

$$P(\varnothing) = 0.$$

If $A, B \in \mathscr{A}$ then

$$P(A \cup B) = P(A) + P(B) - P(A \cap B).$$

If $A, B \in \mathscr{A}$ and $B \subseteq A$ then

$$P(B) \leq P(A).$$

If $A_n \in \mathscr{A}$, $n = 1, 2, \ldots$, and $\bigcup A_n \in \mathscr{A}$, then

$$P(A_1 \cup A_2 \cup \cdots) \leq P(A_1) + P(A_2) + \cdots.$$

The first three properties are evident. To establish the last one it is enough to observe that $\bigcup_{n=1}^{\infty} A_n = \sum_{n=1}^{\infty} B_n$, where $B_1 = A_1$, $B_n = \overline{A}_1 \cap \cdots \cap \overline{A}_{n-1} \cap A_n$, $n \geq 2$, $B_i \cap B_j = \varnothing$, $i \neq j$, and therefore

$$P\left(\bigcup_{n=1}^{\infty} A_n\right) = P\left(\sum_{n=1}^{\infty} B_n\right) = \sum_{n=1}^{\infty} P(B_n) \leq \sum_{n=1}^{\infty} P(A_n).$$

The next theorem, which has many applications, provides conditions under which a finitely additive set function is actually countably additive.

Theorem. *Let P be a finitely additive set function defined over the algebra $\mathscr{A}$, with $P(\Omega) = 1$. The following four conditions are equivalent:*

(1) P *is σ-additive (P is a probability);*

(2) P *is continuous from below, i.e., for any sets $A_1, A_2, \ldots \in \mathscr{A}$ such that $A_n \subseteq A_{n+1}$ and $\bigcup_{n=1}^{\infty} A_n \in \mathscr{A}$,*

$$\lim_n P(A_n) = P\left(\bigcup_{n=1}^{\infty} A_n\right);$$

(3) P *is continuous from above, i.e., for any sets $A_1, A_2, \ldots$ such that $A_n \supseteq A_{n+1}$ and $\bigcap_{n=1}^{\infty} A_n \in \mathscr{A}$,*

$$\lim_n P(A_n) = P\left(\bigcap_{n=1}^{\infty} A_n\right);$$

(4) P *is continuous at $\varnothing$, i.e., for any sets $A_1, A_2, \ldots \in \mathscr{A}$ such that $A_{n+1} \subseteq A_n$ and $\bigcap_{n=1}^{\infty} A_n = \varnothing$,*

$$\lim_n P(A_n) = 0.$$

PROOF. (1) $\Rightarrow$ (2). Since

$$\bigcup_{n=1}^{\infty} A_n = A_1 + (A_2 \backslash A_1) + (A_3 \backslash A_2) + \cdots,$$

we have

$$P\left(\bigcup_{n=1}^{\infty} A_n\right) = P(A_1) + P(A_2 \backslash A_1) + P(A_3 \backslash A_2) + \cdots$$

$$= P(A_1) + P(A_2) - P(A_1) + P(A_3) - P(A_2) + \cdots$$

$$= \lim_n P(A_n).$$

(2) $\rightarrow$ (3). Let $n \geq 1$; then

$$P(A_n) = P(A_1 \backslash (A_1 \backslash A_n)) = P(A_1) - P(A_1 \backslash A_n).$$

The sequence $\{A_1 \backslash A_n\}_{n \geq 1}$ of sets is nondecreasing (see the table below) and

$$\bigcup_{n=1}^{\infty} (A_1 \backslash A_n) = A_1 \backslash \bigcap_{n=1}^{\infty} A_n.$$

Then, by (2)

$$\lim_n P(A_1 \backslash A_n) = P\left(\bigcup_{n=1}^{\infty} (A_1 \backslash A_n)\right)$$

and therefore

$$\lim_n P(A_n) = P(A_1) - \lim_n P(A_1 \backslash A_n)$$

$$= P(A_1) - P\left(\bigcup_{n=1}^{\infty} (A_1 \backslash A_n)\right) = P(A_1) - P\left(A_1 \backslash \bigcap_{n=1}^{\infty} A_n\right)$$

$$= P(A_1) - P(A_1) + P\left(\bigcap_{n=1}^{\infty} A_n\right) = P\left(\bigcap_{n=1}^{\infty} A_n\right).$$

(3) $\Rightarrow$ (4). Obvious.

(4) $\Rightarrow$ (1). Let $A_1, A_2, \ldots \in \mathscr{A}$ be pairwise disjoint and let $\sum_{n=1}^{\infty} A_n \in \mathscr{A}$. Then

$$P\left(\sum_{i=1}^{\infty} A_i\right) = P\left(\sum_{i=1}^{n} A_i\right) + P\left(\sum_{i=n+1}^{\infty} A_i\right),$$

and since $\sum_{i=n+1}^{\infty} A_i \downarrow \varnothing$, $n \rightarrow \infty$, we have

$$\sum_{i=1}^{\infty} P(A_i) = \lim_n \sum_{i=1}^{n} P(A_i) = \lim_n P\left(\sum_{i=1}^{n} A_i\right)$$

$$= \lim_n \left[P\left(\sum_{i=1}^{\infty} A_i\right) - P\left(\sum_{i=n+1}^{\infty} A_i\right)\right]$$

$$= P\left(\sum_{i=1}^{\infty} A_i\right) - \lim_n P\left(\sum_{i=n+1}^{\infty} A_i\right) = P\left(\sum_{i=1}^{\infty} A_i\right). \qquad \square$$

3. We can now formulate the generally accepted *Kolmogorov's axiom system*, which forms the basis for probability models of experiments with outcomes in the set Ω.

Fundamental Definition. An ordered triple $(\Omega, \mathscr{F}, P)$ where

(a) Ω *is a set of points* ω,
(b) $\mathscr{F}$ *is a σ-algebra of subsets of* Ω,
(c) P *is a probability on* $\mathscr{F}$,

is called a *probabilistic model* (of an experiment) or a *probability space*. Here Ω is the *sample space* or *space of elementary events*, the sets A in $\mathscr{F}$ are *events*, and $P(A)$ is the *probability* of the event A.

It is clear from the definition that the axiomatic formulation of probability theory is based on set theory and measure theory. Accordingly, it is useful to have a table (see Table 2.1) displaying the ways in which various concepts are interpreted in the two theories. In the next two sections we shall give examples of the measurable spaces that are most important for probability theory and of how probabilities are assigned on them.

4. PROBLEMS

1. Let $\Omega = \{r: r \in [0,1]\}$ be the set of rational points of $[0,1]$, $\mathscr{A}$ the algebra of sets each of which is a finite sum of disjoint sets A of one of the forms $\{r: a < r < b\}$, $\{r: a \le r < b\}$, $\{r: a < r \le b\}$, $\{r: a \le r \le b\}$, and $P(A) = b - a$. Show that $P(A)$, $A \in \mathscr{A}$, is finitely additive set function but not countably additive.

2. Let Ω be a countable set and $\mathscr{F}$ the collection of all its subsets. Put $\mu(A) = 0$ if A is finite and $\mu(A) = \infty$ if A is infinite. Show that the set function μ is finitely additive but not countably additive.

3. Let μ be a finite measure on a σ-algebra $\mathscr{F}$, $A_n \in \mathscr{F}$, $n = 1, 2, \ldots$, and $A = \lim_n A_n$ (i.e., $A = \liminf_n A_n = \limsup_n A_n$). Show that $\mu(A) = \lim_n \mu(A_n)$.

4. Prove that $P(A \triangle B) = P(A) + P(B) - 2 P(A \cap B)$. (Compare with Problem 4 in Sect. 1, Chap. 1.)

5. Show that the "distances" $\rho_1(A, B)$ and $\rho_2(A, B)$ defined by

$$\rho_1(A, B) = P(A \triangle B),$$

$$\rho_2(A, B) = \begin{cases} \frac{P(A \triangle B)}{P(A \cup B)} & \text{if } P(A \cup B) \neq 0, \\ 0 & \text{if } P(A \cup B) = 0 \end{cases}$$

satisfy the triangle inequality.

Table 2.1

Notation	Set-theoretic interpretation	Interpretation in probability theory
ω	Element or point	Outcome, sample point, elementary event
Ω	Set of points	Sample space; certain event
$\mathscr{F}$	σ-algebra of subsets	σ-algebra of events
$A \in \mathscr{F}$	Set of points	Event (if $\omega \in A$, we say that event A occurs)
$\overline{A} = \Omega \backslash A$	Complement of A, i.e., the set of points ω that are not in A	Event that A does not occur
$A \cup B$	Union of A and B, i.e., the set of points ω belonging either to A or to B (or to both)	Event that either A or B (or both) occurs
$A \cap B$ (or AB)	Intersection of A and B, i.e., the set of points ω belonging to both A and B	Event that both A and B occur
$\varnothing$	Empty set	Impossible event
$A \cap B = \varnothing$	A and B are disjoint	Events A and B are mutually exclusive, i.e., cannot occur simultaneously
$A + B$	Sum of sets, i.e., union of disjoint sets	Event that one of two mutually exclusive events occurs
$A \backslash B$	Difference of A and B, i.e., the set of points that belong to A but not to B	Event that A occurs and B does not
$A \triangle B$	Symmetric difference of sets, i.e., $(A \backslash B) \cup (B \backslash A)$	Event that A or B occurs, but not both
$\bigcup_{n=1}^{\infty} A_n$	Union of the sets $A_1, A_2, \ldots$	Event that at least one of $A_1, A_2, \ldots$ occurs
$\sum_{n=1}^{\infty} A_n$	Sum, i.e., union of pairwise disjoint sets $A_1, A_2, \ldots$	Event that one of the mutually exclusive events $A_1, A_2, \ldots$ occurs
$\bigcap_{n=1}^{\infty} A_n$	Intersection of $A_1, A_2, \ldots$	Event that all the events $A_1, A_2, \ldots$ occur
$A_n \uparrow A$ (or $A = \lim_n \uparrow A_n$)	The increasing sequence of sets A_n converges to A, i.e., $A_1 \subseteq A_2 \subseteq \cdots$ and $A = \bigcup_{n=1}^{\infty} A_n$	The increasing sequence of events converges to event A
$A_n \downarrow A$ (or $A = \lim_n \downarrow A_n$)	The decreasing sequence of sets A_n converges to A, i.e., $A_1 \supseteq A_2 \supseteq \cdots$ and $A = \bigcap_{n=1}^{\infty} A_n$	The decreasing sequence of events converges to event A
$\limsup A_n$ (or *A_n i.o.)	The set $\bigcap_{n=1}^{\infty} \bigcup_{k=n}^{\infty} A_k$	Event that infinitely many of events $A_1, A_2, \ldots$ occur
$\liminf A_n$	The set $\bigcup_{n=1}^{\infty} \bigcap_{k=n}^{\infty} A_k$	Event that all the events $A_1, A_2, \ldots$ occur with the possible exception of a finite number of them

* i.o. = infinitely often

6. Let μ be a finitely additive measure on an algebra $\mathscr{A}$, and let the sets $A_1, A_2, \ldots \in \mathscr{A}$ be pairwise disjoint and satisfy $A = \sum_{i=1}^{\infty} A_i \in \mathscr{A}$. Then $\mu(A) \geq \sum_{i=1}^{\infty} \mu(A_i)$.

7. Prove that

$$\overline{\lim \sup A_n} = \lim \inf \overline{A}_n, \qquad \overline{\lim \inf A_n} = \lim \sup \overline{A}_n,$$
$$\lim \inf A_n \subseteq \lim \sup A_n, \quad \lim \sup(A_n \cup B_n) = \lim \sup A_n \cup \lim \sup B_n,$$
$$\lim \sup A_n \cap \lim \inf B_n \subseteq \lim \sup(A_n \cap B_n) \subseteq \lim \sup A_n \cap \lim \sup B_n.$$

If $A_n \uparrow A$ or $A_n \downarrow A$, then

$$\lim \inf A_n = \lim \sup A_n.$$

8. Let $\{x_n\}$ be a sequence of numbers and $A_n = (-\infty, x_n)$. Show that $x = \lim \sup x_n$ and $A = \lim \sup A_n$ are related in the following way: $(-\infty, x) \subseteq A \subseteq (-\infty, x]$. In other words, A is equal to either $(-\infty, x)$ or to $(-\infty, x]$.

9. Give an example to show that if a measure takes the value $+\infty$, countable additivity in general does not imply continuity at $\varnothing$.

10. Verify the *Boole inequality*: $P(A \cap B) \geq 1 - P(\overline{A}) - P(\overline{B})$.

11. Let $A_1, \ldots, A_n$ be events in $\mathscr{F}$. This system of events is said to be *exchangeable* (or *interchangeable*) if for any $1 \leq l \leq n$ the probabilities $P(A_{i_1} \ldots A_{i_l})$ are the same $(= p_l)$ for any choice of indices $1 \leq i_1 < \cdots < i_l \leq n$. Prove that for such events the following formula holds:

$$P\left(\bigcup_{i=1}^{n} A_i\right) = np_1 - C_n^2 p_2 + C_n^3 p_3 - \cdots + (-1)^{n-1} p_n.$$

12. Let $(A_k)_{k \geq 1}$ be an *infinite* sequence of exchangeable events, i.e., for any $n \geq 1$ the probabilities $P(A_{i_1} \ldots A_{i_n})$ are the same $(= p_n)$ for any set of indices $1 \leq i_1 < \cdots < i_n$. Prove that

$$P\left(\lim \inf_n A_n\right) = P\left(\bigcap_{k=1}^{\infty} A_k\right) = \lim_{j \to \infty} p_j,$$

$$P\left(\lim \sup_n A_n\right) = P\left(\bigcup_{k=1}^{\infty} A_k\right) = 1 - \lim_{j \to \infty} (-1)^j \Delta^j(p_0),$$

where $p_0 = 1$, $\Delta^1(p_n) = p_{n+1} - p_n$, $\Delta^j(p_n) = \Delta^1(\Delta^{j-1}(p_n))$, $j \geq 2$.

13. Let $(A_n)_{n \geq 1}$ be a sequence of sets and let $I(A_n)$ be the indicator of A_n, $n \geq 1$. Show that

$$I\left(\lim \inf_n A_n\right) = \lim \inf_n I(A_n), \qquad I\left(\lim \sup_n A_n\right) = \lim \sup_n I(A_n),$$

$$I\left(\bigcup_{n=1}^{\infty} A_n\right) \leq \sum_{n=1}^{\infty} I(A_n).$$

14. Show that

$$I\left(\bigcup_{n=1}^{\infty} A_n\right) = \max_{n \geq 1} I(A_n), \quad I\left(\bigcap_{n=1}^{\infty} A_n\right) = \min_{n \geq 1} I(A_n).$$

15. Prove that

$$P(\limsup A_n) \geq \limsup P(A_n), \quad P(\liminf A_n) \leq \liminf P(A_n).$$

16. Let $A^* = \limsup A_n$ and $A_* = \liminf A_n$. Show that $P(A_n \backslash A_*) \to 0$ and $P(A^* \backslash A_n) \to 0$.

17. Let (A_n) be a sequence of sets such that $A_n \to A$ (in the sense that $A = A^* = A_*$). Show that $P(A \triangle A_n) \to 0$.

18. Let A_n converge to A in the sense that $P(A \triangle A^*) = P(A \triangle A_*) = 0$. Show that then $P(A \triangle A_n) \to 0$.

19. Prove that the symmetric difference $A \triangle B$ of sets A and B satisfies the following equality:

$$I(A \triangle B) = I(A) + I(B) \pmod 2.$$

Deduce from this equality that $P(A \triangle B) = P(A) + P(B) - 2P(A \cap B)$. (Compare with Problem 4.) Verify also the following properties of the symmetric difference:

$$(A \triangle B) \triangle C = A \triangle (B \triangle C), \quad (A \triangle B) \triangle (B \triangle C) = (A \triangle C),$$
$$A \triangle B = C \Leftrightarrow A = B \triangle C.$$

2 Algebras and σ-Algebras: Measurable Spaces

1. Algebras and σ-algebras are the components out of which probabilistic models are constructed. We shall present some examples and a number of results for these systems.

Let Ω be a sample space. Evidently each of the collections of sets

$$\mathscr{F}_* = \{\varnothing, \Omega\}, \quad \mathscr{F}^* = \{A : A \subseteq \Omega\}$$

is both an algebra and a σ-algebra. In fact, $\mathscr{F}_*$ is trivial, the "poorest" σ-algebra, whereas $\mathscr{F}^*$ is the "richest" σ-algebra, consisting of all subsets of Ω.

When Ω is a finite space, the σ-algebra $\mathscr{F}^*$ is fully surveyable, and commonly serves as the system of events in the elementary theory. However, when the space is uncountable the class $\mathscr{F}^*$ is much too large, since it is impossible to define "probability" on such a system of sets in any consistent way.

If $A \subseteq \Omega$, the system

$$\mathscr{F}_A = \{A, \overline{A}, \varnothing, \Omega\}$$

is another example of an algebra (and a σ-algebra), the *algebra* (or σ-algebra) *generated by* A.

This system of sets is a special case of the systems induced by *decompositions*. In fact, let

$$\mathcal{D} = \{D_1, D_2, \ldots\}$$

be a *countable* decomposition of Ω into nonempty sets:

$$\Omega = D_1 + D_2 + \cdots ; \quad D_i \cap D_j = \varnothing, \quad i \neq j.$$

Then the system $\mathcal{A} = \alpha(\mathcal{D})$, formed by the sets that are finite or countable unions of elements of the decomposition (including the empty set) is an algebra (and a σ-algebra).

The following lemma is particularly important since it establishes that *in principle* there is a smallest algebra, or σ-algebra, containing a given collection of sets.

Lemma 1. *Let $\mathcal{E}$ be a collection of subsets of Ω. Then there are the smallest algebra $\alpha(\mathcal{E})$ and the smallest σ-algebra $\sigma(\mathcal{E})$ containing all the sets that are in $\mathcal{E}$.*

PROOF. The class $\mathcal{F}^*$ of all subsets of Ω is a σ-algebra. Therefore there are at least one algebra and one σ-algebra containing $\mathcal{E}$. We now define $\alpha(\mathcal{E})$ (or $\sigma(\mathcal{E})$) to consist of all sets that belong to every algebra (or σ-algebra) containing $\mathcal{E}$. It is easy to verify that this system is an algebra (or σ-algebra) and indeed the smallest.

$\square$

Remark 1. The algebra $\alpha(E)$ (or $\sigma(E)$, respectively) is often referred to as the smallest *algebra* (or *σ-algebra*) *generated by $\mathcal{E}$*.

As was pointed out, the concept of a σ-algebra plays very important role in probability theory, being a part of the "fundamental definition" of the *probability space* (Subsection 3 of Sect. 1). In this connection it is desirable to provide a constructive way of obtaining the σ-algebra $\sigma(\mathcal{A})$ generated, say, by an algebra $\mathcal{A}$. (Lemma 1 establishes the existence of such σ-algebra, but gives no effective way of its construction.)

One conceivable and seemingly natural way of constructing $\sigma(\mathcal{A})$ from $\mathcal{A}$ is as follows. For a class $\mathcal{E}$ of subsets of Ω, denote by $\hat{\mathcal{E}}$ the class of subsets of Ω consisting of the sets contained in $\mathcal{E}$, their complements, and finite or countable unions of the sets in $\mathcal{E}$. Define $\mathcal{A}_0 = \mathcal{A}$, $\mathcal{A}_1 = \hat{\mathcal{A}}_0$, $\mathcal{A}_2 = \hat{\mathcal{A}}_1$, etc. Clearly, for each n the system $\mathcal{A}_n$ is contained in $\sigma(\mathcal{A})$, and one might expect that $\mathcal{A}_n = \sigma(\mathcal{A})$ for some n or, at least, $\bigcup \mathcal{A}_n = \sigma(\mathcal{A})$.

However this is, in general, not the case. Indeed, let us take $\Omega = (0, 1]$ and consider as the algebra $\mathcal{A}$ the system of subsets of Ω generated by the empty set $\varnothing$ and finite sums of intervals of the form $(a, b]$ with rational end-points a and b. It is not hard to see that in this case the class of sets $\bigcup_{n=1}^{\infty} \mathcal{A}_n$ is *strictly less* than $\sigma(\mathcal{A})$.

In what follows we will be mainly interested not in the problem of constructing the smallest σ-algebra $\sigma(\mathcal{A})$ out of, say, algebra $\mathcal{A}$, but in the question how to establish that some *given* class of sets is a σ-algebra.

In order to answer this question we need the important notion of a "monotonic class."

Definition 1. A collection $\mathcal{M}$ of subsets of Ω is a *monotonic class* if $A_n \in \mathcal{M}$, $n = 1, 2, \ldots$, together with $A_n \uparrow A$ or $A_n \downarrow A$, implies that $A \in \mathcal{M}$.

Let $\mathcal{E}$ be a system of sets. Denote by $\mu(\mathcal{E})$ the smallest monotonic class containing $\mathcal{E}$. (The proof of the existence of this class is like the proof of Lemma 1.)

Lemma 2. *A necessary and sufficient condition for an algebra $\mathcal{A}$ to be a σ-algebra is that it is a monotonic class.*

PROOF. A σ-algebra is evidently a monotonic class. Now let $\mathcal{A}$ be a monotonic class and $A_n \in \mathcal{A}$, $n = 1, 2, \ldots$. It is clear that $B_n = \bigcup_{i=1}^{n} A_i \in \mathcal{A}$ and $B_n \subseteq B_{n+1}$. Consequently, by the definition of a monotonic class, $B_n \uparrow \bigcup_{i=1}^{\infty} A_i \in \mathcal{A}$. Similarly we could show that $\bigcap_{i=1}^{\infty} A_i \in \mathcal{A}$.
$\square$

By using this lemma, we will prove the following result clarifying the relation between the notions of a σ-algebra and a monotonic class.

Theorem 1. *Let $\mathcal{A}$ be an algebra. Then*

$$\mu(\mathcal{A}) = \sigma(\mathcal{A}). \tag{1}$$

PROOF. By Lemma 2, $\mu(\mathcal{A}) \subseteq \sigma(\mathcal{A})$. Hence it is enough to show that $\mu(\mathcal{A})$ is a σ-algebra. But $\mathcal{M} = \mu(\mathcal{A})$ is a monotonic class, and therefore, by Lemma 2 again, it is enough to show that $\mu(\mathcal{A})$ is an *algebra*.

Let $A \in \mathcal{M}$; we show that $\overline{A} \in \mathcal{M}$. For this purpose, we shall apply a principle that will often be used in the future, the *principle of appropriate sets*, which we now illustrate.

Let

$$\tilde{\mathcal{M}} = \{B : B \in \mathcal{M}, \overline{B} \in \mathcal{M}\}$$

be the sets that have the property that concerns us. It is evident that $\mathcal{A} \subseteq \tilde{\mathcal{M}} \subseteq \mathcal{M}$. Let us show that $\tilde{\mathcal{M}}$ is a monotonic class.

Let $B_n \in \tilde{\mathcal{M}}$; then $B_n \in \mathcal{M}, \overline{B}_n \in \mathcal{M}$, and therefore

$$\lim \uparrow B_n \in \mathcal{M}, \quad \lim \uparrow \overline{B}_n \in \mathcal{M}, \quad \lim \downarrow B_n \in \mathcal{M}, \quad \lim \downarrow \overline{B}_n \in \mathcal{M}.$$

Consequently

$$\overline{\lim \uparrow B_n} = \lim \downarrow \overline{B}_n \in \mathcal{M}, \quad \overline{\lim \downarrow B_n} = \lim \uparrow \overline{B}_n \in \mathcal{M},$$

$$\overline{\lim \uparrow \overline{B}_n} = \lim \downarrow B_n \in \mathcal{M}, \quad \overline{\lim \downarrow \overline{B}_n} = \lim \uparrow B_n \in \mathcal{M},$$

and therefore $\tilde{\mathcal{M}}$ is a monotonic class. But $\tilde{\mathcal{M}} \subseteq \mathcal{M}$ and $\mathcal{M}$ is the smallest monotonic class. Therefore $\tilde{\mathcal{M}} = \mathcal{M}$, and if $A \in \mathcal{M} = \mu(\mathcal{A})$, then we also have $\overline{A} \in \mathcal{M}$, i.e., $\mathcal{M}$ is closed under the operation of taking complements.

Let us now show that $\mathcal{M}$ is closed under intersections. Let $A \in \mathcal{M}$ and

$$\mathcal{M}_A = \{B : B \in \mathcal{M}, A \cap B \in \mathcal{M}\}.$$

From the equations

$$\lim \downarrow (A \cap B_n) = A \cap \lim \downarrow B_n,$$
$$\lim \uparrow (A \cap B_n) = A \cap \lim \uparrow B_n$$

it follows that $\mathcal{M}_A$ is a monotonic class.

Moreover, it is easily verified that

$$(A \in \mathcal{M}_B) \Leftrightarrow (B \in \mathcal{M}_A). \tag{2}$$

Now let $A \in \mathscr{A}$; then since $\mathscr{A}$ is an algebra, for every $B \in \mathscr{A}$ the set $A \cap B \in \mathscr{A}$ and therefore

$$\mathscr{A} \subseteq \mathcal{M}_A \subseteq \mathcal{M}.$$

But $\mathcal{M}_A$ is a monotonic class and $\mathcal{M}$ is the smallest monotonic class. Therefore $\mathcal{M}_A = \mathcal{M}$ for all $A \in \mathscr{A}$. But then it follows from (2) that

$$(A \in \mathcal{M}_B) \Leftrightarrow (B \in \mathcal{M}_A = \mathcal{M})$$

whenever $A \in \mathscr{A}$ and $B \in \mathcal{M}$. Consequently if $A \in \mathscr{A}$ then

$$A \in \mathcal{M}_B$$

for every $B \in \mathcal{M}$. Since A is any set in $\mathscr{A}$, it follows that

$$\mathscr{A} \subseteq \mathcal{M}_B \subseteq \mathcal{M}.$$

Therefore for every $B \in \mathcal{M}$

$$\mathcal{M}_B = \mathcal{M},$$

i.e., if $B \in \mathcal{M}$ and $C \in \mathcal{M}$ then $C \cap B \in \mathcal{M}$.

Thus $\mathcal{M}$ is closed under complementation and intersection (and therefore under unions). Consequently $\mathcal{M} = \mu(\mathscr{A})$ is an algebra, and the theorem is established.

$\square$

If we examine the above proof we see that when dealing with systems of sets formed by the *principle of appropriate sets*, the important feature of these systems was that they were *closed* under certain set-theoretic operations.

From this point of view, it turns out that in the "monotonic classes" problems it is expedient to single out the classes of sets called "π-*systems*" and "λ-*systems*," which were actually used in the proof of Theorem 1. These concepts allow us to formulate a number of additional statements (Theorem 2) related to the same topic, which are often more usable than the direct verification that the system of sets at hand is a "monotonic class."

Definition 2 ("π-λ-systems"). Let Ω be a space. A system $\mathscr{P}$ of subsets of Ω is called a π-*system* if it is closed under *finite intersections*, i.e., for any $A_1, \ldots, A_n \in \mathscr{P}$ we have $\bigcap_{1 \le k \le n} A_k \in \mathscr{P}$, $n \ge 1$.

A system $\mathscr{L}$ of subsets of Ω is called a λ-*system*, if

(λ_a) $\Omega \in \mathscr{L}$,

(λ_b) $(A, B \in \mathscr{L}$ and $A \subseteq B)$ $\implies$ $(B \backslash A \in \mathscr{L})$,

(λ_c) $(A_n \in \mathscr{L}, n \geq 1$, and $A_n \uparrow A)$ $\implies$ $(A \in \mathscr{L})$.

A system $\mathscr{D}$ of subsets of Ω that is both a π-system and a λ-system is called a π-λ-*system* or Dynkin's *d-system*.

Remark 2. It is worth to notice that the group of conditions (λ_a), (λ_b), (λ_c) defining a λ-system is equivalent (Problem 3) to the group of conditions (λ_a), (λ'_b), (λ'_c), where

(λ'_b) if $A \in \mathscr{L}$, then $\overline{A} \in \mathscr{L}$,

(λ'_c) if $A_n \in \mathscr{L}, n \geq 1$, $A_n \cap A_m = \varnothing$ for $m \neq n$, then $\bigcup A_n \in \mathscr{L}$.

Note also that any algebra is obviously a π-system.

If $\mathscr{E}$ is a system of sets, then $\pi(\mathscr{E})$, $\lambda(\mathscr{E})$ and $d(\mathscr{E})$ denote, respectively, the smallest π-, λ- and d-systems containing $\mathscr{E}$.

The role of π-λ-systems is clarified in Theorem 2 below. In order to better explain the meaning of this theorem, note that any σ-*algebra is a* λ-*system*, but the converse is, in general, not true. For example, if $\Omega = \{1, 2, 3, 4\}$, then the system

$$\mathscr{L} = \{\varnothing, \Omega, (1, 2), (1, 3), (1, 4), (2, 3), (2, 4), (3, 4)\}$$

is a λ-system, but not a σ-algebra.

It turns out, however, that if we require *additionally* that a λ-system is also a π-system, then this π-λ-system is a σ-algebra.

Theorem 2 (On π-λ-Systems).

(a) *Any π-λ-system $\mathscr{E}$ is a σ-algebra.*

(b) *Let $\mathscr{E}$ be a π-system of sets. Then $\lambda(\mathscr{E}) = d(\mathscr{E}) = \sigma(\mathscr{E})$.*

(c) *Let $\mathscr{E}$ be a π-system of sets, $\mathscr{L}$ a λ-system and $\mathscr{E} \subseteq \mathscr{L}$. Then $\sigma(\mathscr{E}) \subseteq \mathscr{L}$.*

PROOF. (a) The system $\mathscr{E}$ contains Ω (because of (λ_a)) and is closed under taking complements and finite intersections (because of (λ'_b) and the assumption that $\mathscr{E}$ is a π-system). Therefore the system of sets $\mathscr{E}$ is an algebra (according to Definition 1 of Sect. 1). Now to prove that $\mathscr{E}$ is also a σ-algebra we have to show (by Definition 4 of Sect. 1) that if the sets $B_1, B_2, \ldots$ belong to $\mathscr{E}$, then their union $\bigcup_n B_n$ also belongs to $\mathscr{E}$.

Let $A_1 = B_1$ and $A_n = B_n \cap \overline{A}_1 \cap \cdots \cap \overline{A}_{n-1}$. Then (λ'_c) implies $\bigcup A_n \in \mathscr{E}$. But $\bigcup B_n = \bigcup A_n$, consequently $\bigcup B_n \in \mathscr{E}$ as well.

Thus any π-λ-system is a σ-algebra.

(b) Consider a λ-system $\lambda(\mathscr{E})$ and a σ-algebra $\sigma(\mathscr{E})$. As was pointed out, any σ-algebra is a λ-system. Then since $\sigma(\mathscr{E}) \supseteq \mathscr{E}$, we have $\sigma(\mathscr{E}) = \lambda(\sigma(\mathscr{E})) \supseteq \lambda(\mathscr{E})$. Hence $\lambda(\mathscr{E}) \subseteq \sigma(\mathscr{E})$.

Now if we show that the system $\lambda(\mathscr{E})$ is a π-system as well, then by (a) we obtain that $\lambda(\mathscr{E})$ is a σ-algebra containing $\mathscr{E}$. But since $\sigma(\mathscr{E})$ is the minimal σ-algebra containing $\mathscr{E}$, and we have proved that $\lambda(\mathscr{E}) \subseteq \sigma(\mathscr{E})$, we obtain $\lambda(\mathscr{E}) = \sigma(\mathscr{E})$.

Thus we proceed to prove that $\lambda(\mathcal{E})$ is a π-system. As in the proof of Theorem 1, we will use the *principle of appropriate sets*. Let

$$\mathcal{E}_1 = \{B \in \lambda(\mathcal{E}) : B \cap A \in \lambda(\mathcal{E}) \text{ for any } A \in \mathcal{E}\}.$$

If $B \in \mathcal{E}$, then $B \cap A \in \mathcal{E}$ (since $\mathcal{E}$ is a π-system). Hence $\mathcal{E} \subseteq \mathcal{E}_1$. But $\mathcal{E}_1$ is a λ-system (by the very definition of $\mathcal{E}_1$). Therefore $\lambda(\mathcal{E}) \subseteq \lambda(\mathcal{E}_1) = \mathcal{E}_1$. On the other hand, by definition of $\mathcal{E}_1$ we have $\mathcal{E}_1 \subseteq \lambda(\mathcal{E})$. Hence $\mathcal{E}_1 = \lambda(\mathcal{E})$.

Now let

$$\mathcal{E}_2 = \{B \in \lambda(\mathcal{E}) : B \cap A \in \lambda(\mathcal{E}) \text{ for any } A \in \lambda(\mathcal{E})\}.$$

The system $\mathcal{E}_2$, like $\mathcal{E}_1$, is a λ-system. Take a set $B \in \mathcal{E}$. Then by definition of $\mathcal{E}_1$ we find for any $A \in \mathcal{E}_1 = \lambda(\mathcal{E})$ that $B \cap A \in \lambda(\mathcal{E})$. Consequently we see from the definition of $\mathcal{E}_2$ that $\mathcal{E} \subseteq \mathcal{E}_2$ and $\lambda(\mathcal{E}) \subseteq \lambda(\mathcal{E}_2) = \mathcal{E}_2$. But $\lambda(\mathcal{E}) \supseteq \mathcal{E}_2$. Hence $\lambda(\mathcal{E}) = \mathcal{E}_2$ and therefore $A \cap B \in \lambda(\mathcal{E})$ for any A and B in $\lambda(\mathcal{E})$, i.e., $\lambda(\mathcal{E})$ is a π-system. Thus $\lambda(\mathcal{E})$ is a π-λ-system (hence $\lambda(\mathcal{E}) = d(\mathcal{E})$), and, as we pointed out above, this implies that $\lambda(\mathcal{E}) = \sigma(\mathcal{E})$.

This establishes (b).

(c) The facts that $\mathcal{E} \subseteq \mathcal{L}$ and $\mathcal{L}$ is a λ-system imply that $\lambda(\mathcal{E}) \subseteq \lambda(\mathcal{L}) = \mathcal{L}$. It follows from (b) that $\lambda(\mathcal{E}) = \sigma(\mathcal{E})$. Hence $\sigma(\mathcal{E}) \subseteq \mathcal{L}$.

$\square$

Remark 3. Theorem 2 could also be deduced directly from Theorem 1 (Problem 10).

Now we state two lemmas whose proofs illustrate very well the use of the *principle of appropriate sets* and of Theorem 2 on π-λ-systems.

Lemma 3. *Let* P *and* Q *be two probability measures on a measurable space* $(\Omega, \mathcal{F})$. *Let* $\mathcal{E}$ *be a* π-*system of sets in* $\mathcal{F}$ *and the measures* P *and* Q *coincide on the sets which belong to* $\mathcal{E}$. *Then these measures coincide on the* σ-*algebra* $\sigma(\mathcal{E})$. *In particular, if* $\mathcal{A}$ *is an algebra and the measures* P *and* Q *coincide on its sets, then they coincide on the sets of* $\sigma(\mathcal{A})$.

PROOF. We will use the principle of *appropriate sets* taking for these sets $\mathcal{L} = \{A \in \sigma(\mathcal{E}) : P(A) = Q(A)\}$. Clearly, $\Omega \in \mathcal{L}$. If $A \in \mathcal{L}$, then obviously $\overline{A} \in \mathcal{L}$, since $P(\overline{A}) = 1 - P(A) = 1 - Q(A) = Q(\overline{A})$. If $A_1, A_2, \ldots$ is a system of disjoint sets in $\mathcal{L}$, then, since P and Q are *countably additive*,

$$P\left(\bigcup_n A_n\right) = \sum_n P(A_n) = \sum_n Q(A_n) = Q\left(\bigcup_n A_n\right).$$

Therefore the properties (λ_a), (λ'_b), (λ'_c) are fulfilled, hence $\mathcal{L}$ is a λ-system.

By conditions of the lemma $\mathcal{E} \subseteq \mathcal{L}$ and $\mathcal{E}$ is a π-system. Then Theorem 2 (c) implies that $\sigma(\mathcal{E}) \subseteq \mathcal{L}$. Now P and Q coincide on $\sigma(\mathcal{E})$ by the definition of $\mathcal{L}$.

$\square$

Lemma 4. *Let $\mathscr{A}_1, \mathscr{A}_2, \ldots, \mathscr{A}_n$ be algebras of events independent with respect to the measure P. Then the σ-algebras $\sigma(\mathscr{A}_1), \sigma(\mathscr{A}_2), \ldots, \sigma(\mathscr{A}_n)$ are also independent with respect to this measure.*

PROOF. Note first of all that independence of sets and systems of sets (algebras, σ-algebras, etc.) in general probabilistic models is defined in exactly the same way as in elementary probability theory (see Definitions 2–5 in Sect. 3, Chap. 1).

Let $A_2, \ldots, A_n$ be sets in $\mathscr{A}_2, \ldots, \mathscr{A}_n$ respectively and let

$$\mathscr{L}_1 = \left\{ A \in \sigma(\mathscr{A}_1) : \mathsf{P}(A \cap A_2 \cap \cdots \cap A_n) = \mathsf{P}(A) \prod_{k=2}^{n} \mathsf{P}(A_k) \right\}. \qquad (3)$$

We will show that $\mathscr{L}_1$ is a λ-system.

Obviously, $\Omega \in \mathscr{L}_1$, i.e., the property (λ_a) is fulfilled. Let A and B belong to $\mathscr{L}_1$ and $A \subseteq B$. Then since

$$\mathsf{P}(A \cap A_1 \cap \cdots \cap A_n) = \mathsf{P}(A) \prod_{k=2}^{n} \mathsf{P}(A_k)$$

and

$$\mathsf{P}(B \cap A_1 \cap \cdots \cap A_n) = \mathsf{P}(B) \prod_{k=2}^{n} \mathsf{P}(A_k),$$

subtracting the former equality from the latter we find that

$$\mathsf{P}((B \backslash A) \cap A_1 \cap \cdots \cap A_n) = \mathsf{P}(B \backslash A) \prod_{k=2}^{n} \mathsf{P}(A_k).$$

Therefore the property (λ_b) is fulfilled.

Finally, if the sets B_k belong to $\sigma(\mathscr{A}_1)$, $k \geq 1$, and $B_k \uparrow B$, then

$$B_k \cap A_2 \cap \cdots \cap A_n \uparrow B \cap A_2 \cap \cdots \cap A_n.$$

Therefore by continuity from below of the probability P (see Theorem in Sect. 1) we obtain from $\mathsf{P}(B_k \cap A_2 \cap \cdots \cap A_n) = \mathsf{P}(B_k) \prod_{i=2}^{n} \mathsf{P}(A_i)$ taking the limit as $k \to \infty$ that

$$\mathsf{P}(B \cap A_2 \cap \cdots \cap A_n) = \mathsf{P}(B) \prod_{i=2}^{n} \mathsf{P}(A_i),$$

which establishes the property (λ_c).

Hence the system $\mathscr{L}_1$ is a λ-system and $\mathscr{L}_1 \supseteq \mathscr{A}_1$. Applying Theorem 2 (c) we obtain that $\mathscr{L}_1 \supseteq \sigma(\mathscr{A}_1)$.

Thus we have shown that the systems $\sigma(\mathscr{A}_1), \mathscr{A}_2, \ldots, \mathscr{A}_n$ are *independent*. By applying similar arguments to the systems $\mathscr{A}_2, \ldots, \mathscr{A}_n, \sigma(\mathscr{A}_1)$ we arrive at independence of the systems $\sigma(\mathscr{A}_2), \mathscr{A}_3, \ldots, \mathscr{A}_n, \sigma(\mathscr{A}_1)$, or equivalently, of $\mathscr{A}_3, \ldots, \mathscr{A}_n$, $\sigma(\mathscr{A}_1), \sigma(\mathscr{A}_2)$.

Proceeding in this way we obtain that the σ-algebras $\sigma(\mathscr{A}_1)$, $\sigma(\mathscr{A}_2), \ldots, \sigma(\mathscr{A}_n)$ are independent.

$\square$

Remark 4. Let us examine once more the requirements that we have to impose on a system of sets in order that it be a σ-algebra.

To this end we will say that a system of sets $\mathscr{E}$ is a π^*-system if it is closed under *countable* intersections:

$$A_1, A_2, \cdots \in \mathscr{E} \implies \bigcap_{n=1}^{\infty} A_n \in \mathscr{E}.$$

Then it follows from the definition of a σ-algebra that if an algebra $\mathscr{E}$ is at the same time a π^*-system then it is a σ-algebra as well.

The approach based on the notion of a "π-λ-system" is somewhat different. Our starting point here is the notion of a λ-*system* rather than an *algebra*. And Theorem 2 (a) implies that if this λ-system is at the same time a π-system, then it is a σ-algebra.

Let us clarify the difference between these approaches.

If we are to verify that a system of sets is a σ-algebra and establish first that it is an *algebra*, we start thereby our verification by taking into consideration only *finite* sums (or intersections) of sets. And we begin to treat operations on *countably many* sets (which is the key point here) when we proceed to verify that this system is also a π^*-system.

On the other hand, when we employ the "λ-π" approach we start to verify that the system of sets at hand is a σ-algebra with establishing that this is a λ-system whose properties (λ_c) or (λ'_c) involve "countable" operations. In return, when at the second stage we verify that it is a π-system, we deal only with *finite* intersections or sums of sets.

We conclude the exposition of the "monotonic classes" results by stating one of their "functional" versions. (For an example of its application see the proof of the lemma to Theorem 1 in Sect. 2, Chap. 8, Vol. 2.)

Theorem 3. *Let $\mathscr{E}$ be a π-system of sets in $\mathscr{F}$ and $\mathscr{H}$ a class of real-valued $\mathscr{F}$-measurable functions satisfying the following conditions:*

(h_1) *if $A \in \mathscr{E}$, then $I_A \in \mathscr{H}$;*

(h_2) *if $f \in \mathscr{H}$, $h \in \mathscr{H}$, then $f + h \in \mathscr{H}$ and $cf \in \mathscr{H}$ for any real number c;*

(h_3) *if $h_n \in \mathscr{H}$, $n \geq 1$, $0 \leq h_n \uparrow h$, then $h \in \mathscr{H}$.*

Then $\mathscr{H}$ contains all bounded functions measurable with respect to $\sigma(\mathscr{E})$.

PROOF. Let $\mathscr{L} = \{A \in \mathscr{F} : I_A \in \mathscr{H}\}$. Then (h_1) implies that $\mathscr{E} \subseteq \mathscr{L}$. But by (h_2) and (h_3) the system $\mathscr{L}$ is a λ-system (Problem 11). Therefore by Theorem 2 (c) we see that $\sigma(\mathscr{E}) \subseteq \mathscr{L}$. Hence if $A \in \sigma(\mathscr{E})$ then $I_A \in \mathscr{H}$. Consequently (h_2) implies that all the simple functions (i.e., the functions which are finite linear combinations of indicators I_{A_i}, where $A_i \in \sigma(\mathscr{E})$) also belong to $\mathscr{H}$. Finally, we obtain by (h_3) that any bounded $\sigma(\mathscr{E})$-measurable function also belongs to $\mathscr{H}$.

$\square$

Remark 5. Let $X_1, \ldots, X_n$ be random variables on $(\Omega, \mathscr{F})$, $\mathscr{F}^X = \sigma(X_1, \ldots, X_n)$ and $f = f(\omega)$ an $\mathscr{F}^X$-measurable function. Then there is a Borel function $F = F(x_1, \ldots, x_n)$ such that $f(\omega) = F(X_1(\omega), \ldots, X_n(\omega))$.

To prove this statement it suffices to use Theorem 3 taking for the *set of appropriate functions* $\mathscr{H}$ the set of nonnegative Borel functions $F = F(x_1, \ldots, x_n)$ and for $\mathscr{E}$ the class of sets

$$\mathscr{E} = \{\omega: X_1(\omega) \leq x_1, \ldots, X_n(\omega) \leq x_n; x_i \in R, i = 1, \ldots, n\}.$$

Applying Theorem 3 we obtain that any nonnegative $\mathscr{F}^X$-measurable function $f = f(\omega)$ can be represented as $f(\omega) = F(X_1(\omega), \ldots, X_n(\omega))$. The general case of not necessarily nonnegative functions f reduces to the one just considered by using the representation $f = f^+ - f^-$.

We next consider some measurable spaces $(\Omega, \mathscr{F})$ which are extremely important for probability theory.

2. The measurable space $(R, \mathscr{B}(R))$. Let $R = (-\infty, \infty)$ be the real line and

$$(a, b] = \{x \in R: a < x \leq b\}$$

for all a and b, $-\infty \leq a < b < \infty$. The interval $(a, \infty]$ is taken to be (a, ∞). (This convention is required in order to the complement of an interval $(-\infty, b]$ be an interval of the same form, i.e., open on the left and closed on the right.)

Let $\mathscr{A}$ be the system of subsets of R which are *finite* sums of disjoint intervals of the form $(a, b]$:

$$A \in \mathscr{A} \quad \text{if} \quad A = \sum_{i=1}^{n} (a_i, b_i], \quad n < \infty.$$

It is easily verified that this system of sets, in which we also include the empty set $\varnothing$, is an algebra. However, it is not a σ-algebra, since if $A_n = (0, 1 - 1/n] \in \mathscr{A}$, we have $\bigcup_n A_n = (0, 1) \notin \mathscr{A}$.

Let $\mathscr{B}(R)$ be the smallest σ-algebra $\sigma(\mathscr{A})$ containing $\mathscr{A}$. This σ-algebra, which plays an important role in analysis, is called the *Borel σ-algebra* of subsets of the real line, and its sets are called *Borel sets*.

If $\mathscr{I}$ is the system of intervals of the form $I = (a, b]$, and $\sigma(\mathscr{I})$ is the smallest σ-algebra containing $\mathscr{I}$, it is easily verified that $\sigma(\mathscr{I})$ is the Borel σ-algebra. In other words, we can obtain the Borel σ-algebra from $\mathscr{I}$ without going through the algebra $\mathscr{A}$, since $\sigma(\mathscr{I}) = \sigma(\alpha(\mathscr{I}))$.

We observe that

$$(a, b) = \bigcup_{n=1}^{\infty} \left(a, b - \frac{1}{n}\right], \qquad a < b,$$

$$[a, b] = \bigcap_{n=1}^{\infty} \left(a - \frac{1}{n}, b\right], \qquad a < b,$$

$$\{a\} = \bigcap_{n=1}^{\infty} \left(a - \frac{1}{n}, a\right].$$

Thus the Borel σ-algebra contains not only intervals $(a, b]$ but also the singletons $\{a\}$ and all sets of the six forms

$$(a, b), \quad [a, b], \quad [a, b), \quad (-\infty, b), \quad (-\infty, b], \quad (a, \infty). \tag{4}$$

Let us also notice that the construction of $\mathscr{B}(R)$ could have been based on any of the six kinds of intervals instead of on $(a, b]$, since all the minimal σ-algebras generated by systems of intervals of any of the forms (4) are the same as $\mathscr{B}(R)$.

Sometimes it is useful to deal with the σ-algebra $\mathscr{B}(\overline{R})$ of subsets of the extended real line $\overline{R} = [-\infty, \infty]$. This is the smallest σ-algebra generated by intervals of the form

$$(a, b] = \{x \in \overline{R} \colon a < x \le b\}, \quad -\infty \le a < b \le \infty,$$

where $(-\infty, b]$ is to stand for the set $\{x \in \overline{R} \colon -\infty \le x \le b\}$.

Remark 6. The measurable space $(R, \mathscr{B}(R))$ is often denoted by $(R, \mathscr{B})$ or $(R^1, \mathscr{B}_1)$.

Remark 7. Let us introduce the metric

$$\rho_1(x, y) = \frac{|x - y|}{1 + |x - y|}$$

on the real line R (this is equivalent to the usual metric $|x - y|$) and let $\mathscr{B}_0(R)$ be the smallest σ-algebra generated by finite sums of disjoint open sets $S_\rho(x^0) = \{x \in R \colon \rho_1(x, x^0) < \rho\}$, $\rho > 0$, $x^0 \in R$. Then $\mathscr{B}_0(R) = \mathscr{B}(R)$ (see Problem 7).

3. The measurable space $(R^n, \mathscr{B}(R^n))$. Let $R^n = R \times \cdots \times R$ be the direct, or Cartesian, product of n copies of the real line, i.e., the set of *ordered n-tuples* $x = (x_1, \ldots, x_n)$, where $-\infty < x_k < \infty$, $k = 1, \ldots, n$. The set

$$I = I_1 \times \cdots \times I_n,$$

where $I_k = (a_k, b_k]$, i.e., the set $\{x \in R^n \colon x_k \in I_k, \ k = 1, \ldots, n\}$, is called a *rectangle*, and I_k is a *side* of the rectangle. Let $\mathscr{I}$ be the collection of all sets which are finite sums of disjoint rectangles I. The smallest σ-algebra $\sigma(\mathscr{I})$ generated by the system $\mathscr{I}$ is the *Borel σ-algebra* of subsets of R^n and is denoted by $\mathscr{B}(R^n)$. Let us show that we can arrive at this Borel σ-algebra by starting in a different way.

Instead of the rectangles $I = I_1 \times \cdots \times I_n$ let us consider the rectangles $B = B_1 \times \cdots \times B_n$ with *Borel* sides (B_k is the Borel subset of the real line that appears in the kth place in the direct product $R \times \cdots \times R$). The smallest σ-algebra containing all rectangles with Borel sides is denoted by

$$\mathscr{B}(R) \otimes \cdots \otimes \mathscr{B}(R)$$

and called the *direct product* of the σ-algebras $\mathscr{B}(R)$. Let us show that in fact

$$\mathscr{B}(R^n) = \mathscr{B}(R) \otimes \cdots \otimes \mathscr{B}(R).$$

In other words, the smallest σ-algebras generated by the rectangles $I = I_1 \times \cdots \times I_n$ and by the (broader) class of rectangles $B = B_1 \times \cdots \times B_n$ with Borel sides are actually the same.

The proof depends on the following proposition.

Lemma 5. *Let $\mathscr{E}$ be a class of subsets of Ω, let $B \subseteq \Omega$, and define*

$$\mathscr{E} \cap B = \{A \cap B : A \in \mathscr{E}\}. \tag{5}$$

Then

$$\sigma(\mathscr{E} \cap B) = \sigma(\mathscr{E}) \cap B. \tag{6}$$

PROOF. Since $\mathscr{E} \subseteq \sigma(\mathscr{E})$, we have

$$\mathscr{E} \cap B \subseteq \sigma(\mathscr{E}) \cap B. \tag{7}$$

But $\sigma(\mathscr{E}) \cap B$ is a σ-algebra; hence it follows from (7) that

$$\sigma(\mathscr{E} \cap B) \subseteq \sigma(\mathscr{E}) \cap B.$$

To prove the conclusion in the opposite direction, we again use the principle of appropriate sets.

Define

$$\mathscr{C}_B = \{A \in \sigma(\mathscr{E}): A \cap B \in \sigma(\mathscr{E} \cap B)\}.$$

Since $\sigma(\mathscr{E})$ and $\sigma(\mathscr{E} \cap B)$ are σ-algebras, $\mathscr{C}_B$ is also a σ-algebra, and evidently

$$\mathscr{E} \subseteq \mathscr{C}_B \subseteq \sigma(\mathscr{E}),$$

whence $\sigma(\mathscr{E}) \subseteq \sigma(\mathscr{C}_B) = \mathscr{C}_B \subseteq \sigma(\mathscr{E})$ and therefore $\sigma(\mathscr{E}) = \mathscr{C}_B$. Therefore

$$A \cap B \in \sigma(\mathscr{E} \cap B)$$

for every $A \subseteq \sigma(\mathscr{E})$, and consequently $\sigma(\mathscr{E}) \cap B \subseteq \sigma(\mathscr{E} \cap B)$.

This completes the proof of the lemma.

$\square$

The proof that $\mathscr{B}(R^n)$ and $\mathscr{B} \otimes \cdots \otimes \mathscr{B}$ are the same. This is obvious for $n = 1$. We now show that it is true for $n = 2$.

Since $\mathscr{B}(R^2) \subseteq \mathscr{B} \otimes \mathscr{B}$, it is enough to show that the Borel rectangle $B_1 \times B_2$ belongs to $\mathscr{B}(R^2)$.

Let $R^2 = R_1 \times R_2$, where R_1 and R_2 are the "first" and "second" real lines, $\tilde{\mathscr{B}}_1 = \mathscr{B}_1 \times R_2, \tilde{\mathscr{B}}_2 = R_1 \times \mathscr{B}_2$, where $\mathscr{B}_1 \times R_2$ (or $R_1 \times \mathscr{B}_2$) is the collection of sets of the form $B_1 \times R_2$ (or $R_1 \times B_2$), with $B_1 \in \mathscr{B}_1$ (or $B_2 \in \mathscr{B}_2$). Also let $\mathscr{I}_1$ and $\mathscr{I}_2$ be the sets of intervals in R_1 and R_2, and $\tilde{\mathscr{I}}_1 = \mathscr{I}_1 \times R_2, \tilde{\mathscr{I}}_2 = R_1 \times \mathscr{I}_2$. Then, by (6), with $\tilde{B}_1 = B_1 \times R_2, \tilde{B}_2 = R_1 \times B_2$,

$$B_1 \times B_2 = \tilde{B}_1 \cap \tilde{B}_2 \in \tilde{\mathscr{B}}_1 \cap \tilde{\mathscr{B}}_2 = \sigma(\tilde{\mathscr{I}}_1) \cap \tilde{B}_2$$
$$= \sigma(\tilde{\mathscr{I}}_1 \cap \tilde{B}_2) \subseteq \sigma(\tilde{\mathscr{I}}_1 \cap \tilde{\mathscr{I}}_2)$$
$$= \sigma(\mathscr{I}_1 \times \mathscr{I}_2),$$

as was to be proved.

The case of any n, $n > 2$ can be discussed in the same way. $\quad\square$

Remark 8. Let $\mathscr{B}_0(R^n)$ be the smallest σ-algebra generated by the open "balls"

$$S_\rho(x^0) = \{x \in R^n : \rho_n(x, x^0) < \rho\}, \qquad x^0 \in R^n, \ \rho > 0,$$

in the metric

$$\rho_n(x, x^0) = \sum_{k=1}^{n} 2^{-k} \rho_1(x_k, x_k^0),$$

where $x = (x_1, \ldots, x_n)$, $x^0 = (x_1^0, \ldots, x_n^0)$.

Then $\mathscr{B}_0(R^n) = \mathscr{B}(R^n)$ (Problem 7).

4. The measurable space $(R^\infty, \mathscr{B}(R^\infty))$ plays a significant role in probability theory, since it is used as the basis for constructing probabilistic models of experiments with *infinitely many* steps.

The space R^∞ is the space of sequences of real numbers,

$$x = (x_1, x_2, \ldots), \qquad -\infty < x_k < \infty, \ k = 1, 2, \ldots$$

Let I_k and B_k denote, respectively, the intervals $(a_k, b_k]$ and the Borel subsets of the kth real line (with coordinate x_k). We consider the *cylinder sets*

$$\mathscr{I}(I_1 \times \cdots \times I_n) = \{x : x = (x_1, x_2, \ldots), \ x_1 \in I_1, \ldots, x_n \in I_n\}, \tag{8}$$
$$\mathscr{I}(B_1 \times \cdots \times B_n) = \{x : x = (x_1, x_2, \ldots), \ x_1 \in B_1, \ldots, x_n \in B_n\}, \tag{9}$$
$$\mathscr{I}(B^n) = \{x : (x_1, \ldots, x_n) \in B^n\}, \tag{10}$$

where B^n is a Borel set in $\mathscr{B}(R^n)$. Each cylinder $\mathscr{I}(B_1 \times \cdots \times B_n)$, or $\mathscr{I}(B^n)$, can also be thought of as a cylinder with base in R^{n+1}, R^{n+2}, ..., since

$$\mathscr{I}(B_1 \times \cdots \times B_n) = \mathscr{I}(B_1 \times \cdots \times B_n \times R),$$
$$\mathscr{I}(B^n) = \mathscr{I}(B^{n+1}),$$

where $B^{n+1} = B^n \times R$.

The sets that are finite sums of disjoint cylinders $\mathscr{I}(I_1 \times \cdots \times I_n)$ form an algebra. And the finite sums of disjoint cylinders $\mathscr{I}(B_1 \times \cdots \times B_n)$ also form an algebra. The system of cylinders $\mathscr{I}(B^n)$ itself is an algebra. Let $\mathscr{B}(R^\infty)$, $\mathscr{B}_1(R^\infty)$ and $\mathscr{B}_2(R^\infty)$ be the smallest σ-algebras containing, respectively, these three algebras. (The σ-algebra $\mathscr{B}_1(R^\infty)$ is often denoted by $\mathscr{B}(R) \otimes \mathscr{B}(R) \otimes \cdots$.) It is clear that $\mathscr{B}(R^\infty) \subseteq \mathscr{B}_1(R^\infty) \subseteq \mathscr{B}_2(R^\infty)$. As a matter of fact, all three σ-algebras are the same.

To prove this, we put

$$\mathscr{C}_n = \{A \subseteq R^n : \{x : (x_1, \ldots, x_n) \in A\} \in \mathscr{B}(R^\infty)\}$$

for $n = 1, 2, \ldots$. Let $B^n \in \mathscr{B}(R^n)$. Then

$$B^n \in \mathscr{C}_n.$$

But $\mathscr{C}_n$ is a σ-algebra, and therefore

$$\mathscr{B}(R^n) \subseteq \sigma(\mathscr{C}_n) = \mathscr{C}_n;$$

consequently

$$\mathscr{B}_2(R^\infty) \subseteq \mathscr{B}(R^\infty).$$

Thus $\mathscr{B}(R^\infty) = \mathscr{B}_1(R^\infty) = \mathscr{B}_2(R^\infty)$.

From now on we shall describe sets in $\mathscr{B}(R^\infty)$ as Borel sets (in R^∞).

Remark 9. Let $\mathscr{B}_0(R^\infty)$ be the smallest σ-algebra generated by the system of sets formed by finite sums of disjoint open "balls"

$$S_\rho(x^0) = \{x \in R^\infty : \rho_\infty(x, x^0) < \rho\}, \qquad x^0 \in R^\infty, \; \rho > 0,$$

in the metric

$$\rho_\infty(x, x^0) = \sum_{k=1}^\infty 2^{-k} \rho_1(x_k, x_k^0),$$

where $x = (x_1, x_2, \ldots)$ and $x^0 = (x_1^0, x_2^0, \ldots)$. Then $\mathscr{B}(R^\infty) = \mathscr{B}_0(R^\infty)$ (Problem 7).

Here are some examples of Borel sets in R^∞:

(a) $\{x \subset R^\infty : \sup x_n > a\}, \{x \subset R^\infty : \inf x_n < a\}$;

(b) $\{x \in R^\infty : \limsup x_n \le a\}, \{x \in R^\infty : \liminf x_n > a\}$, where, as usual,

$$\limsup_n x_n = \inf_n \sup_{m \ge n} x_m, \qquad \liminf x_n = \sup_n \inf_{m \ge n} x_m;$$

(c) $\{x \in R^\infty : x_n \to\}$, the set of $x \in R^\infty$ for which $\lim x_n$ exists and is finite;

(d) $\{x \in R^\infty : \lim x_n > a\}$;

(e) $\{x \in R^\infty : \sum_{n=1}^\infty |x_n| > a\}$;

(f) $\{x \in R^\infty : \sum_{k=1}^n x_k = 0$ for at least one $n \ge 1\}$.

To be convinced, for example, that sets in (a) belong to the system $\mathscr{B}(R^\infty)$, it is enough to observe that

$$\{x : \sup x_n > a\} = \bigcup_n \{x : x_n > a\} \in \mathscr{B}(R^\infty),$$

$$\{x : \inf x_n < a\} = \bigcup_n \{x : x_n < a\} \in \mathscr{B}(R^\infty).$$

5. The measurable space $(R^T, \mathscr{B}(R^T))$, where T is an *arbitrary* set. The space R^T is the collection of real functions $x = (x_t)$ defined for $t \in T$.* In general we shall be interested in the case when T is an uncountable subset of the real line. For simplicity and definiteness we shall suppose for the present that $T = [0, \infty)$.

We shall consider three types of cylinder sets:

$$\mathscr{I}_{t_1,\ldots,t_n}(I_1 \times \cdots \times I_n) = \{x \colon x_{t_1} \in I_1, \ldots, x_{t_n} \in I_n\}, \tag{11}$$

$$\mathscr{I}_{t_1,\ldots,t_n}(B_1 \times \cdots \times B_n) = \{x \colon x_{t_1} \in B_1, \ldots, x_{t_n} \in B_n\}, \tag{12}$$

$$\mathscr{I}_{t_1,\ldots,\,t_n}(B^n) = \{x \colon (x_{t_1}, \ldots, x_{t_n}) \in B^n\}, \tag{13}$$

where I_k is a set of the form $(a_k, b_k]$, B_k is a Borel set on the line, and B^n is a Borel set in R^n.

The set $\mathscr{I}_{t_1,\ldots,t_n}(I_1 \times \cdots \times I_n)$ is just the set of functions that, at times $t_1, \ldots, t_n$, "get through the windows" $I_1, \ldots, I_n$ and at other times have arbitrary values (Fig. 24).

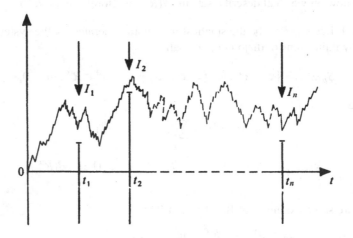

Fig. 24

Let $\mathscr{B}(R^T)$, $\mathscr{B}_1(R^T)$ and $\mathscr{B}_2(R^T)$ be the smallest σ-algebras corresponding respectively to the cylinder sets (11), (12) and (13). It is clear that

$$\mathscr{B}(R^T) \subseteq \mathscr{B}_1(R^T) \subseteq \mathscr{B}_2(R^T). \tag{14}$$

As a matter of fact, all three of these σ-algebras are the same. Moreover, we can give a complete description of the structure of their sets.

Theorem 4. *Let T be any uncountable set. Then $\mathscr{B}(R^T) = \mathscr{B}_1(R^T) = \mathscr{B}_2(R^T)$, and every set $A \in \mathscr{B}(R^T)$ has the following structure: there are a countable set of points $t_1, t_2, \ldots$ of T and a Borel set B in $\mathscr{B}(R^\infty)$ such that*

$$A = \{x \colon (x_{t_1}, x_{t_2}, \ldots) \in B\}. \tag{15}$$

---

* We shall also use the notation $x = (x_t)_{t \in R^T}$ and $x = (x_t)$, $t \in R^T$, for elements of R^T.

PROOF. Let $\mathscr{E}$ denote the collection of sets of the form (15) (for various aggregates $(t_1, t_2, \ldots)$ and Borel sets B in $\mathscr{B}(R^\infty)$). If $A_1, A_2, \ldots \in \mathscr{E}$ and the corresponding aggregates are $T^{(1)} = \left(t_1^{(1)}, t_2^{(1)}, \ldots\right)$, $T^{(2)} = \left(t_1^{(2)}, t_2^{(2)}, \ldots\right), \ldots$, then the set $T^{(\infty)} = \bigcup_k T^{(k)}$ can be taken as a basis, so that every A_i has a representation

$$A_i = \{x \colon (x_{\tau_1}, x_{\tau_2}, \ldots) \in B_i\},$$

where B_i's are sets in one and the same σ-algebra $\mathscr{B}(R^\infty)$, and $\tau_i \in T^{(\infty)}$.

Hence it follows that the system $\mathscr{E}$ is a σ-algebra. Clearly this σ-algebra contains all cylinder sets of the form (13) and, since $\mathscr{B}_2(R^T)$ is the *smallest* σ-algebra containing these sets, and since we have (14), we obtain

$$\mathscr{B}(R^T) \subseteq \mathscr{B}_1(R^T) \subset \mathscr{B}_2(R^T) \subseteq \mathscr{E}. \tag{16}$$

Let us consider a set A from $\mathscr{E}$, represented in the form (15). For a given aggregate $(t_1, t_2, \ldots)$, the same reasoning as for the space $(R^\infty, \mathscr{B}(R^\infty))$ shows that A is an element of the σ-algebra generated by the cylinder sets (11). But this σ-algebra evidently belongs to the σ-algebra $\mathscr{B}(R^T)$; together with (16), this established both conclusions of the theorem.

$\square$

Thus every Borel set A in the σ-algebra $\mathscr{B}(R^T)$ is determined by restrictions imposed on the functions $x = (x_t)$, $t \in T$, on an at most *countable* set of points $t_1, t_2, \ldots$. Hence it follows, in particular, that the sets

$$A_1 = \{x \colon \sup x_t < C \text{ for all } t \in [0, 1]\},$$
$$A_2 = \{x \colon x_t = 0 \text{ for at least one } t \in [0, 1]\},$$
$$A_3 = \{x \colon x_t \text{ is continuous at a given point } t_0 \in [0, 1]\},$$

which depend on the behavior of the function on an uncountable set of points, cannot be Borel sets. And indeed *none of these three sets belongs to $\mathscr{B}(R^{[0,1]})$*.

Let us establish this for A_1. If $A_1 \subset \mathscr{B}(R^{[0,1]})$, then by our theorem there are points $(t_1^0, t_2^0, \ldots)$ and a set $B^0 \in \mathscr{B}(R^\infty)$ such that

$$\left\{x \colon \sup_t x_t < C, \, t \in [0, 1]\right\} = \{x \colon (x_{t_1^0}, x_{t_2^0}, \ldots) \in B^0\}.$$

It is clear that the function $y_t \equiv C - 1$ belongs to A_1, and consequently $(y_{t_1^0}, y_{t_2^0}, \ldots) \in B^0$. Now form the function

$$z_t = \begin{cases} C - 1, & t \in (t_1^0, t_2^0, \ldots), \\ C + 1, & t \notin (t_1^0, t_2^0, \ldots). \end{cases}$$

It is clear that

$$(y_{t_1^0}, y_{t_2^0}, \ldots) = (z_{t_1^0}, z_{t_2^0}, \ldots),$$

and consequently the function $z = (z_t)$ belongs to the set $\{x\colon (x_{t_1^0}, x_{t_2^0}, \cdots) \in B^0\}$. But at the same time it is clear that it does not belong to the set $\{x\colon \sup x_t < C\}$. This contradiction shows that $A_1 \notin \mathcal{B}(R^{[0,1]})$.

Since the sets A_1, A_2 and A_3 are *nonmeasurable* with respect to the σ-algebra $\mathcal{B}(R^{[0,1]})$ in the space of all functions $x = (x_t)$, $t \in [0,1]$, it is natural to consider a smaller class of functions for which these sets are measurable. It is intuitively clear that this will be the case if we take the initial space to be, for example, the space of *continuous* functions.

6. The measurable space $(C, \mathcal{B}(C))$. Let $T = [0,1]$ and let C be the space of *continuous* functions $x = (x_t)$, $0 \leq t \leq 1$. This is a metric space with the metric $\rho(x,y) = \sup_{t\in T} |x_t - y_t|$. We introduce two σ-algebras in C: $\mathcal{B}(C)$ is the σ-algebra generated by the *cylinder* sets, and $\mathcal{B}_0(C)$ is generated by the open sets (open with respect to the metric $\rho(x,y)$). Let us show that in fact these σ-algebras are the same: $\mathcal{B}(C) = \mathcal{B}_0(C)$.

Let $B = \{x\colon x_{t_0} < b\}$ be a cylinder set. It is easy to see that this set is open. Hence it follows that $\{x\colon x_{t_1} < b_1, \ldots, x_{t_n} < b_n\} \in \mathcal{B}_0(C)$, and therefore $\mathcal{B}(C) \subseteq \mathcal{B}_0(C)$.

Conversely, consider a set $B_\rho = \{y\colon y \in S_\rho(x^0)\}$ where x^0 is an element of C and $S_\rho(x^0) = \{x \in C\colon \sup_{t\in T} |x_t - x_t^0| < \rho\}$ is an open ball with center at x^0. Since the functions in C are continuous,

$$B_\rho = \{y \in C\colon y \in S_\rho(x^0)\} = \left\{y \in C\colon \max_t |y_t - x_t^0| < \rho\right\}$$
$$= \bigcap_{t_k} \{y \in C\colon |y_{t_k} - x_{t_k}^0| < \rho\} \in \mathcal{B}(C), \qquad (17)$$

where t_k are the rational points of $[0,1]$. Therefore $\mathcal{B}_0(C) \subseteq \mathcal{B}(C)$.

The space $(C, \mathcal{B}_0(C), \rho)$ is a Polish space, i.e., complete and separable (see [9, 43]).

7. The measurable space $(D, \mathcal{B}(D))$, where D is the space of functions $x = (x_t)$, $t \in [0,1]$, that are *continuous on the right* ($x_t = x_{t+}$ for all $t < 1$) and have limits from the left (at every $t > 0$).

Just as for C, we can introduce a metric $d(x,y)$ on D such that the σ-algebra $\mathcal{B}_0(D)$ generated by the *open* sets will coincide with the σ-algebra $\mathcal{B}(D)$ generated by the *cylinder* sets. The space $(D, \mathcal{B}(D), d)$ is separable (see [9, 43]). The metric $d = d(x,y)$, which was introduced by Skorohod, is defined as follows:

$$d(x,y) = \inf\{\varepsilon > 0\colon \exists \lambda \in \Lambda\colon \sup_t |x_t - y_{\lambda(t)}| + \sup_t |t - \lambda(t)| \leq \varepsilon\}, \qquad (18)$$

where Λ is the set of strictly increasing functions $\lambda = \lambda(t)$ that are continuous on $[0,1]$ and have $\lambda(0) = 0$, $\lambda(1) = 1$.

8. The measurable space $(\prod_{t\in T} \Omega_t, \boxtimes_{t\in T} \mathcal{F}_t)$. Along with the space $(R^T, \mathcal{B}(R^T))$, which is the direct product of T copies of the real line with the system of Borel sets, probability theory also uses the measurable space $(\prod_{t\in T} \Omega_t, \boxtimes_{t\in T} \mathcal{F}_t)$, which is defined in the following way.

Let T be an *arbitrary* set of indices and let $(\Omega_t, \mathscr{F}_t)$ be measurable spaces, $t \in T$. Let $\Omega = \prod_{t \in T} \Omega_t$ be the set of functions $\omega = (\omega_t)$, $t \in T$, such that $\omega_t \in \Omega_t$ for each $t \in T$.

The collection of finite unions of disjoint cylinder sets

$$\mathscr{I}_{t_1,\ldots,t_n}(B_1 \times \cdots \times B_n) = \{\omega : \omega_{t_1} \in B_1, \ldots, \omega_{t_n} \in B_n\},$$

where $B_{t_i} \in \mathscr{F}_{t_i}$, is easily shown to be an algebra. The smallest σ-algebra containing all these cylinder sets is denoted by $\boxtimes_{t \in T} \mathscr{F}_t$, and the measurable space $(\prod \Omega_i, \boxtimes \mathscr{F}_t)$ is called the *direct product* of the measurable spaces $(\Omega_t, \mathscr{F}_t)$, $t \in T$.

9. Problems

1. Let $\mathscr{B}_1$ and $\mathscr{B}_2$ be σ-algebras of subsets of Ω. Are the following systems of sets σ-algebras?

$$\mathscr{B}_1 \cap \mathscr{B}_2 \equiv \{A : A \in \mathscr{B}_1 \text{ and } A \in \mathscr{B}_2\},$$
$$\mathscr{B}_1 \cup \mathscr{B}_2 \equiv \{A : A \in \mathscr{B}_1 \text{ or } A \in \mathscr{B}_2\}.$$

2. Let $\mathscr{D} = \{D_1, D_2, \ldots\}$ be a countable decomposition of Ω and $\mathscr{B} = \sigma(\mathscr{D})$. What is the cardinality of $\mathscr{B}$?

3. Show that
$$\mathscr{B}(R^n) \otimes \mathscr{B}(R) = \mathscr{B}(R^{n+1}).$$

4. Prove that the sets (b)–(f) (see Subsection 4) belong to $\mathscr{B}(R^\infty)$.

5. Prove that the sets A_2 and A_3 (see Subsection 5) do not belong to $\mathscr{B}(R^{[0,1]})$.

6. Prove that the function (18) actually defines a metric.

7. Prove that $\mathscr{B}_0(R^n) = \mathscr{B}(R^n)$, $n \geq 1$, and $\mathscr{B}_0(R^\infty) = \mathscr{B}(R^\infty)$.

8. Let $C = C[0, \infty)$ be the space of continuous functions $x = (x_t)$ defined for $t \geq 0$ endowed with the metric

$$\rho(x, y) = \sum_{n=1}^{\infty} 2^{-n} \min \left[\sup_{0 \leq t \leq n} |x_t - y_t|, 1 \right], \quad x, y \in C.$$

Show that (like $C = C[0, 1]$) this is a Polish space, i.e., a *complete separable metric space*, and that the σ-algebra $\mathscr{B}_0(C)$ generated by the open sets coincides with the σ-algebra $\mathscr{B}(C)$ generated by the cylinder sets.

9. Show that the groups of conditions (λ_a), (λ_b), (λ_c) and (λ_a), (λ_b'), (λ_c') (see Definition 2 and Remark 2) are equivalent.

10. Deduce Theorem 2 from Theorem 1.

11. Prove that the system $\mathscr{L}$ in Theorem 3 is a λ-system.

12. A σ-algebra is said to be *countably generated* or *separable* if it is generated by a countable class of sets.

Show that the σ-algebra $\mathscr{B}$ of Borel subsets of $\Omega = (0, 1]$ is countably generated.

Show by an example that it is possible to have two σ-algebras $\mathscr{F}_1$ and $\mathscr{F}_2$ such that $\mathscr{F}_1 \subseteq \mathscr{F}_2$ and $\mathscr{F}_2$ is countably generated, but $\mathscr{F}_1$ is not.

13. Show that a σ-algebra $\mathcal{G}$ is countably generated if and only if $\mathcal{G} = \sigma(X)$ for some random variable X (for the definition of $\sigma(X)$ see Subsection 4 of Sect. 4).

14. Give an example of a separable σ-algebra having a non-separable sub-σ-algebra.

15. Show that $X_1, X_2, \ldots$ is an independent system of random variables (Sects. 4, 5) if $\sigma(X_n)$ and $\sigma(X_1, \ldots, X_{n-1})$ are independent for each $n \geq 1$.

16. Show by an example that the union of two σ-algebras is not a σ-algebra.

17. Let $\mathcal{A}_1$ and $\mathcal{A}_2$ be two independent systems of sets each of which is a π-system. Show that then $\sigma(\mathcal{A}_1)$ and $\sigma(\mathcal{A}_2)$ are also independent. Give an example of two independent systems $\mathcal{A}_1$ and $\mathcal{A}_2$, which are not π-systems, such that $\sigma(\mathcal{A}_1)$ and $\sigma(\mathcal{A}_2)$ are dependent.

18. Let $\mathcal{L}$ be a λ-system. Then $(A, B \in \mathcal{L}, A \cap B = \varnothing) \implies (A \cup B \in \mathcal{L})$.

19. Let $\mathcal{F}_1$ and $\mathcal{F}_2$ be σ-algebras of subsets of Ω. Set

$$d(\mathcal{F}_1, \mathcal{F}_2) = 4 \sup_{\substack{A_1 \in \mathcal{F}_1 \\ A_2 \in \mathcal{F}_2}} |P(A_1 A_2) - P(A_1) P(A_2)|.$$

Show that this quantity describing the degree of dependence between $\mathcal{F}_1$ and $\mathcal{F}_2$ has the following properties:

(a) $0 \leq d(\mathcal{F}_1, \mathcal{F}_2) \leq 1$;

(b) $d(\mathcal{F}_1, \mathcal{F}_2) = 0$ if and only if $\mathcal{F}_1$ and $\mathcal{F}_2$ are independent.

(c) $d(\mathcal{F}_1, \mathcal{F}_2) = 1$ if and only if $\mathcal{F}_1 \cap \mathcal{F}_2$ contains a set of probability $1/2$.

20. Using the method of the proof of Lemma 1 prove the existence and uniqueness of the classes $\lambda(\mathcal{E})$ and $\pi(\mathcal{E})$ containing a system of sets $\mathcal{E}$.

21. Let $\mathcal{A}$ be an algebra of sets with the property that any sequence $(A_n)_{n\geq1}$ of disjoint sets $A_n \in \mathcal{A}$ satisfies $\bigcup_{n=1}^{\infty} A_n \in \mathcal{A}$. Prove that $\mathcal{A}$ is then a σ-algebra.

22. Let $(\mathcal{F}_n)_{n\geq1}$ be an increasing sequence of σ-algebras, $\mathcal{F}_n \subseteq \mathcal{F}_{n+1}, n \geq 1$. Show that $\bigcup_{n=1}^{\infty} \mathcal{F}_n$ is, in general, only an algebra.

23. Let $\mathcal{F}$ be an algebra (or a σ-algebra) and C a set which is not in $\mathcal{F}$. Consider the smallest algebra (respectively, σ-algebra) generated by the sets in $\mathcal{F} \cup \{C\}$. Show that all the elements of this algebra (respectively, σ-algebra) have the form $(A \cap C) \cup (B \cap \overline{C})$, where $A, B \in \mathcal{F}$.

24. Let $\overline{R} = R \cup \{-\infty\} \cup \{\infty\}$ be the *extended* real line. The Borel σ-algebra $\mathcal{B}(\overline{R})$ can be defined (compare with the definition in Subsection 2) as the σ-algebra generated by sets $[-\infty, x], x \in R$, where $[-\infty, x] = \{-\infty\} \cup (-\infty, x]$. Show that this σ-algebra is the same as any of the σ-algebras generated by the sets

(a) $[-\infty, x), x \in R$, or

(b) $(x, \infty], x \in R$, or

(c) all finite intervals with addition of $\{-\infty\}$ and $\{\infty\}$.

3 Methods of Introducing Probability Measures on Measurable Spaces

1. The measurable space $(R, \mathscr{B}(R))$. Let $\mathsf{P} = \mathsf{P}(A)$ be a probability measure defined on the Borel subsets A of the real line. Take $A = (-\infty, x]$ and put

$$F(x) = \mathsf{P}(-\infty, x], \quad x \in R. \tag{1}$$

This function has the following properties:

(i) $F(x)$ is *nondecreasing*;
(ii) $F(-\infty) = 0$, $F(+\infty) = 1$, *where*

$$F(-\infty) = \lim_{x \downarrow -\infty} F(x), \qquad F(+\infty) = \lim_{x \uparrow \infty} F(x);$$

(iii) $F(x)$ *is continuous on the right and has a limit on the left at each* $x \in R$.

The first property is evident, and the other two follow from the continuity properties of probability measures (Sect. 1).

Definition 1. Every function $F = F(x)$ satisfying conditions (i)–(iii) is called a *distribution function* (on the real line R).

Thus to every probability measure P on $(R, \mathscr{B}(R))$ there corresponds (by (1)) a distribution function. It turns out that the converse is also true.

Theorem 1. *Let* $F = F(x)$ *be a distribution function on the real line* R. *There exists a unique probability measure* P *on* $(R, \mathscr{B}(R))$ *such that*

$$\mathsf{P}(a, b] = F(b) - F(a) \tag{2}$$

for all a, b, $-\infty \leq a < b < \infty$.

PROOF. Let $\mathscr{A}$ be the algebra of the subsets A of R that are finite sums of disjoint intervals of the form $(a, b]$:

$$A = \sum_{k=1}^{n} (a_k, b_k].$$

On these sets we define a set function P_0 by putting

$$\mathsf{P}_0(A) = \sum_{k=1}^{n} [F(b_k) - F(a_k)], \quad A \in \mathscr{A}. \tag{3}$$

This formula defines, evidently uniquely, a *finitely additive* set function on $\mathscr{A}$. Therefore if we show that this function is also *countably additive* on this algebra, the existence and uniqueness of the required measure P on $\mathscr{B}(R)$ will follow immediately from a general result of measure theory (which we quote without proof to be found in [39, 64]).

Carathéodory's Theorem. Let Ω be a space, $\mathscr{A}$ an algebra of its subsets, and $\mathscr{B} = \sigma(\mathscr{A})$ the smallest σ-algebra containing $\mathscr{A}$. Let μ_0 be a σ-finite (Sect. 1) and σ-additive (Sect. 1) measure on $(\Omega, \mathscr{A})$. Then there is a unique measure μ on $(\Omega, \sigma(\mathscr{A}))$ which is an extension of μ_0, i.e., satisfies

$$\mu(A) = \mu_0(A), \qquad A \in \mathscr{A}.$$

We are now to show that P_0 is countably additive (hence is a probability measure) on $\mathscr{A}$. By the theorem from Sect. 1 it is enough to show that P_0 is continuous at $\varnothing$, i.e., to verify that

$$\mathsf{P}_0(A_n) \downarrow 0, \qquad A_n \downarrow \varnothing, \quad A_n \in \mathscr{A}.$$

Let $A_1, A_2, \ldots$ be a sequence of sets from $\mathscr{A}$ with the property $A_n \downarrow \varnothing$. Let us suppose first that the sets A_n belong to a closed interval $[-N, N]$, $N < \infty$. Since each A_n is the sum of finitely many intervals of the form $(a, b]$ and since

$$\mathsf{P}_0(a', b] = F(b) - F(a') \to F(b) - F(a) = \mathsf{P}_0(a, b]$$

as $a' \downarrow a$, because $F(x)$ is continuous on the right, we can find, for every A_n, a set $B_n \in \mathscr{A}$ such that its *closure* $[B_n] \subseteq A_n$ and

$$\mathsf{P}_0(A_n) - \mathsf{P}_0(B_n) \le \varepsilon \cdot 2^{-n},$$

where ε is a preassigned positive number.

By hypothesis, $\bigcap A_n = \varnothing$ and therefore $\bigcap [B_n] = \varnothing$. But the sets $[B_n]$ are closed, and therefore there is a finite $n_0 = n_0(\varepsilon)$ such that

$$\bigcap_{n=1}^{n_0} [B_n] = \varnothing. \tag{4}$$

(In fact, $[-N, N]$ is compact, and the collection of sets $\{[-N, N] \backslash [B_n]\}_{n \ge 1}$ is an open covering of this compact set. By the *Heine–Borel lemma* (see, e.g., [52]) there is a *finite* subcovering:

$$\bigcup_{n=1}^{n_0} ([-N, N] \backslash [B_n]) = [-N, N]$$

and therefore $\bigcap_{n=1}^{n_0} [B_n] = \varnothing$).

Using (4) and the inclusions $A_{n_0} \subseteq A_{n_0-1} \subseteq \cdots \subseteq A_1$, we obtain

$$\mathsf{P}_0(A_{n_0}) = \mathsf{P}_0\left(A_{n_0} \setminus \bigcap_{k=1}^{n_0} B_k\right) + \mathsf{P}_0\left(\bigcap_{k=1}^{n_0} B_k\right)$$

$$= \mathsf{P}_0\left(A_{n_0} \setminus \bigcap_{k=1}^{n_0} B_k\right) \le \mathsf{P}_0\left(\bigcup_{k=1}^{n_0}(A_k \setminus B_k)\right)$$

$$\le \sum_{k=1}^{n_0} \mathsf{P}_0(A_k \setminus B_k) \le \sum_{k=1}^{n_0} \varepsilon \cdot 2^{-k} \le \varepsilon.$$

Therefore $\mathsf{P}_0(A_n) \downarrow 0$, $n \to \infty$.

We now abandon the assumption that all $A_n \subseteq [-N, N]$ for some N. Take an $\varepsilon > 0$ and choose N so that $\mathsf{P}_0[-N, N] > 1 - \varepsilon/2$. Then, since

$$A_n = A_n \cap [-N, N] + A_n \cap \overline{[-N, N]},$$

we have

$$\mathsf{P}_0(A_n) = \mathsf{P}_0(A_n \cap [-N, N]) + \mathsf{P}_0(A_n \overline{[-N, N]})$$
$$\le \mathsf{P}_0(A_n \cap [-N, N]) + \varepsilon/2$$

and, applying the preceding reasoning (replacing A_n by $A_n \cap [-N, N]$), we find that $\mathsf{P}_0(A_n \cap [-N, N]) \le \varepsilon/2$ for sufficiently large n. Hence once again $\mathsf{P}_0(A_n) \downarrow 0$, $n \to \infty$. This completes the proof of the theorem.

$\square$

Thus there is a one-to-one correspondence between probability measures P on $(R, \mathscr{B}(R))$ and distribution functions F on the real line R. The measure P constructed from the function F is usually called the *Lebesgue–Stieltjes probability measure* corresponding to the distribution function F.

The case when

$$F(x) = \begin{cases} 0, & x < 0, \\ x, & 0 \le x \le 1, \\ 1, & x > 1, \end{cases}$$

is particularly important. In this case the corresponding probability measure (denoted by λ) is *Lebesgue measure* on $[0, 1]$. Clearly $\lambda(a, b] = b - a$. In other words, the Lebesgue measure of $(a, b]$ (as well as of any of the intervals (a, b), $[a, b]$ or $[a, b))$ is simply its length $b - a$.

Let

$$\mathscr{B}([0, 1]) = \{A \cap [0, 1] : A \in \mathscr{B}(R)\}$$

be the collection of *Borel* subsets of $[0, 1]$. It is often necessary to consider, besides these sets, the *Lebesgue* measurable subsets of $[0, 1]$. We say that a set $\Lambda \subseteq [0, 1]$ belongs to $\overline{\mathscr{B}}([0, 1])$ if there are Borel sets A and B such that $A \subseteq \Lambda \subseteq B$ and $\lambda(B \setminus A) = 0$. It is easily verified that $\overline{\mathscr{B}}([0, 1])$ is a σ-algebra. It is known as the system of *Lebesgue measurable subsets of* $[0, 1]$. Clearly $\mathscr{B}([0, 1]) \subseteq \overline{\mathscr{B}}([0, 1])$.

The measure λ, defined so far only for sets in $\mathscr{B}([0,1])$, extends in a natural way to the system $\overline{\mathscr{B}}([0,1])$ of Lebesgue measurable sets. Specifically, if $\Lambda \in \overline{\mathscr{B}}([0,1])$ and $A \subseteq \Lambda \subseteq B$, where A and $B \in \overline{\mathscr{B}}([0,1])$ and $\lambda(B \backslash A) = 0$, we define $\overline{\lambda}(\Lambda) = \lambda(A)$. The set function $\overline{\lambda} = \overline{\lambda}(\Lambda)$, $\Lambda \in \overline{\mathscr{B}}([0,1])$, is easily seen to be a probability measure on $([0,1], \overline{\mathscr{B}}([0,1]))$. It is usually called *Lebesgue measure* (on the system of Lebesgue-measurable sets).

Remark 1. This process of completing (or extending) a measure can also be applied, and is useful, in other situations. For example, let $(\Omega, \mathscr{F}, \mathsf{P})$ be a probability space. Let $\bar{\mathscr{F}}^{\mathsf{P}}$ be the collection of all the subsets A of Ω for which there are sets B_1 and B_2 of $\mathscr{F}$ such that $B_1 \subseteq A \subseteq B_2$ and $\mathsf{P}(B_2 \backslash B_1) = 0$. The probability measure can be defined for sets $A \in \bar{\mathscr{F}}^{\mathsf{P}}$ in a natural way (by $\mathsf{P}(A) = \mathsf{P}(B_1)$). The resulting probability space $(\Omega, \bar{\mathscr{F}}^{\mathsf{P}}, \mathsf{P})$ is the *completion* of $(\Omega, \mathscr{F}, \mathsf{P})$ with respect to P.

A probability measure such that $\bar{\mathscr{F}}^{\mathsf{P}} = \mathscr{F}$ is called *complete*, and the corresponding space $(\Omega, \mathscr{F}, \mathsf{P})$ is a *complete probability space*.

Remark 2. Here we briefly outline the idea of the proof of Carathéodory's theorem assuming that $\mu_0(\Omega) = 1$.

Let A be a set in Ω and $A_1, A_2, \ldots$ sets in $\mathscr{A}$ which cover A in the sense that $A \subseteq \bigcup_{n=1}^{\infty} A_n$. Define the *outer measure* $\mu^*(A)$ of the set A as

$$\mu^*(A) = \inf \sum_{n=1}^{\infty} \mu_0(A_n),$$

where the infimum is taken over all coverings $(A_1, A_2, \ldots)$ of A. We define also the *interior measure* $\mu_*(A)$ by

$$\mu_*(A) = 1 - \mu^*(\bar{A}).$$

Denote by $\hat{\mathscr{A}}$ the collection of all sets $A \subseteq \Omega$ such that $\mu^*(A) = \mu_*(A)$. It is not hard to show that the system $\hat{\mathscr{A}}$ is a σ-algebra (Problem 12), and therefore $\mathscr{A} \subseteq \sigma(\mathscr{A}) \subseteq \hat{\mathscr{A}}$. We assign to the sets A in $\hat{\mathscr{A}}$ the "measure" $\mu(A)$ equal to $\mu_*(A)$ ($= \mu^*(A)$). This set function $\mu(A)$, $A \in \hat{\mathscr{A}}$, is a *measure* indeed (Problem 13), i.e., a countably-additive set function (being a probability measure, since $\mu(\Omega) = \mu_0(\Omega) = 1$).

The correspondence between probability measures P and distribution functions F established by the equation $\mathsf{P}(a, b] = F(b) - F(a)$ makes it possible to construct various probability measures by specifying the corresponding distribution functions.

Discrete measures are measures P for which the corresponding distribution functions $F = F(x)$ change their values at the points $x_1, x_2, \ldots$ ($\Delta F(x_i) > 0$, where $\Delta F(x) = F(x) - F(x-)$). In this case the measure is concentrated at the points $x_1, x_2, \ldots$ (Fig. 25):

$$\mathsf{P}(\{x_k\}) = \Delta F(x_k) > 0, \qquad \sum_k \mathsf{P}(\{x_k\}) = 1.$$

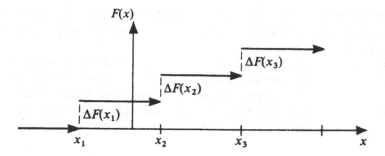

Fig. 25

The set of numbers $(p_1, p_2, \ldots)$, where $p_k = P(\{x_k\})$, is called a *discrete probability distribution* and the corresponding distribution function $F = F(x)$ is called *discrete*.

We present a table of the commonest types of discrete probability distribution, with their names.

Table 2.2

Distribution	Probabilities p_k	Parameters
Discrete uniform	$1/N$, $k = 1, 2, \ldots, N$	$N = 1, 2, \ldots$
Bernoulli	$p_1 = p$, $p_0 = q$	$0 \le p \le 1$, $q = 1 - p$
Binomial	$C_n^k p^k q^{n-k}$, $k = 0, 1, \ldots, n$	$0 \le p \le 1$, $q = 1 - p$, $n = 1, 2, \ldots$
Poisson	$e^{-\lambda} \lambda^k / k!$, $k = 0, 1, \ldots$	$\lambda > 0$
Geometric	$q^{k-1} p$, $k = 1, 2, \ldots$	$0 \le p \le 1$, $q = 1 - p$
Negative binomial (Pascal's distribution)	$C_{k-1}^{r-1} p^r q^{k-r}$, $k = r, r+1, \ldots$	$0 \le p \le 1$, $q = 1 - p$, $r = 1, 2, \ldots$

Remark. The discrete measure presented in Fig. 25 is such that its distribution function is piecewise constant. However one should bear in mind that discrete measures, in general, may have a very complicated structure. For example, such a measure may be concentrated on a countable set of points, which is everywhere dense in R (e.g., on the set of rational numbers).

Absolutely continuous measures. These are measures for which the corresponding distribution functions are such that

$$F(x) = \int_{-\infty}^{x} f(t) dt, \tag{5}$$

where $f = f(t)$ are nonnegative functions integrated to one and the integral is at first taken in the Riemann sense, but later (see Sect. 6) in that of Lebesgue.

(Note that if we have two functions $f(t)$ and $\tilde{f}(t)$ differing only on a set of zero Lebesgue measure then the corresponding distribution functions $F(x) = \int_{-\infty}^{x} f(t) \, dt$ and $\tilde{F}(x) = \int_{-\infty}^{x} \tilde{f}(t) \, dt$ will be the same; it is useful to keep in mind this remark when solving Problem 8 of the next Sect. 4.)

The function $f = f(x)$, $x \in R$, is the *density* of the distribution function $F = F(x)$ (or the density of the probability distribution, or simply the density) and $F = F(x)$ is called *absolutely continuous*.

It is clear that every nonnegative $f = f(x)$ that is Riemann integrable and such that $\int_{-\infty}^{\infty} f(x)dx = 1$ defines a distribution function of some probability measure on $(R, \mathscr{B}(R))$ by (5). Table 2.3 presents some important examples of various kinds of densities $f = f(x)$ with their names and parameters (a density $f(x)$ is taken to be zero for values of x not listed in the table).

Table 2.3

Distribution	Density	Parameters
Uniform on $[a, b]$	$1/(b-a)$, $a \le x \le b$	$a, b \in R$; $a < b$
Normal or Gaussian	$(2\pi\sigma^2)^{-1/2}e^{-(x-m)^2/(2\sigma^2)}$, $x \in R$	$m \in R$, $\sigma > 0$
Gamma	$\frac{x^{\alpha-1}e^{-x/\beta}}{\Gamma(\alpha)\beta^\alpha}$, $x \ge 0$	$\alpha > 0$, $\beta > 0$
Beta	$\frac{x^{r-1}(1-x)^{s-1}}{B(r,s)}$, $0 \le x \le 1$	$r > 0$, $s > 0$
Exponential (gamma with $\alpha = 1$, $\beta = 1/\lambda$)	$\lambda e^{-\lambda x}$, $x \ge 0$	$\lambda > 0$
Bilateral exponential	$\frac{1}{2}\lambda e^{-\lambda|x-\alpha|}$, $x \in R$	$\lambda > 0$, $\alpha \in R$
Chi-square, χ^2 with n degrees of freedom (gamma with $\alpha = n/2$, $\beta = 2$)	$2^{-n/2}x^{n/2-1}e^{-x/2}/\Gamma(n/2)$, $x \ge 0$	$n = 1, 2, \ldots$
Student, t with n degrees of freedom	$\frac{\Gamma(\frac{1}{2}(n+1))}{(n\pi)^{1/2}\Gamma(n/2)}\left(1 + \frac{x^2}{n}\right)^{-(n+1)/2}$, $x \in R$	$n = 1, 2, \ldots$
F	$\frac{(m/n)^{m/2}}{B(m/2,n/2)}\frac{x^{m/2-1}}{(1+mx/n)^{(m+n)/2}}$	$m, n = 1, 2, \ldots$
Cauchy	$\frac{\theta}{\pi(x^2+\theta^2)}$, $x \in R$	$\theta > 0$

Singular measures. These are measures whose distribution functions are continuous but have all their points of increase on sets of *zero Lebesgue measure*. We do not discuss this case in detail; we merely give an example of such a function.

We consider the interval $[0, 1]$ and construct $F(x)$ by the following procedure originated by Cantor.

We divide $[0, 1]$ into thirds and put (Fig. 26)

$$F_1(x) = \begin{cases} \frac{1}{2}, & x \in (\frac{1}{3}, \frac{2}{3}), \\ 0, & x = 0, \\ 1, & x = 1, \end{cases}$$

defining it in the intermediate intervals by linear interpolation.

Then we divide each of the intervals $[0, \frac{1}{3}]$ and $[\frac{2}{3}, 1]$ into three parts and define the function (Fig. 27)

$$F_2(x) = \begin{cases} \frac{1}{2}, & x \in (\frac{1}{3}, \frac{2}{3}), \\ \frac{1}{4}, & x \in (\frac{1}{9}, \frac{2}{9}), \\ \frac{3}{4}, & x \in (\frac{7}{9}, \frac{8}{9}), \\ 0, & x = 0, \\ 1, & x = 1 \end{cases}$$

with its values at other points determined by linear interpolation.

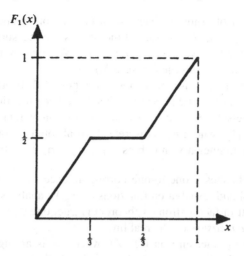

Fig. 26

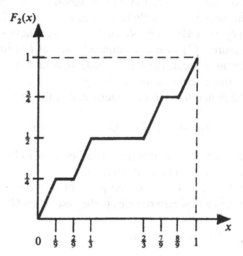

Fig. 27

Continuing this process, we construct a sequence of functions $F_n(x)$, $n = 1, 2, \ldots$, which converges to a nondecreasing continuous function $F(x)$ (the Cantor function), whose points of increase (x is a point of increase of $F(x)$ if $F(x + \varepsilon) - F(x - \varepsilon) > 0$ for every $\varepsilon > 0$) form a set of Lebesgue measure zero. In fact, it is clear from the construction of $F(x)$ that the total length of the intervals $(\frac{1}{3}, \frac{2}{3})$, $(\frac{1}{9}, \frac{2}{9})$, $(\frac{7}{9}, \frac{8}{9})$, $\ldots$ on which the function is constant is

$$\frac{1}{3} + \frac{2}{9} + \frac{4}{27} + \cdots = \frac{1}{3} \sum_{n=0}^{\infty} \left(\frac{2}{3}\right)^n = 1. \tag{6}$$

Let $\mathcal{N}$ be the set of points of increase of the Cantor function $F(x)$. It follows from (6) that $\lambda(\mathcal{N}) = 0$. At the same time, if μ is the measure corresponding to the Cantor function $F(x)$, we have $\mu(\mathcal{N}) = 1$. (We then say that the measure is *singular* with respect to Lebesgue measure λ.)

Without any further discussion of possible types of distribution functions, we merely observe that in fact the *three types* that have been mentioned cover all possibilities. More precisely, every distribution function can be represented in the form $\alpha_1 F_1 + \alpha_2 F_2 + \alpha_3 F_3$, where F_1 is discrete, F_2 is absolutely continuous, and F_3 is singular, and α_i are nonnegative numbers, $\alpha_1 + \alpha_2 + \alpha_3 = 1$ (Problem 18).

2. Theorem 1 establishes a one-to-one correspondence between probability measures on $(R, \mathcal{B}(R))$ and distribution functions on R. An analysis of the proof of the theorem shows that in fact a stronger theorem is true, one that in particular lets us introduce *Lebesgue measure* on the real line.

Let μ be a σ-finite measure on $(\Omega, \mathcal{A})$, where $\mathcal{A}$ is an algebra of subsets of Ω. It turns out that the conclusion of Carathéodory's theorem on the extension of a measure from an algebra $\mathcal{A}$ to a minimal σ-algebra $\sigma(\mathcal{A})$ remains valid with a σ-finite measure; this makes it possible to generalize Theorem 1.

A *Lebesgue–Stieltjes measure* on $(R, \mathcal{B}(R))$ is a (countably additive) measure μ such that the measure $\mu(I)$ of every bounded interval I is finite. A *generalized distribution function* on the real line R is a nondecreasing function $G = G(x)$, with values in $(-\infty, \infty)$, that is continuous on the right.

Theorem 1 can be generalized to the statement that the formula

$$\mu(a, b] = G(b) - G(a), \qquad a < b,$$

again establishes a one-to-one correspondence between Lebesgue–Stieltjes measures μ and generalized distribution functions G.

In fact, if $G(+\infty) - G(-\infty) < \infty$, the proof of Theorem 1 can be taken over without any change, since this case reduces to the case when $G(+\infty) - G(-\infty) = 1$ and $G(-\infty) = 0$.

Now let $G(+\infty) - G(-\infty) = \infty$. Put

$$G_n(x) = \begin{cases} G(x), & |x| \leq n, \\ G(n) & x > n, \\ G(-n), & x < -n. \end{cases}$$

On the algebra $\mathcal{A}$ let us define a finitely additive measure μ_0 such that $\mu_0(a, b] = G(b) - G(a)$, and let μ_n be the countably additive measures previously constructed (by Theorem 1) from $G_n(x)$.

Evidently $\mu_n \uparrow \mu_0$ on $\mathcal{A}$. Now let $A_1, A_2, \ldots$ be disjoint sets in $\mathcal{A}$ and $A \equiv \sum A_n \in \mathcal{A}$. Then (Problem 6 of Sect. 1)

$$\mu_0(A) \geq \sum_{n=1}^{\infty} \mu_0(A_n).$$

If $\sum_{n=1}^{\infty} \mu_0(A_n) = \infty$ then $\mu_0(A) = \sum_{n=1}^{\infty} \mu_0(A_n)$. Let us suppose that $\sum \mu_0(A_n) < \infty$. Then

$$\mu_0(A) = \lim_n \mu_n(A) = \lim_n \sum_{k=1}^{\infty} \mu_n(A_k).$$

By hypothesis, $\sum \mu_0(A_n) < \infty$. Therefore

$$0 \le \mu_0(A) - \sum_{k=1}^{\infty} \mu_0(A_k) = \lim_n \left[\sum_{k=1}^{\infty} (\mu_n(A_k) - \mu_0(A_k))\right] \le 0,$$

since $\mu_n \le \mu_0$.

Thus a σ-finite finitely additive measure μ_0 is countably additive on $\mathscr{A}$, and therefore (by Carathéodory's theorem) it can be extended to a countably additive measure μ on $\sigma(\mathscr{A})$.

The case $G(x) = x$ is particularly important. The measure λ corresponding to this generalized distribution function is *Lebesgue measure* on $(R, \mathscr{B}(R))$. As for the interval $[0, 1]$ of the real line, we can define the system $\overline{\mathscr{B}}(R)$ by writing $\Lambda \in \overline{\mathscr{B}}(R)$ if there are Borel sets A and B such that $A \subseteq \Lambda \subseteq B$, $\lambda(B \backslash A) = 0$. Then Lebesgue measure $\overline{\lambda}$ on $\mathscr{B}(R)$ is defined by $\overline{\lambda}(\Lambda) = \lambda(A)$ if $A \subseteq \Lambda \subseteq B$, $\Lambda \in \overline{\mathscr{B}}(R)$ and $\lambda(B \backslash A) = 0$.

3. The measurable space $(R^n, \mathscr{B}(R^n))$. Let us suppose, as for the real line, that P is a probability measure on $(R^n, \mathscr{B}(R^n))$.

Let us write

$$F_n(x_1, \ldots, x_n) = \mathsf{P}((-\infty, x_1] \times \cdots \times (-\infty, x_n]),$$

or, in a more compact form,

$$F_n(x) = \mathsf{P}(-\infty, x],$$

where $x = (x_1, \ldots, x_n)$, $(-\infty, x] = (-\infty, x_1] \times \cdots \times (-\infty, x_n]$.

Let us introduce the difference operator $\Delta_{a_i, b_i} : R^n \to R$ defined by the formula

$$\begin{aligned} \Delta_{a_i, b_i} F_n(x_1, \ldots, x_n) = &\ F_n(x_1, \ldots, x_{i-1}, b_i, x_{i+1} \ldots, x_n) \\ &-F_n(x_1, \ldots, x_{i-1}, a_i, x_{i+1} \ldots, x_n), \end{aligned}$$

where $a_i \le b_i$. A simple calculation shows that

$$\Delta_{a_1 b_1} \cdots \Delta_{a_n b_n} F_n(x_1, \ldots, x_n) = \mathsf{P}(a, b], \tag{7}$$

where $(a, b] = (a_1, b_1] \times \cdots \times (a_n, b_n]$. Hence it is clear, in particular, that (in contrast to the one-dimensional case) $\mathsf{P}(a, b]$ is in general *not equal* to $F_n(b) - F_n(a)$.

Since $\mathsf{P}(a, b] \ge 0$, it follows from (7) that

$$\Delta_{a_1 b_1} \cdots \Delta_{a_n b_n} F_n(x_1, \ldots, x_n) \ge 0 \tag{8}$$

for arbitrary $a = (a_1, \ldots, a_n)$, $b = (b_1, \ldots, b_n)$, $a_i \leq b_i$.

It also follows from the continuity of P that $F_n(x_1, \ldots, x_n)$ is continuous on the right with respect to the variables collectively, i.e., if $x^{(k)} \downarrow x$, $x^{(k)} = (x_1^{(k)}, \ldots, x_n^{(k)})$, then

$$F_n(x^{(k)}) \downarrow F_n(x), \qquad k \to \infty. \tag{9}$$

It is also clear that

$$F_n(+\infty, \ldots, +\infty) = 1 \tag{10}$$

and

$$\lim_{x \downarrow y} F_n(x_1, \ldots, x_n) = 0, \tag{11}$$

if at least one coordinate of y is $-\infty$.

Definition 2. An *n-dimensional distribution function* (on R^n) is a function $F_n = F_n(x_1, \ldots, x_n)$ with properties (8)–(11).

The following result can be established by the same reasoning as in Theorem 1.

Theorem 2. *Let $F_n = F_n(x_1, \ldots, x_n)$ be a distribution function on R^n. Then there is a unique probability measure P on $(R^n, \mathscr{B}(R^n))$ such that*

$$\mathsf{P}(a, b] = \Delta_{a_1 b_1} \cdots \Delta_{a_n b_n} F_n(x_1, \ldots, x_n). \tag{12}$$

Here are some examples of n-dimensional distribution functions.

Let $F^1, \ldots, F^n$ be one-dimensional distribution functions (on R) and

$$F_n(x_1, \ldots, x_n) = F^1(x_1) \cdots F^n(x_n).$$

It is clear that this function is continuous on the right and satisfies (10) and (11). It is also easy to verify that

$$\Delta_{a_1 b_1} \cdots \Delta_{a_n b_n} F_n(x_1, \ldots, x_n) = \prod [F^k(b_k) - F^k(a_k)] \geq 0.$$

Consequently $F_n(x_1, \ldots, x_n)$ is a distribution function.

The case when

$$F^k(x_k) = \begin{cases} 0, & x_k < 0, \\ x_k, & 0 \leq x_k \leq 1, \\ 1, & x_k > 1, \end{cases}$$

is particularly important. In this case

$$F_n(x_1, \ldots, x_n) = x_1 \cdots x_n.$$

The probability measure corresponding to this n-dimensional distribution function is *n-dimensional Lebesgue measure on* $[0, 1]^n$.

Many n-dimensional distribution functions appear in the form

$$F_n(x_1, \ldots, x_n) = \int_{-\infty}^{x_1} \cdots \int_{-\infty}^{x_n} f_n(t_1, \ldots, t_n) \, dt_1 \cdots dt_n,$$

where $f_n(t_1, \ldots, t_n)$ is a nonnegative function such that

$$\int_{-\infty}^{\infty} \cdots \int_{-\infty}^{\infty} f_n(t_1, \ldots, t_n) \, dt_1 \cdots dt_n = 1,$$

and the integrals are Riemann (more generally, Lebesgue) integrals. The function $f = f_n(t_1, \ldots, t_n)$ is called the *density of the n-dimensional distribution function*, the density of the n-dimensional probability distribution, or simply an *n-dimensional density*.

When $n = 1$, the function

$$f(x) = \frac{1}{\sigma\sqrt{2\pi}} e^{(x-m)^2/(2\sigma^2)}, \qquad x \in R,$$

with $\sigma > 0$ is the density of the (nondegenerate) *Gaussian* or *normal distribution*. There are natural analogs of this density when $n > 1$.

Let $\mathbb{R} = \|r_{ij}\|$ be a positive semi-definite symmetric $n \times n$ matrix:

$$\sum_{i,j=1}^{n} r_{ij}\lambda_i\lambda_j \geq 0, \qquad \lambda_i \in R, \quad i = 1, \ldots, n, \quad r_{ij} = r_{ji}.$$

When $\mathbb{R}$ is a positive definite matrix, $|\mathbb{R}| = \det \mathbb{R} > 0$ and consequently there is an inverse matrix $A = \|a_{ij}\|$. Then the function

$$f_n(x_1, \ldots, x_n) = \frac{|A|^{1/2}}{(2\pi)^{n/2}} \exp\{-\tfrac{1}{2}\sum a_{ij}(x_i - m_i)(x_j - m_j)\}, \qquad (13)$$

where $m_i \in R$, $i = 1, \ldots, n$, has the property that its (Riemann) integral over the whole space equals 1 (this will be proved in Sect. 13) and therefore, since it is also positive, it is a density.

This function is the *density of the n-dimensional* (nondegenerate) *Gaussian* or *normal distribution* (with mean vector $m = (m_1, \ldots, m_n)$ and covariance matrix $\mathbb{R} = A^{-1}$).

When $n = 2$ the density $f_2(x_1, x_2)$ can be put in the form

$$f_2(x_1, x_2) = \frac{1}{2\pi\sigma_1\sigma_2\sqrt{1-\rho^2}}$$

$$\times \exp\left\{-\frac{1}{2(1-\rho^2)}\left[\frac{(x_1-m_1)^2}{\sigma_1^2}\right.\right.$$

$$\left.\left. -2\rho\frac{(x_1-m_1)(x_2-m_2)}{\sigma_1\sigma_2} + \frac{(x_2-m_2)^2}{\sigma_2^2}\right]\right\}, \qquad (14)$$

where $\sigma_i > 0$, $|\rho| < 1$. (The meanings of the parameters m_i, σ_i and ρ will be explained in Sect. 8.)

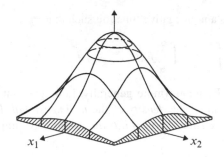

Fig. 28 Density of the two-dimensional Gaussian distribution

Figure 28 indicates the form of the two-dimensional Gaussian density.

Remark 3. As in the case $n = 1$, Theorem 2 can be generalized to (similarly defined) Lebesgue–Stieltjes measures on $(R^n, \mathscr{B}(R^n))$ and generalized distribution functions on R^n. When the generalized distribution function $G_n(x_1, \ldots, x_n)$ is $x_1 \cdots x_n$, the corresponding measure is Lebesgue measure on the Borel sets of R^n. It clearly satisfies

$$\lambda(a, b) = \prod_{i=1}^{n}(b_i - a_i),$$

i.e., the Lebesgue measure of the "rectangle"

$$(a, \ b] = (a_1, b_1] \times \cdots \times (a_n, b_n]$$

is its "volume."

4. The measurable space $(R^\infty, \mathscr{B}(R^\infty))$. For the spaces R^n, $n \geq 1$, the probability measures were constructed in the following way: first for elementary sets (rectangles $(a, b]$), then, in a natural way, for sets $A = \sum(a_i, b_i]$, and finally, by using Carathéodory's theorem, for sets in $\mathscr{B}(R^n)$.

A similar procedure of constructing probability measures also works for the space $(R^\infty, \mathscr{B}(R^\infty))$.

Let

$$\mathscr{I}_n(B) = \{x \in R^\infty : (x_1, \ldots, x_n) \in B\}, \quad B \in \mathscr{B}(R^n),$$

denote a cylinder set in R^∞ with base $B \in \mathscr{B}(R^n)$. As we will see now, it is natural to take the cylinder sets for the *elementary sets* in R^∞ whose probabilities enable us to determine the probability measure on the sets of $\mathscr{B}(R^\infty)$.

Let P be a probability measure on $(R^\infty, \mathscr{B}(R^\infty))$. For $n = 1, 2, \ldots$, we take

$$P_n(B) = \mathsf{P}(\mathscr{I}_n(B)), \quad B \in \mathscr{B}(R^n). \tag{15}$$

The sequence of probability measures $P_1, P_2, \ldots$ defined respectively on $(R, \mathscr{B}(R))$, $(R^2, \mathscr{B}(R^2))$, $\ldots$, has the following evident consistency property: for $n = 1, 2, \ldots$ and $B \in \mathscr{B}(R^n)$,

$$P_{n+1}(B \times R) = P_n(B). \tag{16}$$

It is noteworthy that the *converse* also holds.

Theorem 3 (Kolmogorov's Theorem on the Extension of Measures on $(R^\infty, \mathscr{B}(R^\infty))$). *Let $P_1, P_2, \ldots$ be probability measures on $(R, \mathscr{B}(R))$, $(R^2, \mathscr{B}(R^2))$, $\ldots$ respectively, possessing the consistency property* (16). *Then there is a unique probability measure* P *on* $(R^\infty, \mathscr{B}(R^\infty))$ *such that*

$$\mathsf{P}(\mathscr{I}_n(B)) = P_n(B), \quad B \in \mathscr{B}(R^n), \tag{17}$$

for $n = 1, 2, \ldots$.

PROOF. Let $B^n \in \mathscr{B}(R^n)$ and let $\mathscr{I}_n(B^n)$ be the cylinder with base B^n. We assign the measure $\mathsf{P}(\mathscr{I}_n(B^n))$ to this cylinder by taking $\mathsf{P}(\mathscr{I}_n(B^n)) = P_n(B^n)$.

Let us show that, in virtue of the consistency condition, this definition is consistent, i.e., the value of $\mathsf{P}(\mathscr{I}_n(B^n))$ is independent of the representation of the set $\mathscr{I}_n(B^n)$. In fact, let the same cylinder be represented in two ways:

$$\mathscr{I}_n(B^n) = \mathscr{I}_{n+k}(B^{n+k}).$$

It follows that, if $(x_1, \ldots, x_{n+k}) \in R^{n+k}$, we have

$$(x_1, \ldots, x_n) \in B^n \Leftrightarrow (x_1, \ldots, x_{n+k}) \in B^{n+k}, \tag{18}$$

and therefore, by (16)

$$\begin{aligned}
P_n(B^n) &= P_{n+1}((x_1, \ldots, x_{n+1}) : (x_1, \ldots, x_n) \in B^n) \\
&= \ldots = P_{n+k}((x_1, \ldots, x_{n+k}) : (x_1, \ldots, x_n) \in B^n) \\
&= P_{n+k}(B^{n+k}).
\end{aligned}$$

Let $\mathscr{A}(R^\infty)$ denote the collection of all cylinder sets $\hat{B}^n = \mathscr{I}_n(B^n), B^n \in \mathscr{B}(R^n)$, $n = 1, 2, \ldots$. It is easily seen that $\mathscr{A}(R^\infty)$ is an algebra.

Now let $\hat{B}_1, \ldots, \hat{B}_k$ be disjoint sets in $\mathscr{A}(R^\infty)$. We may suppose without loss of generality that $\hat{B}_i = \mathscr{I}_n(B_i^n)$, $i = 1, \ldots, k$, for some n, where $B_1^n, \ldots, B_k^n$ are disjoint sets in $\mathscr{B}(R^n)$. Then

$$\mathsf{P}\left(\sum_{i=1}^k \hat{B}_i\right) = \mathsf{P}\left(\sum_{i=1}^k \mathscr{I}_n(B_i^n)\right) = P_n\left(\sum_{i=1}^k B_i^n\right) = \sum_{i=1}^n P_n(B_i^n) = \sum_{i=1}^n \mathsf{P}(\hat{B}_i),$$

i.e., the set function P is finitely additive on the algebra $\mathscr{A}(R^\infty)$.

Let us show that P is "continuous at zero" (and therefore σ-additive on $\mathscr{A}(R^\infty)$, see Theorem in Sect. 1), i.e., if a sequence of sets $\hat{B}_n \downarrow \varnothing, n \to \infty$, then $\mathsf{P}(\hat{B}_n) \to 0$,

$n \to \infty$. Suppose the contrary, i.e., let $\lim \mathsf{P}(\hat{B}_n) = \delta > 0$ (the limit exists due to monotonicity). We may suppose without loss of generality that $\{\hat{B}_n\}$ has the form

$$\hat{B}_n = \{x : (x_1, \ldots, x_n) \in B_n\}, \quad B_n \in \mathscr{B}(R^n).$$

We use the following property of probability measures P_n on $(R^n, \mathscr{B}(R^n))$ (see Problem 9): if $B_n \in \mathscr{B}(R^n)$, for a given $\delta > 0$ we can find a compact set $A_n \in \mathscr{B}(R^n)$ such that $A_n \subseteq B_n$ and

$$P_n(B_n \backslash A_n) \le \delta / 2^{n+1}.$$

Therefore if

$$\hat{A}_n = \{x : (x_1, \ldots, x_n) \in A_n\},$$

we have

$$\mathsf{P}(\hat{B}_n \backslash \hat{A}_n) = P_n(B_n \backslash A_n) \le \delta / 2^{n+1}.$$

Form the set $\hat{C}_n = \bigcap_{k=1}^{n} \hat{A}_k$ and let C_n be such that

$$\hat{C}_n = \{x : (x_1, \ldots, x_n) \in C_n\}.$$

Then, since the sets $\hat{B}_n$ decrease, we obtain

$$\mathsf{P}(\hat{B}_n \backslash \hat{C}_n) \le \sum_{k=1}^{n} \mathsf{P}(\hat{B}_n \backslash \hat{A}_k) \le \sum_{k=1}^{n} \mathsf{P}(\hat{B}_k \backslash \hat{A}_k) \le \delta / 2.$$

But by assumption $\lim_n \mathsf{P}(\hat{B}_n) = \delta > 0$, and therefore $\lim_n \mathsf{P}(\hat{C}_n) \ge \delta / 2 > 0$. Let us show that this contradicts the condition $\hat{C}_n \downarrow \varnothing$.

Let us choose a point $\hat{x}^{(n)} = (x_1^{(n)}, x_2^{(n)}, \ldots)$ in $\hat{C}_n$. Then $(x_1^{(n)}, \ldots, x_n^{(n)}) \in C_n$ for $n \ge 1$.

Let (n_1) be a subsequence of (n) such that $x_1^{(n_1)} \to x_1^0$, where x_1^0 is a point in C_1. (Such a sequence exists since $x_1^{(n_1)} \in C_1$ and C_1 is compact.) Then select a subsequence (n_2) of (n_1) such that $(x_1^{(n_2)}, x_2^{(n_2)}) \to (x_1^0, x_2^0) \in C_2$. Similarly let $(x_1^{(n_k)}, \ldots, x_k^{(n_k)}) \to (x_1^0, \ldots, x_k^0) \in C_k$. Finally form the diagonal sequence (m_k), where m_k is the kth term of (n_k). Then $x_i^{(m_k)} \to x_i^0$ as $m_k \to \infty$ for $i = 1, 2, \ldots$, and $(x_1^0, x_2^0, \ldots) \in \hat{C}_n$ for $n = 1, 2, \ldots$, which evidently contradicts the assumption that $\hat{C}_n \downarrow \varnothing$, $n \to \infty$. Thus the set function P is σ-additive on the algebra $\mathscr{A}(R^\infty)$ and hence, by Carathéodory's theorem, it can be extended to a (probability) measure on $(R^\infty, \mathscr{B}(R^\infty))$. This completes the proof of the theorem.

$\square$

Remark 4. In the present case, the space R^∞ is a countable product of *real lines*, $R^\infty = R \times R \times \cdots$. It is natural to ask whether Theorem 3 remains true if $(R^\infty, \mathscr{B}(R^\infty))$ is replaced by a direct product of *measurable spaces* $(\Omega_i, \mathscr{F}_i)$, $i = 1, 2, \ldots$.

We may notice that in the preceding proof the only topological property of the real line that was used was that every set in $\mathscr{B}(R^n)$ contains a *compact subset* whose probability measure is arbitrarily close to the probability measure of the whole set.

It is known, however, that this is a property not only of spaces $(R^n, \mathscr{B}(R^n))$, but also of arbitrary complete separable metric spaces with σ-algebras generated by the open sets.

Consequently Theorem 3 remains valid if we suppose that $P_1, P_2, \ldots$ is a sequence of consistent probability measures on

$$(\Omega_1, \mathscr{F}_1), \ (\Omega_1 \times \Omega_2, \mathscr{F}_1 \otimes \mathscr{F}_2), \ldots,$$

where $(\Omega_i, \mathscr{F}_i)$ are complete separable metric spaces with σ-algebras $\mathscr{F}_i$ generated by open sets, and $(R^\infty, \mathscr{B}(R^\infty))$ is replaced by

$$(\Omega_1 \times \Omega_2 \times \cdots, \mathscr{F}_1 \otimes \mathscr{F}_2 \otimes \cdots).$$

In Sect. 9 (Theorem 2) it will be shown that the result of Theorem 3 remains valid for *arbitrary* measurable spaces $(\Omega_i, \mathscr{F}_i)$ if the measures P_n are constructed in a *particular* way. However, Theorem 3 may fail in the general case (without any hypotheses on the topological nature of the measurable spaces or on the structure of the family of measures $\{P_n\}$). This is shown by the following example.

Let us consider the space $\Omega = (0, 1]$, which is evidently not *complete*, and construct a sequence $\mathscr{F}_1 \subseteq \mathscr{F}_2 \subseteq \cdots$ of σ-algebras in the following way. For $n = 1, 2, \ldots$, let

$$\varphi_n(\omega) = \begin{cases} 1, & 0 < \omega < 1/n, \\ 0, & 1/n \leq \omega \leq 1, \end{cases}$$

$$\mathscr{C}_n = \{A \in \Omega : A = \{\omega : \varphi_n(\omega) \in B\} \quad \text{for some} \quad B \in \mathscr{B}(R)\}$$

and let $\mathscr{F}_n = \sigma\{\mathscr{C}_1, \ldots, \mathscr{C}_n\}$ be the smallest σ-algebra containing the sets $\mathscr{C}_1, \ldots, \mathscr{C}_n$. Clearly $\mathscr{F}_1 \subseteq \mathscr{F}_2 \subseteq \cdots$. Let $\mathscr{F} = \sigma(\cup \mathscr{F}_n)$ be the smallest σ-algebra containing all the $\mathscr{F}_n$. Consider the measurable space $(\Omega, \mathscr{F}_n)$ and define a probability measure P_n on it as follows:

$$P_n\{\omega : (\varphi_1(\omega), \ldots, \varphi_n(\omega)) \in B^n\} = \begin{cases} 1 & \text{if } (1, \ldots, 1) \in B^n, \\ 0 & \text{otherwise,} \end{cases}$$

where $B^n = \mathscr{B}(R^n)$. It is easy to see that the family $\{P_n\}$ is consistent: if $A \in \mathscr{F}_n$ then $P_{n+1}(A) = P_n(A)$. However, we claim that there is *no* probability measure P on $(\Omega, \mathscr{F})$ such that its *restriction* $P \mid \mathscr{F}_n$ (i.e., the measure P considered only on sets in $\mathscr{F}_n$) coincides with P_n for $n = 1, 2, \ldots$. In fact, let us suppose that such a probability measure P exists. Then

$$P\{\omega : \varphi_1(\omega) = \cdots = \varphi_n(\omega) = 1\} = P_n\{\omega : \varphi_1(\omega) = \cdots = \varphi_n(\omega) = 1\} = 1 \tag{19}$$

for $n = 1, 2, \ldots$. But

$$\{\omega : \varphi_1(\omega) = \cdots = \varphi_n(\omega) = 1\} = (0, 1/n) \downarrow \varnothing,$$

which contradicts (19) and the hypothesis of countable additivity (and therefore continuity at the "zero" $\varnothing$) of the set function P.

We now give an example of a probability measure on $(R^\infty, \mathscr{B}(R^\infty))$. Let $F_1(x)$, $F_2(x), \ldots$ be a sequence of one-dimensional distribution functions. Define the functions $G(x) = F_1(x)$, $G_2(x_1, x_2) = F_1(x_1)F_2(x_2), \ldots$, and denote the corresponding probability measures on $(R, \mathscr{B}(R))$, $(R^2, \mathscr{B}(R^2)), \ldots$ by $P_1, P_2, \ldots$. Then it follows from Theorem 3 that there is a measure P on $(R^\infty, \mathscr{B}(R^\infty))$ such that

$$P\{x \in R^\infty : (x_1, \ldots, x_n) \in B\} = P_n(B), \quad B \in \mathscr{B}(R^n),$$

and, in particular,

$$P\{x \in R^\infty : x_1 \leq a_1, \ldots, x_n \leq a_n\} = F_1(a_1) \cdots F_n(a_n).$$

Let us take $F_i(x)$ to be a Bernoulli distribution,

$$F_i(x) = \begin{cases} 0, & x < 0, \\ q, & 0 \leq x < 1, \\ 1, & x \geq 1. \end{cases}$$

Then we can say that there is a probability measure P on the space Ω of sequences of numbers $x = (x_1, x_2, \ldots)$, $x_i = 0$ or 1, together with the σ-algebra of its Borel subsets, such that

$$P\{x : x_1 = a_1, \ldots, x_n = a_n\} = p^{\Sigma a_i} q^{n - \Sigma a_i}.$$

5. The measurable spaces $(R^T, \mathscr{B}(R^T))$. Let T be an *arbitrary* set of indices $t \in T$ and R_t a real line corresponding to the index t. We consider a finite *unordered* set $\tau = [t_1, \ldots, t_n]$ of distinct indices t_i, $t_i \in T$, $n \geq 1$, and let P_τ be a probability measure on $(R^\tau, \mathscr{B}(R^\tau))$, where $R^\tau = R_{t_1} \times \cdots \times R_{t_n}$.

We say that the family $\{P_\tau\}$ of probability measures, where τ runs through all finite unordered sets, is *consistent* if, for any sets $\tau = [t_1, \ldots, t_n]$ and $\sigma = [s_1, \ldots, s_k]$ such that $\sigma \subseteq \tau$ we have

$$P_\sigma\{(x_{s_1}, \ldots, x_{s_k}) : (x_{s_1}, \ldots, x_{s_k}) \in B\} = P_\tau\{(x_{t_1}, \ldots, x_{t_n}) : (x_{s_1}, \ldots, x_{s_k}) \in B\} \tag{20}$$

for every $B \in \mathscr{B}(R^\sigma)$.

Theorem 4 (Kolmogorov's Theorem on the Extension of Measures in $(R^T, \mathscr{B}(R^T))$). *Let $\{P_\tau\}$ be a consistent family of probability measures on $(R^\tau, \mathscr{B}(R^\tau))$. Then there is a unique probability measure P on $(R^T, \mathscr{B}(R^T))$ such that*

$$P(\mathscr{I}_\tau(B)) = P_\tau(B) \tag{21}$$

for all unordered sets $\tau = [t_1, \ldots, t_n]$ of different indices $t_i \in T$, $B \in \mathscr{B}(R^\tau)$ and $\mathscr{I}_\tau(B) = \{x \in R^T : (x_{t_1}, \ldots, x_{t_n}) \in B\}$.

PROOF. Let the set $\hat{B} \in \mathscr{B}(R^T)$. By Theorem 3 of Sect. 2 there is an at most countable set $S = \{s_1, s_2, \ldots\} \subseteq T$ such that $\hat{B} = \{x \colon (x_{s_1}, x_{s_2}, \ldots) \in B\}$, where $B \in \mathscr{B}(R^S)$, $R^S = R_{s_1} \times R_{s_2} \times \cdots$. In other words, $\hat{B} = \mathscr{I}_S(B)$ is a cylinder set with base $B \in \mathscr{B}(R^S)$.

We can define a set function P on such cylinder sets by putting

$$\mathsf{P}(\mathscr{I}_S(B)) = P_S(B), \tag{22}$$

where P_S is the probability measure whose existence is guaranteed by Theorem 3. We claim that P is in fact the measure whose existence is asserted in the theorem. To establish this we first verify that the definition (22) is consistent, i.e., that it leads to a unique value of $\mathsf{P}(\hat{B})$ for all possible representations of $\hat{B}$; and second, that this set function is countably additive.

Let $\hat{B} = \mathscr{I}_{S_1}(B_1)$ and $\hat{B} = \mathscr{I}_{S_2}(B_2)$. It is clear that then $\hat{B} = \mathscr{I}_{S_1 \cup S_2}(B_3)$ with some $B_3 \in \mathscr{B}(R^{S_1 \cup S_2})$; therefore it is enough to show that if $S \subseteq S'$ and $B \in \mathscr{B}(R^S)$, then $P_{S'}(B') = P_S(B)$, where

$$B' = \{(x_{s'_1}, x_{s'_2}, \ldots) \colon (x_{s_1}, x_{s_2}, \ldots) \in B\}$$

with $S' = \{s'_1, s'_2, \ldots\}, S = \{s_1, s_2, \ldots\}$. But by the assumed consistency condition (20) this equation follows immediately from Theorem 3. This establishes that the value of $\mathsf{P}(\hat{B})$ is independent of the representation of $\hat{B}$.

To verify the countable additivity of P, let us suppose that $\{\hat{B}_n\}$ is a sequence of pairwise disjoint sets in $\mathscr{B}(R^T)$. Then there is an at most countable set $S \subseteq T$ such that $\hat{B}_n = \mathscr{I}_S(B_n)$ for all $n \geq 1$, where $B_n \in \mathscr{B}(R^S)$. Since P_S is a probability measure, we have

$$\mathsf{P}\left(\sum \hat{B}_n\right) = \mathsf{P}\left(\sum \mathscr{I}_S(B_n)\right) = P_S\left(\sum B_n\right) = \sum P_S(B_n)$$
$$= \sum \mathsf{P}(I_S(B_n)) = \sum \mathsf{P}(\hat{B}_n).$$

Finally, property (21) follows immediately from the way in which P was constructed. This completes the proof.

$\square$

Remark 5. We emphasize that T is *any* set of indices. Hence, by the remark after Theorem 3, the present theorem remains valid if we replace the real lines R_t by arbitrary complete separable metric spaces Ω_t (with σ-algebras generated by open sets).

Remark 6. The original probability measures $\{P_\tau\}$ were assumed defined for *unordered* sets $\tau = [t_1, \ldots, t_n]$ of different indices. In this connection it is worth to emphasize that these measures $\{P_\tau\}$ as functions of $\tau = [t_1, \ldots, t_n]$ are actually functions of *sets* consisting of n (distinct) points $\{t_1\}, \ldots, \{t_n\}$. (For example, the unordered collections $[a, b]$ and $[b, a]$ should be regarded as the same, since they determine the same set consisting of the points $\{a\}$ and $\{b\}$.) It is also possible to start from a family of probability measures $\{P_\tau\}$ where τ runs through all *ordered* sets $\tau = (t_1, \ldots, t_n)$ of different indices. (Then the collections (a, b) and (b, a) consisting of the same points have to be treated as different, since they differ by the order of their elements.) In this case, in order to have Theorem 4 hold we have to adjoin to (20) a further *consistency condition*:

$$P_{(t_1,\ldots,t_n)}(A_{t_1} \times \cdots \times A_{t_n}) = P_{(t_{i_1},\ldots,t_{i_n})}(A_{t_{i_1}} \times \cdots \times A_{t_{i_n}}), \tag{23}$$

where $(i_1, \ldots, i_n)$ is an arbitrary permutation of $(1, \ldots, n)$ and $A_{t_i} \in \mathscr{B}(R_{t_i})$. The necessity of this condition for the existence of the probability measure P follows from (21) (with $P_{[t_1,\ldots,t_n]}(B)$ replaced by $P_{(t_1,\ldots,t_n)}(B)$).

From now on we shall assume that the sets τ under consideration are *unordered*. If T is a subset of the real line (or some completely ordered set), we may assume without loss of generality that the set $\tau = [t_1, \ldots, t_n]$ satisfies $t_1 < t_2 < \cdots < t_n$. Consequently it is enough to define "finite-dimensional" probabilities only for sets $\tau = [t_1, \ldots, t_n]$ for which $t_1 < t_2 < \cdots < t_n$.

Now consider the case $T = [0, \infty)$. Then R^T is the space of all real functions $x = (x_t)_{t \geq 0}$. A fundamental example of a probability measure on $(R^{[0,\infty)}, \mathscr{B}(R^{[0,\infty)}))$ is Wiener measure, constructed as follows.

Consider the family $\{\varphi_t(y \mid x)\}_{t \geq 0}$ of Gaussian densities (as functions of y for fixed x):

$$\varphi_t(y \mid x) = \frac{1}{\sqrt{2\pi t}} e^{-(y-x)^2/2t}, \quad y \in R,$$

and for each $\tau = [t_1, \ldots, t_n]$, $t_1 < t_2 < \cdots < t_n$, and each set

$$B = I_1 \times \cdots \times I_n, \quad I_k = (a_k, b_k),$$

construct the measure $P_\tau(B)$ according to the formula

$$\begin{aligned}
&P_\tau(I_1 \times \cdots \times I_n) \\
&= \int_{I_1} \cdots \int_{I_n} \varphi_{t_1}(a_1 \mid 0) \varphi_{t_2-t_1}(a_2 \mid a_1) \cdots \varphi_{t_n-t_{n-1}}(a_n \mid a_{n-1}) \, da_1 \cdots da_n
\end{aligned} \tag{24}$$

(integration in the Riemann sense). Now we define the set function P for each cylinder set $\mathscr{I}_{t_1,\ldots,t_n}(I_1 \times \cdots \times I_n) = \{x \in R^T : x_{t_1} \in I_1, \ldots, x_{t_n} \in I_n\}$ by taking

$$P(\mathscr{I}_{t_1,\ldots,t_n}(I_1 \times \cdots \times I_n)) = P_{[t_1,\ldots,t_n]}(I_1 \times \cdots \times I_n).$$

The intuitive meaning of this method of assigning a measure to the cylinder set $\mathscr{I}_{t_1,\ldots,t_n}(I_1 \times \cdots \times I_n)$ is as follows.

The set $\mathscr{I}_{t_1,\ldots,t_n}(I_1 \times \cdots \times I_n)$ is the set of functions that at times $t_1, \ldots, t_n$ pass through the "windows" $I_1, \ldots, I_n$ (see Fig. 24 in Sect. 2). We shall interpret $\varphi_{t_k-t_{k-1}}(a_k \mid a_{k-1}) \, da_k$ as the probability that a particle starting at a_{k-1} arrives in time $t_k - t_{k-1}$ at the da_k-neighborhood of a_k. Then the product of densities that appears in (24) describes a certain independence of the increments of the displacements of the moving "particle" in the time intervals

$$[0, t_1], [t_1, t_2], \ldots, [t_{n-1}, t_n].$$

The family of measures $\{P_\tau\}$ constructed in this way is easily seen to be consistent, and therefore can be extended to a measure on $(R^{[0,\infty)}, \mathscr{B}(R^{[0,\infty)}))$. The measure so obtained plays an important role in probability theory. It was introduced by N. Wiener and is known as *Wiener measure*.

6. PROBLEMS

1. Let $F(x) = P(-\infty, x]$. Verify the following formulas:

$$P(a, b] = F(b) - F(a), \qquad P(a, b) = F(b-) - F(a),$$
$$P[a, b] = F(b) - F(a-), \qquad P[a, b) = F(b-) - F(a-),$$
$$P\{x\} = F(x) - F(x-),$$

where $F(x-) = \lim_{y \uparrow x} F(y)$.

2. Verify (7).

3. Prove Theorem 2.

4. Show that a distribution function $F = F(x)$ on R has at most a countable set of points of discontinuity. Does a corresponding result hold for distribution functions on R^n?

5. Show that each of the functions

$$G(x, y) = \begin{cases} 1, & x + y \geq 0, \\ 0, & x + y < 0, \end{cases}$$
$$G(x, y) = [x + y], \text{ the integral part of } x + y,$$

is continuous on the right and nondecreasing in each argument, but is not a (generalized) distribution function on R^2.

6. Let μ be the Lebesgue–Stieltjes measure generated by a continuous generalized distribution function. Show that if the set Λ is at most countable, then $\mu(\Lambda) = 0$.

7. Let c be the cardinal number of the continuum. Show that the cardinal number of the σ-algebra of Borel sets in R^n is c, whereas that of the σ-algebra of Lebesgue measurable sets is 2^c.

8. Let $(\Omega, \mathscr{F}, P)$ be a probability space and $\mathscr{A}$ an algebra of subsets of Ω such that $\sigma(\mathscr{A}) = \mathscr{F}$. Using the *principle of appropriate sets*, prove that for every $\varepsilon > 0$ and $B \in \mathscr{F}$ there is a set $A \in \mathscr{A}$ such that

$$P(A \triangle B) \leq \varepsilon.$$

9. Let P be a probability measure on $(R^n, \mathscr{B}(R^n))$. Show that, for every $\varepsilon > 0$ and $B \in \mathscr{B}(R^n)$, there are a compact Borel set A_1 and an open Borel set A_2 such that $A_1 \subseteq B \subseteq A_2$ and $P(A_2 \backslash A_1) \leq \varepsilon$. (This was used in the proof of Theorem 3.)

10. Verify the consistency of the family of measures $\{\mathsf{P}_\tau\}$ constructed by means of the formula $\mathsf{P}_\tau(B) = \mathsf{P}(\mathscr{I}_\tau(B))$, where P is a given probability measure. (Compare with (21).)

11. Verify that the "distributions" given in Tables 2.2 and 2.3 are probability distributions indeed.

12. Show that the system $\hat{\mathscr{A}}$ in Remark 2 is a σ-algebra.

13. Show that the set function $\mu(A), A \in \hat{\mathscr{A}}$, introduced in Remark 2, is a *measure*.

14. Give an example showing that a *finitely-additive* (but not countably-additive) measure μ_0 on an algebra $\mathscr{A}$ need not admit an extension to a countably additive measure on $\sigma(\mathscr{A})$.

15. Show that any finitely-additive probability measure defined on an algebra $\mathscr{A}$ of subsets of Ω can be extended to finitely-additive probability on all subsets of Ω.

16. Let P be a probability measure on a σ-algebra $\mathscr{F}$ of subsets of Ω. Suppose a set C is such that $C \subseteq \Omega$, but $C \notin \mathscr{F}$. Show that P can be extended (preserving the countable additivity property) to $\sigma(\mathscr{F} \cup \{C\})$.

17. Show that the support of a *continuous* distribution function F is a *perfect* set (i.e., supp F is a closed set such that for any $x \in$ supp F and $\varepsilon > 0$ there is $y \in$ supp F such that $0 < |x - y| < \varepsilon$). Show that the support of (an arbitrary) distribution function is a *closed* set.

18. Prove the following fundamental result on the structure of distribution functions (see the end of Subsection 1): each distribution function is a convex combination

$$F = \alpha_1 F_{\mathrm{d}} + \alpha_2 F_{\mathrm{abc}} + \alpha_3 F_{\mathrm{sc}}$$

of a discrete (F_{d}), absolutely continuous (F_{abc}) and singular continuous (F_{sc}) distribution functions, $\alpha_i \geq 0, \alpha_1 + \alpha_2 + \alpha_3 = 1$.

19. Let $F = F(x)$ be the Cantor distribution function. Show that any point x in the Cantor set $\mathscr{N}$ of its points of increase (which is the same as the support of F) can be represented as $x = \sum_{k=1}^{\infty} \frac{\alpha_k(x)}{3^k}$, where $\alpha_k(x) = 0$ or 2, and that for such points $F(x) = \sum_{k=1}^{\infty} \frac{\alpha_k(x)}{2^{k+1}}$.

20. Let C be a closed subset of R. Construct a distribution function F with supp $F = C$.

21. Show that the distribution function of the binomial distribution (Subsection 1 of Sect. 2)

$$B_n(m; p) = \mathsf{P}_n\{\nu \leq m\} = \sum_{k=0}^{m} C_n^k p^k q^{n-k}$$

can be expressed in terms of the (incomplete) beta-function:

$$B_n(m; p) = \frac{1}{\mathrm{B}(m + 1, n - m)} \int_p^1 x^m (1 - x)^{n-m-1} \, dx.$$

22. Show that the Poisson distribution function $F(n; \lambda) = \sum_{k=0}^{n} \frac{e^{-\lambda} \lambda^k}{k!}$ is expressed in terms of the (incomplete) gamma-function:

$$F(n; \lambda) = \frac{1}{n!} \int_{\lambda}^{\infty} x^n e^{-x} \, dx.$$

23. Along with the mean value and standard deviation, which serve as location and scale parameters of a distribution, the following two parameters describing the shape of a density function $f = f(x)$ are customarily used: the skewness (related to asymmetry of the distribution)

$$\alpha_3 = \frac{\mu_3}{\sigma^3},$$

and the kurtosis ("peakedness")

$$\alpha_4 = \frac{\mu_4}{\sigma^4},$$

where $\mu_k = \int (x - \mu)^k f(x) \, dx$, $\mu = \int x f(x) \, dx$, $\sigma^2 = \mu_2$.
Find the values of α_3 and α_4 for the distributions listed in Table 2.3 (Subsection 1).

24. Show that for a random variable X having the gamma-distribution (see Table 2.3) with $\beta = 1$

$$\mathsf{E}\, X^k = \frac{\Gamma(k + \alpha)}{\Gamma(\alpha)}.$$

In particular, $\mathsf{E}\, X = \alpha$, $\mathsf{E}\, X^2 = \alpha(\alpha + 1)$ and hence $\mathrm{Var}\, X = \alpha$.
Find the analogs of these formulas when $\beta \neq 1$.

25. Show that for a random variable X with beta-distribution (see Table 2.3)

$$\mathsf{E}\, X^k = \frac{\mathrm{B}(r + k, s)}{\mathrm{B}(r, s)}.$$

26. The *binomial* distribution $\mathsf{P}_n\{\nu = r\} = C_n^r p^r q^{n-r}$, $0 \leq r \leq n$, consists of probabilities that the number ν of "successes" in n trials is r, where p is the probability of "success" in a single trial and the number n of trials is fixed in advance. Now we ask how many trials are needed for r "successes" to occur. Namely, let τ denote the number of trials when r "successes" occur for the first time (unlike the binomial case, r is a given number here and the number of trials is random). We are interested in the probabilities $\mathsf{P}^r\{\tau = k\}$, $k = r, r + 1, \ldots$, which form what is known as the *negative binomial* distribution. Show that for $r = 1, 2, \ldots$

$$\mathsf{P}^r(\tau = k) = C_{k-1}^{r-1} p^r q^{k-r}, \quad k = r, r + 1, \ldots,$$

and $\mathsf{E}^r \tau = rq/p$.

4 Random Variables: I

1. Let $(\Omega, \mathscr{F})$ be a measurable space and let $(R, \mathscr{B}(R))$ be the real line with the system $\mathscr{B}(R)$ of Borel sets.

Definition 1. A real function $\xi = \xi(\omega)$ defined on $(\Omega, \mathscr{F})$ is an $\mathscr{F}$-*measurable function*, or *a random variable*, if

$$\{\omega : \xi(\omega) \in B\} \in \mathscr{F} \tag{1}$$

for every $B \in \mathscr{B}(R)$; or, equivalently, if the inverse image

$$\xi^{-1}(B) \equiv \{\omega : \xi(\omega) \in B\}$$

is a measurable set in Ω.

When $(\Omega, \mathscr{F}) = (R^n, \mathscr{B}(R^n))$, the $\mathscr{B}(R^n)$-measurable functions are called *Borel functions*.

The simplest example of a random variable is the indicator $I_A(\omega)$ of an arbitrary (measurable) set $A \in \mathscr{F}$.

A random variable ξ that has a representation

$$\xi(\omega) = \sum_{i=1}^{\infty} x_i I_{A_i}(\omega), \tag{2}$$

where $\sum A_i = \Omega$, $A_i \in \mathscr{F}$, is called *discrete*. If the sum in (2) is finite, the random variable is called *simple*.

With the same interpretation as in Sect. 4 of Chap. 1, we may say that a random variable is a *numerical property* of an experiment, with a value depending on "chance." Here the requirement (1) of measurability is fundamental, for the following reason. If a probability measure P is defined on $(\Omega, \mathscr{F})$, it then makes sense to speak of the probability of the event $\{\xi(\omega) \in B\}$ that the value of the random variable belongs to a Borel set B.

We introduce the following definitions.

Definition 2. A probability measure P_ξ on $(R, \mathscr{B}(R))$ with

$$P_\xi(B) = P\{\omega : \xi(\omega) \in B\}, \quad B \in \mathscr{B}(R),$$

is called the *probability distribution of ξ* on $(R, \mathscr{B}(R))$.

Definition 3. The function

$$F_\xi(x) = P(\omega : \xi(\omega) \le x), \quad x \in R,$$

is called the *distribution function of ξ*.

For a *discrete* random variable the measure P_ξ is concentrated on an at most countable set and can be represented in the form

$$P_\xi(B) = \sum_{\{k:\ x_k \in B\}} p(x_k), \tag{3}$$

where $p(x_k) = P\{\xi = x_k\} = \Delta F_\xi(x_k)$.

The converse is evidently true: If P_ξ is represented in the form (3) then ξ is a *discrete* random variable.

A random variable ξ is called *continuous* if its distribution function $F_\xi(x)$ is continuous for $x \in R$.

A random variable ξ is called *absolutely continuous* if there is a nonnegative function $f = f_\xi(x)$, called its *density*, such that

$$F_\xi(x) = \int_{-\infty}^{x} f_\xi(y)\, dy, \quad x \in R \tag{4}$$

(the integral can be taken in the Riemann sense, or more generally in that of Lebesgue; see Sect. 6 below).

2. To establish that a function $\xi = \xi(\omega)$ is a random variable, we have to verify property (1) for all sets $B \in \mathscr{F}$. The following lemma shows that the class of such "test" sets can be considerably narrowed.

Lemma 1. *Let $\mathscr{E}$ be a system of sets such that $\sigma(\mathscr{E}) = \mathscr{B}(R)$. A necessary and sufficient condition that a function $\xi = \xi(\omega)$ is $\mathscr{F}$-measurable is that*

$$\{\omega : \xi(\omega) \in E\} \in \mathscr{F} \tag{5}$$

for all $E \in \mathscr{E}$.

PROOF. The necessity is evident. To prove the sufficiency we again use the *principle of appropriate sets* (Sect. 2).

Let $\mathscr{D}$ be the system of those Borel sets D in $\mathscr{B}(R)$ for which $\xi^{-1}(D) \in \mathscr{F}$. The operation "form the inverse image" is easily shown to preserve the set-theoretic operations of union, intersection, and complement:

$$\xi^{-1}\left(\bigcup_\alpha B_\alpha\right) = \bigcup_\alpha \xi^{-1}(B_\alpha),$$

$$\xi^{-1}\left(\bigcap_\alpha B_\alpha\right) = \bigcap_\alpha \xi^{-1}(B_\alpha), \tag{6}$$

$$\overline{\xi^{-1}(B_\alpha)} = \xi^{-1}(\overline{B_\alpha}).$$

It follows that $\mathscr{D}$ is a σ-algebra. Therefore

$$\mathscr{E} \subseteq \mathscr{D} \subseteq \mathscr{B}(R)$$

and
$$\sigma(\mathscr{E}) \subseteq \sigma(\mathscr{D}) = \mathscr{D} \subseteq \mathscr{B}(R).$$
But $\sigma(\mathscr{E}) = \mathscr{B}(R)$ and consequently $\mathscr{D} = \mathscr{B}(R)$.

$\square$

Corollary. *A necessary and sufficient condition for* $\xi = \xi(\omega)$ *to be a random variable is that*
$$\{\omega : \xi(\omega) < x\} \in \mathscr{F}$$
for every $x \in R$, *or that*
$$\{\omega : \xi(\omega) \le x\} \in \mathscr{F}$$
for every $x \in R$.

PROOF. The proof is immediate, since each of the systems
$$\mathscr{E}_1 = \{x : x < c, \, c \in R\},$$
$$\mathscr{E}_2 = \{x : x \le c, \, c \in R\}$$

generates the σ-algebra $\mathscr{B}(R)$, i.e., $\sigma(\mathscr{E}_1) = \sigma(\mathscr{E}_2) = \mathscr{B}(R)$ (see Sect. 2).

$\square$

The following lemma makes it possible to construct random variables as *functions* of other random variables.

Lemma 2. *Let* $\varphi = \varphi(x)$ *be a Borel function and* $\xi = \xi(\omega)$ *a random variable. Then the composition* $\eta = \varphi \circ \xi$, *i.e., the function* $\eta(\omega) = \varphi(\xi(\omega))$, *is also a random variable.*

PROOF. This statement follows from the equations
$$\{\omega : \eta(\omega) \in B\} = \{\omega : \varphi(\xi(\omega)) \in B\} = \{\omega : \xi(\omega) \in \varphi^{-1}(B)\} \in \mathscr{F} \qquad (7)$$
for $B \in \mathscr{B}(R)$, since $\varphi^{-1}(B) \in \mathscr{B}(R)$.

$\square$

Therefore if ξ is a random variable, so are, for examples, ξ^n, $\xi^+ = \max(\xi, 0)$, $\xi^- = -\min(\xi, 0)$, and $|\xi|$, since the functions x^n, x^+, x^- and $|x|$ are Borel functions (Problem 3).

3. Starting from a given collection of random variables $\{\xi_n\}$, we can construct new functions, for example, $\sum_{k=1}^{\infty} |\xi_k|$, $\limsup \xi_n$, $\liminf \xi_n$, etc. Notice that in general such functions take values on the extended real line $\bar{R} = [-\infty, \infty]$. Hence it is advisable to extend the class of $\mathscr{F}$-measurable functions somewhat by allowing them to take the values $\pm\infty$.

Definition 4. A function $\xi = \xi(\omega)$ defined on $(\Omega, \mathscr{F})$ with values in $\bar{R} = [-\infty, \infty]$ will be called an *extended random variable* if condition (1) is satisfied for every Borel set $B \in \mathscr{B}(\bar{R})$, where $\mathscr{B}(\bar{R}) = \sigma(\mathscr{B}(R), \pm\infty)$.

The following theorem, despite its simplicity, is the key to the construction of the Lebesgue integral (Sect. 6).

Theorem 1. (a) *For every random variable* $\xi = \xi(\omega)$ *(extended ones included) there is a sequence of simple random variables* $\xi_1, \xi_2, \ldots$ *such that* $|\xi_n| \leq |\xi|$ *and* $\xi_n(\omega) \to \xi(\omega)$, $n \to \infty$, *for all* $\omega \in \Omega$.
(b) *If also* $\xi(\omega) \geq 0$, *there is a sequence of simple random variables* $\xi_1, \xi_2, \ldots$ *such that* $\xi_n(\omega) \uparrow \xi(\omega)$, $n \to \infty$, *for all* $\omega \in \Omega$.

PROOF. We begin by proving the second statement. For $n = 1, 2, \ldots$, put

$$\xi_n(\omega) = \sum_{k=1}^{n2^n} \frac{k-1}{2^n} I_{k,n}(\omega) + n I_{\{\xi(\omega) \geq n\}}(\omega),$$

where $I_{k,n}$ is the indicator of the set $\{(k-1)/2^n \leq \xi(\omega) < k/2^n\}$. It is easy to verify that the sequence $\xi_n(\omega)$ so constructed is such that $\xi_n(\omega) \uparrow \xi(\omega)$ for all $\omega \in \Omega$. The first statement follows from this if we merely observe that ξ can be represented in the form $\xi = \xi^+ - \xi^-$. This completes the proof of the theorem.
□

We next show that the class of extended random variables is closed under pointwise convergence. For this purpose, we note first that if $\xi_1, \xi_2, \ldots$ is a sequence of extended random variables, then $\sup \xi_n$, $\inf \xi_n$, $\limsup \xi_n$ and $\liminf \xi_n$ are also random variables (possibly extended). This follows immediately from

$$\{\omega : \sup \xi_n > x\} = \bigcup_n \{\omega : \xi_n > x\} \in \mathscr{F},$$

$$\{\omega : \inf \xi_n < x\} = \bigcup_n \{\omega : \xi_n < x\} \in \mathscr{F},$$

and

$$\limsup \xi_n = \inf_n \sup_{m \geq n} \xi_m, \quad \liminf \xi_n = \sup_n \inf_{m \geq n} \xi_m.$$

Theorem 2. *Let* $\xi_1, \xi_2, \ldots$ *be a sequence of extended simple random variables and* $\xi(\omega) = \lim \xi_n(\omega)$, $\omega \in \Omega$. *Then* $\xi(\omega)$ *is also an extended random variable.*

PROOF. It follows immediately from the remark above and the fact that

$$\{\omega : \xi(\omega) < x\} = \{\omega : \lim \xi_n(\omega) < x\}$$
$$= \{\omega : \limsup \xi_n(\omega) = \liminf \xi_n(\omega)\} \cap \{\limsup \xi_n(\omega) < x\}$$
$$= \Omega \cap \{\limsup \xi_n(\omega) < x\} = \{\limsup \xi_n(\omega) < x\} \in \mathscr{F}.$$

□

4. We mention a few more properties of the simplest functions of random variables considered on the measurable space $(\Omega, \mathscr{F})$ and possibly taking values on the extended real line $\overline{R} = [-\infty, \infty]$.*

---

* We shall assume the usual conventions about arithmetic operations in $\overline{R}$: if $a \in R$ then $a \pm \infty = \pm\infty$, $a/\pm\infty = 0$; $a \cdot \infty = \infty$ if $a > 0$, and $a \cdot \infty = -\infty$ if $a < 0$; $0 \cdot (\pm\infty) = 0$, $\infty + \infty = \infty$, $-\infty - \infty = -\infty$.

If ξ and η are random variables, $\xi + \eta$, $\xi - \eta$, $\xi\eta$, and ξ/η are also random variables (assuming that they are defined, i.e., that no indeterminate forms like $\infty - \infty$, ∞/∞, $a/0$ occur).

In fact, let $\{\xi_n\}$ and $\{\eta_n\}$ be sequences of simple random variables converging to ξ and η (see Theorem 1). Then

$$\xi_n \pm \eta_n \to \xi \pm \eta,$$
$$\xi_n\eta_n \to \xi\eta,$$
$$\frac{\xi_n}{\eta_n + \frac{1}{n}I_{\{\eta_n=0\}}(\omega)} \to \frac{\xi}{\eta}.$$

The functions on the left-hand sides of these relations are simple random variables. Therefore, by Theorem 2, the limit functions $\xi \pm \eta$, $\xi\eta$ and ξ/η are also random variables.

5. Let ξ be a random variable. Consider sets from $\mathscr{F}$ of the form $\{\omega : \xi(\omega) \in B\}$, $B \in \mathscr{B}(R)$. It is easily verified that they form a σ-algebra, called the *σ-algebra generated by ξ*, and denoted by $\mathscr{F}_\xi$ or $\sigma(\xi)$.

If φ is a Borel function, it follows from Lemma 2 that the function $\eta = \varphi \circ \xi$ is also a random variable, and in fact $\mathscr{F}_\xi$-measurable, i.e., such that

$$\{\omega : \eta(\omega) \in B\} \in \mathscr{F}_\xi, \quad B \in \mathscr{B}(R)$$

(see (7)). It turns out that the converse is also true.

Theorem 3. *Let $\eta = \eta(\omega)$ be an $\mathscr{F}_\xi$-measurable random variable. Then there is a Borel function φ such that $\eta = \varphi \circ \xi$, i.e. $\eta(\omega) = \varphi(\xi(\omega))$ for every $\omega \in \Omega$.*

PROOF. Let Φ be the class of $\mathscr{F}_\xi$-measurable functions $\eta = \eta(\omega)$ and $\tilde{\Phi}_\xi$ the class of $\mathscr{F}_\xi$-measurable functions representable in the form $\varphi \circ \xi$, where φ is a Borel function. It is clear that $\tilde{\Phi}_\xi \subseteq \Phi_\xi$. The conclusion of the theorem is that in fact $\tilde{\Phi}_\xi = \Phi_\xi$.

Let $A \in \mathscr{F}_\xi$ and $\eta(\omega) = I_A(\omega)$. Let us show that $\eta \in \tilde{\Phi}_\xi$. In fact, if $A \in \mathscr{F}_\xi$ there is a $B \in \mathscr{B}(R)$ such that $A = \{\omega : \xi(\omega) \in B\}$. Let

$$\chi_B(x) = \begin{cases} 1, & x \in B, \\ 0, & x \notin B. \end{cases}$$

Then $I_A(\omega) = \chi_B(\xi(\omega)) \in \tilde{\Phi}_\xi$. Hence it follows that every simple $\mathscr{F}_\xi$-measurable function $\sum_{i=1}^n c_i I_{A_i}(\omega)$, $A_i \in \mathscr{F}_\xi$, also belongs to $\tilde{\Phi}_\xi$.

Now let η be an arbitrary $\mathscr{F}_\xi$-measurable function. By Theorem 1 there is a sequence of simple $\mathscr{F}_\xi$-measurable functions $\{\eta_n\}$ such that $\eta_n(\omega) \to \eta(\omega)$, $n \to \infty$, $\omega \in \Omega$. As we just showed, there are Borel functions $\varphi_n = \varphi_n(x)$ such that $\eta_n(\omega) = \varphi_n(\xi(\omega))$. Moreover $\varphi_n(\xi(\omega)) \to \eta(\omega)$, $n \to \infty$, $\omega \in \Omega$.

Let B denote the set $\{x \in R : \lim_n \varphi_n(x) \text{ exists}\}$. This is a Borel set. Therefore

$$\varphi(x) = \begin{cases} \lim_n \varphi_n(x), & x \in B, \\ 0, & x \notin B \end{cases}$$

is also a Borel function (see Problem 6).

But then it is evident that $\eta(\omega) = \lim_n \varphi_n(\xi(\omega)) = \varphi(\xi(\omega))$ for all $\omega \in \Omega$. Consequently $\tilde{\Phi}_\xi = \Phi_\xi$.

$\square$

6. Consider a measurable space $(\Omega, \mathscr{F})$ and a finite or countably infinite decomposition $\mathscr{D} = \{D_1, D_2, \ldots\}$ of the space Ω: namely, $D_i \in \mathscr{F}$ and $\sum_i D_i = \Omega$. We form the algebra $\mathscr{A}$ containing the empty set $\varnothing$ and the sets of the form $\sum_\alpha D_\alpha$, where the sum is taken in the finite or countably infinite sense. It is evident that the system $\mathscr{A}$ is a monotonic class, and therefore, according to Lemma 2 of Sect. 2, the algebra $\mathscr{A}$ is at the same time a σ-algebra, denoted $\sigma(\mathscr{D})$ and called the σ-algebra generated by the decomposition $\mathscr{D}$. Clearly $\sigma(\mathscr{D}) \subseteq \mathscr{F}$.

Lemma 3. *Let $\xi = \xi(\omega)$ be a $\sigma(\mathscr{D})$-measurable random variable. Then ξ is representable in the form*

$$\xi(\omega) = \sum_{k=1}^\infty x_k I_{D_k}(\omega), \tag{8}$$

where $x_k \in R$, i.e., $\xi(\omega)$ is constant on the elements D_k of the decomposition, $k \geq 1$.

PROOF. Let us choose a set D_k and show that the $\sigma(\mathscr{D})$-measurable function ξ has a constant value on that set. For this purpose, denote

$$x_k = \sup\left[c \colon D_k \cap \{\omega \colon \xi(\omega) < c\} = \varnothing\right].$$

Since $\{\omega \colon \xi(\omega) < x_k\} = \bigcup \{\omega \colon \xi(\omega) < r\}$, where the union is over all rational $r < x_k$, we have

$$D_k \cap \{\omega \colon \xi(\omega) < x_k\} = \varnothing.$$

Now let $c > x_k$. Then $D_k \cap \{\omega \colon \xi(\omega) < c\} \neq \varnothing$, and since the set $\{\omega \colon \xi(\omega) < c\}$ has the form $\sum_\alpha D_\alpha$, where the sum is over a finite or countable collection of indices, we have

$$D_k \cap \{\omega \colon \xi(\omega) < c\} = D_k.$$

Hence, it follows that, for all $c > x_k$,

$$D_k \cap \{\omega \colon \xi(\omega) \geq c\} = \varnothing,$$

and since $\{\omega \colon \xi(\omega) > x_k\} = \bigcup \{\omega \colon \xi(\omega) \geq r\}$, where the union is over all rational $r > x_k$, we have

$$D_k \cap \{\omega \colon \xi(\omega) > x_k\} = \varnothing.$$

Consequently, $D_k \cap \{\omega \colon \xi(\omega) \neq x_k\} = \varnothing$, and therefore

$$D_k \subseteq \{\omega \colon \xi(\omega) = x_k\}$$

as required. $\square$

7. Problems

1. Show that the random variable ξ is continuous if and only if $P(\xi = x) = 0$ for all $x \in R$.
2. If $|\xi|$ is $\mathscr{F}$-measurable, is it true that ξ is also $\mathscr{F}$-measurable?
3. Prove that x^n, $x^+ = \max(x, 0)$, $x^- = -\min(x, 0)$, $|x| = x^+ + x^-$ are Borel functions.
4. If ξ and η are $\mathscr{F}$-measurable, then $\{\omega: \xi(\omega) = \eta(\omega)\} \in \mathscr{F}$.
5. Let ξ and η be random variables on $(\Omega, \mathscr{F})$, and $A \in \mathscr{F}$. Then the function

$$\zeta(\omega) = \xi(\omega) \cdot I_A + \eta(\omega) \cdot I_{\overline{A}}$$

   is also a random variable.
6. Let $\xi_1, \ldots, \xi_n$ be random variables and $\varphi(x_1, \ldots, x_n)$ a Borel function. Show that $\varphi(\xi_1(\omega), \ldots, \xi_n(\omega))$ is also a random variable.
7. Let ξ and η be random variables, both taking the values $1, 2, \ldots, N$. Suppose that $\mathscr{F}_\xi = \mathscr{F}_\eta$. Show that there is a permutation $(i_1, i_2, \ldots, i_N)$ of $(1, 2, \ldots, N)$ such that $\{\omega: \xi = j\} = \{\omega: \eta = i_j\}$ for $j = 1, 2, \ldots, N$.
8. Give an example of a random variable ξ having a density function $f(x)$ such that $\lim_{x \to \infty} f(x)$ does not exist and therefore $f(x)$ does not tend to zero as $n \to \pm\infty$.
   Hint. One solution is suggested by the comment following (5) in Sect. 4. Find another solution, where the required density is continuous.
9. Let ξ and η be bounded random variables ($|\xi| \leq c_1$, $|\eta| \leq c_2$). Prove that if for all $m, n \geq 1$

$$E \xi^m \eta^n = E \xi^m \cdot E \eta^n,$$

   then ξ and η are independent.
10. Let ξ and η be random variables with identical distribution functions F_ξ and F_η. Prove that if $x \in R$ and $\{\omega: \xi(\omega) = x\} \neq \varnothing$, then there exists $y \in R$ such that $\{\omega: \xi(\omega) = x\} = \{\omega: \eta(\omega) = y\}$.
11. Let E be an at most countable subset of R and ξ a mapping $\Omega \to E$. Prove that ξ is a random variable on $(\Omega, \mathscr{F})$ if and only if $\{\omega: \xi(\omega) = x\} \in \mathscr{F}$ for all $x \in E$.

5 Random Elements

1. In addition to random variables, probability theory and its applications involve random objects of more general kinds, for example random points, vectors, functions, processes, fields, sets, measures, etc. In this connection it is desirable to have the concept of a random object of any kind.

Definition 1. Let $(\Omega, \mathscr{F})$ and $(E, \mathscr{E})$ be measurable spaces. We say that a function $X = X(\omega)$, defined on Ω and taking values in E, is $\mathscr{F}/\mathscr{E}$-*measurable*, or is a *random element* (with values in E), if

$$\{\omega : X(\omega) \in B\} \in \mathscr{F} \tag{1}$$

for every $B \in \mathscr{E}$. Random elements (with values in E) are sometimes called *E-valued random variables*.

Let us consider some special cases.

If $(E, \mathscr{E}) = (R, \mathscr{B}(R))$, the definition of a random element is the same as the definition of a random variable (Sect. 4).

Let $(E, \mathscr{E}) = (R^n, \mathscr{B}(R^n))$. Then a random element $X(\omega)$ is a "random point" in R^n. If π_k is the projection of R^n on the kth coordinate axis, $X(\omega)$ can be represented in the form

$$X(\omega) = (\xi_1(\omega), \ldots, \xi_n(\omega)), \tag{2}$$

where $\xi_k = \pi_k \circ X$.

It follows from (1) that ξ_k is an ordinary random variable. In fact, for $B \in \mathscr{B}(R)$ we have

$$\{\omega : \xi_k(\omega) \in B\} = \{\omega : \xi_1(\omega) \in R, \ldots, \xi_{k-1} \in R, \xi_k \in B, \xi_{k+1} \in R, \ldots, \xi_n(\omega) \in R\}$$
$$= \{\omega : X(\omega) \in (R \times \cdots \times R \times B \times R \times \cdots \times R)\} \in \mathscr{F},$$

since $R \times \cdots \times R \times B \times R \times \cdots \times R \in \mathscr{B}(R^n)$.

Definition 2. An ordered set $(\eta_1(\omega), \ldots, \eta_n(\omega))$ of random variables is called an *n-dimensional random vector*.

According to this definition, every random element $X(\omega)$ with values in R^n is an n-dimensional random vector. The converse is also true: every random vector $X(\omega) = (\xi_1(\omega), \ldots, \xi_n(\omega))$ is a random element in R^n. In fact, if $B_k \in \mathscr{B}(R)$, $k = 1, \ldots, n$, then

$$\{\omega : X(\omega) \in (B_1 \times \cdots \times B_n)\} = \bigcap_{k=1}^{n} \{\omega : \xi_k(\omega) \in B_k\} \in \mathscr{F}.$$

But $\mathscr{B}(R^n)$ is the smallest σ-algebra containing the sets $B_1 \times \cdots \times B_n$. Consequently we find immediately, by an evident generalization of Lemma 1 of Sect. 4, that whenever $B \in \mathscr{B}(R^n)$, the set $\{\omega : X(\omega) \in B\}$ belongs to $\mathscr{F}$.

Let $(E, \mathscr{E}) = (\mathbb{Z}, B(\mathbb{Z}))$, where $\mathbb{Z}$ is the set of complex numbers $x + iy$, $x, y \in R$, and $B(\mathbb{Z})$ is the smallest σ-algebra containing the sets $\{z : z = x + iy, a_1 < x \leq b_1, a_2 < y \leq b_2\}$. It follows from the discussion above that a complex-valued random variable $Z(\omega)$ can be represented as $Z(\omega) = X(\omega) + iY(\omega)$, where $X(\omega)$ and $Y(\omega)$ are random variables. Hence we may also call $Z(\omega)$ a *complex random variable*.

Let $(E, \mathscr{E}) = (R^T, \mathscr{B}(R^T))$, where T is a subset of the real line. In this case every random element $X = X(\omega)$ can evidently be represented as $X = (\xi_t)_{t \in T}$ with $\xi_t = \pi_t \circ X$, and is called a *random function* with time domain T.

Definition 3. Let T be a subset of the real line. A set of random variables $X = (\xi_t)_{t \in T}$ is called a *random process** *with time domain* T.

If $T = \{1, 2, \ldots\}$, we call $X = (\xi_1, \xi_2, \ldots)$ a *random process with discrete time*, or *a random sequence*.

If $T = [0, 1]$, $(-\infty, \infty)$, $[0, \infty)$, $\ldots$, we call $X = (\xi_t)_{t \in T}$ a *random process with continuous time*.

It is easy to show, by using the structure of the σ-algebra $\mathcal{B}(R^T)$ (Sect. 2) that every random process $X = (\xi_t)_{t \in T}$ (in the sense of Definition 3) is also a random function (a random element with values in R^T).

Definition 4. Let $X = (\xi_t)_{t \in T}$ be a random process. For each given $\omega \in \Omega$ the function $(\xi_t(\omega))_{t \in T}$ is said to be a *realization* or a *trajectory* of the process corresponding to the outcome ω.

The following definition is a natural generalization of Definition 2 of Sect. 4.

Definition 5. Let $X = (\xi_t)_{t \in T}$ be a random process. The probability measure P_X on $(R^T, \mathcal{B}(R^T))$ defined by

$$P_X(B) = \mathsf{P}\{\omega : X(\omega) \in B\}, \quad B \in \mathcal{B}(R^T),$$

is called the *probability distribution of* X. The probabilities

$$P_{t_1, \ldots, t_n}(B) \equiv \mathsf{P}\{\omega : (\xi_{t_1}, \ldots, \xi_{t_n}) \in B\}, \quad B \in \mathcal{B}(R^n)$$

with $t_1 < t_2 < \cdots < t_n$, $t_i \in T$, are called *finite-dimensional probabilities* (or *probability distributions*). The functions

$$F_{t_1, \ldots, t_n}(x_1, \ldots, x_n) \equiv \mathsf{P}\{\omega : \xi_{t_1} \leq x_1, \ldots, \xi_{t_n} \leq x_n\}$$

with $t_1 < t_2 < \cdots < t_n$, $t_i \in T$, are called *finite-dimensional distribution functions* of the process $X = (\xi_t)_{t \in T}$.

Let $(E, \mathcal{E}) = (C, \mathcal{B}_0(C))$, where C is the space of continuous functions $x = (x_t)_{t \in T}$ on $T = [0, 1]$ and $\mathcal{B}_0(C)$ is the σ-algebra generated by the open sets (Sect. 2). We show that every random element X on $(C, \mathcal{B}_0(C))$ is also a random process with continuous trajectories in the sense of Definition 3.

In fact, according to Sect. 2 the set $A = \{x \in C : x_t < a\}$ is open in $\mathcal{B}_0(C)$. Therefore

$$\{\omega : \xi_t(\omega) < a\} = \{\omega : X(\omega) \in A\} \in \mathcal{F}.$$

On the other hand, let $X = (\xi_t(\omega))_{t \in T}$ be a random process (in the sense of Definition 3) whose trajectories are continuous functions for every $\omega \in \Omega$. According to (17) of Sect. 2

$$\{x \in C : x \in S_\rho(x^0)\} = \bigcap_{t_k} \{x \in C : |x_{t_k} - x_{t_k}^0| < \rho\},$$

---

* Or stochastic process (Translator).

where t_k are the rational points of $[0, 1]$, x^0 is an element of C and

$$S_\rho(x^0) = \{x \in C: \sup_{t \in T} |x_t - x_t^0| < \rho\}.$$

Therefore

$$\{\omega: X(\omega) \in S_\rho(X^0(\omega))\} = \bigcap_{t_k} \{\omega: |\xi_{t_k}(\omega) - \xi_{t_k}^0(\omega)| < \rho\} \in \mathscr{F},$$

and therefore we also have $\{\omega: X(\omega) \in B\} \in \mathscr{F}$ for every $B \in \mathscr{B}_0(C)$.

Similar reasoning will show that every random element of the space $(D, \mathscr{B}_0(D))$ can be considered as a random process with trajectories in the space of functions with no discontinuities of the second kind; and conversely.

2. Let $(\Omega, \mathscr{F}, \mathsf{P})$ be a probability space and $(E_\alpha, \mathscr{E}_\alpha)$ measurable spaces, where α belongs to an (arbitrary) set $\mathfrak{A}$.

Definition 6. We say that the $\mathscr{F}/\mathscr{E}_\alpha$-measurable functions $(X_\alpha(\omega))$, $\alpha \in \mathfrak{A}$, are *independent* (or *mutually independent*) if, for every finite set of indices $\alpha_1, \ldots, \alpha_n$ the random elements $X_{\alpha_1}, \ldots, X_{\alpha_n}$ are independent, i.e.

$$\mathsf{P}(X_{\alpha_1} \in B_{\alpha_1}, \ldots, X_{\alpha_n} \in B_{\alpha_n}) = \mathsf{P}(X_{\alpha_1} \in B_{\alpha_1}) \cdots \mathsf{P}(X_{\alpha_n} \in B_{\alpha_n}), \qquad (3)$$

where $B_\alpha \in \mathscr{E}_\alpha$.

Let $\mathfrak{A} = \{1, 2, \ldots, n\}$, let ξ_α be random variables, $\alpha \in \mathfrak{A}$, and

$$F_\xi(x_1, \ldots, x_n) = \mathsf{P}(\xi_1 \leq x_1, \ldots, \xi_n \leq x_n)$$

be the n-dimensional distribution function of the random vector $\xi = (\xi_1, \ldots, \xi_n)$. Let $F_{\xi_i}(x)$ be the distribution functions of the random variables ξ_i, $i = 1, \ldots, n$.

Theorem. *A necessary and sufficient condition for the random variables $\xi_1, \ldots, \xi_n$ to be independent is that*

$$F_\xi(x_1, \ldots, x_n) = F_{\xi_1}(x_1) \cdots F_{\xi_n}(x_n) \qquad (4)$$

for all $(x_1, \ldots, x_n) \in R^n$.

PROOF. The necessity is evident. To prove the sufficiency we put $a = (a_1, \ldots, a_n)$, $b = (b_1, \ldots, b_n)$,

$$P_\xi(a, b] = \mathsf{P}\{\omega: a_1 < \xi_1 \leq b_1, \ldots, a_n < \xi_n \leq b_n\},$$
$$P_{\xi_i}(a_i, b_i] = \mathsf{P}\{a_i < \xi_i \leq b_i\}.$$

Then

$$P_\xi(a, b] = \prod_{i=1}^n [F_{\xi_i}(b_i) - F_{\xi_i}(a_i)] = \prod_{i=1}^n P_{\xi_i}(a_i, b_i]$$

by (7) of Sect. 3 and (4), and therefore

$$P\{\xi_1 \in I_1, \ldots, \xi_n \in I_n\} = \prod_{i=1}^{n} P\{\xi_i \in I_i\}, \tag{5}$$

where $I_i = (a_i, b_i]$.

We fix $I_2, \ldots, I_n$ and show that

$$P\{\xi_1 \in B_1, \xi_2 \in I_2, \ldots, \xi_n \in I_n\} = P\{\xi_1 \in B_1\} \prod_{i=2}^{n} P\{\xi_i \in I_i\} \tag{6}$$

for all $B_1 \in \mathscr{B}(R)$. Let $\mathscr{M}$ be the collection of sets in $\mathscr{B}(R)$ for which (6) holds (the "principle of appropriate sets," Sect. 2). Then $\mathscr{M}$ evidently contains the algebra $\mathscr{A}$ of sets consisting of sums of disjoint intervals of the form $I_1 = (a_1, b_1]$. Hence $\mathscr{A} \subseteq \mathscr{M} \subseteq \mathscr{B}(R)$. From the countable additivity (and therefore continuity) of probability measures it also follows that $\mathscr{M}$ is a monotonic class. Therefore (see Subsection 1 of Sect. 2)

$$\mu(\mathscr{A}) \subseteq \mathscr{M} \subseteq \mathscr{B}(R).$$

But $\mu(\mathscr{A}) = \sigma(\mathscr{A}) = \mathscr{B}(R)$ by Theorem 1 of Sect. 2. Therefore $\mathscr{M} = \mathscr{B}(R)$.

Thus (6) is established. Now fix $B_1, I_3, \ldots, I_n$; by the same method we can establish (6) with I_2 replaced by the Borel set B_2. Continuing in this way, we can evidently arrive at the required equation,

$$P(\xi_1 \in B_1, \ldots, \xi_n \in B_n) = P(\xi_1 \in B_1) \cdots P(\xi_n \in B_n),$$

where $B_i \in \mathscr{B}(R)$. This completes the proof of the theorem.

$\square$

3. PROBLEMS

1. Let $\xi_1, \ldots, \xi_n$ be discrete random variables. Show that they are independent if and only if

$$P(\xi_1 = x_1, \ldots, \xi_n = x_n) = \prod_{i=1}^{n} P(\xi_i = x_i)$$

   for all real $x_1, \ldots, x_n$.
2. Carry out the proof that every random function $X(\omega) = (\xi_t(\omega))_{t \in T}$ is a random process (in the sense of Definition 3) and conversely.
3. Let $X_1, \ldots, X_n$ be random elements with values in $(E_1, \mathscr{E}_1), \ldots, (E_n, \mathscr{E}_n)$, respectively. In addition let $(E_1', \mathscr{E}_1'), \ldots, (E_n', \mathscr{E}_n')$ be measurable spaces and let $g_1, \ldots, g_n$ be $\mathscr{E}_1/\mathscr{E}_1', \ldots, \mathscr{E}_n/\mathscr{E}_n'$-measurable functions, respectively. Show that if $X_1, \ldots, X_n$ are independent, the random elements $g_1 \circ X_1, \ldots, g_n \circ X_n$ are also independent.
4. Let $X_1, X_2, \ldots$ be an infinite sequence of *exchangeable* random variables (i.e., such that the joint distribution of any collection of k random variables with distinct subscripts, say, $X_{i_1}, \ldots, X_{i_k}$, depends only on k and does not depend on

the specific choice of pairwise distinct $i_1, \ldots, i_k$; compare with the definition in Problem 11 of Sect. 1). Prove that if $\mathsf{E} X_n^2 < \infty, n \geq 1$, then $\mathrm{Cov}(X_1, X_2) \geq 0$.

5. Let ξ, η, ζ be independent random variables. Prove that $\xi + \eta$ and ζ^2 are independent.

6. Let $\xi_1, \ldots, \xi_m, \eta_1, \ldots, \eta_n$ be random variables. Consider the random vectors $X = (\xi_1, \ldots, \xi_m)$ and $Y = (\eta_1, \ldots, \eta_n)$. Suppose the following conditions are fulfilled:

   (i) The random variables $\xi_1, \ldots, \xi_m$ are independent;
   (ii) The random variables $\eta_1, \ldots, \eta_n$ are independent;
   (iii) The random vectors X and Y treated respectively as R^m- and R^n-valued random elements are independent.

   Prove that the random variables $\xi_1, \ldots, \xi_m, \eta_1, \ldots, \eta_n$ are mutually independent.

7. Suppose $X = (\xi_1, \ldots, \xi_m)$ and $Y = (\eta_1, \ldots, \eta_n)$ are random vectors such that the random variables $\xi_1, \ldots, \xi_m, \eta_1, \ldots, \eta_n$ are mutually independent.

   (i) Prove that the random vectors X and Y treated as random elements are independent (compare with Problem 6).
   (ii) Let $f : R^m \to R$, $g : R^n \to R$ be Borel functions. Prove that the random variables $f(\xi_1, \ldots, \xi_m)$ and $g(\eta_1, \ldots, \eta_n)$ are independent.

6 Lebesgue Integral: Expectation

1. When $(\Omega, \mathscr{F}, \mathsf{P})$ is a *finite* probability space and $\xi = \xi(\omega)$ is a simple random variable,

$$\xi(\omega) = \sum_{k=1}^{n} x_k I_{A_k}(\omega), \tag{1}$$

the expectation $\mathsf{E}\xi$ was defined in Sect. 4 of Chap. 1. The same definition of the expectation $\mathsf{E}\xi$ of a *simple* random variable ξ can be used for *any* probability space $(\Omega, \mathscr{F}, \mathsf{P})$. That is, we *define*

$$\mathsf{E}\xi = \sum_{k=1}^{n} x_k \, \mathsf{P}(A_k). \tag{2}$$

This definition is consistent (in the sense that $\mathsf{E}\xi$ is independent of the particular representation of ξ in the form (1)), as can be shown just as for finite probability spaces. The simplest properties of the expectation can be established similarly (see Subsection 5 of Sect. 4, Chap. 1).

   In the present section we shall define and study the properties of the expectation $\mathsf{E}\xi$ of an *arbitrary* random variable. In the language of analysis, $\mathsf{E}\xi$ is merely the Lebesgue integral of the $\mathscr{F}$-measurable function $\xi = \xi(\omega)$ with respect to the measure P. In addition to $\mathsf{E}\xi$ we shall use the notation $\int_\Omega \xi(\omega) \, \mathsf{P}(d\omega)$ or $\int_\Omega \xi \, d\mathsf{P}$.

2. Let $\xi = \xi(\omega)$ be a nonnegative random variable. We construct a sequence of simple nonnegative random variables $\{\xi_n\}_{n \geq 1}$ such that $\xi_n(\omega) \uparrow \xi(\omega)$, $n \to \infty$, for each $\omega \in \Omega$ (see Theorem 1 in Sect. 4).

Since $\mathsf{E}\,\xi_n \leq \mathsf{E}\,\xi_{n+1}$ (cf. Property (3) in Subsection 5 of Sect. 4, Chap. 1), the limit $\lim_n \mathsf{E}\,\xi_n$ exists, possibly with the value $+\infty$.

Definition 1. The *Lebesgue integral* of the nonnegative random variable $\xi = \xi(\omega)$, or its *expectation*, is

$$\mathsf{E}\,\xi \equiv \lim_n \mathsf{E}\,\xi_n. \tag{3}$$

To see that this definition is consistent, we need to show that the limit is independent of the choice of the approximating sequence $\{\xi_n\}$. In other words, we need to show that if $\xi_n \uparrow \xi$ and $\eta_m \uparrow \xi$, where $\{\eta_m\}$ is a sequence of simple functions, then

$$\lim_n \mathsf{E}\,\xi_n = \lim_m \mathsf{E}\,\eta_m. \tag{4}$$

Lemma 1. *Let η and ξ_n be simple nonnegative random variables, $n \geq 1$, and*

$$\xi_n \uparrow \xi \geq \eta.$$

Then

$$\lim_n \mathsf{E}\,\xi_n \geq \mathsf{E}\,\eta. \tag{5}$$

PROOF. Let $\varepsilon > 0$ and

$$A_n = \{\omega : \xi_n \geq \eta - \varepsilon\}.$$

It is clear that $A_n \uparrow \Omega$ and

$$\xi_n = \xi_n I_{A_n} + \xi_n I_{\overline{A}_n} \geq \xi_n I_{A_n} \geq (\eta - \varepsilon) I_{A_n}.$$

Hence by the properties of the expectations of simple random variables we find that

$$\mathsf{E}\,\xi_n \geq \mathsf{E}(\eta - \varepsilon) I_{A_n} = \mathsf{E}\,\eta I_{A_n} - \varepsilon\,\mathsf{P}(A_n)$$
$$= \mathsf{E}\,\eta - \mathsf{E}\,\eta I_{\overline{A}_n} - \varepsilon\,\mathsf{P}(A_n) \geq \mathsf{E}\,\eta - C\,\mathsf{P}(\overline{A}_n) - \varepsilon,$$

where $C = \max_\omega \eta(\omega)$. Since ε is arbitrary, the required inequality (5) follows.
$\square$

It follows from this lemma that $\lim_n \mathsf{E}\,\xi_n \geq \lim_m \mathsf{E}\,\eta_m$ and by symmetry $\lim_m \mathsf{E}\,\eta_m \geq \lim_n \mathsf{E}\,\xi_n$, which proves (4).

The following remark is often useful.

Remark 1. The expectation $\mathsf{E}\,\xi$ of the nonnegative random variable ξ satisfies

$$\mathsf{E}\,\xi = \sup_{\{s \in S\,:\, s \leq \xi\}} \mathsf{E}\,s, \tag{6}$$

where $S = \{s\}$ is the set of simple random variables (Problem 1).

Thus the expectation is well defined for *nonnegative* random variables. We now consider the general case.

Let ξ be a random variable and $\xi^+ = \max(\xi, 0)$, $\xi^- = -\min(\xi, 0)$.

Definition 2. We say that the expectation $\mathsf{E}\,\xi$ of the random variable ξ *exists*, or *is defined*, if at least one of $\mathsf{E}\,\xi^+$ and $\mathsf{E}\,\xi^-$ is finite:

$$\min(\mathsf{E}\,\xi^+, \mathsf{E}\,\xi^-) < \infty.$$

In this case we *define*

$$\mathsf{E}\,\xi \equiv \mathsf{E}\,\xi^+ - \mathsf{E}\,\xi^-.$$

The *expectation* $\mathsf{E}\,\xi$ is also called the *Lebesgue integral* of the function ξ with respect to the probability measure P.

Definition 3. We say that the *expectation of ξ is finite* if $\mathsf{E}\,\xi^+ < \infty$ and $\mathsf{E}\,\xi^- < \infty$.

Since $|\xi| = \xi^+ + \xi^-$, the finiteness of $\mathsf{E}\,\xi$ is equivalent to $\mathsf{E}\,|\xi| < \infty$. (In this sense one says that the Lebesgue integral is absolutely convergent.)

Remark 2. In addition to the expectation $\mathsf{E}\,\xi$, significant numerical characteristics of a random variable ξ are the number $\mathsf{E}\,\xi^r$ (if defined) and $\mathsf{E}\,|\xi|^r$, $r > 0$, which are known as the *moment* of order r (or rth moment) and the *absolute moment* of order r (or absolute rth moment) of ξ.

Remark 3. In the definition of the Lebesgue integral $\int_\Omega \xi(\omega)\,\mathsf{P}(d\omega)$ given above, we supposed that P was a probability measure ($\mathsf{P}(\Omega) = 1$) and that the $\mathscr{F}$-measurable functions (random variables) ξ had values in $R = (-\infty, \infty)$. Suppose now that μ is *any* measure defined on a measurable space $(\Omega, \mathscr{F})$ and possibly taking the value $+\infty$, and that $\xi = \xi(\omega)$ is an $\mathscr{F}$-measurable function with values in $\overline{R} = [-\infty, \infty]$ (an extended random variable). In this case the Lebesgue integral $\int_\Omega \xi(\omega)\,\mu(d\omega)$ is defined in the same way: first, for nonnegative simple ξ (by (2) with P replaced by μ), then for arbitrary nonnegative ξ, and in general by the formula

$$\int_\Omega \xi(\omega)\,\mu(d\omega) = \int_\Omega \xi^+\mu(d\omega) - \int_\Omega \xi^-\mu(d\omega),$$

provided that no indeterminacy of the form $\infty - \infty$ arises.

A case that is particularly important for mathematical analysis is that in which $(\Omega, F) = (R, \mathscr{B}(R))$ and μ is Lebesgue measure. In this case the integral $\int_R \xi(x)\mu(dx)$ is written $\int_R \xi(x)\,dx$, or $\int_{-\infty}^{\infty} \xi(x)\,dx$, or $(\mathrm{L})\int_{-\infty}^{\infty} \xi(x)\,dx$ to emphasize its difference from the Riemann integral $(\mathrm{R})\int_{-\infty}^{\infty} \xi(x)dx$. If the measure μ (Lebesgue–Stieltjes) corresponds to a generalized distribution function $G = G(x)$, the integral $\int_R \xi(x)\,\mu(dx)$ is also called a *Lebesgue–Stieltjes integral* and is denoted by $(\mathrm{L\text{–}S})\int_R \xi(x)\,G(dx)$, a notation that distinguishes it from the corresponding Riemann–Stieltjes integral

$$(\text{R--S}) \int_R \xi(x)\, G(dx)$$

(see Subsection 11 below).

It will be clear from what follows (Property **D**) that if $\mathsf{E}\,\xi$ is defined then so is the expectation $\mathsf{E}(\xi I_A)$ for every $A \in \mathscr{F}$. The notation $\mathsf{E}(\xi; A)$ or $\int_A \xi\, d\mathsf{P}$ is often used for $\mathsf{E}(\xi I_A)$ or its equivalent, $\int_\Omega \xi I_A\, d\,\mathsf{P}$. The integral $\int_A \xi\, d\,\mathsf{P}$ is called the *Lebesgue integral of ξ with respect to P over the set A.*

Similarly, we write $\int_A \xi\, d\mu$ instead of $\int_\Omega \xi \cdot I_A\, d\mu$ for an arbitrary measure μ. In particular, if μ is an n-dimensional Lebesgue–Stieltjes measure, and $A = (a_1 b_1] \times \cdots \times (a_n, b_n]$, the notation

$$\int_{a_1}^{b_1} \cdots \int_{a_n}^{b_n} \xi(x_1, \ldots, x_n)\, \mu(dx_1, \ldots, dx_n) \quad \text{instead of} \quad \int_A \xi\, d\mu,$$

is often used. If μ is Lebesgue measure, we write simply $dx_1 \cdots dx_n$ instead of $\mu(dx_1, \ldots, dx_n)$.

3. Properties of the expectation $\mathsf{E}\,\xi$ of a random variable ξ.

A. *Let c be a constant and let $\mathsf{E}\,\xi$ exist. Then $\mathsf{E}(c\xi)$ exists and*

$$\mathsf{E}(c\xi) = c\,\mathsf{E}\,\xi.$$

B. *Let $\xi \leq \eta$; then*

$$\mathsf{E}\,\xi \leq \mathsf{E}\,\eta$$

with the understanding that

$$\text{if } -\infty < \mathsf{E}\,\xi \text{ then } -\infty < \mathsf{E}\,\eta \text{ and } \mathsf{E}\,\xi \leq \mathsf{E}\,\eta$$

or

$$\text{if } \mathsf{E}\,\eta < \infty \text{ then } \mathsf{E}\,\xi < \infty \text{ and } \mathsf{E}\,\xi \leq \mathsf{E}\,\eta.$$

C. *If $\mathsf{E}\,\xi$ exists then*

$$|\mathsf{E}\,\xi| \leq \mathsf{E}\,|\xi|.$$

D. *If $\mathsf{E}\,\xi$ exists then $\mathsf{E}(\xi I_A)$ exists for each $A \in \mathscr{F}$; if $\mathsf{E}\,\xi$ is finite, $\mathsf{E}(\xi I_A)$ is finite.*
E. *If ξ and η are nonnegative random variables, or such that $\mathsf{E}\,|\xi| < \infty$ and $\mathsf{E}\,|\eta| < \infty$, then*

$$\mathsf{E}(\xi + \eta) = \mathsf{E}\,\xi + \mathsf{E}\,\eta.$$

(See Problem 2 for a generalization.)

Let us establish **A–E**.

A. This is obvious for simple random variables. Let $\xi \geq 0$, $\xi_n \uparrow \xi$, where ξ_n are simple random variables and $c \geq 0$. Then $c\xi_n \uparrow c\xi$ and therefore

$$\mathsf{E}(c\xi) = \lim \mathsf{E}(c\xi_n) = c \lim \mathsf{E}\,\xi_n = c\,\mathsf{E}\,\xi.$$

In the general case we need to use the representation $\xi = \xi^+ - \xi^-$ and notice that $(c\xi)^+ = c\xi^+$, $(c\xi)^- = c\xi^-$ when $c \geq 0$, whereas when $c < 0$, $(c\xi)^+ = -c\xi^-$, $(c\xi)^- = -c\xi^+$.

B. If $0 \leq \xi \leq \eta$, then $\mathsf{E}\,\xi$ and $\mathsf{E}\,\eta$ are defined and the inequality $\mathsf{E}\,\xi \leq \mathsf{E}\,\eta$ follows directly from (6). Now let $\mathsf{E}\,\xi > -\infty$; then $\mathsf{E}\,\xi^- < \infty$. If $\xi \leq \eta$, we have $\xi^+ \leq \eta^+$ and $\xi^- \geq \eta^-$. Therefore $\mathsf{E}\,\eta^- \leq \mathsf{E}\,\xi^- < \infty$; consequently $\mathsf{E}\,\eta$ is defined and $\mathsf{E}\,\xi = \mathsf{E}\,\xi^+ - \mathsf{E}\,\xi^- \leq \mathsf{E}\,\eta^+ - \mathsf{E}\,\eta^- = \mathsf{E}\,\eta$. The case when $\mathsf{E}\,\eta < \infty$ can be discussed similarly.

C. Since $-|\xi| \leq \xi \leq |\xi|$, Properties **A** and **B** imply

$$- \mathsf{E}\,|\xi| \leq \mathsf{E}\,\xi \leq \mathsf{E}\,|\xi|,$$

i.e., $|\mathsf{E}\,\xi| \leq \mathsf{E}\,|\xi|$.

D. This follows from **B** and

$$(\xi I_A)^+ = \xi^+ I_A \leq \xi^+, \qquad (\xi I_A)^- = \xi^- I_A \leq \xi^-.$$

E. Let $\xi \geq 0$, $\eta \geq 0$, and let $\{\xi_n\}$ and $\{\eta_n\}$ be sequences of simple functions such that $\xi_n \uparrow \xi$ and $\eta_n \uparrow \eta$. Then $\mathsf{E}(\xi_n + \eta_n) = \mathsf{E}\,\xi_n + \mathsf{E}\,\eta_n$ and

$$\mathsf{E}(\xi_n + \eta_n) \uparrow \mathsf{E}(\xi + \eta), \quad \mathsf{E}\,\xi_n \uparrow \mathsf{E}\,\xi, \quad \mathsf{E}\,\eta_n \uparrow \mathsf{E}\,\eta$$

and therefore $\mathsf{E}(\xi + \eta) = \mathsf{E}\,\xi + \mathsf{E}\,\eta$. The case when $\mathsf{E}\,|\xi| < \infty$ and $\mathsf{E}\,|\eta| < \infty$ reduces to this if we use the facts that

$$\xi = \xi^+ - \xi^-, \quad \eta = \eta^+ - \eta^-, \quad \xi^+ \leq |\xi|, \quad \xi^- \leq |\xi|,$$

and

$$\eta^+ \leq |\eta|, \qquad \eta^- \leq |\eta|.$$

The following group of statements about expectations involve the notion of "P-almost surely." We say that a property holds "P-*almost surely*" *if there is a set* $N \in \mathscr{F}$ *with* $\mathsf{P}(N) = 0$ *such that the property holds for every point* ω *of* $\Omega \backslash N$. Instead of "P-almost surely" we often say "P-almost everywhere" or simply "almost surely" (a.s.) or "almost everywhere" (a.e.).

F. *If* $\xi = 0$ *(a.s.) then* $\mathsf{E}\,\xi = 0$.

In fact, if ξ is a simple random variable, $\xi = \sum x_k I_{A_k}(\omega)$ and $x_k \neq 0$, we have $\mathsf{P}(A_k) = 0$ by hypothesis and therefore $\mathsf{E}\,\xi = 0$. If $\xi \geq 0$ and $0 \leq s \leq \xi$, where s is a simple random variable, then $s = 0$ (a.s.) and consequently $\mathsf{E}\,s = 0$ and $\mathsf{E}\,\xi = \sup_{\{s \in S: \, s \leq \xi\}} \mathsf{E}\,s = 0$. The general case follows from this by means of the representation $\xi = \xi^+ - \xi^-$ and the facts that $\xi^+ \leq |\xi|$, $\xi^- \leq |\xi|$, and $|\xi| = 0$ (a.s.).

G. *If* $\xi = \eta$ *(a.s.) and* $\mathsf{E}\,|\xi| < \infty$, *then* $\mathsf{E}\,|\eta| < \infty$ *and* $\mathsf{E}\,\xi = \mathsf{E}\,\eta$ (*see also Problem 3*).

In fact, let $N = \{\omega: \xi \neq \eta\}$. Then $\mathsf{P}(N) = 0$ and $\xi = \xi I_N + \xi I_{\overline{N}}$, $\eta = \eta I_N + \eta I_{\overline{N}}$. By properties **E** and **F**, we have $\mathsf{E}\,\xi = \mathsf{E}\,\xi I_N + \mathsf{E}\,\xi I_{\overline{N}} = \mathsf{E}\,\xi I_{\overline{N}} = \mathsf{E}\,\eta I_{\overline{N}}$. But $\mathsf{E}\,\eta I_N = 0$, and therefore $\mathsf{E}\,\xi = \mathsf{E}\,\eta I_{\overline{N}} + \mathsf{E}\,\eta I_N = \mathsf{E}\,\eta$, by Property **E**.

H. *Let* $\xi \geq 0$ *and* $\mathsf{E}\,\xi = 0$. *Then* $\xi = 0$ *(a.s.)*.

For the proof, let $A = \{\omega \colon \xi(\omega) > 0\}$, $A_n = \{\omega \colon \xi(\omega) \geq 1/n\}$. It is clear that $A_n \uparrow A$ and $0 \leq \xi I_{A_n} \leq \xi I_A$. Hence, by Property **B**,

$$0 \leq \mathsf{E}\,\xi I_{A_n} \leq \mathsf{E}\,\xi = 0.$$

Consequently

$$0 = \mathsf{E}\,\xi I_{A_n} \geq \frac{1}{n} \mathsf{P}(A_n)$$

and therefore $\mathsf{P}(A_n) = 0$ for all $n \geq 1$. But $\mathsf{P}(A) = \lim \mathsf{P}(A_n)$ and therefore $\mathsf{P}(A) = 0$.

I. *Let* ξ *and* η *be such that* $\mathsf{E}\,|\xi| < \infty$, $\mathsf{E}\,|\eta| < \infty$ *and* $\mathsf{E}(\xi I_A) \leq \mathsf{E}(\eta I_A)$ *for all* $A \in \mathscr{F}$. *Then* $\xi \leq \eta$ *(a.s.)*.

In fact, let $B = \{\omega \colon \xi(\omega) > \eta(\omega)\}$. Then $\mathsf{E}(\eta I_B) \leq \mathsf{E}(\xi I_B) \leq \mathsf{E}(\eta I_B)$ and therefore $\mathsf{E}(\xi I_B) = \mathsf{E}(\eta I_B)$. By Property **E**, we have $\mathsf{E}((\xi - \eta) I_B) = 0$ and by Property **H** we have $(\xi - \eta) I_B = 0$ (a.s.), whence $\mathsf{P}(B) = 0$.

J. *Let* ξ *be an extended random variable and* $\mathsf{E}\,|\xi| < \infty$. *Then* $|\xi| < \infty$ *(a. s.)*

In fact, let $A = \{\omega \colon |\xi(\omega)| = \infty\}$ and $\mathsf{P}(A) > 0$. Then $\mathsf{E}\,|\xi| \geq \mathsf{E}(|\xi| I_A) = \infty \cdot \mathsf{P}(A) = \infty$, which contradicts the hypothesis $\mathsf{E}\,|\xi| < \infty$. (See also Problem 4.)

4. Here we consider the fundamental theorems on *taking limits* under the expectation sign (or the Lebesgue integral sign).

Theorem 1 (On Monotone Convergence). *Let* η, ξ, ξ_1, ξ_2, $\ldots$ *be random variables.*

(a) *If* $\xi_n \geq \eta$ *for all* $n \geq 1$, $\mathsf{E}\,\eta > -\infty$, *and* $\xi_n \uparrow \xi$, *then*

$$\mathsf{E}\,\xi_n \uparrow \mathsf{E}\,\xi.$$

(b) *If* $\xi_n \leq \eta$ *for all* $n \geq 1$, $\mathsf{E}\,\eta < \infty$, *and* $\xi_n \downarrow \xi$, *then*

$$\mathsf{E}\,\xi_n \downarrow \mathsf{E}\,\xi.$$

PROOF. (a) First suppose that $\eta \geq 0$. For each $k \geq 1$ let $\{\xi_k^{(n)}\}_{n \geq 1}$ be a sequence of simple functions such that $\xi_k^{(n)} \uparrow \xi_k$, $n \to \infty$. Put $\zeta^{(n)} = \max_{1 \leq k \leq n} \xi_k^{(n)}$. Then

$$\zeta^{(n-1)} \leq \zeta^{(n)} = \max_{1 \leq k \leq n} \xi_k^{(n)} \leq \max_{1 \leq k \leq n} \xi_k = \xi_n.$$

Let $\zeta = \lim_n \zeta^{(n)}$. Since

$$\xi_k^{(n)} \leq \zeta^{(n)} \leq \xi_n$$

for $1 \le k \le n$, we find by taking limits as $n \to \infty$ that

$$\xi_k \le \zeta \le \xi$$

for every $k \ge 1$ and therefore $\xi = \zeta$.

The random variables $\zeta^{(n)}$ are simple and $\zeta^{(n)} \uparrow \zeta$. Therefore

$$\mathsf{E}\,\xi = \mathsf{E}\,\zeta = \lim \mathsf{E}\,\zeta^{(n)} \le \lim \mathsf{E}\,\xi_n.$$

On the other hand, it is obvious, since $\xi_n \le \xi_{n+1} \le \xi$, that

$$\lim \mathsf{E}\,\xi_n \le \mathsf{E}\,\xi.$$

Consequently $\lim \mathsf{E}\,\xi_n = \mathsf{E}\,\xi$.

Now let η be any random variable with $\mathsf{E}\,\eta > -\infty$.

If $\mathsf{E}\,\eta = \infty$ then $\mathsf{E}\,\xi_n = \mathsf{E}\,\xi = \infty$ by Property **B**, and our proposition is proved. Let $\mathsf{E}\,\eta < \infty$. Together with the assumption that $\mathsf{E}\,\eta > -\infty$ this implies $\mathsf{E}\,|\eta| < \infty$. It is clear that $0 \le \xi_n - \eta \uparrow \xi - \eta$ for all $\omega \in \Omega$. Therefore by what has been established, $\mathsf{E}(\xi_n - \eta) \uparrow \mathsf{E}(\xi - \eta)$ and therefore (by Property **E** and Problem 2)

$$\mathsf{E}\,\xi_n - \mathsf{E}\,\eta \uparrow \mathsf{E}\,\xi - \mathsf{E}\,\eta.$$

But $\mathsf{E}\,|\eta| < \infty$, and therefore $\mathsf{E}\,\xi_n \uparrow \mathsf{E}\,\xi$, $n \to \infty$.

The proof of (b) follows from (a) if we replace the original variables by their negatives.
$\square$

Corollary. *Let* $\{\eta_n\}_{n \ge 1}$ *be a sequence of nonnegative random variables. Then*

$$\mathsf{E}\sum_{n=1}^{\infty} \eta_n = \sum_{n-1}^{\infty} \mathsf{E}\,\eta_n.$$

The proof follows from Property **E** (see also Problem 2), the monotone convergence theorem, and the remark that

$$\sum_{n=1}^{k} \eta_n \uparrow \sum_{n=1}^{\infty} \eta_n, \quad k \to \infty.$$

$\square$

Theorem 2 (Fatou's Lemma). *Let* η, ξ_1, ξ_2, $\ldots$ *be random variables.*

(a) *If* $\xi_n \ge \eta$ *for all* $n \ge 1$ *and* $\mathsf{E}\,\eta > -\infty$, *then*

$$\mathsf{E} \liminf \xi_n \le \liminf \mathsf{E}\,\xi_n.$$

(b) *If* $\xi_n \le \eta$ *for all* $n \ge 1$ *and* $\mathsf{E}\,\eta < \infty$, *then*

$$\limsup \mathsf{E}\,\xi_n \le \mathsf{E} \limsup \xi_n.$$

(c) *If $|\xi_n| \leq \eta$ for all $n \geq 1$ and $\mathsf{E}\,\eta < \infty$, then*

$$\mathsf{E} \liminf \xi_n \leq \liminf \mathsf{E}\,\xi_n \leq \limsup \mathsf{E}\,\xi_n \leq \mathsf{E} \limsup \xi_n. \tag{7}$$

PROOF. (a) Let $\zeta_n = \inf_{m \geq n} \xi_m$; then

$$\liminf_n \xi_n = \lim_n \inf_{m \geq n} \xi_m = \lim_n \zeta_n.$$

It is clear that $\zeta_n \uparrow \liminf \xi_n$ and $\zeta_n \geq \eta$ for all $n \geq 1$. Then by Theorem 1

$$\mathsf{E} \liminf \xi_n = \mathsf{E} \lim_n \zeta_n = \lim_n \mathsf{E}\,\zeta_n = \liminf_n \mathsf{E}\,\zeta_n \leq \liminf_n \mathsf{E}\,\xi_n,$$

which establishes (a). The second conclusion follows from the first. The third is a corollary of the first two.

□

Theorem 3 (Lebesgue's Theorem on Dominated Convergence). *Let $\eta, \xi, \xi_1, \xi_2, \ldots$ be random variables such that $|\xi_n| \leq \eta$, $\mathsf{E}\,\eta < \infty$ and $\xi_n \to \xi$ (a.s.). Then $\mathsf{E}\,|\xi| < \infty$,*

$$\mathsf{E}\,\xi_n \to \mathsf{E}\,\xi \tag{8}$$

and

$$\mathsf{E}\,|\xi_n - \xi| \to 0 \tag{9}$$

as $n \to \infty$.

PROOF. By hypothesis, $\liminf \xi_n = \limsup \xi_n = \xi$ (a.s.). Therefore by Property **G** and Fatou's lemma (item (c))

$$\mathsf{E}\,\xi = \mathsf{E} \liminf \xi_n \leq \liminf \mathsf{E}\,\xi_n = \limsup \mathsf{E}\,\xi_n = \mathsf{E} \limsup \xi_n = \mathsf{E}\,\xi,$$

which establishes (8). It is also clear that $|\xi| \leq \eta$. Hence $\mathsf{E}\,|\xi| < \infty$.

Conclusion (9) can be proved in the same way if we observe that $|\xi_n - \xi| \leq 2\eta$.

□

Corollary. *Let $\eta, \xi, \xi_1, \xi_2, \ldots$ be random variables such that $|\xi_n| \leq \eta$, $\xi_n \to \xi$ (a.s.) and $\mathsf{E}\,\eta^p < \infty$ for some $p > 0$. Then $\mathsf{E}\,|\xi|^p < \infty$ and $\mathsf{E}\,|\xi - \xi_n|^p \to 0$, $n \to \infty$.*

For the proof, it is sufficient to observe that

$$|\xi| \leq \eta, \quad |\xi - \xi_n|^p \leq (|\xi| + |\xi_n|)^p \leq (2\eta)^p.$$

The condition "$|\xi_n| \leq \eta$, $\mathsf{E}\,\eta < \infty$" that appears in Fatou's lemma and the dominated convergence theorem and ensures the validity of formulas (7)–(9) can be somewhat weakened. In order to be able to state the corresponding result (Theorem 4), we introduce the following definition.

Definition 4. A family $\{\xi_n\}_{n\geq 1}$ of random variables is said to be *uniformly integrable* (with respect to the measure P) if

$$\sup_n \int_{\{|\xi_n|>c\}} |\xi_n|\, P(d\omega) \to 0, \quad c \to \infty, \tag{10}$$

or, in a different notation,

$$\sup_n E[|\xi_n|I_{\{|\xi_n|>c\}}] \to 0, \quad c \to \infty. \tag{11}$$

It is clear that if ξ_n, $n \geq 1$, satisfy $|\xi_n| \leq \eta$, $E\eta < \infty$, then the family $\{\xi_n\}_{n\geq 1}$ is uniformly integrable.

Theorem 4. *Let $\{\xi_n\}_{n\geq 1}$ be a uniformly integrable family of random variables. Then*

(a) $E\liminf \xi_n \leq \liminf E\xi_n \leq \limsup E\xi_n \leq E\limsup \xi_n$.
(b) *If in addition $\xi_n \to \xi$ (a.s.) then ξ is integrable and*

$$E\xi_n \to E\xi, \quad n \to \infty,$$
$$E|\xi_n - \xi| \to 0, \quad n \to \infty.$$

PROOF. (a) For every $c > 0$

$$E\xi_n = E[\xi_n I_{\{\xi_n < -c\}}] + E[\xi_n I_{\{\xi_n \geq -c\}}]. \tag{12}$$

By uniform integrability, for every $\varepsilon > 0$ we can take c so large that

$$\sup_n |E[\xi_n I_{\{\xi_n < -c\}}]| < \varepsilon. \tag{13}$$

By Fatou's lemma,

$$\liminf E[\xi_n I_{\{\xi_n \geq -c\}}] \geq E[\liminf \xi_n I_{\{\xi_n \geq -c\}}].$$

But $\xi_n I_{\{\zeta_n \geq -c\}} \geq \xi_n$ and therefore

$$\liminf E[\xi_n I_{\{\xi_n \geq -c\}}] \geq E[\liminf \xi_n]. \tag{14}$$

From (12)–(14) we obtain

$$\liminf E\xi_n \geq E[\liminf \xi_n] - \varepsilon.$$

Since $\varepsilon > 0$ is arbitrary, it follows that $\liminf E\xi_n \geq E\liminf \xi_n$. The inequality with upper limits, $\limsup E\xi_n \leq E\limsup \xi_n$, is proved similarly.

Conclusion (b) can be deduced from (a) as in Theorem 3.

$\square$

The deeper significance of the concept of uniform integrability is revealed by the following theorem, which gives a necessary and sufficient condition for taking limits under the expectation sign.

Theorem 5. *Let* $0 \leq \xi_n \to \xi$ *(P- a.s.) and* $\mathsf{E}\,\xi_n < \infty$, $n \geq 1$. *Then* $\mathsf{E}\,\xi_n \to \mathsf{E}\,\xi < \infty$ *if and only if the family* $\{\xi_n\}_{n \geq 1}$ *is uniformly integrable.*

PROOF. The sufficiency follows from conclusion (b) of Theorem 4. For the proof of the necessity we consider the (at most countable) set

$$A = \{a \colon \mathsf{P}(\xi = a) > 0\}.$$

Then we have $\xi_n I_{\{\xi_n < a\}} \to \xi I_{\{\xi < a\}}$ for each $a \notin A$, and the family

$$\{\xi_n I_{\{\xi_n < a\}}\}_{n \geq 1}$$

is uniformly integrable. Hence, by the sufficiency part of the theorem, we have $\mathsf{E}\,\xi_n I_{\{\xi_n < a\}} \to \mathsf{E}\,\xi I_{\{\xi < a\}}$, $a \notin A$, and therefore

$$\mathsf{E}\,\xi_n I_{\{\xi_n \geq a\}} \to \mathsf{E}\,\xi I_{\{\xi \geq a\}}, \quad a \notin A, \ n \to \infty. \tag{15}$$

Take an $\varepsilon > 0$ and choose $a_0 \notin A$ so large that $\mathsf{E}\,\xi I_{\{\xi \geq a_0\}} < \varepsilon/2$; then choose N_0 so large that

$$\mathsf{E}\,\xi_n I_{\{\xi_n \geq a_0\}} \leq \mathsf{E}\,\xi I_{\{\xi \geq a_0\}} + \varepsilon/2$$

for all $n \geq N_0$, and consequently $\mathsf{E}\,\xi_n I_{\{\xi_n \geq a_0\}} \leq \varepsilon$. Then choose $a_1 \geq a_0$ so large that $\mathsf{E}\,\xi_n I_{\{\xi_n \geq a_1\}} \leq \varepsilon$ for all $n \leq N_0$. Then we have

$$\sup_n \mathsf{E}\,\xi_n I_{\{\xi_n \geq a_1\}} \leq \varepsilon,$$

which establishes the uniform integrability of the family $\{\xi_n\}_{n \geq 1}$ of random variables.

$\square$

5. Let us notice some tests for uniform integrability.

We first observe that if $\{\xi_n\}$ is a family of uniformly integrable random variables, then

$$\sup_n \mathsf{E}\,|\xi_n| < \infty. \tag{16}$$

In fact, for a given $\varepsilon > 0$ and sufficiently large $c > 0$

$$\sup_n \mathsf{E}\,|\xi_n| = \sup_n [\mathsf{E}(|\xi_n| I_{\{|\xi_n| \geq c\}}) + \mathsf{E}(|\xi_n| I_{\{|\xi_n| < c\}})]$$
$$\leq \sup_n \mathsf{E}(|\xi_n| I_{\{|\xi_n| \geq c\}}) + \sup_n \mathsf{E}(|\xi_n| I_{(|\xi_n| < c)}) \leq \varepsilon + c,$$

which establishes (16).

It turns out that (16) together with a condition of uniform continuity is necessary and sufficient for uniform integrability.

Lemma 2. *A necessary and sufficient condition for a family* $\{\xi_n\}_{n \geq 1}$ *of random variables to be uniformly integrable is that* $\mathsf{E}\,|\xi_n|$, $n \geq 1$, *are uniformly bounded (i.e.,* (16) *holds) and that* $\mathsf{E}\{|\xi_n| I_A\}$, $n \geq 1$, *are uniformly continuous (i.e.* $\sup_n \mathsf{E}\{|\xi_n| I_A\} \to 0$ *when* $\mathsf{P}(A) \to 0$).

PROOF. *Necessity*. Condition (16) was verified above. Moreover,

$$E\{|\xi_n|I_A\} = E\{|\xi_n|I_{A\cap\{|\xi_n|\geq c\}}\} + E\{|\xi_n|I_{A\cap\{|\xi_n|<c\}}\}$$
$$\leq E\{|\xi_n|I_{\{|\xi_n|\geq c\}}\} + cP(A). \tag{17}$$

Take c so large that $\sup_n E\{|\xi_n|I_{\{|\xi_n|\geq c\}}\} \leq \varepsilon/2$. Then if $P(A) \leq \varepsilon/2c$, we have

$$\sup_n E\{|\xi_n|I_A\} \leq \varepsilon$$

by (17). This establishes the uniform continuity.

Sufficiency. Let $\varepsilon > 0$ and let $\delta > 0$ be chosen so that $P(A) < \delta$ implies that $E(|\xi_n|I_A) \leq \varepsilon$, uniformly in n. Since

$$E|\xi_n| \geq E|\xi_n|I_{\{|\xi_n|\geq c\}} \geq cP\{|\xi_n| \geq c\}$$

for every $c > 0$ (cf. Chebyshev's inequality), we have

$$\sup_n P\{|\xi_n| \geq c\} \leq \frac{1}{c}\sup E|\xi_n| \to 0, \qquad c \to \infty,$$

and therefore, when c is sufficiently large, any set $\{|\xi_n| \geq c\}$, $n \geq 1$, can be taken as A. Therefore $\sup E(|\xi_n|I_{\{|\xi_n|\geq c\}}) \leq \varepsilon$, which establishes the uniform integrability. This completes the proof of the lemma.

□

The following proposition provides a simple sufficient condition for uniform integrability.

Lemma 3. *Let $\xi_1, \xi_2, \ldots$ be a sequence of integrable random variables and $G = G(t)$ a nonnegative increasing function, defined for $t \geq 0$, such that*

$$\lim_{t\to\infty} \frac{G(t)}{t} = \infty, \tag{18}$$

$$\sup_n E[G(|\xi_n|)] < \infty. \tag{19}$$

Then the family $\{\xi_n\}_{n\geq 1}$ is uniformly integrable.

PROOF. Let $\varepsilon > 0$, $M = \sup_n E[G(|\xi_n|)]$, $a = M/\varepsilon$. Take c so large that $G(t)/t \geq a$ for $t \geq c$. Then

$$E[|\xi_n|I_{\{|\xi_n|\geq c\}}] \leq \frac{1}{a}E[G(|\xi_n|) \cdot I_{\{|\xi_n|\geq c\}}] \leq \frac{M}{a} = \varepsilon$$

uniformly for $n \geq 1$.

□

6. If ξ and η are independent simple random variables, we can show, as in Subsection 5 of Sect. 4, Chap. 1, that $E\xi\eta = E\xi \cdot E\eta$. Let us now establish a similar proposition in the general case (see also Problem 6).

Theorem 6. *Let ξ and η be independent random variables, $\mathsf{E}\,|\xi| < \infty$, $\mathsf{E}\,|\eta| < \infty$.*
Then $\mathsf{E}\,|\xi\eta| < \infty$ and

$$\mathsf{E}\,\xi\eta = \mathsf{E}\,\xi \cdot \mathsf{E}\,\eta. \tag{20}$$

PROOF. First let $\xi \geq 0$, $\eta \geq 0$. Put

$$\xi_n = \sum_{k=0}^{\infty} \frac{k}{n} I_{\{k/n \leq \xi(\omega) < (k+1)/n\}},$$

$$\eta_n = \sum_{k=0}^{\infty} \frac{k}{n} I_{\{k/n \leq \eta(\omega) < (k+1)/n\}}.$$

Then $\xi_n \leq \xi$, $|\xi_n - \xi| \leq 1/n$ and $\eta_n \leq \eta$, $|\eta_n - \eta| \leq 1/n$. Since $\mathsf{E}\,\xi < \infty$ and
$\mathsf{E}\,\eta < \infty$, it follows from Lebesgue's dominated convergence theorem that

$$\lim \mathsf{E}\,\xi_n = \mathsf{E}\,\xi, \quad \lim \mathsf{E}\,\eta_n = \mathsf{E}\,\eta.$$

Moreover, since ξ and η are independent,

$$\mathsf{E}\,\xi_n\eta_n = \sum_{k,l \geq 0} \frac{kl}{n^2} \mathsf{E}\,I_{\{k/n \leq \xi < (k+1)/n\}} I_{\{l/n \leq \eta < (l+1)/n\}}$$

$$= \sum_{k,l \geq 0} \frac{kl}{n^2} \mathsf{E}\,I_{\{k/n \leq \xi < (k+1)/n\}} \cdot \mathsf{E}\,I_{\{l/n \leq \eta < (l+1)/n\}} = \mathsf{E}\,\xi_n \cdot \mathsf{E}\,\eta_n.$$

Now notice that

$$|\mathsf{E}\,\xi\eta - \mathsf{E}\,\xi_n\eta_n| \leq \mathsf{E}\,|\xi\eta - \xi_n\eta_n| \leq \mathsf{E}[\xi \cdot |\eta - \eta_n|]$$
$$+ \mathsf{E}[\eta_n \cdot |\xi - \xi_n|] \leq \frac{1}{n}\mathsf{E}\,\xi + \frac{1}{n}\mathsf{E}\left(\eta + \frac{1}{n}\right) \to 0, \ n \to \infty.$$

Therefore $\mathsf{E}\,\xi\eta = \lim_n \mathsf{E}\,\xi_n\eta_n = \lim \mathsf{E}\,\xi_n \cdot \lim \mathsf{E}\,\eta_n = \mathsf{E}\,\xi \cdot \mathsf{E}\,\eta$, and $\mathsf{E}\,\xi\eta < \infty$.

The general case reduces to this one if we use the representations $\xi = \xi^+ - \xi^-$, $\eta = \eta^+ - \eta^-$, $\xi\eta = \xi^+\eta^+ - \xi^-\eta^+ - \xi^+\eta^- + \xi^-\eta^-$. This completes the proof.
$\square$

7. The inequalities for expectations that we develop in this subsection are regularly
used both in probability theory and in analysis.

Chebyshev's (Bienaymé–Chebyshev's) Inequality. *Let ξ be a nonnegative ran-*
dom variable. Then for every $\varepsilon > 0$

$$\mathsf{P}(\xi \geq \varepsilon) \leq \frac{\mathsf{E}\,\xi}{\varepsilon}. \tag{21}$$

The proof follows immediately from

$$\mathsf{E}\,\xi \geq \mathsf{E}[\xi \cdot I_{\{\xi \geq \varepsilon\}}] \geq \varepsilon\,\mathsf{E}\,I_{\{\xi \geq \varepsilon\}} = \varepsilon\,\mathsf{P}(\xi \geq \varepsilon).$$

From (21) we can obtain the following versions of Chebyshev's inequality: *If ξ is any random variable then*

$$P(|\xi| \geq \varepsilon) \leq \frac{\mathsf{E}\,\xi^2}{\varepsilon^2} \tag{22}$$

and

$$P(|\xi - \mathsf{E}\,\xi| \geq \varepsilon) \leq \frac{\operatorname{Var}\xi}{\varepsilon^2}, \tag{23}$$

where $\operatorname{Var}\xi = \mathsf{E}(\xi - \mathsf{E}\,\xi)^2$ *is the variance of* ξ.

The Cauchy–Bunyakovskii Inequality. *Let ξ and η satisfy $\mathsf{E}\,\xi^2 < \infty$, $\mathsf{E}\,\eta^2 < \infty$. Then $\mathsf{E}\,|\xi\eta| < \infty$ and*

$$(\mathsf{E}\,|\xi\eta|)^2 \leq \mathsf{E}\,\xi^2 \cdot \mathsf{E}\,\eta^2. \tag{24}$$

PROOF. Suppose that $\mathsf{E}\,\xi^2 > 0$, $\mathsf{E}\,\eta^2 > 0$. Then, with $\tilde{\xi} = \xi/\sqrt{\mathsf{E}\,\xi^2}, \tilde{\eta} = \eta/\sqrt{\mathsf{E}\,\eta^2}$, we find, since $2|\tilde{\xi}\tilde{\eta}| \leq \tilde{\xi}^2 + \tilde{\eta}^2$, that

$$2\,\mathsf{E}\,|\tilde{\xi}\tilde{\eta}| \leq \mathsf{E}\,\tilde{\xi}^2 + \mathsf{E}\,\tilde{\eta}^2 = 2,$$

i.e. $\mathsf{E}\,|\tilde{\xi}\tilde{\eta}| \leq 1$, which establishes (24).

On the other hand if, say, $\mathsf{E}\,\xi^2 = 0$, then $\xi = 0$ (a.s.) by Property **I**, and then $\mathsf{E}\,\xi\eta = 0$ by Property **F**, i.e. (24) is still satisfied.

□

Jensen's Inequality. *Let the Borel function $g = g(x)$ defined on R be convex downward and ξ a random variable such that $\mathsf{E}\,|\xi| < \infty$. Then*

$$g(\mathsf{E}\,\xi) \leq \mathsf{E}\,g(\xi). \tag{25}$$

PROOF. If $g = g(x)$ is convex downward, for each $x_0 \in R$ there is a number $\lambda(x_0)$ such that

$$g(x) \geq g(x_0) + (x - x_0) \cdot \lambda(x_0) \tag{26}$$

for all $x \in R$. Putting $x = \xi$ and $x_0 = \mathsf{E}\,\xi$, we find from (26) that

$$g(\xi) \geq g(\mathsf{E}\,\xi) + (\xi - \mathsf{E}\,\xi) \cdot \lambda(\mathsf{E}\,\xi),$$

and consequently $\mathsf{E}\,g(\xi) \geq g(\mathsf{E}\,\xi)$.

□

Remark 4. Jensen's inequality (25) holds also for vector random variables $\xi = (\xi_1, \ldots, \xi_d)$ with $\mathsf{E}\,|\xi_i| < \infty$, $i = 1, \ldots, d$, and functions $g = g(x)$, $x \in R^d$, convex downward (i.e., functions $g \colon R^d \to R$ such that $g(px + (1 - p)y) \leq pg(x) + (1 - p)g(y)$, $x, y \in R^d$, $p \in [0, 1]$).

A whole series of useful inequalities can be derived from Jensen's inequality. We obtain the following one as an example.

Lyapunov's Inequality. *If* $0 < s < t$,

$$(E |\xi|^s)^{1/s} \leq (E |\xi|^t)^{1/t}. \tag{27}$$

To prove this, let $r = t/s$. Then, putting $\eta = |\xi|^s$ and applying Jensen's inequality to $g(x) = |x|^r$, we obtain $| E \eta|^r \leq E |\eta|^r$, i.e.

$$(E |\xi|^s)^{t/s} \leq E |\xi|^t,$$

which establishes (27).

The following chain of *inequalities among absolute moments* is a consequence of Lyapunov's inequality:

$$E |\xi| \leq (E |\xi|^2)^{1/2} \leq \cdots \leq (E |\xi|^n)^{1/n}. \tag{28}$$

Hölder's Inequality. *Let* $1 < p < \infty$, $1 < q < \infty$, *and* $(1/p) + (1/q) = 1$. *If* $E |\xi|^p < \infty$ *and* $E |\eta|^q < \infty$, *then* $E |\xi\eta| < \infty$ *and*

$$E |\xi\eta| \leq (E |\xi|^p)^{1/p}(E |\eta|^q)^{1/q}. \tag{29}$$

PROOF. If $E |\xi|^p = 0$ or $E |\eta|^q = 0$, (29) follows immediately as for the Cauchy–Bunyakovskii inequality (which is the special case $p = q = 2$ of Hölder's inequality).

Now let $E |\xi|^p > 0$, $E |\eta|^q > 0$ and

$$\tilde{\xi} = \frac{\xi}{(E |\xi|^p)^{1/p}}, \quad \tilde{\eta} = \frac{\eta}{(E |\eta|^q)^{1/q}}.$$

We apply the inequality

$$x^a y^b \leq ax + by, \tag{30}$$

which holds for positive x, y, a, b and $a + b = 1$, and follows immediately from the concavity of the logarithm:

$$\log[ax + by] \geq a \log x + b \log y = \log x^a y^b.$$

Then, putting $x = |\tilde{\xi}|^p$, $y = |\tilde{\eta}|^q$, $a = 1/p$, $b = 1/q$, we find that

$$|\tilde{\xi}\tilde{\eta}| \leq \frac{1}{p}|\tilde{\xi}|^p + \frac{1}{q}|\tilde{\eta}|^q,$$

whence

$$E |\tilde{\xi}\tilde{\eta}| \leq \frac{1}{p} E |\tilde{\xi}|^p + \frac{1}{q} E |\tilde{\eta}|^q = \frac{1}{p} + \frac{1}{q} = 1.$$

This establishes (29).

$\square$

Minkowski's Inequality. *If* $\mathsf{E}\,|\xi|^p < \infty$, $\mathsf{E}\,|\eta|^p < \infty$, $1 \le p < \infty$, *then we have* $\mathsf{E}\,|\xi + \eta|^p < \infty$ *and*

$$(\mathsf{E}\,|\xi + \eta|^p)^{1/p} \le (\mathsf{E}\,|\xi|^p)^{1/p} + (\mathsf{E}\,|\eta|^p)^{1/p}. \tag{31}$$

PROOF. We begin by establishing the following inequality: if a, $b > 0$ and $p \ge 1$, then

$$(a + b)^p \le 2^{p-1}(a^p + b^p). \tag{32}$$

In fact, consider the function $F(x) = (a + x)^p - 2^{p-1}(a^p + x^p)$. Then

$$F'(x) = p(a + x)^{p-1} - 2^{p-1}px^{p-1},$$

and since $p \ge 1$, we have $F'(a) = 0$, $F'(x) > 0$ for $x < a$ and $F'(x) < 0$ for $x > a$. Therefore

$$F(b) \le \max F(x) = F(a) = 0,$$

from which (32) follows.

According to this inequality,

$$|\xi + \eta|^p \le (|\xi| + |\eta|)^p \le 2^{p-1}(|\xi|^p + |\eta|^p) \tag{33}$$

and therefore if $\mathsf{E}\,|\xi|^p < \infty$ and $\mathsf{E}\,|\eta|^p < \infty$ it follows that $\mathsf{E}\,|\xi + \eta|^p < \infty$.

If $p = 1$, inequality (31) follows from (33).

Now suppose that $p > 1$. Take $q > 1$ so that $(1/p) + (1/q) = 1$. Then

$$|\xi + \eta|^p = |\xi + \eta| \cdot |\xi + \eta|^{p-1} \le |\xi| \cdot |\xi + \eta|^{p-1} + |\eta||\xi + \eta|^{p-1}. \tag{34}$$

Notice that $(p - 1)q = p$. Consequently

$$\mathsf{F}(|\xi + \eta|^{p-1})^q = \mathsf{E}\,|\xi + \eta|^p < \infty,$$

and therefore by Hölder's inequality

$$\begin{aligned}
\mathsf{E}(|\xi||\xi + \eta|^{p-1}) &\le (\mathsf{E}\,|\xi|^p)^{1/p}(\mathsf{E}\,|\xi + \eta|^{(p-1)q})^{1/q} \\
&= (\mathsf{E}\,|\xi|^p)^{1/p}(\mathsf{E}\,|\xi + \eta|^p)^{1/q} < \infty.
\end{aligned}$$

In the same way,

$$\mathsf{E}(|\eta||\xi + \eta|^{p-1}) \le (\mathsf{E}\,|\eta|^p)^{1/p}(\mathsf{E}\,|\xi + \eta|^p)^{1/q}.$$

Consequently, by (34),

$$\mathsf{E}\,|\xi + \eta|^p \le (\mathsf{E}\,|\xi + \eta|^p)^{1/q}((\mathsf{E}\,|\xi|^p)^{1/p} + (\mathsf{E}\,|\eta|^p)^{1/p}). \tag{35}$$

If $E|\xi + \eta|^p = 0$, the desired inequality (31) is evident. Now let $E|\xi + \eta|^p > 0$. Then we obtain

$$(E|\xi + \eta|^p)^{1-(1/q)} \leq (E|\xi|^p)^{1/p} + (E|\eta|^p)^{1/p}$$

from (35), and (31) follows since $1 - (1/q) = 1/p$.

$\square$

7. Let ξ be a random variable for which $E\xi$ is defined. Then, by Property **D**, the *set function*

$$Q(A) \equiv \int_A \xi\, dP, \quad A \in \mathscr{F}, \tag{36}$$

is well defined. Let us show that this function is *countably additive*.

First suppose that ξ is nonnegative. If $A_1, A_2, \ldots$ are pairwise disjoint sets from $\mathscr{F}$ and $A = \sum A_n$, the corollary to Theorem 1 implies that

$$Q(A) = E(\xi \cdot I_A) = E(\xi \cdot I_{\Sigma A_n}) = E(\sum \xi \cdot I_{A_n})$$
$$= \sum E(\xi \cdot I_{A_n}) = \sum Q(A_n).$$

If ξ is an arbitrary random variable for which $E\xi$ is defined, the countable additivity of $Q(A)$ follows from the representation

$$Q(A) = Q^+(A) - Q^-(A), \tag{37}$$

where

$$Q^+(A) = \int_A \xi^+ dP, \quad Q^-(A) = \int_A \xi^- dP,$$

together with the countable additivity for nonnegative random variables and the fact that $\min(Q^+(\Omega), Q^-(\Omega)) < \infty$.

Thus if $E\xi$ is defined, the set function $Q = Q(A)$ is a signed measure—a countably additive set function representable as $Q = Q_1 - Q_2$, where at least one of the measures Q_1 and Q_2 is finite.

We now show that $Q = Q(A)$ has the following important property of *absolute continuity* with respect to P:

$$\text{if } P(A) = 0 \text{ then } Q(A) = 0 \quad (A \in \mathscr{F})$$

(this property is denoted by the abbreviation $Q \ll P$).

To prove this property it is sufficient to consider nonnegative random variables. If $\xi = \sum_{k=1}^n x_k I_{A_k}$ is a simple nonnegative random variable and $P(A) = 0$, then

$$Q(A) = E(\xi \cdot I_A) = \sum_{k=1}^n x_k P(A_k \cap A) = 0.$$

If $\{\xi_n\}_{n\geq 1}$ is a sequence of nonnegative simple functions such that $\xi_n \uparrow \xi \geq 0$, then the theorem on monotone convergence shows that

$$Q(A) = \mathsf{E}(\xi \cdot I_A) = \lim \mathsf{E}(\xi_n \cdot I_A) = 0,$$

since $\mathsf{E}(\xi_n \cdot I_A) = 0$ for all $n \geq 1$ and A with $\mathsf{P}(A) = 0$.

Thus the Lebesgue integral $Q(A) = \int_A \xi\,d\mathsf{P}$, considered as a function of sets $A \in \mathscr{F}$, is a *signed measure* that is absolutely continuous with respect to P ($Q \ll \mathsf{P}$). It is quite remarkable that the converse is also valid.

Radon–Nikodým Theorem. *Let $(\Omega, \mathscr{F})$ be a measurable space, μ a σ-finite measure, and λ a signed measure (i.e., $\lambda = \lambda_1 - \lambda_2$, where at least one of the measures λ_1 and λ_2 is finite), which is absolutely continuous with respect to μ. Then there is an $\mathscr{F}$-measurable function $f = f(\omega)$ with values in $\overline{R} = [-\infty, \infty]$ such that*

$$\lambda(A) = \int_A f(\omega)\,\mu(d\omega), \quad A \in \mathscr{F}. \tag{38}$$

The function $f(\omega)$ is unique up to sets of μ-measure zero: if $h = h(\omega)$ is another $\mathscr{F}$-measurable function such that $\lambda(A) = \int_A h(\omega)\mu(d\omega)$, $A \in \mathscr{F}$, then $\mu\{\omega : f(\omega) \neq h(\omega)\} = 0$.

If λ is a measure, then $f = f(\omega)$ has its values in $\overline{R}^+ = [0, \infty]$.

The function $f = f(\omega)$ in the representation (38) is called the *Radon–Nikodým derivative* or the *density* of the measure λ with respect to μ, and denoted by $d\lambda/d\mu$ or $(d\lambda/d\mu)(\omega)$.

Some important properties of these derivatives are presented in the lemma in Subsection 8 of the next Sect. 7. Here we state an especially useful particular case of formula (35) therein, which is often used for recalculation of expectations under a change of measure.

Namely, let P and $\widetilde{\mathsf{P}}$ be two probability measures and E and $\widetilde{\mathsf{E}}$ the corresponding expectations. Suppose that $\widetilde{\mathsf{P}}$ is absolutely continuous with respect to P (denoted $\widetilde{\mathsf{P}} \ll \mathsf{P}$). Then for any nonnegative random variable $\xi = \xi(\omega)$ the following "*formula for recalculation of expectations*" holds:

$$\widetilde{\mathsf{E}}\xi = \mathsf{E}\Big[\xi\frac{d\widetilde{\mathsf{P}}}{d\mathsf{P}}\Big]. \tag{39}$$

This formula remains valid also without assuming that ξ is nonnegative with the following modification of the statement: the random variable ξ is integrable with respect to $\widetilde{\mathsf{P}}$ if and only if $\xi\frac{d\widetilde{\mathsf{P}}}{d\mathsf{P}}$ is integrable with respect to P; then (39) is valid.

The proof of (39) is not hard: for simple functions ξ it follows directly from the definition of the derivative $\frac{d\widetilde{\mathsf{P}}}{d\mathsf{P}}$, and for nonnegative ξ we use Theorem 1 (b) of Sect. 4, which states the existence of simple functions $\xi_n \uparrow \xi$, $n \to \infty$, and then Theorem 1 (a) on monotone convergence. For an arbitrary ξ we have by (39) that $\widetilde{\mathsf{E}}|\xi| = \mathsf{E}|\xi|\frac{d\widetilde{\mathsf{P}}}{d\mathsf{P}}$. This implies that ξ is integrable with respect to $\widetilde{\mathsf{P}}$ if and only if $\xi\frac{d\widetilde{\mathsf{P}}}{d\mathsf{P}}$ is integrable with respect to P. The formula (39) follows then from the representation $\xi = \xi^+ - \xi^-$.

The Radon–Nikodým theorem, which we quote without proof (for the proof see, e.g., [39]), will play a key role in the construction of conditional expectations (Sect. 7).

9. If $\xi = \sum_{i=1}^{n} x_i I_{A_i}$ is a simple random variable, $A_i = \{\omega : \xi = x_i\}$, then

$$\mathsf{E}\, g(\xi) = \sum g(x_i) \mathsf{P}(A_i) = \sum g(x_i) \Delta F_\xi(x_i).$$

In other words, in order to calculate the expectation of a function of the (simple) random variable ξ it is unnecessary to know the probability measure P completely; it is enough to know the probability distribution P_ξ or, equivalently, the distribution function F_ξ of ξ.

The following important theorem generalizes this property.

Theorem 7 (Change of Variables in a Lebesgue Integral). *Let $(\Omega, \mathscr{F})$ and $(E, \mathscr{E})$ be measurable spaces and $X = X(\omega)$ an $\mathscr{F}/\mathscr{E}$-measurable function with values in E. Let P be a probability measure on $(\Omega, \mathscr{F})$ and P_X the probability measure on $(E, \mathscr{E})$ induced by $X = X(\omega)$:*

$$P_X(A) = \mathsf{P}\{\omega : X(\omega) \in A\}, \quad A \in \mathscr{E}. \tag{40}$$

Then

$$\int_A g(x)\, P_X(dx) = \int_{X^{-1}(A)} g(X(\omega))\, \mathsf{P}(d\omega), \quad A \in \mathscr{E}, \tag{41}$$

for every $\mathscr{E}$-measurable function $g = g(x)$, $x \in E$ (in the sense that if one integral exists, the other is well defined, and the two are equal).

PROOF. Let $A \in \mathscr{E}$ and $g(x) = I_B(x)$, where $B \in \mathscr{E}$. Then (41) becomes

$$P_X(AB) = \mathsf{P}(X^{-1}(A) \cap X^{-1}(B)), \tag{42}$$

which follows from (40) and the observation that $X^{-1}(A) \cap X^{-1}(B) = X^{-1}(A \cap B)$.

It follows from (42) that (41) is valid for nonnegative simple functions $g = g(x)$, and therefore, by the monotone convergence theorem, also for all nonnegative $\mathscr{E}$-measurable functions.

In the general case we need only represent g as $g^+ - g^-$. Then, since (41) is valid for g^+ and g^-, if (for example) $\int_A g^+(x)\, P_X(dx) < \infty$, we have

$$\int_{X^{-1}(A)} g^+(X(\omega))\, \mathsf{P}(d\omega) < \infty$$

also, and therefore the existence of $\int_A g(x)\, P_X(dx)$ implies the existence of $\int_{X^{-1}(A)} g(X(\omega))\, \mathsf{P}(d\omega)$. $\quad\square$

Corollary. *Let $(E, \mathscr{E}) = (R, \mathscr{B}(R))$ and let $\xi = \xi(\omega)$ be a random variable with probability distribution P_ξ. Then if $g = g(x)$ is a Borel function and either of the integrals $\int_A g(x)\, P_\xi(dx)$ or $\int_{\xi^{-1}(A)} g(\xi(\omega))\, \mathsf{P}(d\omega)$ exists, we have*

$$\int_A g(x) P_\xi(dx) = \int_{\xi^{-1}(A)} g(\xi(\omega))\, \mathsf{P}(d\omega).$$

In particular, for $A = R$ we obtain

$$\mathsf{E}\, g(\xi(\omega)) = \int_\Omega g(\xi(\omega))\, \mathsf{P}(d\omega) = \int_R g(x)\, P_\xi(dx). \tag{43}$$

The measure P_ξ can be uniquely reconstructed from the distribution function F_ξ (Theorem 1 of Sect. 3). Hence the Lebesgue integral $\int_R g(x)\, P_\xi(dx)$ is often denoted by $\int_R g(x)\, F_\xi(dx)$ and called a *Lebesgue–Stieltjes integral* (with respect to the measure corresponding to the distribution function $F_\xi(x)$).

Let us consider the case when $F_\xi(x)$ has a density $f_\xi(x)$, i.e., let

$$F_\xi(x) = \int_{-\infty}^x f_\xi(y)\, dy, \tag{44}$$

where $f_\xi = f_\xi(x)$ is a nonnegative Borel function and the integral is a Lebesgue integral with respect to Lebesgue measure on the set $(-\infty, x]$ (see Remark 3). With the assumption of (44), formula (43) takes the form

$$\mathsf{E}\, g(\xi(\omega)) = \int_{-\infty}^\infty g(x) f_\xi(x)\, dx, \tag{45}$$

where the integral is the Lebesgue integral of the function $g(x) f_\xi(x)$ with respect to Lebesgue measure. In fact, if $g(x) = I_B(x)$, $B \in \mathscr{B}(R)$, the formula becomes

$$P_\xi(B) = \int_B f_\xi(x)\, dx, \quad B \in \mathscr{B}(R); \tag{46}$$

its correctness follows from Theorem 1 of Sect. 3 and the formula

$$F_\xi(b) - F_\xi(a) = \int_a^b f_\xi(x)\, dx.$$

In the general case, the proof is the same as for Theorem 7.

10. Let us consider the special case of measurable spaces $(\Omega, \mathscr{F})$ with a measure ρ, where $\Omega = \Omega_1 \times \Omega_2$, $\mathscr{F} = \mathscr{F}_1 \otimes \mathscr{F}_2$, and $\rho = \rho_1 \times \rho_2$ is the direct product of measures ρ_1 and ρ_2 (i.e., the measure on $\mathscr{F}$ such that

$$\rho_1 \times \rho_2(A \times B) = \rho_1(A_1)\rho_2(B), \quad A \in \mathscr{F}_1, \quad B \in \mathscr{F}_2;$$

the existence of this measure follows from the proof of Theorem 8.

The following theorem plays the same role as the theorem on the reduction of a double Riemann integral to an iterated integral.

Theorem 8 (Fubini's Theorem). *Let $\xi = \xi(\omega_1, \omega_2)$ be an $\mathscr{F}_1 \otimes \mathscr{F}_2$-measurable function, integrable with respect to the measure $\rho_1 \times \rho_2$:*

$$\int_{\Omega_1 \times \Omega_2} |\xi(\omega_1, \omega_2)| \, d(\rho_1 \times \rho_2) < \infty. \tag{47}$$

Then the integrals $\int_{\Omega_1} \xi(\omega_1, \omega_2) \, \rho_1(d\omega_1)$ and $\int_{\Omega_2} \xi(\omega_1, \omega_2) \, \rho_2(d\omega_2)$

(1) *are defined for ρ_2-almost all ω_2 and ρ_1-almost all ω_1;*

(2) *are respectively $\mathscr{F}_2$- and $\mathscr{F}_1$-measurable functions with*

$$\rho_2 \left\{ \omega_2 : \int_{\Omega_1} |\xi(\omega_1, \omega_2)| \, \rho_1(d\omega_1) = \infty \right\} = 0,$$

$$\tag{48}$$

$$\rho_1 \left\{ \omega_1 : \int_{\Omega_2} |\xi(\omega_1, \omega_2)| \, \rho_2(d\omega_2) = \infty \right\} = 0,$$

   and

(3)

$$\int_{\Omega_1 \times \Omega_2} \xi(\omega_1, \omega_2) \, d(\rho_1 \times \rho_2) = \int_{\Omega_1} \left[\int_{\Omega_2} \xi(\omega_1, \omega_2) \, \rho_2(d\omega_2) \right] \rho_1(d\omega_1)$$

$$\tag{49}$$

$$= \int_{\Omega_2} \left[\int_{\Omega_1} \xi(\omega_1, \omega_2) \, \rho_1(d\omega_1) \right] \rho_2(d\omega_2).$$

PROOF. We first show that $\xi_{\omega_1}(\omega_2) = \xi(\omega_1, \omega_2)$ is $\mathscr{F}_2$-measurable with respect to ω_2, for each $\omega_1 \in \Omega_1$.

Let $F \in \mathscr{F}_1 \otimes \mathscr{F}_2$ and $\xi(\omega_1, \omega_2) = I_F(\omega_1, \omega_2)$. Let

$$F_{\omega_1} = \{\omega_2 \in \Omega_2 : (\omega_1, \omega_2) \in F\}$$

be the *cross-section* of F at ω_1, and let $\mathscr{C}_{\omega_1} = \{F \in \mathscr{F} : F_{\omega_1} \in \mathscr{F}_2\}$. We must show that $\mathscr{C}_{\omega_1} = \mathscr{F}$ for every ω_1.

If $F = A \times B$, $A \in \mathscr{F}_1$, $B \in \mathscr{F}_2$, then

$$(A \times B)_{\omega_1} = \begin{cases} B & \text{if } \omega_1 \in A, \\ \varnothing & \text{if } \omega_1 \notin A. \end{cases}$$

Hence rectangles with measurable sides belong to $\mathscr{C}_{\omega_1}$. In addition, if $F \in \mathscr{F}$, then $(\overline{F})_{\omega_1} = \overline{F_{\omega_1}}$, and if $\{F^n\}_{n \geq 1}$ are sets in $\mathscr{F}$, then $(\bigcup F^n)_{\omega_1} = \bigcup F^n_{\omega_1}$. It follows that $\mathscr{C}_{\omega_1} = \mathscr{F}$.

Now let $\xi(\omega_1, \omega_2) \geq 0$. Then, since the function $\xi(\omega_1, \omega_2)$ is $\mathscr{F}_2$-measurable for each ω_1, the integral $\int_{\Omega_2} \xi(\omega_1, \omega_2) \, \rho_2(d\omega_2)$ is defined. Let us show that this integral is an $\mathscr{F}_1$-measurable function and

$$\int_{\Omega_1} \left[\int_{\Omega_2} \xi(\omega_1, \omega_2) \, \rho_2(d\omega_2) \right] \rho_1(d\omega_1) = \int_{\Omega_1 \times \Omega_2} \xi(\omega_1, \omega_2) \, d(\rho_1 \times \rho_2). \quad (50)$$

Let us suppose that $\xi(\omega_1, \omega_2) = I_{A \times B}(\omega_1, \omega_2)$, $A \in \mathscr{F}_1$, $B \in \mathscr{F}_2$. Then since $I_{A \times B}(\omega_1, \omega_2) = I_A(\omega_1) I_B(\omega_2)$, we have

$$\int_{\Omega_2} I_{A \times B}(\omega_1, \omega_2) \, \rho_2(d\omega_2) = I_A(\omega_1) \int_{\Omega_2} I_B(\omega_2) \, \rho_2(d\omega_2) \quad (51)$$

and consequently the integral on the left of (51) is an $\mathscr{F}_1$-measurable function.

Now let $\xi(\omega_1, \omega_2) = I_F(\omega_1, \omega_2)$, $F \in \mathscr{F} = \mathscr{F}_1 \otimes \mathscr{F}_2$. Let us show that the integral $f(\omega_1) = \int_{\Omega_2} I_F(\omega_1, \omega_2) \, \rho_2(d\omega_2)$ is $\mathscr{F}_1$-measurable. For this purpose we put $\mathscr{C} = \{F \in \mathscr{F} : f(\omega_1) \text{ is } \mathscr{F}_1\text{-measurable}\}$. According to what has been proved, the set $A \times B$ (where $A \in \mathscr{F}_1$, $B \in \mathscr{F}_2$) belongs to $\mathscr{C}$ and therefore the algebra $\mathscr{A}$ consisting of finite sums of disjoint sets of this form also belongs to $\mathscr{C}$. It follows from the monotone convergence theorem that $\mathscr{C}$ is a monotonic class, $\mathscr{C} = \mu(\mathscr{C})$. Therefore, because of the inclusions $\mathscr{A} \subseteq \mathscr{C} \subseteq \mathscr{F}$ and Theorem 1 of Sect. 2, we have $\mathscr{F} = \sigma(\mathscr{A}) = \mu(\mathscr{A}) \subseteq \mu(\mathscr{C}) = \mathscr{C} \subseteq \mathscr{F}$, i.e., $\mathscr{C} = \mathscr{F}$.

Finally, if $\xi(\omega_1, \omega_2)$ is an arbitrary nonnegative $\mathscr{F}$-measurable function, the $\mathscr{F}_1$-measurability of the integral $\int_{\Omega_2} \xi(\omega_1, \omega_2) \, \rho_2(d\omega)$ follows from the monotone convergence theorem and Theorem 2 of Sect. 4.

Let us now show that the measure $\rho = \rho_1 \times \rho_2$ defined on $\mathscr{F} = \mathscr{F}_1 \otimes \mathscr{F}_2$, with the property $(\rho_1 \times \rho_2)(A \times B) = \rho_1(A) \cdot \rho_2(B)$, $A \in \mathscr{F}_1$, $B \in \mathscr{F}_2$, actually exists and is unique.

For $F \in \mathscr{F}$ we put

$$\rho(F) = \int_{\Omega_1} \left[\int_{\Omega_2} I_{F_{\omega_1}}(\omega_2) \, \rho_2(d\omega_2) \right] \rho_1(d\omega_1).$$

As we have shown, the inner integral is an $\mathscr{F}_1$-measurable function, and consequently the set function $\rho(F)$ is actually defined for $F \in \mathscr{F}$. It is clear that if $F = A \times B$, then $\rho(A \times B) = \rho_1(A)\rho_2(B)$. Now let $\{F^n\}$ be disjoint sets from $\mathscr{F}$. Then

$$\rho\left(\sum_n F^n \right) = \int_{\Omega_1} \left[\int_{\Omega_2} I_{(\Sigma F^n)_{\omega_1}}(\omega_2) \, \rho_2(d\omega_2) \right] \rho_1(d\omega_1)$$

$$= \int_{\Omega_1} \sum_n \left[\int_{\Omega_2} I_{F^n_{\omega_1}}(\omega_2) \, \rho_2(d\omega_2) \right] \rho_1(d\omega_1)$$

$$= \sum_n \int_{\Omega_1} \left[\int_{\Omega_2} I_{F^n_{\omega_1}}(\omega_2) \, \rho_2(d\omega_2) \right] \rho_1(d\omega_1) = \sum_n \rho(F^n),$$

i.e., ρ is a (σ-finite) measure on $\mathscr{F}$.

It follows from Carathéodory's theorem that this measure ρ is the unique measure with the property that $\rho(A \times B) = \rho_1(A)\rho_2(B)$.

We can now establish (50). If $\xi(\omega_1, \omega_2) = I_{A \times B}(\omega_1, \omega_2)$, $A \in \mathscr{F}_1$, $B \in \mathscr{F}_2$, then

$$\int_{\Omega_1 \times \Omega_2} I_{A \times B}(\omega_1, \omega_2) \, d(\rho_1 \times \rho_2) = (\rho_1 \times \rho_2)(A \times B), \qquad (52)$$

and since $I_{A \times B}(\omega_1, \omega_2) = I_A(\omega_1)I_B(\omega_2)$, we have

$$\int_{\Omega_1} \left[\int_{\Omega_2} I_{A \times B}(\omega_1, \omega_2) \, \rho_2(d\omega_2) \right] \rho_1(d\omega_1)$$
$$= \int_{\Omega_1} \left[I_A(\omega_1) \int_{\Omega_2} I_B(\omega_2) \, \rho_2(d\omega_2) \right] \rho_1(d\omega_1) = \rho_1(A)\rho_2(B). \qquad (53)$$

But, by the definition of $\rho_1 \times \rho_2$,

$$(\rho_1 \times \rho_2)(A \times B) = \rho_1(A)\rho_2(B).$$

Hence it follows from (52) and (53) that (50) is valid for $\xi(\omega_1, \omega_2) = I_{A \times B}(\omega_1, \omega_2)$.

Now let $\xi(\omega_1, \omega_2) = I_F(\omega_1, \omega_2)$, $F \in \mathscr{F}$. The set function

$$\lambda(F) = \int_{\Omega_1 \times \Omega_2} I_F(\omega_1, \omega_2) \, d(\rho_1 \times \rho_2), \qquad F \in \mathscr{F},$$

is evidently a σ-finite measure. It is also easily verified that the set function

$$\nu(F) = \int_{\Omega_1} \left[\int_{\Omega_2} I_F(\omega_1, \omega_2) \, \rho_2(d\omega_2) \right] \rho_1(d\omega_1)$$

is a σ-finite measure. As was shown above, λ and ν coincide on sets of the form $F = A \times B$, and therefore on the algebra $\mathscr{A}$. Hence it follows by Carathéodory's theorem that λ and ν coincide for all $F \in \mathscr{F}$.

We turn now to the proof of the full conclusion of Fubini's theorem. By (47),

$$\int_{\Omega_1 \times \Omega_2} \xi^+(\omega_1, \omega_2) \, d(\rho_1 \times \rho_2) < \infty, \qquad \int_{\Omega_1 \times \Omega_2} \xi^-(\omega_1, \omega_2) \, d(\rho_1 \times \rho_2) < \infty.$$

By what has already been proved, the integral $\int_{\Omega_2} \xi^+(\omega_1, \omega_2) \, \rho_2(d\omega_2)$ is an $\mathscr{F}_1$-measurable function of ω_1 and

$$\int_{\Omega_1} \left[\int_{\Omega_2} \xi^+(\omega_1, \omega_2) \, \rho_2(d\omega_2) \right] \rho_1(d\omega_1) = \int_{\Omega_1 \times \Omega_2} \xi^+(\omega_1, \omega_2) \, d(\rho_1 \times \rho_2) < \infty.$$

Consequently by Problem 4 (see also Property **J** in Subsection 3)

$$\int_{\Omega_2} \xi^+(\omega_1, \omega_2) \, \rho_2(d\omega_2) < \infty \quad (\rho_1\text{-a.s.}).$$

In the same way

$$\int_{\Omega_2} \xi^-(\omega_1, \omega_2) \, \rho_2(d\omega_2) < \infty \quad (\rho_1\text{-a.s.}),$$

and therefore

$$\int_{\Omega_2} |\xi(\omega_1, \omega_2)| \, \rho_2(d\omega_2) < \infty \quad (\rho_1\text{-a.s.}).$$

It is clear that, except on a set $\mathcal{N}$ of ρ_1-measure zero,

$$\int_{\Omega_2} \xi(\omega_1, \omega_2) \, \rho_2(d\omega_2) = \int_{\Omega_2} \xi^+(\omega_1, \omega_2) \, \rho_2(d\omega_2) - \int_{\Omega_2} \xi^-(\omega_1, \omega_2) \, \rho_2(d\omega_2).$$
$$(54)$$

Taking the integrals to be zero for $\omega_1 \in \mathcal{N}$, we may suppose that (54) holds for *all* $\omega_1 \in \Omega_1$. Then, integrating (54) with respect to ρ_1 and using (50), we obtain

$$\int_{\Omega_1} \left[\int_{\Omega_2} \xi(\omega_1, \omega_2) \, \rho_2(d\omega_2) \right] \rho_1(d\omega_1) = \int_{\Omega_1} \left[\int_{\Omega_2} \xi^+(\omega_1, \omega_2) \, \rho_2(d\omega_2) \right] \rho_1(d\omega_1)$$
$$- \int_{\Omega_1} \left[\int_{\Omega_2} \xi^-(\omega_1, \omega_2) \, \rho_2(d\omega_2) \right] \rho_1(d\omega_1)$$
$$= \int_{\Omega_1 \times \Omega_2} \xi^+(\omega_1, \omega_2) \, d(\rho_1 \times \rho_2) - \int_{\Omega_1 \times \Omega_2} \xi^-(\omega_1, \omega_2) \, d(\rho_1 \times \rho_2)$$
$$= \int_{\Omega_1 \times \Omega_2} \xi(\omega_1, \omega_2) \, d(\rho_1 \times \rho_2).$$

Similarly we can establish the first equation in (48) and the equation

$$\int_{\Omega_1 \times \Omega_2} \xi(\omega_1, \omega_2) \, d(\rho_1 \times \rho_2) = \int_{\Omega_2} \left[\int_{\Omega_1} \xi(\omega_1, \omega_2) \, \rho_1(d\omega_1) \right] \rho_2(d\omega_2).$$

This completes the proof of the theorem.

□

Corollary. *If* $\int_{\Omega_1} [\int_{\Omega_2} |\xi(\omega_1, \omega_2)| \, \rho_2(d\omega_2)] \, \rho_1(d\omega_1) < \infty$, *the conclusion of Fubini's theorem is still valid.*

In fact, under this hypothesis (47) follows from (50), and consequently the conclusions of Fubini's theorem hold.

Example. Let (ξ, η) be a pair of random variables whose distribution has a two-dimensional density $f_{\xi, \eta}(x, y)$, i.e.

$$P((\xi, \eta) \in B) = \int_B f_{\xi, \eta}(x, y) \, dx \, dy, \quad B \in \mathcal{B}(R^2),$$

where $f_{\xi, \eta}(x, y)$ is a nonnegative $\mathcal{B}(R^2)$-measurable function, and the integral is a Lebesgue integral with respect to two-dimensional Lebesgue measure.

Let us show that the one-dimensional distributions for ξ and η have densities $f_\xi(x)$ and $f_\eta(y)$, where

$$f_\xi(x) = \int_{-\infty}^{\infty} f_{\xi,\eta}(x, y)\, dy \quad \text{and} \quad f_\eta(y) = \int_{-\infty}^{\infty} f_{\xi,\eta}(x, y)\, dx. \tag{55}$$

In fact, if $A \in \mathscr{B}(R)$, then by Fubini's theorem

$$P(\xi \in A) = P((\xi, \eta) \in A \times R) = \int_{A \times R} f_{\xi,\eta}(x, y)\, dx\, dy = \int_A \left[\int_R f_{\xi,\eta}(x, y)\, dy \right] dx.$$

This establishes both the existence of a density for the probability distribution of ξ and the first formula in (55). The second formula is established similarly.

According to the theorem in Sect. 5, a necessary and sufficient condition that ξ and η are independent is that

$$F_{\xi,\eta}(x, y) = F_\xi(x) F_\eta(y), \quad (x, y) \in R^2.$$

Let us show that when there is a two-dimensional density $f_{\xi,\eta}(x, y)$, the variables ξ and η are independent if and only if

$$f_{\xi,\eta}(x, y) = f_\xi(x) f_\eta(y) \tag{56}$$

(where the equation is to be understood in the sense of holding almost surely with respect to two-dimensional Lebesgue measure).

In fact, if (56) holds, then by Fubini's theorem

$$F_{\xi,\eta}(x, y) = \int_{(-\infty, x] \times (-\infty, y]} f_{\xi,\eta}(u, v)\, du\, dv = \int_{(-\infty, x] \times (-\infty, y]} f_\xi(u) f_\eta(v)\, du\, dv$$

$$= \int_{(-\infty, x]} f_\xi(u)\, du \left(\int_{(-\infty, y]} f_\eta(v)\, dv \right) = F_\xi(x) F_\eta(y)$$

and consequently ξ and η are independent.

Conversely, if they are independent and have a density $f_{\xi,\eta}(x, y)$, then again by Fubini's theorem

$$\int_{(-\infty, x] \times (-\infty, y]} f_{\xi,\eta}(u, v)\, du\, dv = \left(\int_{(-\infty, x]} f_\xi(u)\, du \right) \left(\int_{(-\infty, y]} f_\eta(v)\, dv \right)$$

$$= \int_{(-\infty, x] \times (-\infty, y]} f_\xi(u) f_\eta(v)\, du\, dv.$$

It follows that

$$\int_B f_{\xi,\eta}(x, y)\, dx\, dy = \int_B f_\xi(x) f_\eta(y)\, dx\, dy$$

for every $B \in \mathscr{B}(R^2)$, and it is easily deduced from Property **I** that (56) holds.

11. In this subsection we discuss the relation between the Lebesgue and Riemann integrals.

We first observe that the construction of the Lebesgue integral is independent of the measurable space $(\Omega, \mathscr{F})$ on which the integrands are given. On the other hand, the Riemann integral is not defined on abstract spaces in general, and for $\Omega = R^n$ it is defined sequentially: first for R^1, and then extended, with corresponding changes, to the case $n > 1$.

We emphasize that the constructions of the Riemann and Lebesgue integrals are based on different ideas. The first step in the construction of the Riemann integral is to group the points $x \in R^1$ according to their distances along the x axis. On the other hand, in Lebesgue's construction (for $\Omega = R^1$) the points $x \in R^1$ are grouped according to a different principle: by the distances between the values of the integrand. It is a consequence of these different approaches that the Riemann approximating sums have limits only for "mildly" discontinuous functions, whereas the Lebesgue sums converge to limits for a much wider class of functions.

Let us recall the definition of the *Riemann–Stieltjes* integral. Let $G = G(x)$ be a generalized distribution function on R (see Subsection 2 of Sect. 3) and μ its corresponding Lebesgue–Stieltjes measure, and let $g = g(x)$ be a bounded function that vanishes outside $[a, b]$.

Consider a decomposition $\mathscr{P} = \{x_0, \ldots, x_n\}$,

$$a = x_0 < x_1 < \cdots < x_n = b,$$

of $[a, b]$, and form the *upper* and *lower* sums

$$\overline{\sum_{\mathscr{P}}} = \sum_{i=1}^{n} \overline{g}_i[G(x_{i+1}) - G(x_i)], \quad \underline{\sum_{\mathscr{P}}} = \sum_{i=1}^{n} \underline{g}_i[G(x_{i+1}) - G(x_i)],$$

where

$$\overline{g}_i = \sup_{x_{i-1} < y \leq x_i} g(y), \quad \underline{g}_i = \inf_{x_{i-1} < y \leq x_i} g(y).$$

Define simple functions $\overline{g}_{\mathscr{P}}(x)$ and $\underline{g}_{\mathscr{P}}(x)$ by taking

$$\overline{g}_{\mathscr{P}}(x) = \overline{g}_i, \quad \underline{g}_{\mathscr{P}}(x) = \underline{g}_i$$

on $x_{i-1} < x \leq x_i$, and define $\overline{g}_{\mathscr{P}}(a) = \underline{g}_{\mathscr{P}}(a) = g(a)$. Then it is clear that, according to the construction of the *Lebesgue–Stieltjes* integral (see Remark 3 in Subsection 2),

$$\overline{\sum_{\mathscr{P}}} = \text{(L–S)} \int_a^b \overline{g}_{\mathscr{P}}(x)\, G(dx)$$

and

$$\underline{\sum_{\mathscr{P}}} = \text{(L–S)} \int_a^b \underline{g}_{\mathscr{P}}(x)\, G(dx).$$

Now let $\{\mathscr{P}_k\}$ be a sequence of decompositions such that $\mathscr{P}_k \subseteq \mathscr{P}_{k+1}$ and $\mathscr{P}_k = \{x_0^{(k)}, \ldots, x_{n_k}^{(k)}\}$ are such that $\max_{0 \leq i \leq n_k} |x_{i+1}^{(k)} - x_i^{(k)}| \to 0, k \to \infty$. Then

$$\overline{g}_{\mathscr{P}_1} \geq \overline{g}_{\mathscr{P}_2} \geq \cdots \geq g \geq \cdots \geq \underline{g}_{\mathscr{P}_2} \geq \underline{g}_{\mathscr{P}_1},$$

and if $|g(x)| \leq C$ we have, by the dominated convergence theorem,

$$\lim_{k \to \infty} \sum_{\mathscr{P}_k} = (\text{L--S}) \int_a^b \overline{g}(x)\, G(dx),$$

(57)

$$\lim_{k \to \infty} \sum_{\mathscr{P}_k} = (\text{L--S}) \int_a^b \underline{g}(x)\, G(dx),$$

where $\overline{g}(x) = \lim_k \overline{g}_{\mathscr{P}_k}(x)$, $\underline{g}(x) = \lim_k \underline{g}_{\mathscr{P}_k}(x)$.

If the limits $\lim_k \sum_{\mathscr{P}_k}$ and $\lim_k \sum_{\mathscr{P}_k}$ are *finite and equal, and their common value is independent of the sequence of decompositions* $\{\mathscr{P}_k\}$, we say that $g = g(x)$ is *Riemann–Stieltjes integrable*, and the common value of the limits is denoted by

$$(\text{R--S}) \int_a^b g(x)\, G(dx) \quad \text{or} \quad (\text{R--S}) \int_a^b g(x)\, dG(x).$$

(58)

When $G(x) = x$, the integral is called a *Riemann integral* and denoted by

$$(\text{R}) \int_a^b g(x)\, dx.$$

Now let $(\text{L-S}) \int_a^b g(x)G(dx)$ be the corresponding *Lebesgue–Stieltjes* integral (see Remark 3).

Theorem 9. *If $g = g(x)$ is continuous on $[a, b]$, it is Riemann–Stieltjes integrable and*

$$(\text{R--S}) \int_a^b g(x)\, G(dx) = (\text{L--S}) \int_a^b g(x)\, G(dx).$$

(59)

PROOF. Since $g(x)$ is continuous, we have $\overline{g}(x) = g(x) = \underline{g}(x)$. Hence by (57) $\lim_{k \to \infty} \sum_{\mathscr{P}_k} = \lim_{k \to \infty} \sum_{\mathscr{P}_k}$. Consequently a continuous function $g = g(x)$ is Riemann–Stieltjes integrable and its Riemann–Stieltjes integral equals the Lebesgue–Stieltjes integral (again by (57)).

□

Let us consider in more detail the question of the correspondence between the Riemann and Lebesgue integrals for the case of *Lebesgue* measure on the *line R*.

Theorem 10. *Let $g(x)$ be a bounded function on $[a, b]$.*

(a) *The function $g = g(x)$ is Riemann integrable on $[a, b]$ if and only if it is continuous almost everywhere (with respect to Lebesgue measure $\overline{\lambda}$ on $\mathscr{B}([a, b])$).*

(b) *If $g = g(x)$ is Riemann integrable, it is Lebesgue integrable and*

$$(R) \int_a^b g(x)\, dx = (L) \int_a^b g(x)\, \overline{\lambda}(dx). \tag{60}$$

PROOF. (a) Let $g = g(x)$ be Riemann integrable. Then, by (57),

$$(L) \int_a^b \overline{g}(x)\, \overline{\lambda}(dx) = (L) \int_a^b \underline{g}(x)\, \overline{\lambda}(dx).$$

But $\underline{g}(x) \le g(x) \le \overline{g}(x)$, and hence by Property **H**

$$\underline{g}(x) = g(x) = \overline{g}(x) \quad (\overline{\lambda}\text{-a. s.}), \tag{61}$$

from which it is easy to see that $g(x)$ is continuous almost everywhere (with respect to $\overline{\lambda}$).

Conversely, let $g = g(x)$ be continuous almost everywhere (with respect to $\overline{\lambda}$). Then (61) is satisfied and consequently $g(x)$ differs from the (Borel) measurable function $\overline{g}(x)$ only on a set $\mathcal{N}$ with $\overline{\lambda}(\mathcal{N}) = 0$. But then

$$\{x : g(x) \le c\} = \{x : g(x) \le c\} \cap \overline{\mathcal{N}} + \{x : g(x) \le c\} \cap \mathcal{N}$$
$$= \{x : \overline{g}(x) \le c\} \cap \overline{\mathcal{N}} + \{x : g(x) \le c\} \cap \mathcal{N}.$$

It is clear that the set $\{x : \overline{g}(x) \le c\} \cap \overline{\mathcal{N}} \in \mathscr{B}([a, b])$, and that

$$\{x : g(x) \le c\} \cap \mathcal{N}$$

is a subset of $\mathcal{N}$ having Lebesgue measure $\overline{\lambda}$ equal to zero and therefore also belonging to $\overline{\mathscr{B}}([a, b)]$. Therefore $g(x)$ is $\overline{\mathscr{B}}([a, b])$-measurable and, as a bounded function, is Lebesgue integrable. Therefore by Property **G**,

$$(L) \int_a^b \overline{g}(x)\, \overline{\lambda}(dx) = (L) \int_a^b \underline{g}(x)\, \overline{\lambda}(dx) = (L) \int_a^b g(x)\, \overline{\lambda}(dx),$$

which completes the proof of (a).

(b) If $g = g(x)$ is Riemann integrable, then according to (a) it is continuous ($\overline{\lambda}$-a. s.). It was shown above than then $g(x)$ is Lebesgue integrable and its Riemann and Lebesgue integrals are equal.

This completes the proof of the theorem.

□

Remark 5. Let μ be a Lebesgue–Stieltjes measure on $\mathscr{B}([a, b])$. Let $\overline{\mathscr{B}}_\mu([a, b])$ be the system consisting of those subsets $\Lambda \subseteq [a, b]$ for which there are sets A and B in $\mathscr{B}([a, b])$ such that $A \subseteq \Lambda \subseteq B$ and $\mu(B \backslash A) = 0$. Let $\overline{\mu}$ be the *extension* of μ to $\overline{\mathscr{B}}_\mu([a, b])$ ($\overline{\mu}(\Lambda) = \mu(A)$ for Λ such that $A \subseteq \Lambda \subseteq B$ and $\mu(B \backslash A) = 0$). Then the conclusion of the theorem remains valid if we consider $\overline{\mu}$ instead of Lebesgue measure $\overline{\lambda}$, and the Riemann–Stieltjes and Lebesgue–Stieltjes integrals with respect to $\overline{\mu}$ instead of the Riemann and Lebesgue integrals.

Remark 6. The definition of Lebesgue integral (see Definitions 1 and 2 and formulas (3) and (6)) differs both conceptually and "in outward appearance" from those of Riemann and Riemann–Stieltjes integrals, which employ *upper* and *lower* sums (see (57)).

Now we *compare* these definitions in more detail.

Let $(\Omega, \mathscr{F}, \mu)$ be a measurable space with measure μ. For any $\mathscr{F}$-measurable nonnegative function $f = f(\omega)$ define two integrals, lower L_*f and upper L^*f (denoted also by $\int_* f \, d\mu$ and $\int^* f \, d\mu$), by putting

$$L_*f = \sup \sum_i \left(\inf_{\omega \in A_i} f(\omega) \right) \mu(A_i),$$

$$L^*f = \inf \sum_i \left(\sup_{\omega \in A_i} f(\omega) \right) \mu(A_i),$$

where sup and inf are taken over all finite decompositions $(A_1, A_2, \ldots, A_n)$ of Ω into $\mathscr{F}$-measurable sets $A_1, A_2, \ldots, A_n$ $\left(\sum_{i=1}^n A_i = \Omega \right), n \geq 1$.

It can be shown that $L_*f \leq L^*f$ and $L_*f = L^*f$ provided that f is bounded and μ is finite (Problem 20).

One approach (Darboux–Young) to the definition of the integral Lf of f with respect to μ consists in saying that f is μ-integrable if $L_*f = L^*f$, letting then $Lf = L_*f (= L^*f)$.

If we now consider Definition 1 of Lebesgue integral $\mathsf{E}f$ (Subsection 1), we can see (Problem 21) that

$$\mathsf{E}f = L_*f.$$

Thus for *bounded nonnegative* functions $f = f(\omega)$ the Lebesgue and Darboux–Young approaches give the same result ($\mathsf{E}f = Lf = L^*f = L_*f$).

But these approaches to integration become different when we deal with *unbounded* functions or *infinite* measure μ.

For example, the Lebesgue integrals $\int_{(0,1]} \frac{dx}{x^{1/2}}$ and $\int_{(1,\infty)} \frac{dx}{x^2}$ are well defined and equal to L_*f for $f(x) = x^{-1/2}I(0, 1]$ and $f(x) = x^{-2}I(1, \infty)$ respectively. However in both cases $L^*f = \infty$.

Thus $L_*f < L^*f$ here and the functions at hand are not integrable in the Darboux–Young sense, whereas they are integrable in the Lebesgue sense.

Consider now integration in Riemann's sense in terms of the above approach dealing with lower, L_*f, and upper, L^*f, integrals.

Suppose that $\Omega = (0, 1]$, $\mathscr{F} = \mathscr{B}$ (Borel σ-algebra) and $\mu = \lambda$ (Lebesgue measure). Let $f = f(x)$, $x \in \Omega$, be a bounded function (for the present we do not assume its measurability).

By analogy with L_*f and L^*f define *lower* and *upper* Riemann integrals R_*f, R^*f by putting

$$R_*f = \sup \sum_i \left(\inf_{\omega \in B_i} f(\omega) \right) \lambda(B_i),$$

$$R^*f = \inf \sum_i \left(\sup_{\omega \in B_i} f(\omega) \right) \lambda(B_i),$$

where $(B_1, B_2, \ldots, B_n)$ is a finite decomposition of $\Omega = (0, 1]$ with B_i's of the form $(a_i, b_i]$ (unlike arbitrary $\mathscr{F}$-measurable A_i's in the definition of $L_* f$ and $L^* f$).

Obviously, the above definitions imply that

$$R_* f \leq L_* f \leq L^* f \leq R^* f.$$

The Riemann integrability properties given in Theorems 9 and 10 can be restated and complemented in terms of the following conditions:

(a) $R^* f = R_* f$;
(b) The set D_f of discontinuity points of f has zero Lebesgue measure ($\lambda(D_f) = 0$);
(c) There esists a constant $R(f)$ such that for any $\varepsilon > 0$ there is $\delta > 0$ such that

$$\left| R(f) - \sum_i f(\omega_i) \lambda((a_i, b_i]) \right| < \varepsilon, \quad \omega_i \in (a_i, b_i],$$

for any finite system of disjoint intervals $(a_i, b_i]$ satisfying $\sum (a_i, b_i] = (0, 1]$ and $\lambda((a_i, b_i]) < \delta$.

Arguing as in the proofs of Theorems 9 and 10 one can show (Problem 22) that *for a bounded function f*

(A) *conditions* (a), (b), (c) *are equivalent and*
(B) *under either of conditions* (a), (b), (c)

$$R(f) = R_* f = R^* f.$$

11. In this part we present a useful theorem on integration by parts for the Lebesgue–Stieltjes integral.

Let two generalized distribution functions $F = F(x)$ and $G = G(x)$ be given on $(R, \mathscr{B}(R))$.

Theorem 11. *The following formulas are valid for all real a and b, $a < b$*:

$$F(b)G(b) - F(a)G(a) = \int_a^b F(s-) \, dG(s) + \int_a^b G(s) \, dF(s), \tag{62}$$

or equivalently

$$F(b)G(b) - F(a)G(a) = \int_a^b F(s-) \, dG(s) + \int_a^b G(s-) \, dF(s)$$
$$+ \sum_{a < s \leq b} \Delta F(s) \cdot \Delta G(s), \tag{63}$$

where $F(s-) = \lim_{t \uparrow s} F(t)$, $\Delta F(s) = F(s) - F(s-)$.

Remark 7. Formula (62) can be written symbolically in "differential" form

$$d(FG) = F_- \, dG + G \, dF. \tag{64}$$

Remark 8. The conclusion of the theorem remains valid for functions F and G of bounded variation on $[a, b]$. (Every such function that is continuous on the right and has limits on the left can be represented as the difference of two monotone nondecreasing functions.)

PROOF. We first recall that in accordance with Subsection 1 an integral $\int_a^b (\cdot)$ means $\int_{(a,b]} (\cdot)$. Then (see formula (2) in Sect. 3)

$$(F(b) - F(a))(G(b) - G(a)) = \int_a^b dF(s) \cdot \int_a^b dG(t).$$

Let $F \times G$ denote the direct product of the measures corresponding to F and G. Then by Fubini's theorem

$$(F(b) - F(a))(G(b) - G(a)) = \int_{(a,b] \times (a,b]} d(F \times G)(s,t)$$

$$= \int_{(a,b] \times (a,b]} I_{\{s \geq t\}}(s,\, t)\, d(F \times G)(s,t) + \int_{(a,b] \times (a,b]} I_{\{s < t\}}(s,t)\, d(F \times G)(s,\, t)$$

$$= \int_{(a,b]} (G(s) - G(a))\, dF(s) + \int_{(a,b]} (F(t-) - F(a))\, dG(t)$$

$$= \int_a^b G(s)\, dF(s) + \int_a^b F(s-)\, dG(s) - G(a)(F(b) - F(a)) - F(a)(G(b) - G(a)),$$

$$\tag{65}$$

where I_A is the indicator of the set A.

Formula (62) follows immediately from (65). In turn, (63) follows from (62) if we observe that

$$\int_a^b (G(s) - G(s-))\, dF(s) = \sum_{a < s \leq b} \Delta G(s) \cdot \Delta F(s). \tag{66}$$

$\square$

Corollary 1. *If $F(x)$ and $G(x)$ are distribution functions, then*

$$F(x)G(x) = \int_{-\infty}^x F(s-)\, dG(s) + \int_{-\infty}^x G(s)\, dF(s). \tag{67}$$

If also

$$F(x) = \int_{-\infty}^x f(s)\, ds,$$

then

$$F(x)G(x) = \int_{-\infty}^x F(s)\, dG(s) + \int_{-\infty}^x G(s)f(s)\, ds. \tag{68}$$

Corollary 2. *Let ξ be a random variable with distribution function $F(x)$ and* $\mathsf{E}\,|\xi|^n < \infty$. *Then*

$$\int_0^\infty x^n\,dF(x) = n\int_0^\infty x^{n-1}[1 - F(x)]\,dx, \tag{69}$$

$$\int_{-\infty}^0 |x|^n\,dF(x) = -\int_0^\infty x^n\,dF(-x) = n\int_0^\infty x^{n-1}F(-x)\,dx \tag{70}$$

and

$$\mathsf{E}\,|\xi|^n = \int_{-\infty}^\infty |x|^n\,dF(x) = n\int_0^\infty x^{n-1}[1 - F(x) + F(-x)]\,dx. \tag{71}$$

To prove (69) we observe that

$$\int_0^b x^n\,dF(x) = -\int_0^b x^n\,d(1 - F(x))$$

$$= -b^n(1 - F(b)) + n\int_0^b x^{n-1}(1 - F(x))\,dx. \tag{72}$$

Let us show that since $\mathsf{E}\,|\xi|^n < \infty$,

$$b^n(1 - F(b) + F(-b)) \leq b^n\mathsf{P}(|\xi| \geq b) \to 0, \quad b \to \infty. \tag{73}$$

In fact,

$$\mathsf{E}\,|\xi|^n = \sum_{k=1}^\infty \int_{k-1}^k |x|^n dF(x) < \infty$$

and therefore

$$\sum_{k \geq b\,|\,1} \int_{k-1}^k |x|^n dF(x) \to 0, \quad b \to \infty.$$

But

$$\sum_{k \geq b+1} \int_{k-1}^k |x|^n\,dF(x) \geq b^n\,\mathsf{P}(|\xi| \geq b),$$

which establishes (73).

Taking the limit as $b \to \infty$ in (72), we obtain (69). Formula (70) is proved similarly, and (71) follows from (69) and (70).

13. Let $A = A(t)$, $t \geq 0$, be a function of locally bounded variation (i.e., of bounded variation on each finite interval $[a, b]$), which is continuous on the right and has limits on the left. Consider the equation

$$Z_t = 1 + \int_0^t Z_{s-}\,dA(s), \tag{74}$$

which can be written in differential form as

$$dZ_t = Z_{t-}\, dA(t), \qquad Z_0 = 1. \tag{75}$$

The formula that we have proved for integration by parts lets us solve (74) explicitly in the class of functions of locally bounded variation.

We introduce the function (called the *stochastic exponent*, see [43])

$$\mathscr{E}_t(A) = e^{A(t)-A(0)} \prod_{0 \le s \le t} (1 + \Delta A(s)) e^{-\Delta A(s)}, \tag{76}$$

where $\Delta A(s) = A(s) - A(s-)$ for $s > 0$, and $\Delta A(0) = 0$.

The function $A(s)$, $0 \le s \le t$, has bounded variation and therefore has at most countably many discontinuities and the series $\sum_{0 \le s \le t} |\Delta A(s)|$ converges. It follows that

$$\prod_{0 \le s \le t} (1 + \Delta A(s)) e^{-\Delta A(s)}, \quad t \ge 0,$$

is a function of locally bounded variation.

If $A^c(t) = A(t) - \sum_{0 \le s \le t} \Delta A(s)$ is the continuous component of $A(t)$, we can rewrite (76) in the form

$$\mathscr{E}_t(A) = e^{A^c(t)-A^c(0)} \prod_{0 < s \le t} (1 + \Delta A(s)). \tag{77}$$

Let us write

$$F(t) = e^{A^c(t)-A^c(0)}, \qquad G(t) = \prod_{0 < s \le t} (1 + \Delta A(s)), \quad G(0) = 1.$$

Then by (62)

$$\mathscr{E}_t(A) = F(t)G(t) = 1 + \int_0^t F(s)\, dG(s) + \int_0^t G(s-)\, dF(s)$$

$$= 1 + \sum_{0 < s \le t} F(s)G(s-)\Delta A(s) + \int_0^t G(s-)F(s)\, dA^c(s)$$

$$= 1 + \int_0^t \mathscr{E}_{s-}(A)\, dA(s).$$

Therefore $\mathscr{E}_t(A)$, $t \ge 0$, is a (locally bounded) *solution of* (74). Let us show that this is the only locally bounded solution.

Suppose that there are two such solutions and let $Y = Y(t)$, $t \ge 0$, be their difference. Then

$$Y(t) = \int_0^t Y(s-)\, dA(s).$$

Put

$$T = \inf\{t \geq 0 : Y(t) \neq 0\},$$

where we take $T = \infty$ if $Y(t) = 0$ for $t \geq 0$.

Since $A(t)$, $t \geq 0$, is a function of locally bounded variation, there are two generalized distribution functions $A_1(t)$ and $A_2(t)$ such that $A(t) = A_1(t) - A_2(t)$. If we suppose that $T < \infty$, we can find a finite $T' > T$ such that

$$[A_1(T') + A_2(T')] - [A_1(T) + A_2(T)] \leq \tfrac{1}{2}.$$

Then it follows from the equation

$$Y(t) = \int_T^t Y(s-) \, dA(s), \quad t \geq T,$$

that

$$\sup_{t \leq T'} |Y(t)| \leq \tfrac{1}{2} \sup_{t \leq T'} |Y(t)|$$

and since $\sup_{t \leq T'} |Y(t)| < \infty$, we have $Y(t) = 0$ for $T < t \leq T'$, contradicting the assumption that $T < \infty$.

Thus we have proved the following theorem.

Theorem 12. *There is a unique locally bounded solution of* (74), *and it is given by* (76).

14. PROBLEMS

1. Establish the representation (6).
2. Prove the following extension of Property E. Let ξ and η be random variables for which $\mathsf{E}\,\xi$ and $\mathsf{E}\,\eta$ are defined and the sum $\mathsf{E}\,\xi + \mathsf{E}\,\eta$ is meaningful (does not have the form $\infty - \infty$ or $-\infty + \infty$). Then

$$\mathsf{E}(\xi + \eta) = \mathsf{E}\,\xi + \mathsf{E}\,\eta.$$

3. Generalize Property **G** by showing that if $\xi = \eta$ (a. s.) and $\mathsf{E}\,\xi$ exists, then $\mathsf{E}\,\eta$ exists and $\mathsf{E}\,\xi = \mathsf{E}\,\eta$.
4. Let ξ be an extended random variable, μ a σ-finite measure, and $\int_\Omega |\xi| d\mu < \infty$. Show that $|\xi| < \infty$ (μ-a. s.) (cf. Property **J**).
5. Let μ be a σ-finite measure, ξ and η extended random variables for which $\mathsf{E}\,\xi$ and $\mathsf{E}\,\eta$ are defined. If $\int_A \xi \, d\mu \leq \int_A \eta \, d\mu$ for all $A \in \mathscr{F}$ then $\xi \leq \eta$ (μ-a. s.). (Cf. Property **I**.)
6. Let ξ and η be independent nonnegative random variables. Show that $\mathsf{E}\,\xi\eta = \mathsf{E}\,\xi \cdot \mathsf{E}\,\eta$.
7. Using Fatou's lemma, show that

$$\mathsf{P}(\liminf A_n) \leq \liminf \mathsf{P}(A_n), \quad \mathsf{P}(\limsup A_n) \geq \limsup \mathsf{P}(A_n).$$

8. Find an example to show that in general it is impossible to weaken the hypothesis "$|\xi_n| \leq \eta$, $\mathsf{E}\,\eta < \infty$" in the dominated convergence theorem.

9. Find an example to show that in general the hypothesis "$\xi_n \leq \eta$, $\mathsf{E}\,\eta > -\infty$" in Fatou's lemma cannot be omitted.

10. Prove the following variant of Fatou's lemma. Let the family $\{\xi_n^+\}_{n\geq 1}$ of random variables be uniformly integrable. Then

$$\limsup \mathsf{E}\,\xi_n \leq \mathsf{E} \limsup \xi_n.$$

11. Dirichlet's function

$$d(x) = \begin{cases} 1, & x \text{ irrational,} \\ 0, & x \text{ rational,} \end{cases}$$

is defined on $[0,1]$, Lebesgue integrable, but not Riemann integrable. Why?

12. Find an example of a sequence of Riemann integrable functions $\{f_n\}_{n\geq 1}$, defined on $[0,1]$, such that $|f_n| \leq 1$, $f_n \to f$ almost everywhere (with Lebesgue measure), but f is not Riemann integrable.

13. Let $(a_{ij};\ i,j \geq 1)$ be a sequence of real numbers such that $\sum_{i,j} |a_{ij}| < \infty$. Deduce from Fubini's theorem that

$$\sum_{i,j} a_{ij} = \sum_i \left(\sum_j a_{ij} \right) = \sum_j \left(\sum_i a_{ij} \right). \tag{78}$$

14. Find an example of a sequence $(a_{ij};\ i,j \geq 1)$ for which $\sum_{i,j} |a_{ij}| = \infty$ and the equations in (78) do not hold.

15. Starting from simple functions and using the theorem on taking limits under the Lebesgue integral sign, prove the following result on *integration by substitution*.

Let $h = h(y)$ be a nondecreasing continuously differentiable function on $[a,b]$, and let $f(x)$ be (Lebesgue) integrable on $[h(a), h(b)]$. Then the function $f(h(y))h'(y)$ is integrable on $[a,b]$ and

$$\int_{h(a)}^{h(b)} f(x)\,dx = \int_a^b f(h(y))h'(y)\,dy.$$

16. Prove formula (70).

17. Let $\xi, \xi_1, \xi_2, \ldots$ be nonnegative integrable random variables such that $\mathsf{E}\,\xi_n \to \mathsf{E}\,\xi$ and $\mathsf{P}(|\xi - \xi_n| > \varepsilon) \to 0$ for every $\varepsilon > 0$. Show that then $\mathsf{E}\,|\xi_n - \xi| \to 0$, $n \to \infty$.

18. Let ξ be an integrable random variable ($\mathsf{E}\,|\xi| < \infty$). Prove that for any $\varepsilon > 0$ there exists $\delta > 0$ such that $\mathsf{E}\,I_A|\xi| < \varepsilon$ for any $A \in \mathscr{F}$ with $\mathsf{P}(A) < \delta$ (the property of "absolute continuity of Lebesgue integral").

19. Let ξ, η, ζ and ξ_n, η_n, ζ_n, $n \geq 1$, be random variables such that[*]

---

[*] *Convergence in probability* $\xi_n \overset{\mathsf{P}}{\to} \xi$ means that $\mathsf{P}\{|\xi_n - \xi| > \varepsilon\} \to 0$ as $n \to \infty$ for any $\varepsilon > 0$ (for more details see Sect. 10).

$$\xi_n \xrightarrow{P} \xi, \quad \eta_n \xrightarrow{P} \eta, \quad \zeta_n \xrightarrow{P} \zeta, \quad \eta_n \le \xi_n \le \zeta_n, \quad n \ge 1,$$
$$\mathsf{E}\,\zeta_n \to \mathsf{E}\,\zeta, \qquad \mathsf{E}\,\eta_n \to \mathsf{E}\,\eta,$$

and the expectations $\mathsf{E}\,\xi$, $\mathsf{E}\,\eta$, $\mathsf{E}\,\zeta$ are finite. Show that then $\mathsf{E}\,\xi_n \to \mathsf{E}\,\xi$ (*Pratt's lemma*).

If also $\eta_n \le 0 \le \zeta_n$ then $\mathsf{E}\,|\xi_n - \xi| \to 0$.

Deduce that if $\xi_n \xrightarrow{P} \xi$, $\mathsf{E}\,|\xi_n| \to \mathsf{E}\,|\xi|$ and $\mathsf{E}\,|\xi| < \infty$, then $\mathsf{E}\,|\xi_n - \xi| \to 0$.

Give an example showing that, in general, under the conditions of Pratt's lemma $\mathsf{E}\,|\xi_n - \xi| \not\to 0$.

20. Prove that $L_* f \le L^* f$ and if f is bounded and μ is finite then $L_* f = L^* f$ (see Remark 6).

21. Prove that $\mathsf{E}\, f = L_* f$ for bounded f (see Remark 6).

22. Prove the final statement of Remark 6.

23. Let X be a random variable and $F(x)$ its distribution function. Show that

$$\mathsf{E}\,X^+ < \infty \iff \int_a^\infty \log \frac{1}{F(x)}\, dx < \infty \quad \text{for some } a.$$

24. Show that if $\lim_{x\to\infty} x^p \,\mathsf{P}\{|\xi| > x\} = 0$ for $p > 0$ then $\mathsf{E}\,|\xi|^r < \infty$ for all $r < p$. Give an example showing that for $r = p$ it is possible that $\mathsf{E}\,|\xi|^p = \infty$.

25. Find an example of a density $f(x)$, which is not an even function but has zero odd moments, $\int_{-\infty}^\infty x^k f(x)\, dx = 0$, $k = 1, 3, \ldots$

26. Give an example of random variables ξ_n, $n \ge 1$, such that

$$\mathsf{E}\,\sum_{n=1}^\infty \xi_n \ne \sum_{n=1}^\infty \mathsf{E}\,\xi_n.$$

27. Let a random variable X be such that

$$\frac{\mathsf{P}\{|X| > \alpha n\}}{\mathsf{P}\{|X| > n\}} \to 0, \quad n \to \infty,$$

for any $\alpha > 1$. Prove that in this case all the moments of X are finite. *Hint*: use the formula

$$\mathsf{E}\,|X|^N = N \int_0^\infty x^{N-1}\, \mathsf{P}(|X| > x)\, dx.$$

28. Let X be a random variable taking values $k = 0, 1, 2, \ldots$ with probabilities p_k. By the definition in Sect. 13 of Chap. 1 the function $F(s) = \sum_{k=0}^\infty p_k s^k$, $|s| \le 1$, is called the *generating function* of X. Establish the following formulas:

   (i) If X is a Poisson random variable, i.e., $p_k = e^{-\lambda}\lambda^k/k!$, $\lambda > 0$, $k = 0, 1, 2, \ldots$, then

$$F(s) = e^{-\lambda(1-s)}, \quad |s| \le 1;$$

   (ii) If X has a geometric distribution, i.e., $p_k = pq^k$, $0 < p < 1$, $q = 1 - p$, $k = 0, 1, 2, \ldots$, then

$$F(s) = \frac{p}{1 - sq}, \quad |s| \le 1.$$

29. Besides the generating function $F(s)$, it is often expedient to consider the *moment generating function* $M(s) = \mathsf{E}\, e^{sX}$ (for s such that $\mathsf{E}\, e^{sX} < \infty$).
    (a) Show that if the moment generating function $M(s)$ is defined for all s in a neighborhood of zero ($s \in [-a, a]$, $a > 0$), then $M(s)$ has derivatives $M^{(k)}(s)$ of any order $k = 1, 2, \ldots$ at $s = 0$ and

$$M^{(k)}(0) = \mathsf{E}\, X^k$$

    (which explains the term for $M(s)$).
    (b) Give an example of a random variable for which $M(s) = \infty$ for all $s > 0$.
    (c) Show that the moment generating function of a Poisson random variable X with $\lambda > 0$ is $M(s) = e^{-\lambda(1 - e^s)}$ for all $s \in R$.

30. Let $X_n \in L^r$, $0 < r < \infty$, and $X_n \overset{\mathsf{P}}{\to} X$. Then the following conditions are equivalent:
    (i) The family $\{|X_n|^r, n \ge 1\}$ is uniformly integrable;
    (ii) $X_n \to X$ in L^r;
    (iii) $\mathsf{E}\,|X_n|^r \to \mathsf{E}\,|X|^r < \infty$.

31. *Spitzer's identity.* Let $X_1, X_2, \ldots$ be independent identically distributed random variables with $\mathsf{P}\{X_1 \le 1\} = 1$ and let $S_n = X_1 + \cdots + X_n$. Then for $|u|, |t| < 1$

$$\sum_{n=0}^{\infty} t^n \, \mathsf{E}\, e^{uM_n} = \exp\left(\sum_{n=1}^{\infty} \frac{t^n}{n} \, \mathsf{E}\, e^{uS_n^+} \right),$$

    where $M_n = \max(0, X_1, X_2, \ldots, X_n)$, $S_n^+ = \max(0, S_n)$.

32. Let $S_0 = 0$, $S_n = X_1 + \cdots + X_n$, $n \ge 1$, be a simple symmetric random walk (i.e., X_i, $i = 1, 2, \ldots$, are independent and take the values ± 1 with probabilities $1/2$) and $\tau = \min\{n > 0 : S_n \ge 0\}$. Show that

$$\mathsf{E} \min(\tau, 2m) = 2\,\mathsf{E}\,|S_{2m}| = 4m\,\mathsf{P}\{S_{2m} = 0\}, \quad m \ge 0.$$

33. Let ξ be a standard Gaussian random variable ($\xi \sim \mathcal{N}(0, 1)$). Using integration by parts, show that $\mathsf{E}\,\xi^k = (k - 1)\,\mathsf{E}\,\xi^{k-2}$. Derive from this the formulas:

$$\mathsf{E}\,\xi^{2k-1} = 0 \quad \text{and} \quad \mathsf{E}\,\xi^{2k} = 1 \cdot 3 \cdot \ldots \cdot (2k - 3)(2k - 1) \ (= (2k - 1)!!).$$

34. Show that the function $x^{-1} \sin x$, $x \in R$, is Riemann integrable, but not Lebesgue integrable (with Lebesgue measure on $(R, \mathscr{B}(R))$).

35. Show that the function

$$\xi(\omega_1, \omega_2) = e^{-\omega_1 \omega_2} - 2e^{-2\omega_1 \omega_2}, \quad \omega_1 \in \Omega_1 = [1, \infty), \ \omega_2 \in \Omega_2 = (0, 1],$$

is such that
    (a) it is Lebesgue integrable with respect to $\omega_1 \in \Omega_1$ for every ω_2 and
    (b) it is Lebesgue integrable with respect to $\omega_2 \in \Omega_2$ for every ω_1,
but Fubibi's theorem does not hold.

36. Prove the *Beppo Levi theorem*: If random variables $\xi_1, \xi_2, \ldots$ are integrable
($E |\xi_n| < \infty$ for all $n \geq 1$), $\sup_n E \xi_n < \infty$ and $\xi_n \uparrow \xi$, then ξ is integrable
and $E \xi_n \uparrow E \xi$ (cf. Theorem 1 (a)).

37. Prove the following version of *Fatou's lemma*: if $0 \leq \xi_n \to \xi$ (P-a. s.) and
$E \xi_n \leq A < \infty$, $n \geq 1$, then ξ is integrable and $E \xi \leq A$.

38. (*On relation of Lebesgue and Riemann integration.*) Let a Borel function $f = f(x)$ be Lebesgue integrable on R: $\int_R |f(x)| \, dx < \infty$. Prove that for any $\varepsilon > 0$ there are:
    (a) a step function $f_\varepsilon(x) = \sum_{i=1}^n f_i I_{A_i}(x)$ with bounded intervals A_i such
that $\int_R |f(x) - f_\varepsilon(x)| \, dx < \varepsilon$;
    (b) an integrable continuous function $g_\varepsilon(x)$ with bounded support such that
$\int_R |f(x) - g_\varepsilon(x)| \, dx < \varepsilon$.

39. Show that if ξ is an integrable random variable, then

$$E \xi = \int_0^\infty P\{\xi > x\} \, dx - \int_{-\infty}^0 P\{\xi < x\} \, dx.$$

40. Let ξ and η be integrable random variables. Show that

$$E \xi - E \eta = \int_{-\infty}^\infty [P\{\eta < x \leq \xi\} - P\{\xi < x \leq \eta\}] \, dx.$$

41. Let ξ be a nonnegative random variable ($\xi \geq 0$) with Laplace transform
$\varphi_\xi(\lambda) = E e^{-\lambda \xi}$, $\lambda \geq 0$.
    (a) Show that for any $0 < r < 1$

$$E \xi^r = \frac{r}{\Gamma(1-r)} \int_0^\infty \frac{1 - \varphi_\xi(\lambda)}{\lambda^{r+1}} \, d\lambda.$$

Hint: use that for $s \geq 0$, $0 < r < 1$

$$\frac{1}{r} \Gamma(1-r) s^r = \int_0^\infty \frac{1 - e^{-s\lambda}}{\lambda^{r+1}} \, d\lambda.$$

    (b) Show that if $\xi > 0$ then for any $r > 0$

$$E \xi^{-r} = \frac{1}{r \Gamma(r)} \int_0^\infty \varphi_\xi(\lambda^{1/r}) \, d\lambda.$$

Hint: use that for $s \geq 0$, $r > 0$

$$s = \frac{r}{\Gamma(1/r)} \int_0^\infty \exp\{-(\lambda/s)^r\} \, d\lambda.$$

7 Conditional Probabilities and Conditional Expectations with Respect to a σ-Algebra

1. Let $(\Omega, \mathscr{F}, \mathsf{P})$ be a probability space and $A \in \mathscr{F}$ an event such that $\mathsf{P}(A) > 0$. As for finite probability spaces, the *conditional probability of B with respect to A* (denoted by $\mathsf{P}(B \mid A)$) means $\mathsf{P}(BA)/\mathsf{P}(A)$, and the *conditional probability of B with respect to the finite or countable decomposition* $\mathscr{D} = \{D_1, D_2, \ldots\}$ such that $\mathsf{P}(D_i) > 0$, $i \geq 1$ (denoted by $\mathsf{P}(B \mid \mathscr{D})$) is the random variable equal to $\mathsf{P}(B \mid D_i)$ for $\omega \in D_i$, $i \geq 1$:

$$\mathsf{P}(B \mid \mathscr{D}) = \sum_{i \geq 1} \mathsf{P}(B \mid D_i) I_{D_i}(\omega).$$

In a similar way, if ξ is a random variable for which $\mathsf{E}\,\xi$ is defined, the *conditional expectation of ξ with respect to the event A* with $\mathsf{P}(A) > 0$ (denoted by $\mathsf{E}(\xi \mid A)$) is $\mathsf{E}(\xi I_A)/\mathsf{P}(A)$ (cf. (10) in Sect. 8 of Chap. 1).

The random variable $\mathsf{P}(B \mid \mathscr{D})$ is evidently measurable with respect to the σ-algebra $\mathscr{G} = \sigma(\mathscr{D})$, and is consequently also denoted by $\mathsf{P}(B \mid \mathscr{G})$ (see Sect. 8 of Chap. 1).

However, in probability theory we may have to consider conditional probabilities with respect to events whose probabilities are *zero*.

Consider, for example, the following experiment. Let ξ be a random variable uniformly distributed on $[0, 1]$. If $\xi = x$, toss a coin for which the probability of head is x, and the probability of tail is $1 - x$. Let ν be the number of heads in n independent tosses of this coin. What is the "conditional probability $\mathsf{P}(\nu = k \mid \xi = x)$"? Since $\mathsf{P}(\xi = x) = 0$, the conditional probability $\mathsf{P}(\nu = k \mid \xi = x)$ is undefined, although it is intuitively plausible that "it ought to be $C_n^k x^k (1 - x)^{n-k}$."

Let us now give a general definition of conditional expectation (and, in particular, of conditional probability) with respect to a σ-algebra $\mathscr{G}$, $\mathscr{G} \subseteq \mathscr{F}$, and compare it with the definition given in Sect. 8 of Chap. 1 for finite probability spaces.

2. Let $(\Omega, \mathscr{F}, \mathsf{P})$ be a probability space, $\mathscr{G}$ a σ-algebra, $\mathscr{G} \subseteq \mathscr{F}$ ($\mathscr{G}$ is a σ-subalgebra of $\mathscr{F}$), and $\xi = \xi(\omega)$ a random variable. Recall that, according to Sect. 6, the expectation $\mathsf{E}\,\xi$ was defined in two stages: first for a nonnegative random variable ξ, then in the general case by

$$\mathsf{E}\,\xi = \mathsf{E}\,\xi^+ - \mathsf{E}\,\xi^-,$$

and only under the assumption that

$$\min(\mathsf{E}\,\xi^-, \mathsf{E}\,\xi^+) < \infty$$

(in order to avoid an indeterminacy of the form $\infty - \infty$). A similar two-stage construction is also used to define conditional expectations $\mathsf{E}(\xi \mid \mathscr{G})$.

Definition 1.

(1) The *conditional expectation of a nonnegative* random variable ξ *with respect to the σ-algebra $\mathscr{G}$* is a nonnegative extended random variable, denoted by $\mathsf{E}(\xi \mid \mathscr{G})$ or $\mathsf{E}(\xi \mid \mathscr{G})(\omega)$, such that

(a) $\mathsf{E}(\xi\,|\,\mathscr{G})$ is $\mathscr{G}$-measurable;

(b) for every $A \in \mathscr{G}$

$$\int_A \xi\,d\mathsf{P} = \int_A \mathsf{E}(\xi\,|\,\mathscr{G})\,d\mathsf{P}. \tag{1}$$

(2) *The conditional expectation* $\mathsf{E}(\xi\,|\,\mathscr{G})$, *or* $\mathsf{E}(\xi\,|\,\mathscr{G})(\omega)$, *of any* random variable ξ *with respect to the* σ-*algebra* $\mathscr{G}$ is considered to be defined if

$$\min(\mathsf{E}(\xi^+\,|\,\mathscr{G}), \mathsf{E}(\xi^-\,|\,\mathscr{G})) < \infty \quad (\mathsf{P}\text{-a. s.}),$$

and it is given by the formula

$$\mathsf{E}(\xi\,|\,\mathscr{G}) \equiv \mathsf{E}(\xi^+\,|\,\mathscr{G}) - \mathsf{E}(\xi^-\,|\,\mathscr{G}),$$

where, on the set (of probability zero) of sample points for which $\mathsf{E}(\xi^+\,|\,\mathscr{G}) = \mathsf{E}(\xi^-\,|\,\mathscr{G}) = \infty$, the difference $\mathsf{E}(\xi^+\,|\,\mathscr{G}) - \mathsf{E}(\xi^-\,|\,\mathscr{G})$ is given an arbitrary value, for example zero.

We begin by showing that, for nonnegative random variables, $\mathsf{E}(\xi\,|\,\mathscr{G})$ actually exists. By Subsection 8 of Sect. 6 the set function

$$Q(A) = \int_A \xi\,d\mathsf{P}, \quad A \in \mathscr{G}, \tag{2}$$

is a measure on $(\Omega, \mathscr{G})$, and is absolutely continuous with respect to P (considered on $(\Omega, \mathscr{G})$, $\mathscr{G} \subseteq \mathscr{F}$). Therefore (by the Radon–Nikodým theorem) there is a nonnegative $\mathscr{G}$-measurable extended random variable $\mathsf{E}(\xi\,|\,\mathscr{G})$ such that

$$Q(A) = \int_A \mathsf{E}(\xi\,|\,\mathscr{G})\,d\mathsf{P}. \tag{3}$$

Then (1) follows from (2) and (3).

Remark 1. In accordance with the Radon–Nikodým theorem, the conditional expectation $\mathsf{E}(\xi\,|\,\mathscr{G})$ is defined only up to sets of P-measure zero. In other words, $\mathsf{E}(\xi\,|\,\mathscr{G})$ can be taken to be any $\mathscr{G}$-measurable function $f(\omega)$ for which $Q(A) = \int_A f(\omega)\,d\mathsf{P}$, $A \in \mathscr{G}$ (a "version" of the conditional expectation).

Let us observe that, in accordance with the remark on the Radon–Nikodým theorem,

$$\mathsf{E}(\xi\,|\,\mathscr{G}) \equiv \frac{dQ}{d\mathsf{P}}(\omega), \tag{4}$$

i.e., the conditional expectation is just the Radon–Nikodým derivative of the measure Q with respect to P (considered on $(\Omega, \mathscr{G})$).

It is worth to note that if a nonnegative random variable ξ is such that $\mathsf{E}\,\xi < \infty$, then $\mathsf{E}(\xi\,|\,\mathscr{G}) < \infty$ (P-a. s.), which directly follows from (1). Similarly, if $\xi \leq 0$ and $\mathsf{E}\,\xi > -\infty$, then $\mathsf{E}(\xi\,|\,\mathscr{G}) > -\infty$ (P-a. s.).

Remark 2. In connection with (1), we observe that we cannot in general put $\mathsf{E}(\xi\,|\,\mathscr{G}) = \xi$, since ξ is *not necessarily* $\mathscr{G}$-measurable.

Remark 3. Suppose that ξ is a random variable for which $\mathsf{E}\,\xi$ exists. Then $\mathsf{E}(\xi\,|\,\mathscr{G})$ could be defined as a $\mathscr{G}$-measurable function for which (1) holds. This is usually just what happens. Our definition $\mathsf{E}(\xi\,|\,\mathscr{G}) \equiv \mathsf{E}(\xi^+\,|\,\mathscr{G}) - \mathsf{E}(\xi^-\,|\,\mathscr{G})$ has the advantage that for the trivial σ-algebra $\mathscr{G} = \{\varnothing,\,\Omega\}$ it reduces to the definition of $\mathsf{E}\,\xi$ but does not presuppose the existence of $\mathsf{E}\,\xi$. (For example, if ξ is a random variable with $\mathsf{E}\,\xi^+ = \infty$, $\mathsf{E}\,\xi^- = \infty$, and $\mathscr{G} = \mathscr{F}$, then $\mathsf{E}\,\xi$ is not defined, but in terms of Definition 1, $\mathsf{E}(\xi\,|\,\mathscr{G})$ exists and is simply $\xi = \xi^+ - \xi^-$).

Remark 4. Let the random variable ξ have a conditional expectation $\mathsf{E}(\xi\,|\,\mathscr{G})$ with respect to the σ-algebra $\mathscr{G}$. The *conditional variance* $\mathrm{Var}(\xi\,|\,\mathscr{G})$ of ξ with respect to $\mathscr{G}$ is the random variable

$$\mathrm{Var}(\xi\,|\,\mathscr{G}) = \mathsf{E}[(\xi - \mathsf{E}(\xi\,|\,\mathscr{G}))^2\,|\,\mathscr{G}].$$

(Cf. the definition of the conditional variance $\mathrm{Var}(\xi\,|\,\mathscr{D})$ with respect to a decomposition $\mathscr{D}$, as given in Problem 2 in Sect. 8, Chap. 1, and the definition of the *variance* in Sect. 8).

Definition 2. Let $B \in \mathscr{F}$. The conditional expectation $\mathsf{E}(I_B\,|\,\mathscr{G})$ is denoted by $\mathsf{P}(B\,|\,\mathscr{G})$, or $\mathsf{P}(B\,|\,\mathscr{G})(\omega)$, and is called the *conditional probability of the event B with respect to the σ-algebra $\mathscr{G}$*, $\mathscr{G} \subseteq \mathscr{F}$.

It follows from Definitions 1 and 2 that, for a given $B \in \mathscr{F}$, $\mathsf{P}(B\,|\,\mathscr{G})$ is a random variable such that

(a) $\mathsf{P}(B\,|\,\mathscr{G})$ is $\mathscr{G}$-measurable,
(b) for every $A \in \mathscr{G}$

$$\mathsf{P}(A \cap B) = \int_A \mathsf{P}(B\,|\,\mathscr{G})\,d\mathsf{P}. \tag{5}$$

Definition 3. Let ξ be a random variable and $\mathscr{G}_\eta$ the σ-algebra generated by a random element η. Then $\mathsf{E}(\xi\,|\,\mathscr{G}_\eta)$, if defined, is denoted by $\mathsf{E}(\xi\,|\,\eta)$ or $\mathsf{E}(\xi\,|\,\eta)(\omega)$, and is called the *conditional expectation of ξ with respect to η*.

The conditional probability $\mathsf{P}(B\,|\,\mathscr{G}_\eta)$ is denoted by $\mathsf{P}(B\,|\,\eta)$ or $\mathsf{P}(B\,|\,\eta)(\omega)$, and is called the *conditional probability of B with respect to η*.

3. Let us show that the definition of $\mathsf{E}(\xi\,|\,\mathscr{G})$ given here agrees with the definition of conditional expectation in Sect. 8 of Chap. 1.

Let $\mathscr{D} = \{D_1, D_2, \ldots\}$ be a finite or countable decomposition with atoms D_i $(\sum_i D_i = \Omega)$ such that $\mathsf{P}(D_i) > 0, i \geq 1$.

Theorem 1. *If $\mathscr{G} = \sigma(\mathscr{D})$ and ξ is a random variable for which $\mathsf{E}\,\xi$ is defined, then*

$$\mathsf{E}(\xi\,|\,\mathscr{G}) = \mathsf{E}(\xi\,|\,D_i) \quad (\text{P-a. s. on } D_i) \tag{6}$$

or equivalently

$$\mathsf{E}(\xi\,|\,\mathscr{G}) = \frac{\mathsf{E}(\xi I_{D_i})}{\mathsf{P}(D_i)} \quad (\text{P-a. s. on } D_i).$$

(The notation "$\xi = \eta$ (P-a. s. on A)" or "$\xi = \eta$ (A; P-a. s.)" means that $\mathsf{P}(A \cap \{\xi \neq \eta\}) = 0$.)

PROOF. According to Lemma 3 of Sect. 4, $\mathsf{E}(\xi \mid \mathscr{G}) = K_i$ on D_i, where K_i are constants. But

$$\int_{D_i} \xi \, d\mathsf{P} = \int_{D_i} \mathsf{E}(\xi \mid \mathscr{G}) \, d\mathsf{P} = K_i \, \mathsf{P}(D_i),$$

whence

$$K_i = \frac{1}{\mathsf{P}(D_i)} \int_{D_i} \xi \, d\mathsf{P} = \frac{\mathsf{E}(\xi I_{D_i})}{\mathsf{P}(D_i)} = \mathsf{E}(\xi \mid D_i).$$

This completes the proof of the theorem.

□

Consequently the concept of the conditional expectation $\mathsf{E}(\xi \mid \mathscr{D})$ with respect to a finite decomposition $\mathscr{D} = \{D_1, \ldots, D_n\}$, as introduced in Chap. 1, is a special case of the concept of conditional expectation with respect to the σ-algebra $\mathscr{G} = \sigma(D)$.

4. Properties of conditional expectations. We shall suppose that the expectations are defined for all the random variables that we consider and that $\mathscr{G} \subseteq \mathscr{F}$.

A*. *If C is a constant and $\xi = C$ (a. s.), then $\mathsf{E}(\xi \mid \mathscr{G}) = C$ (a. s.).*

B*. *If $\xi \leq \eta$ (a. s.) then $\mathsf{E}(\xi \mid \mathscr{G}) \leq \mathsf{E}(\eta \mid \mathscr{G})$ (a. s.).*

C*. *$|\mathsf{E}(\xi \mid \mathscr{G})| \leq \mathsf{E}(|\xi| \mid \mathscr{G})$ (a. s.).*

D*. *If a, b are constants and $a \mathsf{E}\, \xi + b \mathsf{E}\, \eta$ is defined, then*

$$\mathsf{E}(a\xi + b\eta \mid \mathscr{G}) = a\, \mathsf{E}(\xi \mid \mathscr{G}) + b\, \mathsf{E}(\eta \mid \mathscr{G}) \quad (a.\, s.).$$

E*. *Let $\mathscr{F}_* = \{\varnothing, \Omega\}$ be the trivial σ-algebra. Then*

$$\mathsf{E}(\xi \mid \mathscr{F}_*) = \mathsf{E}\, \xi \quad (a.\, s.).$$

F*. $\mathsf{E}(\xi \mid \mathscr{F}) = \xi$ *(a. s.).*

G*. $\mathsf{E}(\mathsf{E}(\xi \mid \mathscr{G})) = \mathsf{E}\, \xi$

H*. *If $\mathscr{G}_1 \subseteq \mathscr{G}_2$ then the (first) "telescopic property" holds:*

$$\mathsf{E}[\mathsf{E}(\xi \mid \mathscr{G}_2) \mid \mathscr{G}_1] = \mathsf{E}(\xi \mid \mathscr{G}_1) \quad (a.\, s.).$$

I*. *If $\mathscr{G}_1 \supseteq \mathscr{G}_2$ then the (second) "telescopic property" holds:*

$$\mathsf{E}[\mathsf{E}(\xi \mid \mathscr{G}_2) \mid \mathscr{G}_1)] = \mathsf{E}(\xi \mid \mathscr{G}_2) \quad (a.\, s.).$$

J*. *Let a random variable ξ for which $\mathsf{E}\, \xi$ is defined be independent of the σ-algebra $\mathscr{G}$ (i.e., independent of I_B, $B \in \mathscr{G}$). Then*

$$\mathsf{E}(\xi \mid \mathscr{G}) = \mathsf{E}\, \xi \quad (a.\, s.).$$

K*. *Let η be a $\mathscr{G}$-measurable random variable, $\mathsf{E}\, |\xi| < \infty$ and $\mathsf{E}\, |\xi\eta| < \infty$. Then*

$$\mathsf{E}(\xi\eta \mid \mathscr{G}) = \eta\, \mathsf{E}(\xi \mid \mathscr{G}) \quad (a.\, s.).$$

Let us establish these properties.

A*. A constant function is measurable with respect to $\mathscr{G}$. Therefore we need only verify that

$$\int_A \xi \, dP = \int_A C \, dP, \quad A \in \mathscr{G}.$$

But, by the hypothesis $\xi = C$ (a. s.) and Property **G** of Sect. 6, this equation is obviously satisfied.

B*. If $\xi \leq \eta$ (a. s.), then by Property **B** of Sect. 6

$$\int_A \xi \, dP \leq \int_A \eta \, dP, \quad A \in \mathscr{G},$$

and therefore

$$\int_A \mathsf{E}(\xi \mid \mathscr{G}) \, dP \leq \int_A \mathsf{E}(\eta \mid \mathscr{G}) \, dP, \quad A \in \mathscr{G}.$$

The required inequality now follows from Property **I** (Sect. 6).

C*. This follows from the preceding property if we observe that $-|\xi| \leq \xi \leq |\xi|$.

D*. If $A \in \mathscr{G}$ then by Problem 2 of Sect. 6,

$$\int_A (a\xi + b\eta) \, dP = \int_A a\xi \, dP + \int_A b\eta \, dP = \int_A a \mathsf{E}(\xi \mid \mathscr{G}) \, dP$$

$$+ \int_A b \mathsf{E}(\eta \mid \mathscr{G}) \, dP = \int_A [a \mathsf{E}(\xi \mid \mathscr{G}) + b \mathsf{E}(\eta \mid \mathscr{G})] \, dP,$$

which establishes **D***.

E*. This property follows from the remark that $\mathsf{E}\,\xi$ is an $\mathscr{F}_*$-measurable function and the evident fact that if $A = \Omega$ or $A = \varnothing$ then

$$\int_A \xi \, dP = \int_A \mathsf{E}\,\xi \, dP.$$

F*. Since ξ if $\mathscr{F}$-measurable and

$$\int_A \xi \, dP = \int_A \mathsf{E}(\xi \mid \mathscr{F}) \, dP, \quad A \in \mathscr{F},$$

we have $\mathsf{E}(\xi \mid \mathscr{F}) = \xi$ (a. s.).

G*. This follows from **E*** and **H*** by taking $\mathscr{G}_1 = \{\varnothing, \Omega\}$ and $\mathscr{G}_2 = \mathscr{G}$.

H*. Let $A \in \mathscr{G}_1$; then

$$\int_A \mathsf{E}(\xi \mid \mathscr{G}_1) \, dP = \int_A \xi \, dP.$$

Since $\mathscr{G}_1 \subseteq \mathscr{G}_2$, we have $A \in \mathscr{G}_2$ and therefore

$$\int_A \mathsf{E}[\mathsf{E}(\xi \mid \mathscr{G}_2) \mid \mathscr{G}_1] \, dP = \int_A \mathsf{E}(\xi \mid \mathscr{G}_2) \, dP = \int_A \xi \, dP.$$

Consequently, when $A \in \mathcal{G}_1$,

$$\int_A \mathsf{E}(\xi \mid \mathcal{G}_1)\, d\mathsf{P} = \int_A \mathsf{E}[\mathsf{E}(\xi \mid \mathcal{G}_2) \mid \mathcal{G}_1]\, d\mathsf{P}$$

and arguing as in the proof of Property **I** (Sect. 6) (see also Problem 5 of Sect. 6)

$$\mathsf{E}(\xi \mid \mathcal{G}_1) = \mathsf{E}[\mathsf{E}(\xi \mid \mathcal{G}_2) \mid \mathcal{G}_1] \quad \text{(a. s.)}.$$

I*. If $A \in \mathcal{G}_1$, then by the definition of $\mathsf{E}[\mathsf{E}(\xi \mid \mathcal{G}_2) \mid \mathcal{G}_1]$

$$\int_A \mathsf{E}[\mathsf{E}(\xi \mid \mathcal{G}_2) \mid \mathcal{G}_1]\, d\mathsf{P} = \int_A \mathsf{E}(\xi \mid \mathcal{G}_2)\, d\mathsf{P}.$$

The function $\mathsf{E}(\xi \mid \mathcal{G}_2)$ is $\mathcal{G}_2$-measurable and, since $\mathcal{G}_2 \subseteq \mathcal{G}_1$, also $\mathcal{G}_1$-measurable. It follows that $\mathsf{E}(\xi \mid \mathcal{G}_2)$ is a variant of the expectation $\mathsf{E}[\mathsf{E}(\xi \mid \mathcal{G}_2) \mid \mathcal{G}_1]$, which proves Property **I***.

J*. Since $\mathsf{E}\,\xi$ is a $\mathcal{G}$-measurable function, we have only to verify that

$$\int_B \xi\, d\mathsf{P} = \int_B \mathsf{E}\,\xi\, d\mathsf{P} \qquad \text{for any} \quad B \in \mathcal{G},$$

i.e., that $\mathsf{E}[\xi \cdot I_B] = \mathsf{E}\,\xi \cdot \mathsf{E}\,I_B$. If $\mathsf{E}\,|\xi| < \infty$, this follows immediately from Theorem 6 of Sect. 6. In the general case use Problem 6 of Sect. 6 instead of this theorem.

The proof of Property **K*** will be given a little later; it depends on conclusion (a) of the following theorem.

Theorem 2 (On Taking Limits Under the Conditional Expectation Sign). *Let* $\{\xi_n\}_{n \geq 1}$ *be a sequence of extended random variables.*

(a) *If* $|\xi_n| \leq \eta$, $\mathsf{E}\,\eta < \infty$, *and* $\xi_n \to \xi$ (*a. s.*), *then*

$$\mathsf{E}(\xi_n \mid \mathcal{G}) \to \mathsf{E}(\xi \mid \mathcal{G}) \quad (a. s.)$$

and

$$\mathsf{E}(|\xi_n - \xi| \mid \mathcal{G}) \to 0 \quad (a. s.).$$

(b) *If* $\xi_n \geq \eta$, $\mathsf{E}\,\eta > -\infty$, *and* $\xi_n \uparrow \xi$ (*a. s.*), *then*

$$\mathsf{E}(\xi_n \mid \mathcal{G}) \uparrow \mathsf{E}(\xi \mid \mathcal{G}) \quad (a. s.).$$

(c) *If* $\xi_n \leq \eta$, $\mathsf{E}\,\eta < \infty$, *and* $\xi_n \downarrow \xi$ (*a. s.*), *then*

$$\mathsf{E}(\xi_n \mid \mathcal{G}) \downarrow \mathsf{E}(\xi \mid \mathcal{G}) \quad (a. s.).$$

(d) *If* $\xi_n \geq \eta$, $\mathsf{E}\,\eta > -\infty$, *then*

$$\mathsf{E}(\liminf \xi_n \mid \mathcal{G}) \leq \liminf \mathsf{E}(\xi_n \mid \mathcal{G}) \quad (a. s.).$$

(e) *If $\xi_n \leq \eta$, $E\eta < \infty$, then*

$$\limsup E(\xi_n \mid \mathcal{G}) \leq E(\limsup \xi_n \mid \mathcal{G}) \quad (a.s.).$$

(f) *If $\xi_n \geq 0$ then*

$$E\left(\sum \xi_n \mid \mathcal{G}\right) = \sum E(\xi_n \mid \mathcal{G}) \quad (a.s.).$$

PROOF. (a) Let $\zeta_n = \sup_{m \geq n} |\xi_m - \xi|$. Since $\xi_n \to \xi$ (a.s.), we have $\zeta_n \downarrow 0$ (a.s.). The expectations $E\xi_n$ and $E\xi$ are finite; therefore by Properties D^* and C^* (a.s.)

$$|E(\xi_n \mid \mathcal{G}) - E(\xi \mid \mathcal{G})| = |E(\xi_n - \xi \mid \mathcal{G})| \leq E(|\xi_n - \xi| \mid \mathcal{G}) \leq E(\zeta_n \mid \mathcal{G}).$$

Since $E(\zeta_{n+1} \mid \mathcal{G}) \leq E(\zeta_n \mid \mathcal{G})$ (a.s.), the limit $h = \lim_n E(\zeta_n \mid \mathcal{G})$ exists (a.s.). Then

$$0 \leq \int_\Omega h \, dP \leq \int_\Omega E(\zeta_n \mid \mathcal{G}) \, dP = \int_\Omega \zeta_n \, dP \to 0, \quad n \to \infty,$$

where the last statement follows from the dominated convergence theorem, since $0 \leq \zeta_n \leq 2\eta$, $E\eta < \infty$. Consequently $\int_\Omega h \, dP = 0$ and then $h = 0$ (a.s.) by Property **H**.

(b) First let $\eta \equiv 0$. Since $E(\xi_n \mid \mathcal{G}) \leq E(\xi_{n+1} \mid \mathcal{G})$ (a.s.), the limit $\zeta(\omega) = \lim_n E(\xi_n \mid \mathcal{G})$ exists (a.s.). Then by the equation

$$\int_A \xi_n \, dP = \int_A E(\xi_n \mid \mathcal{G}) \, dP, \quad A \in \mathcal{G},$$

and the theorem on monotone convergence,

$$\int_A \xi \, dP = \int_A E(\xi \mid \mathcal{G}) \, dP = \int_A \zeta \, dP, \quad A \in \mathcal{G}.$$

Consequently $\xi = \zeta$ (a.s.) by Property **I** and Problem 5 of Sect. 6.

For the proof in the general case, we observe that $0 \leq \xi_n^+ \uparrow \xi^+$, and by what has been proved,

$$E(\xi_n^+ \mid \mathcal{G}) \uparrow E(\xi^+ \mid \mathcal{G}) \quad (a.s.). \tag{7}$$

But $0 \leq \xi_n^- \leq \xi^-$, $E\xi^- < \infty$, and therefore by (a)

$$E(\xi_n^- \mid \mathcal{G}) \to E(\xi^- \mid \mathcal{G}),$$

which, with (7), proves (b).

Conclusion (c) follows from (b).

(d) Let $\zeta_n = \inf_{m \geq n} \xi_m$; then $\zeta_n \uparrow \zeta$, where $\zeta = \liminf \xi_n$. According to (b), $E(\zeta_n \mid \mathcal{G}) \uparrow E(\zeta \mid \mathcal{G})$ (a.s.). Therefore (a.s.) $E(\liminf \xi_n \mid \mathcal{G}) = E(\zeta \mid \mathcal{G}) = \lim_n E(\zeta_n \mid \mathcal{G}) = \liminf E(\zeta_n \mid \mathcal{G}) \leq \liminf E(\xi_n \mid \mathcal{G})$.

Conclusion (e) follows from (d).

(f) If $\xi_n \geq 0$, by Property **D*** we have

$$\mathsf{E}\left(\sum_{k=1}^{n} \xi_k \mid \mathscr{G}\right) = \sum_{k=1}^{n} \mathsf{E}(\xi_k \mid \mathscr{G}) \quad \text{(a. s.)}$$

which, with (b), establishes the required result.

This completes the proof of the theorem.

$\square$

We can now establish Property **K***. Let $\eta = I_B$, $B \in \mathscr{G}$. Then, for every $A \in \mathscr{G}$,

$$\int_A \xi \eta \, d\mathsf{P} = \int_{A \cap B} \xi \, d\mathsf{P} = \int_{A \cap B} \mathsf{E}(\xi \mid \mathscr{G}) \, d\mathsf{P} = \int_A I_B \mathsf{E}(\xi \mid \mathscr{G}) \, d\mathsf{P} = \int_A \eta \mathsf{E}(\xi \mid \mathscr{G}) \, d\mathsf{P}.$$

By the additivity of the Lebesgue integral, the equation

$$\int_A \xi \eta \, d\mathsf{P} = \int_A \eta \mathsf{E}(\xi \mid \mathscr{G}) \, d\mathsf{P}, \quad A \in \mathscr{G}, \tag{8}$$

remains valid for the simple random variables $\eta = \sum_{k=1}^{n} y_k I_{B_k}$, $B_k \in \mathscr{G}$. Therefore, by Property **I** (Sect. 6), we have

$$\mathsf{E}(\xi \eta \mid \mathscr{G}) = \eta \mathsf{E}(\xi \mid \mathscr{G}) \quad \text{(a. s.)} \tag{9}$$

for these random variables.

Now let η be any $\mathscr{G}$-measurable random variable, and let $\{\eta_n\}_{n \geq 1}$ be a sequence of simple $\mathscr{G}$-measurable random variables such that $|\eta_n| \leq \eta$ and $\eta_n \to \eta$. Then by (9)

$$\mathsf{E}(\xi \eta_n \mid \mathscr{G}) = \eta_n \mathsf{E}(\xi \mid \mathscr{G}) \quad \text{(a. s.)}.$$

It is clear that $|\xi \eta_n| \leq |\xi \eta|$, where $\mathsf{E}|\xi \eta| < \infty$. Therefore $\mathsf{E}(\xi \eta_n \mid \mathscr{G}) \to \mathsf{E}(\xi \eta \mid \mathscr{G})$ (a. s.) by Property (a). In addition, since $\mathsf{E}|\xi| < \infty$, we have $\mathsf{E}(\xi \mid \mathscr{G})$ finite (a. s.) (see Property **C*** and Property **J** of Sect. 6). Therefore $\eta_n \mathsf{E}(\xi \mid \mathscr{G}) \to \eta \mathsf{E}(\xi \mid \mathscr{G})$ (a. s.). (The hypothesis that $\mathsf{E}(\xi \mid \mathscr{G})$ is finite, almost surely, is essential, since, according to the footnote in Subsection 4 of Section 4, $0 \cdot \infty = 0$, but if $\eta_n = 1/n$, $\eta \equiv 0$, we have $1/n \cdot \infty = \infty \nrightarrow 0 \cdot \infty = 0$.)

Remark 5. For the property **K*** the following conditions suffice: η *is $\mathscr{G}$-measurable and* $\mathsf{E}(\xi \mid \mathscr{G})$ *is well defined.*

5. Here we consider the more detailed structure of conditional expectations $\mathsf{E}(\xi \mid \mathscr{G}_\eta)$, which we also denote, as usual, by $\mathsf{E}(\xi \mid \eta)$.

Since $\mathsf{E}(\xi \mid \eta)$ is a $\mathscr{G}_\eta$-measurable function, then by Theorem 3 of Sect. 4 (more precisely, by its obvious modification for extended random variables) there is a Borel function $m = m(y)$ from $\overline{R}$ to $\overline{R}$ such that

$$m(\eta(\omega)) = \mathsf{E}(\xi \mid \eta)(\omega) \tag{10}$$

for all $\omega \in \Omega$. We denote this function $m(y)$ by $\mathsf{E}(\xi \mid \eta = y)$ and call it the *conditional expectation of ξ with respect to the event $\{\eta = y\}$*, or *the conditional expectation of ξ under the condition that $\eta = y$*, or *given that $\eta = y$.*

By definition,

$$\int_A \xi \, d\mathsf{P} = \int_A \mathsf{E}(\xi \mid \eta) \, d\mathsf{P} = \int_A m(\eta) \, d\mathsf{P}, \quad A \in \mathscr{G}_\eta. \tag{11}$$

Therefore by Theorem 7 of Sect. 6 (on change of variable under the Lebesgue integral sign)

$$\int_{\{\omega \,:\, \eta \in B\}} m(\eta) \, d\mathsf{P} = \int_B m(y) \, P_\eta(dy), \quad B \in \mathscr{B}(\overline{R}), \tag{12}$$

where P_η is the probability distribution of η. Consequently $m = m(y)$ is a Borel function such that

$$\int_{\{\omega \,:\, \eta \in B\}} \xi \, d\mathsf{P} = \int_B m(y) \, P_\eta(dy) \tag{13}$$

for every $B \in \mathscr{B}(R)$.

This remark shows that we can give a different definition of the conditional expectation $\mathsf{E}(\xi \mid \eta = y)$.

Definition 4. Let ξ and η be random variables (possibly, extended) and let $\mathsf{E}\,\xi$ be defined. The conditional expectation of the random variable ξ under the condition that $\eta = y$ is any $\mathscr{B}(\overline{R})$-measurable function $m = m(y)$ for which

$$\int_{\{\omega \,:\, \eta \in B\}} \xi \, d\mathsf{P} = \int_B m(y) \, P_\eta(dy), \quad B \in \mathscr{B}(\overline{R}). \tag{14}$$

That such a function exists follows again from the Radon–Nikodým theorem if we observe that the set function

$$Q(B) = \int_{\{\omega \,:\, \eta \in B\}} \xi \, d\mathsf{P}$$

is a signed measure absolutely continuous with respect to the measure P_η.

Now suppose that $m(y)$ is a conditional expectation in the sense of Definition 4. Then if we again apply the theorem on change of variable under the Lebesgue integral sign, we obtain

$$\int_{\{\omega \,:\, \eta \in B\}} \xi \, d\mathsf{P} = \int_B m(y) \, P_\eta(dy) = \int_{\{\omega \,:\, \eta \in B\}} m(\eta) \, d\mathsf{P}, \quad B \in \mathscr{B}(\overline{R}).$$

The function $m(\eta)$ is $\mathscr{G}_\eta$-measurable, and the sets $\{\omega \colon \eta \in B\}$, $B \in \mathscr{B}(\overline{R})$, exhaust the subsets of $\mathscr{G}_\eta$.

Hence it follows that $m(\eta)$ is the expectation $\mathsf{E}(\xi \mid \eta)$. Consequently if we know $\mathsf{E}(\xi \mid \eta = y)$ we can reconstruct $\mathsf{E}(\xi \mid \eta)$, and conversely from $\mathsf{E}(\xi \mid \eta)$ we can find $\mathsf{E}(\xi \mid \eta = y)$.

From an intuitive point of view, the conditional expectation $E(\xi \mid \eta = y)$ is simpler and more natural than $E(\xi \mid \eta)$. However, $E(\xi \mid \eta)$, considered as a $\mathcal{G}_\eta$-measurable random variable, is more convenient to work with.

Observe that Properties $\mathbf{A^*}$–$\mathbf{K^*}$ above and the conclusions of Theorem 2 can easily be transferred to $E(\xi \mid \eta = y)$ (replacing "almost surely" by "P_η-almost surely"). Thus, for example, Property $\mathbf{K^*}$ transforms as follows: if $E \mid \xi \mid < \infty$ and $E \mid \xi f(\eta) \mid < \infty$, where $f = f(y)$ is a $\mathcal{B}(\overline{R})$ measurable function, then

$$E(\xi f(\eta) \mid \eta = y) = f(y) E(\xi \mid \eta = y) \quad (P_\eta\text{-a.s.}). \tag{15}$$

In addition (cf. Property $\mathbf{J^*}$), if ξ and η are independent, then

$$E(\xi \mid \eta = y) = E\xi \quad (P_\eta\text{-a.s.}).$$

We also observe that if $B \in \mathcal{B}(R^2)$ and ξ and η are independent, then

$$E[I_B(\xi, \eta) \mid \eta = y] = E I_B(\xi, y) \quad (P_\eta\text{-a.s.}), \tag{16}$$

and if $\varphi = \varphi(x, y)$ is a $\mathcal{B}(R^2)$-measurable function such that $E \mid \varphi(\xi, \eta) \mid < \infty$, then

$$E[\varphi(\xi, \eta) \mid \eta = y] = E[\varphi(\xi, y)] \quad (P_\eta\text{-a.s.}).$$

To prove (16) we make the following observation. If $B = B_1 \times B_2$, the validity of (16) will follow from

$$\int_{\{\omega: \, \eta \in A\}} I_{B_1 \times B_2}(\xi, \eta) \, P(d\omega) = \int_{(y \in A)} E I_{B_1 \times B_2}(\xi, y) \, P_\eta(dy).$$

But the left-hand side here is $P\{\xi \in B_1, \, \eta \in A \cap B_2\}$, and the right-hand side is $P(\xi \in B_1) P(\eta \in A \cap B_2)$; their equality follows from the independence of ξ and η. In the general case the proof depends on an application of Theorem 1 of Sect. 2 on monotone classes (cf. the corresponding part of the proof of Fubini's theorem).

Definition 5. The conditional probability of the event $A \in \mathcal{F}$ under the condition that $\eta = y$ (notation: $P(A \mid \eta = y)$) is $E(I_A \mid \eta = y)$.

It is clear that $P(A \mid \eta = y)$ could be defined as a $\mathcal{B}(\overline{R})$-measurable function such that

$$P(A \cap \{\eta \in B\}) = \int_B P(A \mid \eta = y) P_\eta(dy), \quad B \in \mathcal{B}(\overline{R}). \tag{17}$$

6. Let us calculate some examples of conditional probabilities and conditional expectations.

Example 1. Let η be a discrete random variable with $P(\eta = y_k) > 0$, $\sum_{k=1}^{\infty} P(\eta = y_k) = 1$. Then

$$P(A \mid \eta = y_k) = \frac{P(A \cap \{\eta = y_k\})}{P(\eta = y_k)}, \quad k \geq 1.$$

For $y \notin \{y_1, y_2, \ldots\}$ the conditional probability $P(A \mid \eta = y)$ can be defined in any way, for example as zero.

If ξ is a random variable for which $E \xi$ exists, then

$$E(\xi \mid \eta = y_k) = \frac{1}{P(\eta = y_k)} \int_{\{\omega \,:\, \eta = y_k\}} \xi \, dP.$$

When $y \notin \{y_1, y_2, \ldots\}$ the conditional expectation $E(\xi \mid \eta = y)$ can be defined in any way (for example, as zero).

Example 2. Let (ξ, η) be a pair of random variables whose distribution has a density $f_{\xi, \eta}(x, y)$:

$$P\{(\xi, \eta) \in B\} = \int_B f_{\xi, \eta}(x, y) \, dx \, dy, \quad B \in \mathscr{B}(R^2).$$

Let $f_\xi(x)$ and $f_\eta(y)$ be the densities of the probability distributions of ξ and η (see (46), (55), and (56) in Sect. 6).

Let us put

$$f_{\xi \mid \eta}(x \mid y) = \frac{f_{\xi, \eta}(x, y)}{f_\eta(y)}, \tag{18}$$

taking $f_{\xi \mid \eta}(x \mid y) = 0$ if $f_\eta(y) = 0$.

Then

$$P(\xi \in C \mid \eta = y) = \int_C f_{\xi \mid \eta}(x \mid y) \, dx, \quad C \in \mathscr{B}(R), \tag{19}$$

i.e., $f_{\xi \mid \eta}(x \mid y)$ is the density of a conditional probability distribution.

In fact, to prove (19) it is enough to verify (17) for $B \in \mathscr{B}(R)$, $A = \{\xi \in C\}$. By (43) and (45) of Sect. 6 and Fubini's theorem,

$$
\begin{aligned}
\int_B \left[\int_C f_{\xi \mid \eta}(x \mid y) \, dx \right] P_\eta(dy) &= \int_B \left[\int_C f_{\xi \mid \eta}(x \mid y) \, dx \right] f_\eta(y) \, dy \\
&= \int_{C \times B} f_{\xi \mid \eta}(x \mid y) f_\eta(y) \, dx \, dy \\
&= \int_{C \times B} f_{\xi, \eta}(x, y) \, dx \, dy \\
&= P\{(\xi, \eta) \in C \times B\} = P\{(\xi \in C) \cap (\eta \in B)\},
\end{aligned}
$$

which proves (17).

In a similar way we can show that if $\mathsf{E}\,\xi$ exists, then

$$\mathsf{E}(\xi \mid \eta = y) = \int_{-\infty}^{\infty} x f_{\xi \mid \eta}(x \mid y)\,dx. \tag{20}$$

Example 3. Let the length of time that a piece of apparatus will continue to operate be described by a nonnegative random variable $\eta = \eta(\omega)$ whose distribution function $F_\eta(y)$ has a density $f_\eta(y)$ (naturally, $F_\eta(y) = f_\eta(y) = 0$ for $y < 0$). Find the conditional expectation $\mathsf{E}(\eta - a \mid \eta \geq a)$, i.e., the average time for which the apparatus will continue to operate given that it has already been operating for time a.

Let $\mathsf{P}(\eta \geq a) > 0$. Then according to the definition of conditional probability given in Subsection 1 and (45) of Sect. 6 we have

$$\mathsf{E}(\eta - a \mid \eta \geq a) = \frac{\mathsf{E}[(\eta - a)I_{\{\eta \geq a\}}]}{\mathsf{P}(\eta \geq a)} = \frac{\int_\Omega (\eta - a)I_{\{\eta \geq a\}}\,\mathsf{P}(d\omega)}{\mathsf{P}(\eta \geq a)}$$
$$= \frac{\int_a^\infty (y - a)f_\eta(y)\,dy}{\int_a^\infty f_\eta(y)\,dy}.$$

It is interesting to observe that if η is exponentially distributed, i.e.

$$f_\eta(y) = \begin{cases} \lambda e^{-\lambda y}, & y \geq 0, \\ 0, & y < 0, \end{cases} \tag{21}$$

then $\mathsf{E}\,\eta = \mathsf{E}(\eta \mid \eta \geq 0) = 1/\lambda$ and $\mathsf{E}(\eta - a \mid \eta \geq a) = 1/\lambda$ for every $a > 0$. In other words, in this case the average time for which the apparatus continues to operate, assuming that it has already operated for time a, is independent of a and simply equals the average time $\mathsf{E}\,\eta$.

Let us find the conditional distribution $\mathsf{P}(\eta - a \leq x \mid \eta \geq a)$ assuming (21). We have

$$\mathsf{P}(\eta - a \leq x \mid \eta \geq a) = \frac{\mathsf{P}(u \leq \eta \leq a + x)}{\mathsf{P}(\eta \geq a)}$$
$$= \frac{F_\eta(a + x) - F_\eta(a) + \mathsf{P}(\eta = a)}{1 - F_\eta(a) + \mathsf{P}(\eta = a)}$$
$$= \frac{[1 - e^{-\lambda(a+x)}] - [1 - e^{-\lambda a}]}{1 - [1 - e^{-\lambda a}]}$$
$$= \frac{e^{-\lambda a}[1 - e^{-\lambda x}]}{e^{-\lambda a}} = 1 - e^{-\lambda x}.$$

Therefore the conditional distribution $\mathsf{P}(\eta - a \leq x \mid \eta \geq a)$ is the same as the unconditional distribution $\mathsf{P}(\eta \leq x)$. This remarkable property is characteristic for the exponential distribution: there are no other distributions that have densities and possess the property $\mathsf{P}(\eta - a \leq x \mid \eta \geq a) = \mathsf{P}(\eta \leq x)$, $a \geq 0, 0 \leq x < \infty$.

Example 4 (Buffon's Needle). Suppose that we toss a needle of unit length "at random" onto a pair of parallel straight lines, a unit distance apart, in a plane (see Fig. 29). What is the probability that the needle will intersect at least one of the lines?

To solve this problem we must first define what it means to toss the needle "at random." Let ξ be the distance from the midpoint of the needle to the left-hand line. We shall suppose that ξ is uniformly distributed on $[0, 1]$, and (see Fig. 29) that the angle θ is uniformly distributed on $[-\pi/2, \pi/2]$. In addition, we shall assume that ξ and θ are independent.

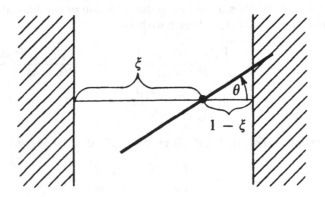

Fig. 29

Let A be the event that the needle intersects one of the lines. It is easy to see that if

$$B = \{(a, x) \colon |a| \leq \pi/2, \; x \in [0, \tfrac{1}{2} \cos a] \cup [1 - \tfrac{1}{2} \cos a, 1]\},$$

then $A = \{\omega \colon (\theta, \xi) \in B\}$, and therefore the probability in question is

$$P(A) = \mathsf{E}\, I_A(\omega) = \mathsf{E}\, I_B(\theta(\omega), \xi(\omega)).$$

By Property $\mathbf{G}^*$ and formula (16),

$$
\begin{aligned}
\mathsf{E}\, I_B(\theta(\omega), \xi(\omega)) &= \mathsf{E}(\mathsf{E}[I_B(\theta(\omega), \xi(\omega)) \mid \theta(\omega)]) \\
&= \int_{\Omega} \mathsf{E}[I_B(\theta(\omega), \xi(\omega)) \mid \theta(\omega)]\, \mathsf{P}(d\omega) \\
&= \int_{-\pi/2}^{\pi/2} \mathsf{E}[I_B(\theta(\omega), \xi(\omega)) \mid \theta(\omega) = a]\, P_\theta(da) \\
&= \frac{1}{\pi} \int_{-\pi/2}^{\pi/2} \mathsf{E}\, I_B(a, \xi(\omega))\, da = \frac{1}{\pi} \int_{-\pi/2}^{\pi/2} \cos a\, da = \frac{2}{\pi},
\end{aligned}
$$

where we have used the fact that

$$\mathsf{E}\, I_B(a, \xi(\omega)) = \mathsf{P}\{\xi \in [0, \tfrac{1}{2} \cos a] \cup [1 - \tfrac{1}{2} \cos a]\} = \cos a.$$

Thus the probability that a "random" toss of the needle intersects one of the lines is $2/\pi$. This result can be used as the basis for an experimental evaluation of π. In fact, let the needle be tossed N times independently. Define ξ_i to be 1 if the needle intersects a line on the ith toss, and 0 otherwise. Then by the law of large numbers (see, for example, (5) in Sect. 5 of Chap. 1)

$$\mathsf{P}\left\{\left|\frac{\xi_1 + \cdots + \xi_N}{N} - \mathsf{P}(A)\right| > \varepsilon\right\} \to 0, \quad N \to \infty,$$

for every $\varepsilon > 0$.

In this sense the frequency satisfies

$$\frac{\xi_1 + \cdots + \xi_N}{N} \approx \mathsf{P}(A) = \frac{2}{\pi}$$

and therefore

$$\frac{2N}{\xi_1 + \cdots + \xi_N} \approx \pi.$$

This formula has actually been used for a statistical evaluation of π. In 1850, R. Wolf (an astronomer in Zurich) threw a needle 5000 times and obtained the value 3.1596 for π. Apparently this problem was one of the first applications (now known as "Monte Carlo methods") of probabilistic-statistical regularities to numerical analysis.

Remark 6. Example 4 (Buffon's problem) is a typical example of a problem on *geometric probabilities*. In such problems one often can see how to assign probabilities to "elementary events" from geometric considerations based, for example, on a "symmetry." (Cf. Subsections 3 and 4 in Sect. 1 of Chap. 1 and Sect. 3 of the present Chapter.) Problems 9 to 12 below deal with geometric probabilities.

7. If $\{\xi_n\}_{n\geq 1}$ is a sequence of nonnegative random variables, then according to conclusion (f) of Theorem 2,

$$\mathsf{E}\left(\sum \xi_n \,|\, \mathscr{G}\right) = \sum \mathsf{E}(\xi_n \,|\, \mathscr{G}) \quad \text{(a. s.)}.$$

In particular, if $B_1, B_2, \ldots$ is a sequence of pairwise disjoint sets,

$$\mathsf{P}\left(\sum B_n \,|\, \mathscr{G}\right) = \sum \mathsf{P}\left(B_n \,|\, \mathscr{G}\right) \quad \text{(a. s.)}. \tag{22}$$

It must be emphasized that this equation is satisfied only almost surely and that consequently the conditional probability $\mathsf{P}(B \,|\, \mathscr{G})(\omega)$ cannot be considered as a measure in B for given ω. One might suppose that, except for a set $\mathscr{N}$ of measure zero, $\mathsf{P}(\cdot \,|\, \mathscr{G})(\omega)$ would still be a measure for $\omega \in \overline{\mathscr{N}}$. However, in general this is not the case, for the following reason. Let $\mathscr{N}(B_1, B_2, \ldots)$ be the set of sample points ω such that the countable additivity property (22) fails for these $B_1, B_2, \ldots$. Then the excluded set $\mathscr{N}$ is

$$\mathcal{N} = \bigcup \mathcal{N}(B_1, B_2, \ldots), \tag{23}$$

where the union is taken over all $B_1, B_2, \ldots$ in $\mathscr{F}$. Although the P-measure of each set $\mathcal{N}(B_1, B_2, \ldots)$ is zero, the P-measure of $\mathcal{N}$ can be different from zero (because of an uncountable union in (23)). (Recall that the Lebesgue measure of a single point is zero, but the measure of the set $\mathcal{N} = [0, 1]$, which is an uncountable sum of the individual points $\{x\}, 0 \le x < 1$, is 1.)

However, it would be convenient if the conditional probability $P(\cdot \mid \mathscr{G})(\omega)$ were a measure for each $\omega \in \Omega$, since then, for example, the calculation of conditional expectations $E(\xi \mid \mathscr{G})(\omega)$ could be carried out (see Theorem 3 below) in a simple way by averaging with respect to the measure $P(\cdot \mid \mathscr{G})(\omega)$:

$$E(\xi \mid \mathscr{G}) = \int_\Omega \xi(\tilde\omega) \, P(d\tilde\omega \mid \mathscr{G})(\omega) \quad \text{(a. s.)}$$

(cf. (10) in Sect. 8 of Chap. 1).

We introduce the following definition.

Definition 6. A function $P(\omega; B)$, defined for all $\omega \in \Omega$ and $B \in \mathscr{F}$, is a *regular conditional probability* with respect to $\mathscr{G} \subseteq \mathscr{F}$ if

(a) $P(\omega; \cdot)$ is a probability measure on $\mathscr{F}$ for every $\omega \in \Omega$;
(b) For each $B \in \mathscr{F}$ the function $P(\omega; B)$, as a function of ω, is a version of the conditional probability $P(B \mid \mathscr{G})(\omega)$, i.e., $P(\omega; B) = P(B \mid \mathscr{G})(\omega)$ (a. s.).

Theorem 3. *Let $P(\omega; B)$ be a regular conditional probability with respect to $\mathscr{G}$ and let ξ be an integrable random variable. Then*

$$E(\xi \mid \mathscr{G})(\omega) = \int_\Omega \xi(\tilde\omega) P(\omega; d\tilde\omega) \quad \text{(a. s.)}. \tag{24}$$

PROOF. If $\xi = I_B, B \in \mathscr{F}$, the required formula (24) becomes

$$P(B \mid \mathscr{G})(\omega) = P(\omega; B) \quad \text{(a. s.)},$$

which holds by Definition 6 (b). Consequently (24) holds for simple functions.

Now let $\xi \ge 0$ and $\xi_n \uparrow \xi$, where ξ_n are simple functions. Then by (b) of Theorem 2 we have $E(\xi \mid \mathscr{G})(\omega) = \lim_n E(\xi_n \mid \mathscr{G})(\omega)$ (a. s.). But since $P(\omega; \cdot)$ is a measure for every $\omega \in \Omega$, we have

$$\lim_n E(\xi_n \mid \mathscr{G})(\omega) = \lim_n \int_\Omega \xi_n(\tilde\omega) P(\omega; d\tilde\omega) = \int_\Omega \xi(\tilde\omega) P(\omega; d\tilde\omega)$$

by the monotone convergence theorem.

The general case reduces to this one if we use the representation $\xi = \xi^+ - \xi^-$. This completes the proof.
$\square$

Corollary. *Let $\mathscr{G} = \mathscr{G}_\eta$, where η is a random variable, and let the pair (ξ, η) have a probability distribution with density $f_{\xi,\eta}(x, y)$. Let $\mathsf{E}\,|g(\xi)| < \infty$. Then*

$$\mathsf{E}(g(\xi)\,|\,\eta = y) = \int_{-\infty}^{\infty} g(x) f_{\xi|\eta}(x\,|\,y)\,dx,$$

where $f_{\xi|\eta}(x\,|\,y)$ is the density of the conditional distribution (see (18)).

In order to be able to state the basic result on the *existence* of regular conditional probabilities, we need the following definitions.

Definition 7. Let $(E, \mathscr{E})$ be a measurable space, $X = X(\omega)$ a random element with values in E, and $\mathscr{G}$ a σ-subalgebra of $\mathscr{F}$. A function $Q(\omega; B)$, defined for $\omega \in \Omega$ and $B \in \mathscr{E}$, is a *regular conditional distribution of X with respect to $\mathscr{G}$* if

(a) for each $\omega \in \Omega$ the function $Q(\omega; B)$ is a probability measure on $(E, \mathscr{E})$;
(b) for each $B \in \mathscr{E}$ the function $Q(\omega; B)$, as a function of ω, is a version of the conditional probability $\mathsf{P}(X \in B\,|\,\mathscr{G})(\omega)$, i.e.

$$Q(\omega; B) = \mathsf{P}(X \in B\,|\,\mathscr{G})(\omega) \quad \text{(a. s.).}$$

Definition 8. Let ξ be a random variable. A function $F = F(\omega; x)$, $\omega \in \Omega$, $x \in R$, is a *regular distribution function for ξ with respect to $\mathscr{G}$* if:

(a) $F(\omega; x)$ is, for each $\omega \in \Omega$, a distribution function on R;
(b) $F(\omega; x) = \mathsf{P}(\xi \le x\,|\,\mathscr{G})(\omega)$ (a. s.), for each $x \in R$.

Theorem 4. *A regular distribution function and a regular conditional distribution always exist for the random variable ξ with respect to a σ-algebra $\mathscr{G} \subseteq \mathscr{F}$.*

PROOF. For each rational number $r \in R$, define $F_r(\omega) = \mathsf{P}(\xi \le r\,|\,\mathscr{G})(\omega)$, where $\mathsf{P}(\xi \le r\,|\,\mathscr{G})(\omega) = \mathsf{E}(I_{\{\xi \le r\}}\,|\,\mathscr{G})(\omega)$ is any version of the conditional probability, with respect to $\mathscr{G}$, of the event $\{\xi \le r\}$. Let $\{r_i\}$ be the set of rational numbers in R. If $r_i < r_j$, Property $\mathbf{B^*}$ implies that $\mathsf{P}(\xi \le r_i\,|\,\mathscr{G}) \le \mathsf{P}(\xi \le r_j\,|\,\mathscr{G})$ (a. s.), and therefore if $A_{ij} = \{\omega: F_{r_j}(\omega) < F_{r_i}(\omega)\}$, $A = \bigcup A_{ij}$, we have $\mathsf{P}(A) = 0$. In other words, the set of points ω at which the distribution function $F_r(\omega)$, $r \in \{r_i\}$, fails to be monotonic has measure zero.

Now let

$$B_i = \left\{\omega: \lim_{n \to \infty} F_{r_i + (1/n)}(\omega) \ne F_{r_i}(\omega)\right\}, \quad B = \bigcup_{i=1}^{\infty} B_i.$$

It is clear that $I_{\{\xi \le r_i + (1/n)\}} \downarrow I_{\{\xi \le r_i\}}$, $n \to \infty$. Therefore, by (a) of Theorem 2, $F_{r_i + (1/n)}(\omega) \to F_{r_i}(\omega)$ (a. s.), and therefore the set B on which continuity on the right (along the rational numbers) fails also has measure zero, $\mathsf{P}(B) = 0$.

In addition, let

$$C = \left\{\omega: \lim_{n \to \infty} F_n(\omega) \ne 1\right\} \cup \left\{\omega: \lim_{n \to -\infty} F_n(\omega) \ne 0\right\}.$$

Then, since $\{\xi \le n\} \uparrow \Omega$, $n \to \infty$, and $\{\xi \le n\} \downarrow \varnothing$, $n \to -\infty$, we have $\mathsf{P}(C) = 0$.

Now put

$$F(\omega; x) = \begin{cases} \lim_{r \downarrow x} F_r(\omega), & \omega \notin A \cup B \cup C, \\ G(x), & \omega \in A \cup B \cup C, \end{cases}$$

where $G(x)$ is any distribution function on R; we show that $F(\omega; x)$ satisfies the conditions of Definition 8.

Let $\omega \notin A \cup B \cup C$. Then it is clear that $F(\omega; x)$ is a nondecreasing function of x. If $x < x' \le r$, then $F(\omega; x) \le F(\omega; x') \le F(\omega; r) = F_r(\omega) \downarrow F(\omega, x)$ when $r \downarrow x$. Consequently $F(\omega; x)$ is continuous on the right. Similarly $\lim_{x \to \infty} F(\omega; x) = 1$, $\lim_{x \to -\infty} F(\omega; x) = 0$. Since $F(\omega; x) = G(x)$ when $\omega \in A \cup B \cup C$, it follows that $F(\omega; x)$ is a distribution function on R for every $\omega \in \Omega$, i.e., condition (a) of Definition 6 is satisfied.

By construction, $\mathsf{P}(\xi \le r \,|\, \mathscr{G})(\omega) = F_r(\omega) = F(\omega; r)$. If $r \downarrow x$, we have $F(\omega; r) \downarrow F(\omega; x)$ for all $\omega \in \Omega$ by the continuity on the right that we just established. But by conclusion (a) of Theorem 2, we have $\mathsf{P}(\xi \le r \,|\, \mathscr{G})(\omega) \to \mathsf{P}(\xi \le x \,|\, \mathscr{G})(\omega)$ (a.s.). Therefore $F(\omega; x) = \mathsf{P}(\xi \le x \,|\, \mathscr{G})(\omega)$ (a.s.), which establishes condition (b) of Definition 8.

We now turn to the proof of the existence of a regular conditional distribution of ξ with respect to $\mathscr{G}$.

Let $F(\omega; x)$ be the function constructed above. Put

$$Q(\omega; B) = \int_B F(\omega; dx),$$

where the integral is a Lebesgue–Stieltjes integral. From the properties of the integral (see Subsection 8 in Sect. 6), it follows that $Q(\omega; B)$ is a measure in B for each given $\omega \in \Omega$. To establish that $Q(\omega; B)$ is a *version* of the conditional probability $\mathsf{P}(\xi \in B \,|\, \mathscr{G})(\omega)$, we use the *principle of appropriate sets*.

Let $\mathscr{C}$ be the collection of sets B in $\mathscr{B}(R)$ for which $Q(\omega; B) = \mathsf{P}(\xi \in B \,|\, \mathscr{G})(\omega)$ (a.s.). Since $F(\omega; x) = \mathsf{P}(\xi \le x \,|\, \mathscr{G})(\omega)$ (a.s.), the system $\mathscr{C}$ contains the sets B of the form $B = (-\infty, x]$, $x \in R$. Therefore $\mathscr{C}$ also contains the intervals of the form $(a, b]$, and the algebra $\mathscr{A}$ consisting of finite sums of disjoint sets of the form $(a, b]$. Then it follows from the continuity properties of $Q(\omega; B)$ (ω fixed) and from conclusion (b) of Theorem 2 that $\mathscr{C}$ is a monotonic class, and since $\mathscr{A} \subseteq \mathscr{C} \subseteq \mathscr{B}(R)$, we have, from Theorem 1 of Sect. 2,

$$\mathscr{B}(R) = \sigma(\mathscr{A}) \subseteq \sigma(\mathscr{C}) = \mu(\mathscr{C}) = \mathscr{C} \subseteq \mathscr{B}(R),$$

whence $\mathscr{C} = \mathscr{B}(R)$.

This completes the proof of the theorem.

$\square$

By using topological considerations we can extend the conclusion of Theorem 4 on the existence of a regular conditional distribution to random elements with values in what are known as Borel spaces. We need the following definition.

Definition 9. A measurable space $(E, \mathscr{E})$ is a *Borel space* if it is Borel equivalent to a Borel subset of the real line, i.e., there is a one-to-one mapping $\varphi = \varphi(e) \colon (E, \mathscr{E}) \rightarrow (R, \mathscr{B}(R))$ such that

(1) $\varphi(E) \equiv \{\varphi(e) \colon e \in E\}$ is a set in $\mathscr{B}(R)$;
(2) φ is $\mathscr{E}$-measurable ($\varphi^{-1}(A) \in \mathscr{E}, A \in \varphi(E) \cap \mathscr{B}(R)$),
(3) φ^{-1} is $\mathscr{B}(R)/\mathscr{E}$-measurable ($\varphi(B) \in \varphi(E) \cap \mathscr{B}(R), B \in \mathscr{E}$).

Theorem 5. *Let $X = X(\omega)$ be a random element with values in the Borel space $(E, \mathscr{E})$. Then there is a regular conditional distribution of X with respect to $\mathscr{G} \subseteq \mathscr{F}$.*

PROOF. Let $\varphi = \varphi(e)$ be the function in Definition 9. By (2) in this definition $\varphi(X(\omega))$ is a random variable. Hence, by Theorem 4, we can define the conditional distribution $Q(\omega; A)$ of $\varphi(X(\omega))$ with respect to $\mathscr{G}, A \in \varphi(E) \cap \mathscr{B}(R)$.

We introduce the function $\tilde{Q}(\omega; B) = Q(\omega; \varphi(B)), B \in \mathscr{E}$. By (3) of Definition 9, $\varphi(B) \in \varphi(E) \cap \mathscr{B}(R)$ and consequently $\tilde{Q}(\omega; B)$ is defined. Evidently $\tilde{Q}(\omega; B)$ is a measure in $B \in \mathscr{E}$ for every ω. Now fix $B \in \mathscr{E}$. By the one-to-one character of the mapping $\varphi = \varphi(e)$,

$$\tilde{Q}(\omega; B) = Q(\omega; \varphi(B)) = \mathsf{P}\{\varphi(X) \in \varphi(B) \,|\, \mathscr{G}\}(\omega) = \mathsf{P}\{X \in B \,|\, \mathscr{G}\}(\omega) \quad \text{(a. s.)}.$$

Therefore $\tilde{Q}(\omega; B)$ is a regular conditional distribution of X with respect to $\mathscr{G}$. This completes the proof of the theorem.

□

Corollary. *Let $X = X(\omega)$ be a random element with values in a complete separable metric space $(E, \mathscr{E})$. Then there is a regular conditional distribution of X with respect to $\mathscr{G}$. In particular, such a distribution exists for the spaces $(R^n, \mathscr{B}(R^n))$ and $(R^\infty, \mathscr{B}(R^\infty))$.*

The proof follows from Theorem 5 and the well-known topological result that such spaces $(E, \mathscr{E})$ are Borel spaces.

8. The theory of conditional expectations developed above makes it possible to give a generalization of Bayes's theorem; this has applications in statistics.

Recall that if $\mathscr{D} = \{A_1, \ldots, A_n\}$ is a partition of the space Ω with $\mathsf{P}(A_i) > 0$, Bayes's theorem, see (9) in Sect. 3 of Chap. 1, states that

$$\mathsf{P}(A_i \,|\, B) = \frac{\mathsf{P}(A_i) \, \mathsf{P}(B \,|\, A_i)}{\sum_{j=1}^n \mathsf{P}(A_j) \, \mathsf{P}(B \,|\, A_j)} \tag{25}$$

for every B with $\mathsf{P}(B) > 0$. Therefore if $\theta = \sum_{i=1}^n a_i I_{A_i}$ is a discrete random variable then, according to (10) in Sect. 8 of Chap. 1,

$$\mathsf{E}[g(\theta) \,|\, B] = \frac{\sum_{i=1}^n g(a_i) \, \mathsf{P}(A_i) \, \mathsf{P}(B \,|\, A_i)}{\sum_{j=1}^n \mathsf{P}(A_j) \, \mathsf{P}(B \,|\, A_j)}, \tag{26}$$

or

$$E[g(\theta) \mid B] = \frac{\int_{-\infty}^{\infty} g(a) \, P(B \mid \theta = a) \, P_\theta(da)}{\int_{-\infty}^{\infty} P(B \mid \theta = a) \, P_\theta(da)}, \tag{27}$$

where $P_\theta(A) = P\{\theta \in A\}$.

On the basis of the definition of $E[g(\theta) \mid B]$ given at the beginning of this section, it is easy to establish that (27) holds for all events B with $P(B) > 0$, random variables θ and functions $g = g(a)$ with $E \, |g(\theta)| < \infty$.

We now consider an analog of (27) for conditional expectations $E[g(\theta) \mid \mathscr{G}]$ with respect to a σ-algebra $\mathscr{G}$, $\mathscr{G} \subseteq \mathscr{F}$.

Let

$$Q(B) = \int_B g(\theta(\omega)) \, P(d\omega), \quad B \in \mathscr{G}. \tag{28}$$

Then by (4)

$$E[g(\theta) \mid \mathscr{G}](\omega) = \frac{dQ}{dP}(\omega). \tag{29}$$

We also consider the σ-algebra $\mathscr{G}_\theta$. Then, by (5),

$$P(B) = \int_\Omega P(B \mid \mathscr{G}_\theta) \, dP \tag{30}$$

or, by the formula for change of variable in Lebesgue integrals,

$$P(B) = \int_{-\infty}^{\infty} P(B \mid \theta = a) \, P_\theta(da). \tag{31}$$

Since

$$Q(B) = E[g(\theta) I_B] = E[g(\theta) \cdot E(I_B \mid \mathscr{G}_\theta)],$$

we have

$$Q(B) = \int_{-\infty}^{\infty} g(a) \, P(B \mid \theta = a) \, P_\theta(da). \tag{32}$$

Now suppose that the conditional probability $P(B \mid \theta = a)$ is regular and admits the representation

$$P(B \mid \theta = a) = \int_B \rho(\omega; a) \, \lambda(d\omega), \tag{33}$$

where $\rho = \rho(\omega; a)$ is nonnegative and measurable in the two variables jointly, and λ is a σ-finite measure on $(\Omega, \mathscr{G})$.

Let $E \, |g(\theta)| < \infty$. Let us show that (P-a. s.)

$$E[g(\theta) \mid \mathscr{G}](\omega) = \frac{\int_{-\infty}^{\infty} g(a) \rho(\omega; a) \, P_\theta(da)}{\int_{-\infty}^{\infty} \rho(\omega, a) \, P_\theta(da)} \tag{34}$$

(*generalized Bayes theorem*).

In proving (34) we shall need the following lemma.

Lemma. *Let $(\Omega, \mathscr{F})$ be a measurable space.*

(a) *Let μ and λ be σ-finite measures, $\mu \ll \lambda$, and $f = f(\omega)$ an $\mathscr{F}$-measurable function. Then*

$$\int_\Omega f \, d\mu = \int_\Omega f \frac{d\mu}{d\lambda} \, d\lambda \tag{35}$$

   (in the sense that if either integral exists, the other exists and they are equal).

(b) *If v is a signed measure and μ, λ are σ-finite measures, $v \ll \mu$, $\mu \ll \lambda$, then*

$$\frac{dv}{d\lambda} = \frac{dv}{d\mu} \cdot \frac{d\mu}{d\lambda} \quad (\lambda\text{-}a.\,s.) \tag{36}$$

and

$$\frac{dv}{d\mu} = \frac{dv}{d\lambda} \Big/ \frac{d\mu}{d\lambda} \quad (\mu\text{-}a.\,s.). \tag{37}$$

PROOF. (a) Since

$$\mu(A) = \int_A \left(\frac{d\mu}{d\lambda} \right) d\lambda, \quad A \in \mathscr{F},$$

(35) is evidently satisfied for simple functions $f = \sum f_i I_{A_i}$. The general case follows from the representation $f = f^+ - f^-$ and the monotone convergence theorem (cf. the proof of (39) in Sect. 6).

(b) From (a) with $f = dv/d\mu$ we obtain

$$v(A) = \int_A \left(\frac{dv}{d\mu} \right) d\mu = \int_A \left(\frac{dv}{d\mu} \right) \left(\frac{d\mu}{d\lambda} \right) d\lambda.$$

Then $v \ll \lambda$ and therefore

$$v(A) = \int_A \frac{dv}{d\lambda} \, d\lambda,$$

whence (36) follows since A is arbitrary, by Property I (Sect. 6).

   Property (37) follows from (36) and the remark that

$$\mu \left\{ \omega : \frac{d\mu}{d\lambda} = 0 \right\} = \int_{\{\omega \,:\, d\mu/d\lambda = 0\}} \frac{d\mu}{d\lambda} \, d\lambda = 0$$

(on the set $\{\omega : d\mu/d\lambda = 0\}$ the right-hand side of (37) can be defined arbitrarily, for example as zero). This completes the proof of the lemma.
□

   To prove (34) we observe that by Fubini's theorem and (33),

$$Q(B) = \int_B \left[\int_{-\infty}^{\infty} g(a) \rho(\omega; a) \, P_\theta(da) \right] \lambda(d\omega), \tag{38}$$

$$P(B) = \int_B \left[\int_{-\infty}^{\infty} \rho(\omega; a) \, P_\theta(da) \right] \lambda(d\omega). \tag{39}$$

Then by the lemma

$$\frac{d\mathsf{Q}}{d\mathsf{P}} = \frac{d\mathsf{Q}/d\lambda}{d\mathsf{P}/d\lambda} \quad \text{(P-a. s.).}$$

Taking account of (38), (39) and (29), we have (34).

Remark 7. Formula (34) remains valid if we replace θ by a random element with values in some measurable space $(E, \mathscr{E})$ (and replace integration over R by integration over E).

Let us consider some special cases of (34).

Let the σ-algebra $\mathscr{G}$ be generated by the random variable ξ, $\mathscr{G} = \mathscr{G}_\xi$. Suppose that

$$\mathsf{P}(\xi \in A \mid \theta = a) = \int_A q(x; a)\, \lambda(dx), \quad A \in \mathscr{B}(R), \tag{40}$$

where $q = q(x; a)$ is a nonnegative function, measurable with respect to both variables jointly, and λ is a σ-finite measure on $(R, \mathscr{B}(R))$. Then by the formula for change of variable in Lebesgue integrals and (34) we obtain

$$\mathsf{E}[g(\theta) \mid \xi = x] = \frac{\int_{-\infty}^{\infty} g(a)\, q(x; a)\, P_\theta(da)}{\int_{-\infty}^{\infty} q(x; a)\, P_\theta(da)}. \tag{41}$$

In particular, let (θ, ξ) be a pair of discrete random variables, $\theta = \sum a_i I_{A_i}$, $\xi = \sum x_j I_{B_j}$. Then, taking λ to be the *counting* measure $(\lambda(\{x_i\}) = 1$, $i = 1, 2, \ldots)$ we find from (41) that

$$\mathsf{E}[g(\theta) \mid \xi = x_j] = \frac{\sum_i g(a_i)\, \mathsf{P}(\xi = x_j \mid \theta = a_i)\, \mathsf{P}(\theta = a_i)}{\sum_i \mathsf{P}(\xi = x_j \mid \theta = a_i)\, \mathsf{P}(\theta = a_i)}. \tag{42}$$

(Compare (26).)

Now let (θ, ξ) be a pair of absolutely continuous random variables with density $f_{\theta,\xi}(a, x)$. Then by (19) the representation (40) applies with $q(x; a) = f_{\xi|\theta}(x \mid a)$ and Lebesgue measure λ. Therefore

$$\mathsf{E}[g(\theta) \mid \xi = x] = \frac{\int_{-\infty}^{\infty} g(a) f_{\xi|\theta}(x \mid a) f_\theta(a)\, da}{\int_{-\infty}^{\infty} f_{\xi|\theta}(x \mid a) f_\theta(a)\, da}. \tag{43}$$

9. Here we give one more version of the generalized Bayes theorem (see (34)), which is especially appropriate in problems related to the change of probability measures.

Theorem 6. *Let* P *and* $\tilde{\mathsf{P}}$ *be two probability measures on a measurable space* $(\Omega, \mathscr{F})$ *with* $\tilde{\mathsf{P}}$ *being absolutely continuous with respect to* P *(denoted* $\tilde{\mathsf{P}} \ll \mathsf{P}$*) and let* $\frac{d\tilde{\mathsf{P}}}{d\mathsf{P}}$ *be the Radon–Nikodým derivative of* $\tilde{\mathsf{P}}$ *with respect to* P. *Let* $\mathscr{G}$ *be a* σ-subalgebra of $\mathscr{F}$ $(\mathscr{G} \subseteq \mathscr{F})$ *and* $\mathsf{E}(\cdot \mid \mathscr{G})$ *and* $\tilde{\mathsf{E}}(\cdot \mid \mathscr{G})$ *be conditional expectations with respect to* P *and* $\tilde{\mathsf{P}}$ *given* $\mathscr{G}$. *Let* ξ *be a nonnegative* $(\mathscr{F}$-measurable) *random variable. Then the following "recalculation formula of conditional expectations" holds:*

$$\tilde{\mathsf{E}}(\xi \mid \mathcal{G}) = \frac{\mathsf{E}\left(\xi \frac{d\tilde{\mathsf{P}}}{d\mathsf{P}} \mid \mathcal{G}\right)}{\mathsf{E}\left(\frac{d\tilde{\mathsf{P}}}{d\mathsf{P}} \mid \mathcal{G}\right)} \quad (\tilde{\mathsf{P}}\text{-a. s.}). \tag{44}$$

Formula (44) is valid also for any random variable ξ whose conditional expectation $\tilde{\mathsf{E}}(\xi \mid \mathcal{G})$ is well defined.

PROOF. Note first of all that the $\tilde{\mathsf{P}}$-measure (as well as the P-measure) of the event $\left\{\omega \colon \mathsf{E}\left(\frac{d\tilde{\mathsf{P}}}{d\mathsf{P}} \mid \mathcal{G}\right) = 0\right\}$ equals zero. Indeed, if $A \in \mathcal{G}$, then

$$\int_A \mathsf{E}\left(\frac{d\tilde{\mathsf{P}}}{d\mathsf{P}} \mid \mathcal{G}\right) d\mathsf{P} = \int_A \frac{d\tilde{\mathsf{P}}}{d\mathsf{P}} \, d\mathsf{P} = \int_A d\tilde{\mathsf{P}} = \tilde{\mathsf{P}}(A),$$

and therefore the set $A = \left\{\omega \colon \mathsf{E}\left(\frac{d\tilde{\mathsf{P}}}{d\mathsf{P}} \mid \mathcal{G}\right) = 0\right\}$ has zero $\tilde{\mathsf{P}}$-measure.

Let $\xi \geq 0$. By the definition of conditional expectation, $\tilde{\mathsf{E}}(\xi \mid \mathcal{G})$ is a $\mathcal{G}$-measurable random variable such that

$$\tilde{\mathsf{E}}[I_A \tilde{\mathsf{E}}(\xi \mid \mathcal{G})] = \tilde{\mathsf{E}}[I_A \xi] \tag{45}$$

for any $A \in \mathcal{G}$. Hence for the proof of (44) we only need to establish that the ($\mathcal{G}$-measurable) random variable in the right-hand side of (44) satisfies the equality

$$\tilde{\mathsf{E}}\left[I_A \cdot \frac{1}{\mathsf{E}\left(\frac{d\tilde{\mathsf{P}}}{d\mathsf{P}} \mid \mathcal{G}\right)} \cdot \mathsf{E}\left(\xi \frac{d\tilde{\mathsf{P}}}{d\mathsf{P}} \mid \mathcal{G}\right)\right] = \tilde{\mathsf{E}}[I_A \xi]. \tag{46}$$

Using the properties of conditional expectations and (39) of Sect. 6 we find that

$$\tilde{\mathsf{E}}\left[I_A \cdot \frac{1}{\mathsf{E}\left(\frac{d\tilde{\mathsf{P}}}{d\mathsf{P}} \mid \mathcal{G}\right)} \cdot \mathsf{E}\left(\xi \frac{d\tilde{\mathsf{P}}}{d\mathsf{P}} \mid \mathcal{G}\right)\right] = \mathsf{E}\left[I_A \cdot \frac{1}{\mathsf{E}\left(\frac{d\tilde{\mathsf{P}}}{d\mathsf{P}} \mid \mathcal{G}\right)} \cdot \mathsf{E}\left(\xi \frac{d\tilde{\mathsf{P}}}{d\mathsf{P}} \mid \mathcal{G}\right) \cdot \frac{d\tilde{\mathsf{P}}}{d\mathsf{P}}\right]$$

$$= \mathsf{E}\left[I_A \cdot \frac{1}{\mathsf{E}\left(\frac{d\tilde{\mathsf{P}}}{d\mathsf{P}} \mid \mathcal{G}\right)} \cdot \mathsf{E}\left(\xi \frac{d\tilde{\mathsf{P}}}{d\mathsf{P}} \mid \mathcal{G}\right) \cdot \mathsf{E}\left(\frac{d\tilde{\mathsf{P}}}{d\mathsf{P}} \mid \mathcal{G}\right)\right]$$

$$= \mathsf{E}\left[I_A \, \mathsf{E}\left(\xi \frac{d\tilde{\mathsf{P}}}{d\mathsf{P}} \mid \mathcal{G}\right)\right] = \mathsf{E}\left[I_A \xi \frac{d\tilde{\mathsf{P}}}{d\mathsf{P}}\right] = \tilde{\mathsf{E}}[I_A \xi],$$

which proves (45) for nonnegative ξ. The general case is treated similarly to the proof of (39) in Sect. 6 for arbitrary integrable random variables ξ.

□

10. The generalized Bayes theorem stated above (see (34), (41), and (43)), which is one of the basic tools in the "Bayes's approach" in mathematical statistics, answers the question of how our knowledge about the distribution of a random variable θ redistributes depending on the results of observations on a random variable ξ statistically related to θ.

Now we will consider one more application of the concept of conditional expectation to the estimation problem of an *unknown parameter* θ based on the observational data. (We emphasize that unlike the above case, where θ was a random variable, we now treat θ simply as a parameter taking values in a parameter set Θ specified *a priori*, cf. Sect. 7 of Chap. 1).

Actually we will consider an important concept of mathematical statistics, namely, that of *sufficient σ-subalgebra*.

Let $\mathscr{P} = \{\mathsf{P}_\theta, \theta \in \Theta\}$ be a family of probability measures on a measurable space $(\Omega, \mathscr{F})$, these measures depending on a parameter θ running over a *parameter set* Θ. The triple $\mathscr{E} = (\Omega, \mathscr{F}, \mathscr{P})$ is often said to specify a *probabilistic-statistical model* or a *probabilistic-statistical experiment*.

To clarify Definition 10 to be given below assume that we have an $\mathscr{F}$-measurable function $T = T(\omega)$ (a statistic), depending on the outcome ω, and a σ-algebra $\mathscr{G} = \sigma(T(\omega))$ generated by this function. It is clear that $\mathscr{G} \subseteq \mathscr{F}$ and, in general, $\mathscr{F}$ may contain events which do not belong to $\mathscr{G}$ (i.e., $\mathscr{F}$ is "richer" than $\mathscr{G}$). But it may happen that with regard to determining which value of θ is in fact "acting" we do not need anything but $T = T(\omega)$. In this sense it would be natural to call the statistic T "*sufficient*."

Remark 8. When $T(\omega) = \omega$, i.e., when we know the *outcomes* themselves (rather then a function of them), we can single out the following two extreme cases.

One of them is when the probabilities P_θ are the same for all $\theta \in \Theta$. Clearly neither outcome ω can then give any information about θ.

Another case is when the supports of all the measures P_θ, $\theta \in \Theta$, are contained in different subsets of $\mathscr{F}$ (i.e., for any two values θ_1 and θ_2 the measures P_{θ_1} and P_{θ_2} are *singular*, in which case there are two sets (supports) Ω_{θ_1} and Ω_{θ_2} such that $\mathsf{P}_{\theta_1}(\Omega \backslash \Omega_{\theta_1}) = 0$, $\mathsf{P}_{\theta_2}(\Omega \backslash \Omega_{\theta_2}) = 0$ and $\Omega_{\theta_1} \cap \Omega_{\theta_2} = \varnothing$). In this case the outcome ω uniquely determines θ.

Both these cases are of little interest. The cases of interest are, say, when all the measures P_θ are equivalent to each other (and their supports are then the same), or these measures are *dominated*, which is a weaker property than equivalence, namely, there exists a σ-finite measure λ such that $\mathsf{P}_\theta \ll \lambda$ for all $\theta \in \Theta$. In the general statistical theory it is customarily assumed that the family at hand is dominated (which allows one to exclude some measure-theoretic pathologies). The role of this property is fully revealed by the Factorization Theorem 7 to be stated below.

The following definition may be regarded as one of the ways to formalize the concept of sufficiency of "information" contained in a σ-subalgebra $\mathscr{G} \subseteq \mathscr{F}$.

Definition 10. Let $(\Omega, \mathscr{F}, \mathscr{P})$ be a probabilistic-statistical model, $\mathscr{P} = \{\mathsf{P}_\theta, \theta \in \Theta\}$, and $\mathscr{G}$ be a σ-subalgebra of $\mathscr{F}$ ($\mathscr{G} \subseteq \mathscr{F}$). Then $\mathscr{G}$ is said to be *sufficient* for the family $\mathscr{P}$ if there exist versions of conditional probabilities $\mathsf{P}_\theta(\,\cdot\,|\,\mathscr{G})(\omega)$, $\theta \in \Theta$, $\omega \in \Omega$, independent of θ, i.e., there is a function $P(A;\omega)$, $A \in \mathscr{F}$, $\omega \in \Omega$, such that

$$\mathsf{P}_\theta(A\,|\,\mathscr{G})(\omega) = P(A;\omega) \quad (\mathsf{P}_\theta\text{-a.s.}) \tag{47}$$

for all $A \in \mathscr{F}$ and $\theta \in \Theta$; in other words, $P(A;\omega)$ is a version of $\mathsf{P}_\theta(A\,|\,\mathscr{G})(\omega)$ for all $\theta \in \Theta$.

If $\mathscr{G} = \sigma(T(\omega))$, then the statistic $T = T(\omega)$ is called *sufficient* for the family $\mathscr{P}$.

Remark 9. As was pointed out above, we are interested in finding *sufficient* statistics in our statistical research because we try to obtain functions $T = T(\omega)$ of outcomes ω which provide the *reduction of data preserving the information* (about θ). For example, suppose that $\omega = (x_1, x_2, \ldots, x_n)$, where $x_i \in R$ and n is very large. Then finding "good" estimators for θ (as, e.g., in Sect. 7 of Chap. 1) may be a complicated problem because of large dimensionality of the data $x_1, x_2, \ldots, x_n$. However it may happen (as we observed in Sect. 7 of Chap. 1) that for obtaining "good" estimators it suffices to know only the value of a "summarizing" statistic like $T(\omega) = x_1 + x_2 + \cdots + x_n$ rather than individual values $x_1, x_2, \ldots, x_n$.

Clearly such a statistic provides an essential reduction of data (and computational complexity) being at the same time sufficient for obtaining "good" estimators for θ.

The following *factorization theorem* provides conditions that ensure sufficiency of a σ-subalgebra $\mathscr{G}$ for the family $\mathscr{P}$.

Theorem 7. *Let $\mathscr{P} = \{\mathsf{P}_\theta, \theta \in \Theta\}$ be a dominated family, i.e., there exists a σ-finite measure λ on $(\Omega, \mathscr{F})$ such that the measures P_θ are absolutely continuous with respect to λ ($\mathsf{P}_\theta \ll \lambda$) for all $\theta \in \Theta$.*

Let $g_\theta^{(\lambda)}(\omega) = \dfrac{d\mathsf{P}_\theta}{d\lambda}(\omega)$ be a Radon–Nikodým derivative of P_θ with respect to λ.

The σ-subalgebra $\mathscr{G}$ is sufficient for the family $\mathscr{P}$ if and only if the functions $g_\theta^{(\lambda)}(\omega)$ admit the following factorization: there are nonnegative functions $\hat{g}_\theta^{(\lambda)}(\omega)$ and $h(\omega)$ such that $\hat{g}_\theta^{(\lambda)}(\omega)$ are $\mathscr{G}$-measurable, $h(\omega)$ is $\mathscr{F}$-measurable and

$$g_\theta^{(\lambda)}(\omega) = \hat{g}_\theta^{(\lambda)}(\omega)\, h(\omega) \quad (\lambda\text{-a.s.}) \tag{48}$$

for all $\theta \in \Theta$.

If we can take the measure P_{θ_0} for λ, where θ_0 is a parameter in Θ, then $\mathscr{G}$ is sufficient if and only if the derivative $g_\theta^{(\theta_0)}(\omega) = \dfrac{d\mathsf{P}_\theta}{d\mathsf{P}_{\theta_0}}$ itself is $\mathscr{G}$-measurable.

PROOF. *Sufficiency.* By assumption, the dominating measure λ is σ-finite. This means that there are $\mathscr{F}$-measurable disjoint sets Ω_k, $k \geq 1$, such that $\Omega = \sum_{k\geq 1} \Omega_k$ and $0 < \lambda(\Omega_k) < \infty$, $k \geq 1$.

Form the measure

$$\tilde{\lambda}(\cdot) = \sum_{k\geq 1} \frac{1}{2^k} \frac{\lambda(\Omega_k \cap \cdot)}{1 + \lambda(\Omega_k)}.$$

This measure if finite, $\tilde{\lambda}(\Omega) < \infty$, and $\tilde{\lambda}(\Omega) > 0$. Without loss of generality it can be taken to be a probability measure, $\tilde{\lambda}(\Omega) = 1$.

Then by the formula (44) of recalculating conditional expectations we find that, for any $\mathscr{F}$-measurable bounded random variable $X = X(\omega)$,

$$E_\theta(X \mid \mathscr{G}) = \frac{E_{\tilde\lambda}\left(X \frac{dP_\theta}{d\tilde\lambda} \mid \mathscr{G}\right)}{E_{\tilde\lambda}\left(\frac{dP_\theta}{d\tilde\lambda} \mid \mathscr{G}\right)} \quad (P_\theta\text{-a.s.}). \tag{49}$$

By (48) we have

$$g_\theta^{(\tilde\lambda)} = \frac{dP_\theta}{d\tilde\lambda} = \frac{dP_\theta}{d\lambda} \cdot \frac{d\lambda}{d\tilde\lambda} = g_\theta^{(\lambda)} \frac{d\lambda}{d\tilde\lambda} = \hat{g}_\theta^{(\lambda)} h \frac{d\lambda}{d\tilde\lambda}. \tag{50}$$

Therefore (49) takes the form

$$E_\theta(X \mid \mathscr{G}) = \frac{E_{\tilde\lambda}\left(X \hat{g}_\theta^{(\lambda)} h \frac{d\lambda}{d\tilde\lambda} \mid \mathscr{G}\right)}{E_{\tilde\lambda}\left(\hat{g}_\theta^{(\lambda)} h \frac{d\lambda}{d\tilde\lambda} \mid \mathscr{G}\right)} \quad (P_\theta\text{-a.s.}). \tag{51}$$

But $\hat{g}_\theta^{(\lambda)}$ are $\mathscr{G}$-measurable and

$$E_\theta(X \mid \mathscr{G}) = \frac{E_{\tilde\lambda}\left(X h \frac{d\lambda}{d\tilde\lambda} \mid \mathscr{G}\right)}{E_{\tilde\lambda}\left(h \frac{d\lambda}{d\tilde\lambda} \mid \mathscr{G}\right)} \quad (P_\theta\text{-a.s.}). \tag{52}$$

The right-hand side here does not depend on θ, hence the property (47) holds. Thus by Definition 10 the σ-algebra $\mathscr{G}$ is sufficient.

The *necessity* will be proved only under the additional assumption that the family $\mathscr{P} = \{P_\theta, \theta \in \Theta\}$ is not only dominated by a σ-finite measure λ, but also that there is a $\theta_0 \in \Theta$ such that all the measures $P_\theta \ll P_{\theta_0}$, i.e., that for any $\theta \in \Theta$ the measure P_θ absolutely continuous with respect to P_{θ_0}. (In the general case the proof becomes more complicated, see Theorem 34.6 in [10].)

So, let $\mathscr{G}$ be a sufficient σ-algebra, i.e., (47) holds. We will show that under the assumption that $P_\theta \ll P_{\theta_0}$, $\theta \in \Theta$, the derivative $g_\theta^{(\theta_0)} = \frac{dP_\theta}{dP_{\theta_0}}$ is $\mathscr{G}$-measurable for any $\theta \in \Theta$.

Let $A \in \mathscr{F}$. Then using the properties of conditional expectations we find for any $\theta \in \Theta$ that $\left(\text{with } g_\theta^{(\theta_0)} = dP_\theta / dP_{\theta_0}\right)$

$$P_\theta(A) = E_\theta I_A = E_\theta E_\theta(I_A \mid \mathscr{G}) = E_\theta E_{\theta_0}(I_A \mid \mathscr{G}) = E_{\theta_0}[g_\theta^{(\theta_0)} E_{\theta_0}(I_A \mid \mathscr{G})]$$

$$= E_{\theta_0} E_{\theta_0}(g_\theta^{(\theta_0)} E_{\theta_0}(I_A \mid \mathscr{G}) \mid \mathscr{G}) = E_{\theta_0}([E_{\theta_0}(g_\theta^{(\theta_0)} \mid \mathscr{G})] \cdot [E_{\theta_0}(I_A \mid \mathscr{G})])$$

$$= E_{\theta_0} E_{\theta_0}(I_A E_{\theta_0}(g_\theta^{(\theta_0)} \mid \mathscr{G}) \mid \mathscr{G}) = E_{\theta_0} I_A E_{\theta_0}(g_\theta^{(\theta_0)} \mid \mathscr{G}) = \int_A E_{\theta_0}(g_\theta^{(\theta_0)} \mid \mathscr{G}) \, dP_{\theta_0}.$$

Therefore the $\mathscr{G}$-measurable function $E_{\theta_0}(g_\theta^{(\theta_0)} \mid \mathscr{G})$ is a version of the derivative $g_\theta^{(\theta_0)} = \frac{dP_\theta}{dP_{\theta_0}}$.

Hence if $\lambda = \mathsf{P}_{\theta_0}$, sufficiency of $\mathscr{G}$ implies the factorization property (48) with $\hat{g}_\theta^{(\theta_0)} = g_\theta^{(\theta_0)}$ and $h \equiv 1$.

In the general case (again under the additional assumption $\mathsf{P}_\theta \ll \mathsf{P}_{\theta_0}, \theta \in \Theta$) we find that

$$g_\theta^{(\lambda)} = \frac{d\mathsf{P}_\theta}{d\lambda} = \frac{d\mathsf{P}_\theta}{d\mathsf{P}_{\theta_0}} \cdot \frac{d\mathsf{P}_{\theta_0}}{d\lambda} = g_\theta^{(\theta_0)} \frac{d\mathsf{P}_{\theta_0}}{d\lambda}.$$

Denoting

$$\hat{g}_\theta^{(\theta_0)} = g_\theta^{(\theta_0)}, \quad h = \frac{d\mathsf{P}_{\theta_0}}{d\lambda},$$

we obtain the desired factorization representation (48).

$\square$

Remark 10. It is worth to emphasize that there exists a sufficient σ-algebra for any family $\mathscr{P} = \{\mathsf{P}_\theta, \theta \in \Theta\}$ (without any assumptions like $\mathscr{P}$ being dominated). We can always take the "richest" σ-algebra $\mathscr{F}$ for this σ-algebra.

Indeed, in this case $\mathsf{E}_\theta(X \mid \mathscr{F}) = X$ (P_θ-a. s.) for any integrable random variable X and therefore (47) is fulfilled.

It is clear that such sufficient σ-algebra is not of much interest because it does not provide any "reduction of data." What we are really interested in is to find the *minimal* sufficient σ-algebra $\mathscr{G}_{\min}$, i. e. the σ-algebra which is the intersection of all sufficient σ-subalgebras (cf. the proof of Lemma 1 in Sect. 2, which implies that such a σ-algebra exists). But regretfully an explicit construction of such σ-algebras is, as a rule, rather complicated (see, however, Sects. 13–15, Chap. 2 in Borovkov's book [13]).

Remark 11. Suppose that $\mathscr{P} = \{\mathsf{P}_\theta, \theta \in \Theta\}$ is a dominated family ($\mathsf{P}_\theta \ll \lambda, \theta \in \Theta$, with λ a σ-finite measure) and the density $g_\theta^{(\lambda)} = \frac{d\mathsf{P}_\theta}{d\lambda}(\omega)$ is representable as

$$g_\theta^{(\lambda)}(\omega) = G_\theta^{(\lambda)}(T(\omega)) h(\omega) \quad (\lambda\text{-a. s.}) \tag{53}$$

for all $\theta \in \Theta$, where $T = T(\omega)$ is an $\mathscr{F}/\mathscr{E}$-measurable function (random element, see Sect. 5) taking values in a set E with a σ-algebra $\mathscr{E}$ of its subsets. The functions $G_\theta^{(\lambda)}(t), t \in E$, and $h(\omega), \omega \in \Omega$, are assumed to be nonnegative and $\mathscr{E}$-and $\mathscr{F}$-measurable respectively.

By comparing (48) and (53) we see that σ-algebra $\mathscr{G} = \sigma(T(\omega))$ is sufficient and the function $T = T(\omega)$ is a sufficient statistic (in the sense of Definition 10).

Note that in dominated settings it is the factorization representation (53) that is usually taken for the definition of the sufficient statistic $T = T(\omega)$ involved in this equation.

Example 5 (Exponential Family). Assume that $\Omega = R^n$, $\mathscr{F} = \mathscr{B}(R^n)$ and the measure P_θ is such that

$$\mathsf{P}_\theta(d\omega) = \mathsf{P}_\theta(dx_1) \cdots \mathsf{P}_\theta(dx_n) \tag{54}$$

for $\omega = (x_1, \ldots, x_n)$, where the measure $\mathsf{P}_\theta(dx), x \in R$, has the following structure:

$$P_\theta(dx) = \alpha(\theta)\, e^{\beta(\theta)s(x)}\gamma(x)\,\lambda(dx). \tag{55}$$

Here $s = s(x)$ is a $\mathscr{B}$-measurable function and the meaning of $\alpha(\theta)$, $\beta(\theta)$, $\gamma(x)$, $\lambda(dx)$ is obvious. (The family of measures P_θ, $\theta \in \Theta$, presents the simplest example of an *exponential family*.) It follows from (54) and (55) that

$$P_\theta(d\omega) = \alpha^n(\theta) e^{\beta(\theta)[s(x_1)+\cdots+s(x_n)]}\gamma(x_1)\cdots\gamma(x_n)\,dx_1 \ldots dx_n. \tag{56}$$

Comparing (56) with (53) we see that $T(\omega) = s(x_1) + \cdots + s(x_n)$, $\omega = (x_1, \ldots, x_n)$, is a sufficient statistic (for the exponential family at hand).

If we denote $X_1(\omega) = x_1, \ldots, X_n(\omega) = x_n$ for $\omega = (x_1, \ldots, x_n)$, then the structure of the measures P_θ (which have the form of the *direct product of measures P_θ*) implies that relative to them $X_1, \ldots, X_n$ are a sequence of *independent identically distributed* random variables. Thus the statistic $T(\omega) = s(X_1(\omega)) + \cdots + s(X_n(\omega))$ is a sufficient statistic related to such a sequence $X_1(\omega), \ldots, X_n(\omega)$. (In Problem 20 we ask whether this statistic is *minimal sufficient*.)

Example 6. Let $\Omega = R^n$, $\mathscr{F} = \mathscr{B}(R^n)$, $\omega = (x_1, \ldots, x_n)$, and the distributions P_θ, $\theta > 0$, have densities (with respect to the Lebesgue measure λ)

$$\frac{dP_\theta}{d\lambda}(\omega) = \begin{cases} \theta^{-n}, & \text{if } 0 \le x_i \le \theta \text{ for all } i = 1, \ldots, n, \\ 0 & \text{otherwise.} \end{cases}$$

Putting

$$T(\omega) = \max_{1 \le i \le n} x_i,$$

$$h(\omega) = \begin{cases} 1, & \text{if } x_i \ge 0 \text{ for all } i = 1, \ldots, n, \\ 0 & \text{otherwise,} \end{cases}$$

$$G_\theta^{(\lambda)}(t) = \begin{cases} \theta^{-n}, & \text{if } 0 \le t \le \theta, \\ 0 & \text{otherwise,} \end{cases}$$

we obtain

$$\frac{dP_\theta}{d\lambda}(\omega) = G_\theta^{(\lambda)}(T(\omega))h(\omega). \tag{57}$$

Thus $T(\omega) = \max_{1 \le i \le n} x_i$ is a sufficient statistic.

11. Let Θ be a subset of the real line and $\mathscr{E} = (\Omega, \mathscr{F}, \mathscr{P} = \{P_\theta, \theta \in \Theta\})$ a probabilistic-statistical model. We are now interested in construction of "good" estimators for the parameter θ.

By an estimator we mean any random variable $\hat\theta = \hat\theta(\omega)$ (cf. Sect. 7 of Chap. 1).

The theorem to be stated below shows how the use of a sufficient σ-algebra enables us to *improve* the "quality" of an estimator measured by the mean-square deviation of $\hat\theta$ from θ. More precisely, we say that $\hat\theta$ is an *unbiased* estimator of θ if $E_\theta\,|\hat\theta| < \infty$ and $E_\theta\,\hat\theta = \theta$ for all $\theta \in \Theta$ (cf. the property (2) in Sect. 7 of Chap. 1).

Theorem 8 (Rao–Blackwell). *Let $\mathcal{G}$ be a sufficient σ-algebra for the family $\mathcal{P}$ and $\hat{\theta} = \hat{\theta}(\omega)$ an estimator.*

(a) *If $\hat{\theta}$ is an unbiased estimator, then the estimator*

$$T = \mathsf{E}_\theta(\hat{\theta} \,|\, \mathcal{G}) \tag{58}$$

is also unbiased.

(b) *The estimator T is "better" than $\hat{\theta}$ in the sense that*

$$\mathsf{E}_\theta(T - \theta)^2 \leq \mathsf{E}_\theta(\hat{\theta} - \theta)^2, \quad \theta \in \Theta. \tag{59}$$

PROOF. Conclusion (a) follows from

$$\mathsf{E}_\theta T = \mathsf{E}_\theta \mathsf{E}_\theta(\hat{\theta} \,|\, \mathcal{G}) = \mathsf{E}_\theta \hat{\theta} = \theta.$$

For the proof of (b) we have only to note that by Jensen's inequality (see Problem 5 with $g(x) = (x - \theta)^2$),

$$(\mathsf{E}_\theta(\hat{\theta} \,|\, \mathcal{G}) - \theta)^2 \leq \mathsf{E}_\theta[(\hat{\theta} - \theta)^2 \,|\, \mathcal{G}].$$

Taking the expectation $\mathsf{E}_\theta(\cdot)$ of both sides we obtain (59). □

12. PROBLEMS

1. Let ξ and η be independent identically distributed random variables with $\mathsf{E}\,\xi$ defined. Show that

$$\mathsf{E}(\xi \,|\, \xi + \eta) = \mathsf{E}(\eta \,|\, \xi + \eta) = \frac{\xi + \eta}{2} \quad \text{(a. s.)}.$$

2. Let $\xi_1, \xi_2, \ldots$ be independent identically distributed random variables with $\mathsf{E}\,|\xi_i| < \infty$. Show that

$$\mathsf{E}(\xi_1 \,|\, S_n, S_{n+1}, \ldots) = \frac{S_n}{n} \quad \text{(a. s.)},$$

where $S_n = \xi_1 + \cdots + \xi_n$.

3. Suppose that the random elements (X, Y) are such that there is a regular conditional distribution $P_x(B) = \mathsf{P}(Y \in B \,|\, X = x)$. Show that if $\mathsf{E}\,|g(X, Y)| < \infty$ then

$$\mathsf{E}[g(X, Y) \,|\, X = x] = \int g(x, y)\, P_x(dy) \quad (P_x\text{-a.s.}).$$

4. Let ξ be a random variable with distribution function $F_\xi(x)$. Show that

$$\mathsf{E}(\xi \,|\, a < \xi \leq b) = \frac{\int_a^b x\, dF_\xi(x)}{F_\xi(b) - F_\xi(a)}$$

(assuming that $F_\xi(b) - F_\xi(a) > 0$).

5. Let $g = g(x)$ be a convex Borel function with $\mathsf{E}\,|g(\xi)| < \infty$. Show that Jensen's inequality

$$g(\mathsf{E}(\xi\,|\,\mathscr{G})) \leq \mathsf{E}(g(\xi)\,|\,\mathscr{G})$$

holds (a.s.) for the conditional expectations.

6. Show that a necessary and sufficient condition for the random variable ξ and the σ-algebra $\mathscr{G}$ to be independent (i.e., the random variables ξ and $I_B(\omega)$ are independent for every $B \in \mathscr{G}$) is that $\mathsf{E}(g(\xi)\,|\,\mathscr{G}) = \mathsf{E}\,g(\xi)$ for every Borel function $g(x)$ with $\mathsf{E}\,|g(\xi)| < \infty$.

7. Let ξ be a nonnegative random variable and $\mathscr{G}$ a σ-algebra, $\mathscr{G} \subseteq \mathscr{F}$. Show that $\mathsf{E}(\xi\,|\,\mathscr{G}) < \infty$ (a. s.) if and only if the measure Q, defined on sets $A \in \mathscr{G}$ by $\mathsf{Q}(A) = \int_A \xi\,d\mathsf{P}$, is σ-finite.

8. Show that the conditional probabilities $\mathsf{P}(A\,|\,B)$ are "continuous" in the sense that $\lim_n \mathsf{P}(A_n\,|\,B_n) = \mathsf{P}(A\,|\,B)$ whenever $\lim_n A_n = A$, $\lim_n B_n = B$, $\mathsf{P}(B_n) > 0$, $\mathsf{P}(B) > 0$.

9. Let $\Omega = (0,1)$, $\mathscr{F} = \mathscr{B}((0,1))$ and P the Lebesgue measure. Let $X(\omega)$ and $Y(\omega)$ be two independent random variables uniformly distributed on $(0,1)$. Consider the random variable $Z(\omega) = |X(\omega) - Y(\omega)|$ (the distance between the "points" $X(\omega)$ and $Y(\omega)$). Prove that the distribution function $F_Z(z)$ has a density $f_Z(z)$ and $f_Z(z) = 2(1 - z)$, $0 \leq z \leq 1$. (This, of course, implies that $\mathsf{E}\,Z = 1/3$.)

10. Suppose that two points A_1 and A_2 are chosen "at random" in the circle of radius R ($\{(x,y): x^2 + y^2 \leq R^2\}$), i. e. these points are chosen independently with probabilities (in polar coordinates, $A_i = (\rho_i, \theta_i)$, $i = 1, 2$)

$$\mathsf{P}(\rho_i \in dr, \theta_i \in d\theta) = \frac{r\,dr\,d\theta}{\pi R^2}, \quad i = 1, 2.$$

Show that the distance ρ between A_1 and A_2 has a density $f_\rho(r)$ and

$$f_\rho(r) = \frac{2r}{\pi R^2}\left[2\arccos\left(\frac{r}{2R}\right) - \frac{r}{R}\sqrt{1 - \left(\frac{r}{2R}\right)^2}\right],$$

where $0 < r < 2R$.

11. A point $P = (x, y)$ is chosen "at random" (explain what it means!) in the unit square (with vertices $(0,0)$, $(0,1)$, $(1,1)$, $(1,0)$). Find the probability that this point will be closer to the point $(1, 1)$ than to $(1/2, 1/2)$.

12. Two people A and B made an appointment to meet between 7 and 8 p.m. But both of them forgot the exact time of the meeting and come between 7 and 8 "at random" waiting at most 10 minutes. Show that the probability for them to meet is $11/36$.

13. Let $X_1, X_2, \ldots$ be a sequence of independent random variables and $S_n = \sum_{i=1}^n X_i$. Show that S_1 and S_3 are conditionally independent relative to the σ-algebra $\sigma(S_2)$ generated by S_2.

14. Two σ-algebras $\mathscr{G}_1$ and $\mathscr{G}_2$ are said to be conditionally independent relative to a σ-algebra $\mathscr{G}_3$ if

$$P(A_1 A_2 \mid \mathscr{G}_3) = P(A_1 \mid \mathscr{G}_3) \, P(A_2 \mid \mathscr{G}_3) \quad \text{for all } A_i \in \mathscr{G}_i, \ i = 1, 2.$$

Show that conditional independence of $\mathscr{G}_1$ and $\mathscr{G}_2$ relative to $\mathscr{G}_3$ holds (P-a. s.) if and only if any of the following conditions is fulfilled:
   (a) $P(A_1 \mid \sigma(\mathscr{G}_2 \cup \mathscr{G}_3)) = P(A_1 \mid \mathscr{G}_3)$ for all $A_1 \in \mathscr{G}_1$;
   (b) $P(B \mid \sigma(\mathscr{G}_2 \cup \mathscr{G}_3)) = P(B \mid \mathscr{G}_3)$ for any set B in a system $\mathscr{P}_1$, which is a π-system such that $\mathscr{G}_1 = \sigma(\mathscr{P}_1)$;
   (c) $P(B_1 B_2 \mid \sigma(\mathscr{G}_2 \cup \mathscr{G}_3)) = P(B_1 \mid \mathscr{G}_3) \, P(B_2 \mid \mathscr{G}_3)$ for any B_1 and B_2 in π-systems $\mathscr{P}_1$ and $\mathscr{P}_2$ respectively such that $\mathscr{G}_1 = \sigma(\mathscr{P}_1)$ and $\mathscr{G}_2 = \sigma(\mathscr{P}_2)$;
   (d) $E(X \mid \sigma(\mathscr{G}_2 \cup \mathscr{G}_3)) = E(X \mid \mathscr{G}_3)$ for any $\sigma(\mathscr{G}_2 \cup \mathscr{G}_3)$-measurable random variable X for which the expectation EX is well defined (see Definition 2 in Sect. 6).
15. Prove the following extended version of Fatou's lemma for conditional expectations (cf. (d) in Theorem 2):
   Let $(\Omega, \mathscr{F}, P)$ be a probability space and $(\xi_n)_{n \geq 1}$ a sequence of random variables such that the expectations $E\,\xi_n$, $n \geq 1$, and $E \liminf \xi_n$ (which may take the values $\pm\infty$, see Definition 2 in Sect. 6) are well defined. Let $\mathscr{G}$ be a σ-subalgebra of $\mathscr{F}$ and

$$\sup_{n \geq 1} E(\xi_n^- I(\xi_n \geq a) \mid \mathscr{G}) \to 0 \quad \text{(P-a. s.)}, \quad a \to \infty.$$

Then

$$E(\liminf \xi_n \mid \mathscr{G}) \leq \liminf E(\xi_n \mid \mathscr{G}) \quad \text{(P-a. s.)}.$$

16. Let, as in the previous problem, $(\xi_n)_{n \geq 1}$ be a sequence of random variables such that the expectations $E\,\xi_n$, $n \geq 1$, are well defined and $\mathscr{G}$ a σ-subalgebra of $\mathscr{F}$ such that

$$\sup_{n} \lim_{k \to \infty} E(|\xi_n| I(|\xi_n| \geq k) \mid \mathscr{G}) = 0 \quad \text{(P-a. s.)}. \tag{60}$$

Then

$$E(\xi_n \mid \mathscr{G}) \to E(\xi \mid \mathscr{G}) \quad \text{(P-a. s.)}$$

provided that $\xi_n \to \xi$ (P-a. s.) and $E\,\xi$ is well defined.
17. In the previous problem replace (60) by $\sup_n E(|\xi_n|^\alpha \mid \mathscr{G}) < \infty$ (P-a. s.) for some $\alpha > 1$. Then

$$E(\xi_n \mid \mathscr{G}) \to E(\xi \mid \mathscr{G}) \quad \text{(P-a. s.)}.$$

18. Let $\xi_n \xrightarrow{L^p} \xi$ for some $p \geq 1$. Then $E(\xi_n \mid \mathscr{G}) \xrightarrow{L^p} E(\xi \mid \mathscr{G})$.

19. (a) Let $\mathrm{Var}(X \mid Y) \equiv \mathsf{E}[(X - \mathsf{E}(X \mid Y))^2 \mid Y]$. Show that $\mathrm{Var}\,X = \mathsf{E}\,\mathrm{Var}(X \mid Y) + \mathrm{Var}\,\mathsf{E}(X \mid Y)$.
   (b) Show that $\mathrm{Cov}(X, Y) = \mathrm{Cov}(X, \mathsf{E}(Y \mid X))$.
20. Determine whether the sufficient statistic $T(\omega) = s(X_1(\omega)) + \cdots + s(X_n(\omega))$ in Example 5 is minimal.
21. Prove the factorization representation (57).
22. In Example 6 (Subsection 10), show that $\mathsf{E}_\theta(X_i \mid T) = \frac{n+1}{2n} T$, where $X_i(\omega) = x_i$ for $\omega = (x_1, \ldots, x_n)$, $i = 1, \ldots, n$.

8 Random Variables: II

1. In the first chapter we introduced characteristics of simple random variables, such as the variance, covariance, and correlation coefficient. These extend similarly to the general case. Let $(\Omega, \mathscr{F}, \mathsf{P})$ be a probability space and $\xi = \xi(\omega)$ a random variable for which $\mathsf{E}\,\xi$ is defined.

The *variance* of ξ is

$$\mathrm{Var}\,\xi = \mathsf{E}(\xi - \mathsf{E}\,\xi)^2.$$

The number $\sigma = +\sqrt{\mathrm{Var}\,\xi}$ is the *standard deviation* (Cf. Definition 5 in Sect. 4, Chap. 1).

If ξ is a random variable with a Gaussian (normal) density

$$f_\xi(x) = \frac{1}{\sqrt{2\pi}\sigma} e^{-(x-m)^2/2\sigma^2}, \quad \sigma > 0, \ -\infty < m < \infty, \tag{1}$$

the meaning of the parameters m and σ in (1) is very simple:

$$m = \mathsf{E}\,\xi, \quad \sigma^2 = \mathrm{Var}\,\xi.$$

Hence the probability distribution of this random variable ξ, which we call *Gaussian*, or *normally distributed*, is completely determined by its mean value m and variance σ^2. (It is often convenient to write $\xi \sim \mathscr{N}(m, \sigma^2)$.)

Now let (ξ, η) be a pair of random variables. Their covariance is

$$\mathrm{Cov}(\xi, \eta) = \mathsf{E}(\xi - \mathsf{E}\,\xi)(\eta - \mathsf{E}\,\eta) \tag{2}$$

(assuming that the expectations are defined).

If $\mathrm{Cov}(\xi, \eta) = 0$ we say that ξ and η are *uncorrelated*.

If $0 < \mathrm{Var}\,\xi < \infty$ and $0 < \mathrm{Var}\,\eta < \infty$, the number

$$\rho(\xi, \eta) \equiv \frac{\mathrm{Cov}(\xi, \eta)}{\sqrt{\mathrm{Var}\,\xi \cdot \mathrm{Var}\,\eta}} \tag{3}$$

is the *correlation coefficient* of ξ and η.

The properties of variance, covariance, and correlation coefficient were investigated in Sect. 4 of Chap. 1 for simple random variables. In the general case these properties can be stated in a completely analogous way.

Let $\xi = (\xi_1, \ldots, \xi_n)$ be a random vector whose components have finite second moments. The *covariance matrix* of ξ is the $n \times n$ matrix $\mathbb{R} = \|R_{ij}\|$, where $R_{ij} = \text{Cov}(\xi_i, \xi_j)$. It is clear that $\mathbb{R}$ is *symmetric*. Moreover, it is *positive semi-definite*, i.e.

$$\sum_{i,j=1}^{n} R_{ij}\lambda_i\lambda_j \geq 0$$

for all $\lambda_i \in R, i = 1, \ldots, n$, since

$$\sum_{i,j=1}^{n} R_{ij}\lambda_i\lambda_j = \mathsf{E}\left[\sum_{i=1}^{n}(\xi_i - \mathsf{E}\,\xi_i)\lambda_i\right]^2 \geq 0.$$

The following lemma shows that the converse is also true.

Lemma. *A necessary and sufficient condition that an $n \times n$ matrix $\mathbb{R}$ is the covariance matrix of a vector $\xi = (\xi_1, \ldots, \xi_n)$ is that the matrix $\mathbb{R}$ is symmetric and positive semi-definite, or, equivalently, that there is an $n \times k$ matrix A $(1 \leq k \leq n)$ such that*

$$\mathbb{R} = AA^*,$$

*where * denotes the transpose.*

PROOF. We showed above that every covariance matrix is symmetric and positive semi-definite.

Conversely, let $\mathbb{R}$ be a matrix with these properties. We know from matrix theory that corresponding to every symmetric positive semi-definite matrix $\mathbb{R}$ there is an orthogonal matrix $\mathcal{O}$ (i.e., $\mathcal{O}\mathcal{O}^* = E$, the identity matrix) such that

$$\mathcal{O}^*\mathbb{R}\mathcal{O} = D,$$

where

$$D = \begin{pmatrix} d_1 & & 0 \\ & \ddots & \\ 0 & & d_n \end{pmatrix}$$

is a diagonal matrix with nonnegative elements $d_i, i = 1, \ldots, n$.

It follows that

$$\mathbb{R} = \mathcal{O}D\mathcal{O}^* = (\mathcal{O}B)(B^*\mathcal{O}^*),$$

where B is the diagonal matrix with elements $b_i = +\sqrt{d_i}, i = 1, \ldots, n$. Consequently if we put $A = \mathcal{O}B$ we have the required representation $\mathbb{R} = AA^*$ for $\mathbb{R}$.

It is clear that every matrix AA^* is symmetric and positive semi-definite. Consequently we have only to show that $\mathbb{R}$ is the covariance matrix of some random vector.

Let $\eta_1, \eta_2, \ldots, \eta_n$ be a sequence of independent normally distributed random variables, $\mathcal{N}(0,1)$. (The existence of such a sequence follows, for example, from Corollary 1 of Theorem 1, Sect. 9, and in principle could easily be derived from Theorem 2 of Sect. 3.) Then the random vector $\xi = A\eta$ (vectors are thought of as column vectors) has the required properties. In fact,

$$\mathsf{E}\,\xi\xi^* = \mathsf{E}(A\eta)(A\eta)^* = A \cdot \mathsf{E}\,\eta\eta^* \cdot A^* = AEA^* = AA^*.$$

(If $\zeta = \|\zeta_{ij}\|$ is a matrix whose elements are random variables, $\mathsf{E}\,\zeta$ means the matrix $\|\mathsf{E}\,\xi_{ij}\|$).

This completes the proof of the lemma.

$\square$

We now turn our attention to the two-dimensional Gaussian (normal) density

$$f_{\xi,\eta}(x,y) = \frac{1}{2\pi\sigma_1\sigma_2\sqrt{1-\rho^2}} \exp\left\{-\frac{1}{2(1-\rho^2)}\left[\frac{(x-m_1)^2}{\sigma_1^2}\right.\right.$$
$$\left.\left.-2\rho\frac{(x-m_1)(y-m_2)}{\sigma_1\sigma_2} + \frac{(y-m_2)^2}{\sigma_2^2}\right]\right\}, \tag{4}$$

characterized by the five parameters $m_1, m_2, \sigma_1, \sigma_2$ and ρ (cf. (14) in Sect. 3), where $|m_1| < \infty$, $|m_2| < \infty$, $\sigma_1 > 0$, $\sigma_2 > 0$, $|\rho| < 1$. (See Fig. 28 in Sect. 3.) An easy calculation identifies these parameters:

$$m_1 = \mathsf{E}\,\xi, \quad \sigma_1^2 = \mathrm{Var}\,\xi,$$
$$m_2 = \mathsf{E}\,\eta, \quad \sigma_2^2 = \mathrm{Var}\,\eta,$$
$$\rho = \rho(\xi, \eta).$$

In Sect. 4 of Chap. 1 we explained that if ξ and η are uncorrelated ($\rho(\xi, \eta) = 0$), it does not follow that they are independent. However, if the pair (ξ, η) is Gaussian, it does follow that if ξ and η are uncorrelated then they are independent.

In fact, if $\rho = 0$ in (4), then

$$f_{\xi,\eta}(x,y) = \frac{1}{2\pi\sigma_1\sigma_2}e^{-(x-m_1)^2/2\sigma_1^2} \cdot e^{-(y-m_2)^2/2\sigma_2^2}.$$

But by (55) in Sect. 6 and (4),

$$f_\xi(x) = \int_{-\infty}^{\infty} f_{\xi,\eta}(x,y)\,dy = \frac{1}{\sqrt{2\pi}\sigma_1}e^{-(x-m_1)^2/2\sigma_1^2},$$
$$f_\eta(y) = \int_{-\infty}^{\infty} f_{\xi,\eta}(x,y)\,dx = \frac{1}{\sqrt{2\pi}\sigma_2}e^{-(y-m_2)^2/2\sigma_2^2}.$$

Consequently

$$f_{\xi,\eta}(x,y) = f_\xi(x) \cdot f_\eta(y),$$

from which it follows that ξ and η are independent (see the end of Subsection 9 of Sect. 6).

2. A striking example of the utility of the concept of conditional expectation (introduced in Sect. 7) is its application to the solution of the following problem which is connected with *estimation theory* (cf. Subsection 8 of Sect. 4 of Chap. 1).

Let (ξ, η) be a pair of random variables such that ξ is observable but η is not. We ask how the unobservable component η can be "estimated" from the knowledge of observation of ξ.

To state the problem more precisely, we need to define the concept of an *estimator*. Let $\varphi = \varphi(x)$ be a Borel function. We call the random variable $\varphi(\xi)$ an *estimator* of η in terms of ξ, and $E[\eta - \varphi(\xi)]^2$ the *(mean-square) error* of this estimator. An estimator $\varphi^*(\xi)$ is called *optimal* (in the mean-square sense) if

$$\Delta \equiv E[\eta - \varphi^*(\xi)]^2 = \inf_{\varphi} E[\eta - \varphi(\xi)]^2, \tag{5}$$

where inf is taken over all Borel functions $\varphi = \varphi(x)$.

Theorem 1. *Let* $E\eta^2 < \infty$. *Then there is an optimal estimator* $\varphi^* = \varphi^*(\xi)$ *and* $\varphi^*(x)$ *can be taken to be the function*

$$\varphi^*(x) = E(\eta \mid \xi = x). \tag{6}$$

PROOF. Without loss of generality we may consider only estimators $\varphi(\xi)$ for which $E\varphi^2(\xi) < \infty$. Then if $\varphi(\xi)$ is such an estimator, and $\varphi^*(\xi) = E(\eta \mid \xi)$, we have

$$\begin{aligned}
E[\eta - \varphi(\xi)]^2 &= E[(\eta - \varphi^*(\xi)) + (\varphi^*(\xi) - \varphi(\xi))]^2 \\
&= E[\eta - \varphi^*(\xi)]^2 + E[\varphi^*(\xi) - \varphi(\xi)]^2 \\
&\quad + 2E[(\eta - \varphi^*(\xi))(\varphi^*(\xi) - \varphi(\xi))] \geq E[\eta - \varphi^*(\xi)]^2,
\end{aligned}$$

since $E[\varphi^*(\xi) - \varphi(\xi)]^2 > 0$ and, by the properties of conditional expectations,

$$\begin{aligned}
E[(\eta - \varphi^*(\xi))(\varphi^*(\xi) - \varphi(\xi))] &= E\{E[(\eta - \varphi^*(\xi))(\varphi^*(\xi) - \varphi(\xi)) \mid \xi]\} \\
&= E\{(\varphi^*(\xi) - \varphi(\xi)) E(\eta - \varphi^*(\xi) \mid \xi)\} = 0.
\end{aligned}$$

This completes the proof of the theorem.
□

Remark 1. It is clear from the proof that the conclusion of the theorem is still valid when ξ is not merely a random variable but any random element with values in a measurable space $(E, \mathscr{E})$. We would then assume that the estimator $\varphi = \varphi(x)$ is an $\mathscr{E}/\mathscr{B}(R)$-measurable function.

Let us consider the form of $\varphi^*(x)$ on the hypothesis that (ξ, η) is a Gaussian pair with density given by (4).

From (1), (4) and (18) of Sect. 7 we find that the density $f_{\eta|\xi}(y \mid x)$ of the conditional probability distribution is given by

$$f_{\eta|\xi}(y \mid x) = \frac{1}{\sqrt{2\pi(1 - \rho^2)\sigma_2^2}} e^{-(y - m(x))^2/[2\sigma_2^2(1 - \rho^2)]}, \tag{7}$$

where

$$m(x) = m_2 + \frac{\sigma_2}{\sigma_1}\rho \cdot (x - m_1).$$ (8)

Then by the Corollary of Theorem 3, Sect. 7,

$$\mathsf{E}(\eta \mid \xi = x) = \int_{-\infty}^{\infty} y f_{\eta \mid \xi}(y \mid x)\, dy = m(x)$$ (9)

and

$$\mathrm{Var}(\eta \mid \xi = x) \equiv \mathsf{E}[(\eta - \mathsf{E}(\eta \mid \xi = x))^2 \mid \xi = x]$$
$$= \int_{-\infty}^{\infty} (y - m(x))^2 f_{\eta \mid \xi}(y \mid x)\, dy$$
$$= \sigma_2^2(1 - \rho^2).$$ (10)

Notice that the conditional variance $\mathrm{Var}(\eta \mid \xi = x)$ is independent of x and therefore

$$\Delta = \mathsf{E}[\eta - \mathsf{E}(\eta \mid \xi)]^2 = \sigma_2^2(1 - \rho^2).$$ (11)

Formulas (9) and (11) were obtained under the assumption that $\mathrm{Var}\,\xi > 0$ and $\mathrm{Var}\,\eta > 0$. However, if $\mathrm{Var}\,\xi > 0$ and $\mathrm{Var}\,\eta = 0$ they are still evidently valid.

Hence we have the following result (cf. (16), (17) in Sect. 4 of Chap. 1).

Theorem 2 (Theorem on the Normal Correlation). *Let (ξ, η) be a Gaussian vector with $\mathrm{Var}\,\xi > 0$. Then the optimal estimator of η in terms of ξ is*

$$\mathsf{E}(\eta \mid \xi) = \mathsf{E}\,\eta + \frac{\mathrm{Cov}(\xi, \eta)}{\mathrm{Var}\,\xi}(\xi - \mathsf{E}\,\xi),$$ (12)

and its error is

$$\Delta \equiv \mathsf{E}[\eta - \mathsf{E}(\eta \mid \xi)]^2 = \mathrm{Var}\,\eta - \frac{\mathrm{Cov}^2(\xi, \eta)}{\mathrm{Var}\,\xi}.$$ (13)

Remark 2. The curve $y(x) = \mathsf{E}(\eta \mid \xi = x)$ is the *regression curve of η on ξ or of η with respect to ξ*. In the Gaussian case $\mathsf{E}(\eta \mid \xi = x) = a + bx$ and consequently the regression of η and ξ is *linear*. Hence it is not surprising that the right-hand sides of (12) and (13) agree with the corresponding parts of (16) and (17) in Sect. 4 of Chap. 1 for the optimal linear estimator and its error.

Corollary. *Let ε_1 and ε_2 be independent Gaussian random variables with mean zero and unit variance, and*

$$\xi = a_1\varepsilon_1 + a_2\varepsilon_2, \quad \eta = b_1\varepsilon_1 + b_2\varepsilon_2.$$

Then $\mathsf{E}\,\xi = \mathsf{E}\,\eta = 0$, $\mathrm{Var}\,\xi = a_1^2 + a_2^2$, $\mathrm{Var}\,\eta = b_1^2 + b_2^2$, $\mathrm{Cov}(\xi, \eta) = a_1 b_1 + a_2 b_2$, and if $a_1^2 + a_2^2 > 0$, then

$$E(\eta \mid \xi) = \frac{a_1 b_1 + a_2 b_2}{a_1^2 + a_2^2} \xi, \tag{14}$$

$$\Delta = \frac{(a_1 b_2 - a_2 b_1)^2}{a_1^2 + a_2^2}. \tag{15}$$

3. Let us consider the problem of determining the distribution functions of random variables that are functions of other random variables.

Let ξ be a random variable with distribution function $F_\xi(x)$ (and density $f_\xi(x)$, if it exists), let $\varphi = \varphi(x)$ be a Borel function and $\eta = \varphi(\xi)$. Letting $I_y = (-\infty, y)$, we obtain

$$F_\eta(y) = P(\eta \leq y) = P(\varphi(\xi) \in I_y) = P(\xi \in \varphi^{-1}(I_y)) = \int_{\varphi^{-1}(I_y)} F_\xi(dx), \tag{16}$$

which expresses the distribution function $F_\eta(y)$ in terms of $F_\xi(x)$ and φ.

For example, if $\eta = a\xi + b, a > 0$, we have

$$F_\eta(y) = P\left(\xi \leq \frac{y-b}{a}\right) = F_\xi\left(\frac{y-b}{a}\right). \tag{17}$$

If $\eta = \xi^2$, it is evident that $F_\eta(y) = 0$ for $y < 0$, while for $y \geq 0$

$$\begin{aligned} F_\eta(y) = P(\xi^2 \leq y) &= P(-\sqrt{y} \leq \xi \leq \sqrt{y}) \\ &= F_\xi(\sqrt{y}) - F_\xi(-\sqrt{y}) + P(\xi = -\sqrt{y}). \end{aligned} \tag{18}$$

We now turn to the problem of determining $f_\eta(y)$.

Let us suppose that the range of ξ is a (finite or infinite) open interval $I = (a, b)$, and that the function $\varphi = \varphi(x)$, with domain (a, b), is continuously differentiable and either strictly increasing or strictly decreasing. We also suppose that $\varphi'(x) \neq 0$, $x \in I$. Let us write $h(y) = \varphi^{-1}(y)$ and suppose for definiteness that $\varphi(x)$ is strictly increasing. Then when $y \in \{\varphi(x) : x \in I\}$,

$$F_\eta(y) = P(\eta \leq y) = P(\varphi(\xi) \leq y) = P(\xi \leq \varphi^{-1}(y))$$

$$= P(\xi \leq h(y)) = \int_{-\infty}^{h(y)} f_\xi(x)\, dx. \tag{19}$$

By Problem 15 of Sect. 6,

$$\int_{-\infty}^{h(y)} f_\xi(x)\, dx = \int_{-\infty}^{y} f_\xi(h(z)) h'(z)\, dz \tag{20}$$

and therefore

$$f_\eta(y) = f_\xi(h(y)) h'(y). \tag{21}$$

Similarly, if $\varphi(x)$ is strictly decreasing,

$$f_\eta(y) = f_\xi(h(y))(-h'(y)).$$

Hence in either case

$$f_\eta(y) = f_\xi(h(y))|h'(y)|. \tag{22}$$

For example, if $\eta = a\xi + b$, $a \neq 0$, we have

$$h(y) = \frac{y-b}{a} \quad \text{and} \quad f_\eta(y) = \frac{1}{|a|}f_\xi\left(\frac{y-b}{a}\right).$$

If $\xi \sim \mathcal{N}(m, \sigma^2)$ and $\eta = e^\xi$, we find from (22) that

$$f_\eta(y) = \begin{cases} \frac{1}{\sqrt{2\pi}\sigma y} \exp\left[-\frac{\log(y/M)^2}{2\sigma^2}\right], & y > 0, \\ 0 & y \leq 0, \end{cases} \tag{23}$$

with $M = e^m$. A probability distribution with the density (23) is said to be *logarithmically normal* or *lognormal*.

If $\varphi = \varphi(x)$ is neither strictly increasing nor strictly decreasing, formula (22) is inapplicable. However, the following generalization suffices for many applications.

Let $\varphi = \varphi(x)$ be defined on the set $\sum_{k=1}^n [a_k, b_k]$, continuously differentiable and either strictly increasing or strictly decreasing on each open interval $I_k = (a_k, b_k)$, and with $\varphi'(x) \neq 0$ for $x \in I_k$. Let $h_k = h_k(y)$ be the inverse of $\varphi(x)$ for $x \in I_k$. Then we have the following generalization of (22):

$$f_\eta(y) = \sum_{k=1}^n f_\xi(h_k(y))|h'_k(y)| \cdot I_{D_k}(y), \tag{24}$$

where D_k is the domain of $h_k(y)$.

For example, if $\eta = \xi^2$ we can take $I_1 = (-\infty, 0)$, $I_2 = (0, \infty)$, and find that $h_1(y) = -\sqrt{y}$, $h_2(y) = \sqrt{y}$, and therefore

$$f_\eta(y) = \begin{cases} \frac{1}{2\sqrt{y}}[f_\xi(\sqrt{y}) + f_\xi(-\sqrt{y})], & y > 0, \\ 0, & y \leq 0. \end{cases} \tag{25}$$

We can observe that this result also follows from (18), since $\mathsf{P}(\xi = -\sqrt{y}) = 0$. In particular, if $\xi \sim \mathcal{N}(0, 1)$,

$$f_{\xi^2}(y) = \begin{cases} \frac{1}{\sqrt{2\pi y}}e^{-y/2}, & y > 0, \\ 0, & y \leq 0. \end{cases} \tag{26}$$

A straightforward calculation shows that

$$f_{|\xi|}(y) = \begin{cases} f_\xi(y) + f_\xi(-y), & y > 0, \\ 0, & y \leq 0. \end{cases} \tag{27}$$

$$f_{+\sqrt{|\xi|}}(y) = \begin{cases} 2y(f_\xi(y^2) + f_\xi(-y^2)), & y > 0, \\ 0, & y \leq 0. \end{cases} \tag{28}$$

4. We now consider functions of several random variables.

If ξ and η are random variables with joint distribution $F_{\xi,\eta}(x, y)$, and $\varphi = \varphi(x, y)$ is a Borel function, then if we put $\zeta = \varphi(\xi, \eta)$ we see at once that

$$F_\zeta(z) = \int_{\{x,y:\ \varphi(x,y)\leq z\}} dF_{\xi,\eta}(x, y). \qquad (29)$$

For example, if $\varphi(x, y) = x + y$, and ξ and η are independent (and therefore $F_{\xi,\eta}(x, y) = F_\xi(x) \cdot F_\eta(y)$) then Fubini's theorem shows that

$$\begin{aligned}
F_\zeta(z) &= \int_{\{x,y:\ x+y\leq z\}} dF_\xi(x)\, dF_\eta(y) \\
&= \int_{R^2} I_{\{x+y\leq z\}}(x, y)\, dF_\xi(x)\, dF_\eta(y) \\
&= \int_{-\infty}^{\infty} dF_\xi(x) \left\{ \int_{-\infty}^{\infty} I_{\{x+y\leq z\}}(x, y)\, dF_\eta(y) \right\} = \int_{-\infty}^{\infty} F_\eta(z - x)\, dF_\xi(x) \quad (30)
\end{aligned}$$

and similarly

$$F_\zeta(z) = \int_{-\infty}^{\infty} F_\xi(z - y)\, dF_\eta(y). \qquad (31)$$

If F and G are distribution functions, the function

$$H(z) = \int_{-\infty}^{\infty} F(z - x)\, dG(x)$$

is denoted by $F * G$ and called the *convolution* of F and G.

Thus *the distribution function F_ζ of the sum of two independent random variables ξ and η is the convolution of their distribution functions F_ξ and F_η:*

$$F_\zeta = F_\xi * F_\eta.$$

It is clear that $F_\xi * F_\eta = F_\eta * F_\xi$.

Now suppose that the independent random variables ξ and η have densities f_ξ and f_η. Then we find from (31), with another application of Fubini's theorem, that

$$\begin{aligned}
F_\zeta(z) &= \int_{-\infty}^{\infty} \left[\int_{-\infty}^{z-y} f_\xi(u)\, du \right] f_\eta(y)\, dy \\
&= \int_{-\infty}^{\infty} \left[\int_{-\infty}^{z} f_\xi(u - y)\, du \right] f_\eta(y)\, dy = \int_{-\infty}^{z} \left[\int_{-\infty}^{\infty} f_\xi(u - y) f_\eta(y)\, dy \right] du,
\end{aligned}$$

whence

$$f_\zeta(z) = \int_{-\infty}^{\infty} f_\xi(z - y) f_\eta(y)\, dy, \qquad (32)$$

and similarly

$$f_\zeta(z) = \int_{-\infty}^{\infty} f_\eta(z-x) f_\xi(x)\, dx. \tag{33}$$

Let us see some examples of the use of these formulas.

Let $\xi_1, \xi_2, \ldots, \xi_n$ be a sequence of independent identically distributed random variables with the uniform density on $[-1, 1]$:

$$f(x) = \begin{cases} \frac{1}{2}, & |x| \leq 1, \\ 0, & |x| > 1. \end{cases}$$

Then by (32) we have

$$f_{\xi_1+\xi_2}(x) = \begin{cases} \frac{2-|x|}{4}, & |x| \leq 2, \\ 0, & |x| > 2, \end{cases}$$

$$f_{\xi_1+\xi_2+\xi_3}(x) = \begin{cases} \frac{(3-|x|)^2}{16}, & 1 \leq |x| \leq 3, \\ \frac{3-x^2}{8}, & 0 \leq |x| \leq 1, \\ 0, & |x| > 3, \end{cases}$$

and by induction

$$f_{\xi_1+\cdots+\xi_n}(x) = \begin{cases} \frac{1}{2^n(n-1)!} \displaystyle\sum_{k=0}^{[(n+x)/2]} (-1)^k C_n^k (n+x-2k)^{n-1}, & |x| \leq n, \\ 0, & |x| > n. \end{cases}$$

Now let $\xi \sim \mathcal{N}(m_1, \sigma_1^2)$ and $\eta \sim \mathcal{N}(m_2, \sigma_2^2)$. If we write

$$\varphi(x) = \frac{1}{\sqrt{2\pi}} e^{-x^2/2},$$

then

$$f_\xi(x) = \frac{1}{\sigma_1} \varphi\left(\frac{x-m_1}{\sigma_1}\right), \quad f_\eta(x) = \frac{1}{\sigma_2} \varphi\left(\frac{x-m_2}{\sigma_2}\right),$$

and the formula

$$f_{\xi+\eta}(x) = \frac{1}{\sqrt{\sigma_1^2+\sigma_2^2}} \varphi\left(\frac{x-(m_1+m_2)}{\sqrt{\sigma_1^2+\sigma_2^2}}\right)$$

follows easily from (32).

Therefore the *sum of two independent Gaussian random variables is again a Gaussian random variable with mean* $m_1 + m_2$ *and variance* $\sigma_1^2 + \sigma_2^2$.

Let $\xi_1, \ldots, \xi_n$ be independent random variables each of which is normally distributed with mean 0 and variance 1. Then it follows easily from (26) (by induction) that

$$f_{\xi_1^2+\cdots+\xi_n^2}(x) = \begin{cases} \frac{1}{2^{n/2}\Gamma(n/2)} x^{(n/2)-1} e^{-x/2}, & x > 0, \\ 0, & x \leq 0. \end{cases} \tag{34}$$

The variable $\xi_1^2 + \cdots + \xi_n^2$ is usually denoted by χ_n^2, and its distribution (with density (34)) is the χ^2-*distribution* ("chi-square distribution") with n degrees of freedom (cf. Table 2.3 in Sect. 3).

If we write $\chi_n = +\sqrt{\chi_n^2}$, it follows from (28) and (34) that

$$f_{\chi_n}(x) = \begin{cases} \frac{2x^{n-1}e^{-x^2/2}}{2^{n/2}\Gamma(n/2)}, & x \geq 0, \\ 0, & x < 0. \end{cases} \tag{35}$$

The probability distribution with this density is the χ-*distribution* (chi-distribution) with n degrees of freedom. When $n = 2$ it is called the Rayleigh distribution.

Again let ξ and η be independent random variables with densities f_ξ and f_η. Then

$$F_{\xi\eta}(z) = \iint\limits_{\{x,y:\, xy \leq z\}} f_\xi(x) f_\eta(y)\, dx\, dy,$$

$$F_{\xi/\eta}(z) = \iint\limits_{\{x,y:\, x/y \leq z\}} f_\xi(x) f_\eta(y)\, dx\, dy.$$

Hence we easily obtain

$$f_{\xi\eta}(z) = \int_{-\infty}^{\infty} f_\xi\left(\frac{z}{y}\right) f_\eta(y) \frac{dy}{|y|} = \int_{-\infty}^{\infty} f_\eta\left(\frac{z}{x}\right) f_\xi(x) \frac{dx}{|x|} \tag{36}$$

and

$$f_{\xi/\eta}(z) = \int_{-\infty}^{\infty} f_\xi(zy) f_\eta(y) |y|\, dy. \tag{37}$$

Applying (37) with independent $\xi \sim \mathcal{N}(0,1)$ and $\eta \stackrel{d}{=} \sqrt{\chi_n^2/n}$ and using (35), we find that

$$f_{\xi/\eta}(x) = \frac{1}{\sqrt{\pi n}} \frac{\Gamma\left(\frac{n+1}{2}\right)}{\Gamma\left(\frac{n}{2}\right)} \frac{1}{\left(1 + \frac{x^2}{n}\right)^{(n+1)/2}}.$$

This is the density of the *t-distribution*, or *Student's distribution*, with n degrees of freedom (cf. Table 2.3 in Sect. 3). See Problem 17 showing how this distribution arises in mathematical statistics.

5. PROBLEMS

1. Verify formulas (9), (10), (24), (27), (28), (34)–(38).

2. Let $\xi_1, \ldots, \xi_n$, $n \geq 2$, be independent identically distributed random variables with distribution function $F(x)$ (and density $f(x)$, if it exists), and let $\bar{\xi} = \max(\xi_1, \ldots, \xi_n)$, $\underline{\xi} = \min(\xi_1, \ldots, \xi_n)$, $\rho = \bar{\xi} - \underline{\xi}$. Show that

$$F_{\bar{\xi},\underline{\xi}}(y, x) = \begin{cases} (F(y))^n - (F(y) - F(x))^n, & y > x, \\ (F(y))^n, & y \leq x, \end{cases}$$

$$f_{\bar{\xi},\underline{\xi}}(y, x) = \begin{cases} n(n-1)[F(y) - F(x)]^{n-2}f(x)f(y), & y > x, \\ 0, & y < x, \end{cases}$$

$$F_\rho(x) = \begin{cases} n\int_{-\infty}^{\infty}[F(y) - F(y - x)]^{n-1}f(y)\,dy, & x \geq 0, \\ 0, & x < 0, \end{cases}$$

$$f_\rho(x) = \begin{cases} n(n-1)\int_{-\infty}^{\infty}[F(y) - F(y - x)]^{n-2}f(y - x)f(y)\,dy, & x > 0, \\ 0, & x < 0. \end{cases}$$

3. Let ξ_1 and ξ_2 be independent Poisson random variables with respective parameters λ_1 and λ_2. Show that $\xi_1 + \xi_2$ has the Poisson distribution with parameter $\lambda_1 + \lambda_2$.

4. Let $m_1 = m_2 = 0$ in (4). Show that

$$f_{\xi/\eta}(z) = \frac{\sigma_1\sigma_2\sqrt{1 - \rho^2}}{\pi(\sigma_2^2 z^2 - 2\rho\sigma_1\sigma_2 z + \sigma_1^2)}.$$

5. The *maximal correlation coefficient* of ξ and η is $\rho^*(\xi, \eta) = \sup_{u,v} \rho(u(\xi), v(\xi))$, where the supremum is taken over the Borel functions $u = u(x)$ and $v = v(x)$ for which the correlation coefficient $\rho(u(\xi), v(\xi))$ is defined. Show that ξ and η are independent if and only if $\rho^*(\xi, \eta) = 0$.

6. Let $\tau_1, \tau_2, \ldots, \tau_k$ be independent nonnegative identically distributed random variables with the exponential density

$$f(t) = \lambda e^{-\lambda t}, \quad t \geq 0.$$

Show that the distribution of $\tau_1 + \cdots + \tau_k$ has the density

$$\frac{\lambda^k t^{k-1} e^{-\lambda t}}{(k - 1)!}, \quad t \geq 0,$$

and that

$$P(\tau_1 + \cdots + \tau_k > t) = \sum_{i=0}^{k-1} e^{-\lambda t}\frac{(\lambda t)^i}{i!}.$$

7. Let $\xi \sim \mathcal{N}(0, \sigma^2)$. Show that, for every $p \geq 1$,

$$E\,|\xi|^p = C_p\sigma^p,$$

where

$$C_p = \frac{2^{p/2}}{\pi^{1/2}} \Gamma\left(\frac{p+1}{2}\right)$$

and $\Gamma(s) = \int_0^\infty e^{-x} x^{s-1}\, dx$ is the Euler's gamma function. In particular, for each integer $n \geq 1$,

$$\mathsf{E}\,\xi^{2n} = (2n-1)!!\,\sigma^{2n}.$$

8. Let ξ and η be independent random variables such that the distributions of $\xi + \eta$ and ξ are the same. Show that $\eta = 0$ a. s.

9. Let (X, Y) be uniformly distributed on the unit circle $\{(x, y) : x^2 + y^2 \leq 1\}$ and $W = X^2 + Y^2$. Put

$$U = X\sqrt{-\frac{2\log W}{W}}, \quad V = Y\sqrt{-\frac{2\log W}{W}}.$$

Show that U and V are independent $\mathcal{N}(0, 1)$-distributed random variables.

10. Let X and Y be independent random variables uniformly distributed on $(0, 1)$. Put

$$U = \sqrt{-\log Y}\cos(2\pi X), \quad V = \sqrt{-\log Y}\sin(2\pi X).$$

Show that U and V are independent and $\mathcal{N}(0, 1)$ distributed.

11. Give an example of Gaussian random variables ξ and η such that the distribution of their sum $\xi + \eta$ is not Gaussian.

12. Let $X_1, \ldots, X_n$ be independent identically distributed random variables with density $f = f(x)$. Let $\mathcal{R}_n = \max(X_1, \ldots, X_n) - \min(X_1, \ldots, X_n)$ be the sample range of $X_1, \ldots, X_n$. Show that the density $f_{\mathcal{R}_n}(x)$, $x > 0$, of $\mathcal{R}_n$ is

$$f_{\mathcal{R}_n}(x) = n(n-1)\int_{-\infty}^\infty [F(y) - F(y-x)]^{n-2} f(y) f(y-x)\, dx,$$

where $F(y) = \int_{-\infty}^y f(z)\, dz$. In particular, when $X_1, \ldots, X_n$ are uniformly distributed on $[0, 1]$,

$$f_{\mathcal{R}_n}(x) = \begin{cases} n(n-1)x^{n-2}(1-x), & 0 \leq x \leq 1, \\ 0, & x < 0 \text{ or } x > 1. \end{cases}$$

13. Let $F(x)$ be a distribution function. Show that for any $a > 0$ the following functions are also distribution functions:

$$G_1(x) = \frac{1}{a}\int_x^{x+a} F(u)\, du, \quad G_2(x) = \frac{1}{2a}\int_{x-a}^{x+a} F(u)\, du.$$

14. Let a random variable X have the exponential distribution with parameter $\lambda > 0$ ($f_X(x) = \lambda e^{-\lambda x}$, $x \geq 0$). Find the density of the random variable $Y = X^{1/\alpha}$, $\alpha > 0$. (The corresponding distribution is called the *Weibull distribution*.)

Let $\lambda = 1$. Find the density of the random variable $Y = \log X$ (its distribution is called *double exponential*).

15. Let random variables X and Y have the joint density function $f(x, y)$ of the form $f(x, y) = g(\sqrt{x^2 + y^2})$. Find the joint density function of $\rho = \sqrt{X^2 + Y^2}$ and $\theta = \arctan(Y/X)$. Show that ρ and θ are independent.
Let $U = (\cos \alpha)X + (\sin \alpha)Y$ and $V = (-\sin \alpha)X + (\cos \alpha)Y$. Show that the joint density of U and V is again $f(x, y)$. (This property is due to invariance of the distribution of (X, Y) with respect to "rotation.")

16. Let $X_1, \ldots, X_n$ be independent identically distributed random variables with distribution function $F = F(x)$ and density $f = f(x)$. Denote (cf. Problem 12) by $X^{(1)} = \min(X_1, \ldots, X_n)$ the smallest of $X_1, \ldots, X_n$, by $X^{(2)}$ the second smallest, and so on, and by $X^{(n)} = \max(X_1, \ldots, X_n)$ the largest of $X_1, \ldots, X_n$ (the variables $X^{(1)}, \ldots, X^{(n)}$ so defined are called *order statistics* of $X_1, \ldots, X_n$).
Show that: (a) the density function of $X^{(k)}$ has the form

$$nf(x)C_{n-1}^{k-1}[F(x)]^{k-1}[1 - F(x)]^{n-k};$$

(b) the joint density $f(x^1, \ldots, x^n)$ of $X^{(1)}, \ldots, X^{(n)}$ is given by

$$f(x^1, \ldots, x^n) = \begin{cases} n!f(x^1) \cdots f(x^n), & \text{if } x^1 < \cdots < x^n, \\ 0 & \text{otherwise.} \end{cases}$$

17. Let $X_1, \ldots, X_n$ be independent identically distributed Gaussian $\mathcal{N}(\mu, \sigma^2)$ random variables. The statistic

$$S^2 = \frac{1}{n-1} \sum_{i=1}^{n} (X_i - \overline{X})^2, \quad \text{where } n > 1, \ \overline{X} = \frac{1}{n} \sum_{i=1}^{n} X_i,$$

is called the *sample variance*. Show that:
  (a) $\mathsf{E}\, S^2 = \sigma^2$;
  (b) The *sample mean* $\overline{X}$ and sample variance S^2 are independent;
  (c) $\overline{X} \sim \mathcal{N}(\mu, \sigma^2/n)$ and $(n-1)S^2/\sigma^2$ has the χ^2-distribution with $(n-1)$ degrees of freedom.
  (d) The statistic $T = \sqrt{n}(\overline{X} - \mu)/\sqrt{S^2}$ has the Student distribution with $n - 1$ degrees of freedom (independently of μ and σ). In mathematical statistics T is used for testing hypotheses and setting confidence intervals for μ.

18. Let $X_1, \ldots, X_n, \ldots$ be independent identically distributed random variables and N a random variable independent of X_i's ($N = 1, 2, \ldots$) such that $\mathsf{E}\, N < \infty$, $\mathrm{Var}\, N < \infty$. Put $S_N = X_1 + \cdots + X_N$. Show that

$$\mathrm{Var}\, S_N = \mathrm{Var}\, X_1 \ \mathsf{E}\, N + (\mathsf{E}\, X_1)^2 \, \mathrm{Var}\, N, \qquad \frac{\mathrm{Var}\, S_N}{\mathsf{E}\, S_N} = \frac{\mathrm{Var}\, X_1}{\mathsf{E}\, X_1} + \mathsf{E}\, X_1 \frac{\mathrm{Var}\, N}{\mathsf{E}\, N}.$$

19. Let $M(t) = \mathsf{E}\,e^{tX}$ be the generating function of a random variable X. Show that $\mathsf{P}(X \geq 0) \leq M(t)$ for any $t > 0$.

20. Let $X, X_1, \ldots, X_n$ be independent identically distributed random variables, $S_n = \sum_{i=1}^{n} X_i$, $S_0 = 0$, $\overline{M}_n = \max_{0 \leq j \leq n} S_j$, $\overline{M} = \sup_{n \geq 0} S_n$. Show that (the notation $\xi \overset{d}{=} \eta$ means that ξ and η have the same distribution):

   (a) $\overline{M}_n \overset{d}{=} (\overline{M}_{n-1} + X)^+$, $n \geq 1$;

   (b) if $S_n \to -\infty$ (P-a. s.), then $\overline{M} \overset{d}{=} (\overline{M} + X)^+$;

   (c) if $-\infty < \mathsf{E}\,X < 0$ and $\mathsf{E}\,X^2 < \infty$, then

$$\mathsf{E}\,\overline{M} = \frac{\operatorname{Var} X - \operatorname{Var}(S + X)^-}{-2\,\mathsf{E}\,X}.$$

21. Under the conditions of the previous problem, let $\overline{M}(\varepsilon) = \sup_{n \geq 0}(S_n - n\varepsilon)$ for $\varepsilon > 0$. Show that $\lim_{\varepsilon \downarrow 0} \varepsilon \overline{M}(\varepsilon) = (\operatorname{Var} X)/2$.

9 Construction of a Process with Given Finite-Dimensional Distributions

1. Let $\xi = \xi(\omega)$ be a random variable defined on a probability space $(\Omega, \mathscr{F}, \mathsf{P})$, and let

$$F_\xi(x) = \mathsf{P}\{\omega : \xi(\omega) \leq x\}$$

be its distribution function. It is clear that $F_\xi(x)$ is a distribution function on the real line in the sense of Definition 1 of Sect. 3.

We now ask the following question. Let $F = F(x)$ be *some* distribution function on R. *Does there exist* a random variable whose distribution function is $F(x)$?

One reason for asking this question is as follows. Many statements in probability theory begin, "Let ξ be a random variable with the distribution function $F(x)$; then ...." Consequently if a statement of this kind is to be meaningful we need to be certain that the object under consideration actually exists. Since to know a random variable we first have to know its domain $(\Omega, \mathscr{F})$, and in order to speak of its distribution we need to have a probability measure P on $(\Omega, \mathscr{F})$, a correct way of phrasing the question of the existence of a random variable with a given distribution function $F(x)$ is this:

Do there exist a probability space $(\Omega, \mathscr{F}, \mathsf{P})$ and a random variable $\xi = \xi(\omega)$ on it, such that

$$\mathsf{P}\{\omega : \xi(\omega) \leq x\} = F(x)?$$

Let us show that the answer is positive, and essentially contained in Theorem 1 of Sect. 3.

In fact, let us put

$$\Omega = R, \quad \mathscr{F} = \mathscr{B}(R).$$

It follows from Theorem 1 of Sect. 3 that there is a probability measure P (and only one) on $(R, \mathscr{B}(R))$ for which $P(a, b] = F(b) - F(a)$, $a < b$.

Put $\xi(\omega) \equiv \omega$. Then

$$P\{\omega : \xi(\omega) \leq x\} = P\{\omega : \omega \leq x\} = P(-\infty, x] = F(x).$$

Consequently we have constructed the required probability space and the random variable on it.

2. Let us now ask a similar question for random processes.

Let $X = (\xi_t)_{t \in T}$ be a random process (in the sense of Definition 3 in Sect. 5) defined on the probability space $(\Omega, \mathscr{F}, P)$, with $t \in T \subseteq R$.

From a physical point of view, the most fundamental characteristic of a random process is the set $\{F_{t_1, \ldots, t_n}(x_1, \ldots, x_n)\}$ of its *finite-dimensional distribution functions*

$$F_{t_1, \ldots, t_n}(x_1, \ldots, x_n) = P\{\omega : \xi_{t_1} \leq x_1, \ldots, \xi_{t_n} \leq x_n\}, \tag{1}$$

defined for all sets $t_1, \ldots, t_n$ with $t_1 < t_2 < \cdots < t_n$.

We see from (1) that, for each set $t_1, \ldots, t_n$ with $t_1 < t_2 < \cdots < t_n$ the functions $F_{t_1, \ldots, t_n}(x_1, \ldots, x_n)$ are n-dimensional distribution functions (in the sense of Definition 2 in Sect. 3) and that the collection $\{F_{t_1, \ldots, t_n}(x_1, \ldots, x_n)\}$ has the following *consistency* property (cf. (20) in Sect. 3):

$$F_{t_1, \ldots, t_k, \ldots, t_n}(x_1, \ldots, \infty, \ldots, x_n)$$
$$= F_{t_1, \ldots, t_{k-1}, t_{k+1}, \ldots, t_n}(x_1, \ldots, x_{k-1}, x_{k+1}, \ldots, x_n). \tag{2}$$

Now it is natural to ask the following question: under what conditions can a given family $\{F_{t_1, \ldots, t_n}(x_1, \ldots, x_n)\}$ of distribution functions $F_{t_1, \ldots, t_n}(x_1, \ldots, x_n)$ (in the sense of Definition 2 in Sect. 3) be the family of finite-dimensional distribution functions of a random process? It is quite remarkable that all such conditions are covered by the consistency condition (2).

Theorem 1 (Kolmogorov's Theorem on the Existence of a Process). *Let* $\{F_{t_1, \ldots, t_n}(x_1, \ldots, x_n)\}$, *with* $t_i \in T \subseteq R$, $t_1 < t_2 < \cdots < t_n$, $n \geq 1$, *be a given family of finite-dimensional distribution functions, satisfying the consistency condition* (2). *Then there are a probability space* $(\Omega, \mathscr{F}, P)$ *and a random process* $X = (\xi_t)_{t \in T}$ *such that*

$$P\{\omega : \xi_{t_1} \leq x_1, \ldots, \xi_{t_n} \leq x_n\} = F_{t_1, \ldots, t_n}(x_1, \ldots, x_n). \tag{3}$$

PROOF. Put

$$\Omega = R^T, \quad \mathscr{F} = \mathscr{B}(R^T),$$

i.e., take Ω to be the space of real functions $\omega = (\omega_t)_{t \in T}$ with the σ-algebra generated by the cylindrical sets.

Let $\tau = [t_1, \ldots, t_n]$, $t_1 < t_2 < \cdots < t_n$. Then by Theorem 2 of Sect. 3 we can construct on the space $(R^n, \mathscr{B}(R^n))$ a unique probability measure P_τ such that

$$P_\tau\{(\omega_{t_1}, \ldots, \omega_{t_n}) : \omega_{t_1} \leq x_1, \ldots, \omega_{t_n} \leq x_n\} = F_{t_1, \ldots, t_n}(x_1, \ldots, x_n). \tag{4}$$

It follows from the consistency condition (2) that the family $\{P_\tau\}$ is also consistent (see (20) in Sect. 3). According to Theorem 4 of Sect. 3 there is a probability measure P on $(R^T, \mathscr{B}(R^T))$ such that

$$P\{\omega: (\omega_{t_1}, \ldots, \omega_{t_n}) \in B\} = P_\tau(B)$$

for every set $\tau = [t_1, \ldots, t_n], t_1 < \cdots < t_n$.

From this, it also follows that (4) is satisfied. Therefore the required random process $X = (\xi_t(\omega))_{t \in T}$ can be taken to be the process defined by

$$\xi_t(\omega) = \omega_t, \quad t \in T. \tag{5}$$

This completes the proof of the theorem.

$\square$

Remark 1. The probability space $(R^T, \mathscr{B}(R^T), P)$ that we have constructed is called *canonical*, and the construction given by (5) is called the *coordinate method* of constructing the process.

Remark 2. Let $(E_\alpha, \mathscr{E}_\alpha)$ be complete separable metric spaces, where α belongs to some set $\mathfrak{A}$ of indices. Let $\{P_\tau\}$ be a set of consistent finite-dimensional distribution functions P_τ, $\tau = [\alpha_1, \ldots, \alpha_n]$, on

$$(E_{\alpha_1} \times \cdots \times E_{\alpha_n}, \; \mathscr{E}_{\alpha_1} \otimes \cdots \otimes \mathscr{E}_{\alpha_n}).$$

Then there are a probability space $(\Omega, \mathscr{F}, P)$ and a family of $\mathscr{F}/\mathscr{E}_\alpha$-measurable functions $(X_\alpha(\omega))_{\alpha \in \mathfrak{A}}$ such that

$$P\{(X_{\alpha_1}, \ldots, X_{\alpha_n}) \in B\} = P_\tau(B)$$

for all $\tau = [\alpha_1, \ldots, \alpha_n]$ and $B \in \mathscr{E}_{\alpha_1} \otimes \cdots \otimes \mathscr{E}_{\alpha_n}$.

This result, which generalizes Theorem 1, follows from Theorem 4 of Sect. 3 if we put $\Omega = \prod_\alpha E_\alpha$, $\mathscr{F} = \boxtimes_\alpha \mathscr{E}_\alpha$ and $X_\alpha(\omega) = \omega_\alpha$ for each $\omega = (\omega_\alpha)$, $\alpha \in \mathfrak{A}$.

Corollary 1. *Let $F_1(x), F_2(x), \ldots$ be a sequence of one-dimensional distribution functions. Then there exist a probability space $(\Omega, \mathscr{F}, P)$ and a sequence of independent random variables $\xi_1, \xi_2, \ldots$ such that*

$$P\{\omega: \xi_i(\omega) \le x\} = F_i(x). \tag{6}$$

In particular, there is a probability space $(\Omega, \mathscr{F}, P)$ on which an infinite sequence of Bernoulli random variables is defined. Notice that Ω can be taken to be the space

$$\Omega = \{\omega: \omega = (a_1, a_2, \ldots), \; a_i = 0 \text{ or } 1\}$$

(cf. also Theorem 2 below).

To establish the corollary it is enough to put $F_{1,\ldots,n}(x_1, \ldots, x_n) = F_1(x_1) \cdots F_n(x_n)$ and apply Theorem 1.

Corollary 2. *Let $T = [0, \infty)$ and let $\{P(s, x; t, B\}$ be a family of nonnegative functions defined for $s, t \in T, t > s, x \in R, B \in \mathscr{B}(R)$, and satisfying the following conditions:*

(a) *$P(s, x; t, B)$ is a probability measure in B for given s, x and t;*
(b) *for given s, t and B, the function $P(s, x; t, B)$ is a Borel function of x;*
(c) *for all $0 \leq s < t < \tau$ and $B \in \mathscr{B}(R)$, the Kolmogorov–Chapman equation*

$$P(s, x; \tau, B) = \int_R P(s, x; t, dy)P(t, y; \tau, B) \tag{7}$$

is satisfied.

Also let $\pi = \pi(\cdot)$ be a probability measure on $(R, \mathscr{B}(R))$. Then there are a probability space $(\Omega, \mathscr{F}, \mathsf{P})$ and a random process $X = (\xi_t)_{t \geq 0}$ defined on it, such that

$$\mathsf{P}\{\xi_{t_0} \leq x_0, \xi_{t_1} \leq x_1, \ldots, \xi_{t_n} \leq x_n\} = \int_{-\infty}^{x_0} \pi(dy_0) \int_{-\infty}^{x_1} P(0, y_0; t_1, dy_1)$$

$$\ldots \int_{-\infty}^{x_n} P(t_{n-1}, y_{n-1}; t_n, dy_n) \tag{8}$$

for $0 = t_0 < t_1 < \cdots < t_n$.

The process X so constructed is a Markov process with initial distribution π and transition probabilities $\{P(s, x; t, B\}$.

Corollary 3. *Let $T = \{0, 1, 2, \ldots\}$ and let $\{P_k(x; B)\}$ be a family of nonnegative functions defined for $k \geq 1, x \in R, B \in \mathscr{B}(R)$, and such that $P_k(x; B)$ is a probability measure in B (for given k and x) and measurable in x (for given k and B). In addition, let $\pi = \pi(B)$ be a probability measure on $(R, \mathscr{B}(R))$.*

Then there is a probability space $(\Omega, \mathscr{F}, \mathsf{P})$ with a family of random variables $X = \{\xi_0, \xi_1, \ldots\}$ defined on it, such that

$$\mathsf{P}\{\xi_0 \leq x_0, \xi_1 \leq x_1, \ldots, \xi_n \leq x_n\}$$

$$= \int_{-\infty}^{x_0} \pi(dy_0) \int_{-\infty}^{x_1} P_1(y_0; dy_1) \cdots \int_{-\infty}^{x_n} P_n(y_{n-1}; dy_n).$$

3. In the situation of Corollary 1, there is a sequence of independent random variables $\xi_1, \xi_2, \ldots$ whose one-dimensional distribution functions are $F_1, F_2, \ldots$, respectively.

Now let $(E_1, \mathscr{E}_1), (E_2, \mathscr{E}_2), \ldots$ be complete separable metric spaces and let $P_1, P_2, \ldots$ be probability measures on them. Then it follows from Remark 2 that there are a probability space $(\Omega, \mathscr{F}, \mathsf{P})$ and a sequence of independent elements $X_1, X_2, \ldots$ such that X_n is $\mathscr{F}/\mathscr{E}_n$-measurable and $\mathsf{P}(X_n \in B) = P_n(B), B \in \mathscr{E}_n$.

It turns out that this result remains valid when the spaces $(E_n, \mathscr{E}_n)$ are *arbitrary measurable spaces.*

Theorem 2 (Ionescu Tulcea's Theorem on Extending a Measure and the Existence of a Random Sequence). *Let $(\Omega_n, \mathscr{F}_n)$, $n = 1, 2, \ldots$, be arbitrary measurable spaces and $\Omega = \prod \Omega_n$, $\mathscr{F} = \bigotimes \mathscr{F}_n$. Suppose that a probability measure P_1 is given on $(\Omega_1, \mathscr{F}_1)$ and that, for every set $(\omega_1, \ldots, \omega_n) \in \Omega_1 \times \cdots \times \Omega_n$, $n \geq 1$, probability measures $P(\omega_1, \ldots, \omega_n; \cdot)$ are given on $(\Omega_{n+1}, \mathscr{F}_{n+1})$. Suppose that for every $B \in \mathscr{F}_{n+1}$ the functions $P(\omega_1, \ldots, \omega_n; B)$ are $\mathscr{F}^n \equiv \mathscr{F}_1 \otimes \cdots \otimes \mathscr{F}_n$-measurable functions of $(\omega_1, \ldots, \omega_n)$ and let, for $A_i \in \mathscr{F}_i$, $n \geq 1$,*

$$P_n(A_1 \times \cdots \times A_n) = \int_{A_1} P_1(d\omega_1) \int_{A_2} P(\omega_1; d\omega_2)$$
$$\cdots \int_{A_n} P(\omega_1, \ldots, \omega_{n-1}; d\omega_n). \tag{9}$$

Then there is a unique probability measure P on $(\Omega, \mathscr{F})$ such that

$$\mathsf{P}\{\omega: \omega_1 \in A_1, \ldots, \omega_n \in A_n\} = P_n(A_1 \times \cdots \times A_n) \tag{10}$$

for every $n \geq 1$, and there is a random sequence $X = (X_1(\omega), X_2(\omega), \ldots)$ such that

$$\mathsf{P}\{\omega: X_1(\omega) \in A_1, \ldots, X_n(\omega) \in A_n\} = P_n(A_1 \times \cdots \times A_n), \tag{11}$$

where $A_i \in \mathscr{E}_i$.

PROOF. The first step is to establish that for each $n > 1$ the set function P_n defined by (9) on the rectangles $A_1 \times \cdots \times A_n$ can be extended to the σ-algebra $\mathscr{F}^n$.

For each $n \geq 2$ and $B \in \mathscr{F}^n$ we put

$$P_n(B) = \int_{\Omega_1} P_1(d\omega_1) \int_{\Omega_2} P(\omega_1; d\omega_2 \cdots) \int_{\Omega_{n-1}} P(\omega_1, \ldots, \omega_{n-2}; d\omega_{n-1})$$
$$\times \int_{\Omega_n} I_B(\omega_1, \ldots, \omega_n) P(\omega_1, \ldots, \omega_{n-1}; d\omega_n). \tag{12}$$

It is easily seen that when $B = A_1 \times \cdots \times A_n$ the right-hand side of (12) is the same as the right-hand side of (9). Moreover, when $n = 2$ it can be shown, just as in Theorem 8 of Sect. 6, that P_2 is a measure. Consequently it is easily established by induction that P_n is a measure for all $n \geq 2$.

The next step is the same as in Kolmogorov's theorem on the extension of a measure in $(R^\infty, \mathscr{B}(R^\infty))$ (Theorem 3, Sect. 3). Namely, for every cylindrical set $J_n(B) = \{\omega \in \Omega: (\omega_1, \ldots, \omega_n) \in B\}$, $B \in \mathscr{F}^n = \mathscr{F}_1 \otimes \cdots \otimes \mathscr{F}_n$, we define the set function P by

$$\mathsf{P}(J_n(B)) = P_n(B). \tag{13}$$

If we use (12) and the fact that $P(\omega_1, \ldots, \omega_k; \cdot)$ are measures, it is easy to establish that the definition (13) is consistent, in the sense that the value of $\mathsf{P}(J_n(B))$ is independent of the representation of the cylindrical set.

It follows that the set function P defined in (13) for cylindrical sets, and in an obvious way on the algebra that contains all the cylindrical sets, is a finitely additive measure on this algebra. It remains to verify its countable additivity and apply Carathéodory's theorem.

In Theorem 3 of Sect. 3 the corresponding verification was based on the property of $(R^n, \mathscr{B}(R^n))$ that for every Borel set B there is a compact set $A \subseteq B$ whose probability measure is arbitrarily close to the measure of B. In the present case this part of the proof needs to be modified in the following way.

As in Theorem 3 of Sect. 3, let $\{\hat{B}_n\}_{n \geq 1}$ be a sequence of cylindrical sets

$$\hat{B}_n = \{\omega : (\omega_1, \ldots, \omega_n) \in B_n\}$$

that decrease to the empty set $\varnothing$, but have

$$\lim_{n \to \infty} P(\hat{B}_n) > 0. \tag{14}$$

For $n > 1$, we have from (12)

$$P(\hat{B}_n) = \int_{\Omega_1} f_n^{(1)}(\omega_1) \, P_1(d\omega_1),$$

where

$$f_n^{(1)}(\omega_1) = \int_{\Omega_2} P(\omega_1; d\omega_2) \ldots \int_{\Omega_n} I_{B_n}(\omega_1, \ldots, \omega_n) \, P(\omega_1, \ldots, \omega_{n-1}; d\omega_n).$$

Since $\hat{B}_{n+1} \subseteq \hat{B}_n$, we have $B_{n+1} \subseteq B_n \times \Omega_{n+1}$ and therefore

$$I_{B_{n+1}}(\omega_1, \ldots, \omega_{n+1}) \leq I_{B_n}(\omega_1, \ldots, \omega_n) I_{\Omega_{n+1}}(\omega_{n+1}).$$

Hence the sequence $\{f_n^{(1)}(\omega_1)\}_{n \geq 1}$ decreases. Let $f^{(1)}(\omega_1) = \lim_n f_n^{(1)}(\omega_1)$. By the dominated convergence theorem

$$\lim_n P(\hat{B}_n) = \lim_n \int_{\Omega_1} f_n^{(1)}(\omega_1) \, P_1(d\omega_1) = \int_{\Omega_1} f^{(1)}(\omega_1) \, P_1(d\omega_1).$$

By hypothesis, $\lim_n P(\hat{B}_n) > 0$. It follows that there is an $\omega_1^0 \in B_1$ such that $f^{(1)}(\omega_1^0) > 0$, since if $\omega_1 \notin B_1$ then $f_n^{(1)}(\omega_1) = 0$ for $n \geq 1$.

Further, for $n > 2$,

$$f_n^{(1)}(\omega_1^0) = \int_{\Omega_2} f_n^{(2)}(\omega_2) \, P(\omega_1^0; d\omega_2), \tag{15}$$

where

$$f_n^{(2)}(\omega_2) = \int_{\Omega_3} P(\omega_1^0, \omega_2; d\omega_3)$$

$$\cdots \int_{\Omega_n} I_{B_n}(\omega_1^0, \omega_2, \ldots, \omega_n) \, P(\omega_1^0, \omega_2, \ldots, \omega_{n-1}; d\omega_n).$$

We can establish, as for $\{f_n^{(1)}(\omega_1)\}$, that $\{f_n^{(2)}(\omega_2)\}$ is decreasing. Let $f^{(2)}(\omega_2) = \lim_{n\to\infty} f_n^{(2)}(\omega_2)$. Then it follows from (15) that

$$0 < f^{(1)}(\omega_1^0) = \int_{\Omega_2} f^{(2)}(\omega_2) \, P(\omega_1^0; d\omega_2),$$

and there is a point $\omega_2^0 \in \Omega_2$ such that $f^{(2)}(\omega_2^0) > 0$. Then $(\omega_1^0, \omega_2^0) \in B_2$. Continuing this process, we find a point $(\omega_1^0, \ldots, \omega_n^0) \in B_n$ for each n. Consequently $(\omega_1^0, \ldots, \omega_n^0, \ldots) \in \bigcap \hat{B}_n$, but by hypothesis we have $\bigcap \hat{B}_n = \varnothing$. This contradiction shows that $\lim_n P(\hat{B}_n) = 0$.

Thus we have proved the part of the theorem about the existence of the probability measure P. The other part follows from this by putting $X_n(\omega) = \omega_n, n \geq 1$.

$\square$

Corollary 4. *Let $(E_n, \mathscr{E}_n)_{n\geq 1}$ be any measurable spaces and $(P_n)_{n\geq 1}$ measures on them. Then there are a probability space $(\Omega, \mathscr{F}, P)$ and a family of independent random elements $X_1, X_2, \ldots$ with values in $(E_1, \mathscr{E}_1), (E_2, \mathscr{E}_2), \ldots$, respectively, such that*

$$P\{\omega : X_n(\omega) \in B\} = P_n(B), \quad B \in \mathscr{E}_n, \, n \geq 1.$$

Corollary 5. *Let $E = \{1, 2, \ldots\}$, and let $\{p_k(x, y)\}$ be a family of nonnegative functions, $k \geq 1, x, y \in E$, such that $\sum_{y\in E} p_k(x; y) = 1, x \in E, k \geq 1$. Also let $\pi = \pi(x)$ be a probability distribution on E $\big($that is, $\pi(x) \geq 0, \sum_{x\in E} \pi(x) = 1\big)$.*

Then there are a probability space $(\Omega, \mathscr{F}, P)$ and a family $X = \{\xi_0, \xi_1, \ldots\}$ of random variables on it such that

$$P\{\xi_0 = x_0, \, \xi_1 = x_1, \ldots, \xi_n = x_n\} = \pi(x_0) p_1(x_0, x_1) \cdots p_n(x_{n-1}, x_n) \quad (16)$$

(cf. (4) in Sect. 12 of Chapter 1) for all $x_i \in E$ and $n \geq 1$. We may take Ω to be the space

$$\Omega = \{\omega : \omega = (x_0, x_1, \ldots), \, x_i \in E\}.$$

A sequence $X = \{\xi_0, \xi_1, \ldots\}$ of random variables satisfying (16) is a *Markov chain* with a countable set E of states, transition matrices $\{p_k(x, y)\}$ and initial probability distribution π. (Cf. the definition in Sect. 12 of Chap. 1 and the definitions in Sect. 1 of Chapter 8, Vol. 2).

4. Kolmogorov's theorem (Theorem 1) states the existence of a process with a given set of consistent finite-dimensional distribution functions. Its proof exploits the *canonical* probability space and the processes are constructed in a *coordinate-wise* manner, which is due to the complexity of the structure of their paths.

From this point of view of much interest are the instances, where random processes having desired properties can be built *constructively*, and with minimal use of "probabilistic structures."

To demonstrate such possibilities, consider the so-called renewal processes. (A particular case of them is the Poisson process; see Sect. 10 of Chap. 7, Vol. 2.)

Let $(\sigma_1, \sigma_2, \dots)$ be a sequence of independent identically distributed positive random variables with distribution function $F = F(x)$. (The existence of such a sequence follows from Corollary 1 to Theorem 1.)

Based on $(\sigma_1, \sigma_2, \dots)$, we form a new sequence $(T_0, T_1, \dots)$ with $T_0 = 0$ and

$$T_n = \sigma_1 + \cdots + \sigma_n, \quad n \geq 1.$$

For illustrative purposes, let us think of T_n as the time instant of, say, the nth telephone call. Then σ_n is the time between the $(n-1)$th and nth calls.

The random process $N = (N_t)_{t \geq 0}$ with constructively specified random variables

$$N_t = \sum_{n=1}^{\infty} I(T_n \leq t) \tag{17}$$

is referred to as a *renewal process*.

Clearly, N_t could also be defined as

$$N_t = \max\{n : T_n \leq t\}, \tag{18}$$

i. e., N_t is the number of calls that occur in the time interval $(0, t]$, and it is obvious that

$$\{N_t \geq n\} = \{T_n \leq t\}. \tag{19}$$

This simple formula is very useful because it reduces the study of probabilistic properties of the process $N = (N_t)_{t \geq 0}$ to the treatment of the variables $T_n = \sigma_1 + \cdots + \sigma_n$, which are sums of independent random variables $\sigma_1, \dots, \sigma_n, n \geq 1$ (see Subsection 4 in Sect. 3 of Chapter 4 and Subsection 4 in Sect. 2 of Chapter 7 (vol. 2)).

Formula (17) implies that the *renewal function* $m(t) = \mathsf{E}\, N_t, t \geq 0$, is connected with the distribution function $F_n(t) = \mathsf{P}(T_n \leq t)$ by the equation

$$m(t) = \sum_{n=1}^{\infty} F_n(t). \tag{20}$$

4. Problems

1. Let $\Omega = [0, 1]$, let $\mathscr{F}$ be the class of Borel subsets of $[0, 1]$, and let P be Lebesgue measure on $[0, 1]$. Show that the space $(\Omega, \mathscr{F}, \mathsf{P})$ is universal in the following sense. For every distribution function $F(x)$, $x \in R$, there is a random variable $\xi = \xi(\omega)$ such that its distribution function $F_\xi(x) = \mathsf{P}(\xi \leq x)$ coincides with $F(x)$. (*Hint.* Let $\xi(\omega) = F^{-1}(\omega)$, $0 < \omega < 1$, where $F^{-1}(\omega) = \sup\{x : F(x) < \omega\}$, when $0 < \omega < 1$, and $\xi(0), \xi(1)$ can be chosen arbitrarily.)

2. Verify the consistency of the families of distributions in the corollaries to Theorems 1 and 2.
3. Deduce Corollary 2, Theorem 2, from Theorem 1.
4. Let F_n denote the distribution function of T_n, $n \geq 1$ (see Subsection 4). Show that $F_{n+1}(t) = \int_0^t F_n(t-s)\, dF(s)$, $n \geq 1$, where $F_1 = F$.
5. Show that $\mathsf{P}\{N_t = n\} = F_n(t) - F_{n+1}(t)$ (see (17)).
6. Show that the renewal function $m(t)$ defined in Subsection 4 satisfies the *renewal equation*

$$m(t) = F(t) + \int_0^t m(t-x)\, dF(x). \tag{21}$$

7. Show that the function defined by (20) is a unique solution of the equation (21) within the class of functions bounded on finite intervals.
8. Let T be an arbitrary set.
    (i) Suppose that for every $t \in T$ a probability space $(\Omega_t, \mathscr{F}_t, \mathsf{P}_t)$ is given. Put $\Omega = \prod_{t \in T} \Omega_t$, $\mathscr{F} = \bigotimes_{t \in T} \mathscr{F}_t$. Prove that there is a unique probability measure P on $(\Omega, \mathscr{F})$ such that

$$\mathsf{P}\left(\prod_{t \in T} B_t\right) = \prod_{t \in T} \mathsf{P}(B_t),$$

    where $B_t \in \mathscr{F}_t$, $t \in T$, and $B_t = \Omega_t$ for all but finitely many t. (*Hint.* Specify P on an appropriate algebra and use the method of the proof of Ionescu Tulcea's theorem.)
    (ii) Let for every $t \in T$ a measurable space $(E_t, \mathscr{E}_t)$ and a probability measure P_t on it be given. Show that there is a probability space $(\Omega, \mathscr{F}, \mathsf{P})$ and independent random elements $(X_t)_{t \in T}$ such that X_t are $\mathscr{F}/\mathscr{E}_t$-measurable and $\mathsf{P}\{X_t \in B\} = \mathsf{P}_t(B)$, $B \in \mathscr{E}_t$.

10 Various Kinds of Convergence of Sequences of Random Variables

1. Just as in analysis, in probability theory we need to use various kinds of convergence of random variables. Four of these are particularly important: *in probability, with probability one, in the mean of order p, in distribution.*

First some definitions. Let $\xi, \xi_1, \xi_2, \ldots$ be random variables defined on a probability space $(\Omega, \mathscr{F}, \mathsf{P})$.

Definition 1. The sequence $\xi_1, \xi_2, \ldots$ of random variables (denoted by (ξ_n) or $(\xi_n)_{n \geq 1}$) converges *in probability* to the random variable ξ (notation: $\xi_n \xrightarrow{\mathsf{P}} \xi$) if for every $\varepsilon > 0$

$$\mathsf{P}\{|\xi_n - \xi| > \varepsilon\} \to 0, \quad n \to \infty. \tag{1}$$

We have already encountered this convergence in connection with the law of large numbers for a Bernoulli scheme, which stated that

$$P\left(\left|\frac{S_n}{n} - p\right| > \varepsilon\right) \to 0, \quad n \to \infty$$

(see Sect. 5 of Chap. 1). In analysis this is known as *convergence in measure*.

Definition 2. The sequence $\xi_1, \xi_2, \ldots$ of random variables converges *with probability one* (*almost surely, almost everywhere*) to the random variable ξ if

$$P\{\omega: \xi_n \nrightarrow \xi\} = 0, \tag{2}$$

i.e., if the set of sample points ω for which $\xi_n(\omega)$ does not converge to ξ has probability zero.

This convergence is denoted by $\xi_n \to \xi$ (P-a. s.), or $\xi_n \to \xi$ (a. s.), or $\xi_n \xrightarrow{\text{a. s.}} \xi$ or $\xi_n \xrightarrow{\text{a.e.}} \xi$.

Definition 3. The sequence $\xi_1, \xi_2, \ldots$ of random variables converges *in the mean of order p*, $0 < p < \infty$, to the random variable ξ if

$$E |\xi_n - \xi|^p \to 0, \quad n \to \infty. \tag{3}$$

In analysis this is known as *convergence in L^p*, and denoted by $\xi_n \xrightarrow{L^p} \xi$. In the special case $p = 2$ it is called *mean square convergence* and denoted by $\xi = \text{l.i.m.} \, \xi_n$ (for "limit in the mean").

Definition 4. The sequence $\xi_1, \xi_2, \ldots$ of random variables *converges in distribution* to the random variable ξ (notation: $\xi_n \xrightarrow{d} \xi$ or $\xi_n \xrightarrow{law} \xi$) if

$$Ef(\xi_n) \to Ef(\xi), \quad n \to \infty, \tag{4}$$

for every bounded continuous function $f = f(x)$. The reason for the terminology is that, according to what will be proved in Sect. 1 of Chap. 3 condition (4) is equivalent to the convergence of the distribution functions $F_{\xi_n}(x)$ to $F_\xi(x)$ at each *point x of continuity* of $F_\xi(x)$. This convergence is denoted by $F_{\xi_n} \Rightarrow F_\xi$.

We emphasize that the convergence of random variables in distribution is defined only in terms of the convergence of their distribution functions. Therefore it makes sense to discuss this mode of convergence even when the random variables are defined on different probability spaces. This convergence will be studied in detail in Chapter 3, where, in particular, we shall explain why in the definition of $F_{\xi_n} \Rightarrow F_\xi$ we require only convergence at points of continuity of $F_\xi(x)$ and not at all x.

2. In solving problems of analysis on the convergence (in one sense or another) of a given sequence of functions, it is useful to have the concept of a fundamental sequence (or Cauchy sequence). We can introduce a similar concept for each of the first three kinds of convergence of a sequence of random variables.

Let us say that a sequence $\{\xi_n\}_{n \geq 1}$ of random variables is *fundamental in probability*, or *with probability* 1, or *in the mean of order p*, $0 < p < \infty$, if the corresponding one of the following properties is satisfied: $P\{|\xi_n - \xi_m| > \varepsilon\} \to 0$ as $m, n \to \infty$

for every $\varepsilon > 0$; the sequence $\{\xi_n(\omega)\}_{n\geq 1}$ is fundamental for almost all $\omega \in \Omega$; the sequence $\{\xi_n(\omega)\}_{n\geq 1}$ is fundamental in L^p, i.e., $\mathsf{E}\,|\xi_n - \xi_m|^p \to 0$ as $n, m \to \infty$.

3. Theorem 1.

(a) *A necessary and sufficient condition that $\xi_n \to \xi$ (P-a.s.) is that*

$$\mathsf{P}\left\{\sup_{k\geq n} |\xi_k - \xi| \geq \varepsilon\right\} \to 0, \quad n \to \infty, \tag{5}$$

for every $\varepsilon > 0$.

(b) *The sequence $\{\xi_n\}_{n\geq 1}$ is fundamental with probability 1 if and only if*

$$\mathsf{P}\left\{\sup_{k\geq n, l\geq n} |\xi_k - \xi_l| \geq \varepsilon\right\} \to 0, \quad n \to \infty, \tag{6}$$

for every $\varepsilon > 0$; or equivalently

$$\mathsf{P}\left\{\sup_{k\geq 0} |\xi_{n+k} - \xi_n| \geq \varepsilon\right\} \to 0, \quad n \to \infty. \tag{7}$$

PROOF. (a) Let $A_n^\varepsilon = \{\omega: |\xi_n - \xi| \geq \varepsilon\}$, $A^\varepsilon = \limsup A_n^\varepsilon \equiv \bigcap_{n=1}^\infty \bigcup_{k\geq n} A_k^\varepsilon$. Then

$$\{\omega: \xi_n \nrightarrow \xi\} = \bigcup_{\varepsilon \geq 0} A^\varepsilon = \bigcup_{m=1}^\infty A^{1/m}.$$

But

$$\mathsf{P}(A^\varepsilon) = \lim_n \mathsf{P}\left(\bigcup_{k>n} A_k^\varepsilon\right),$$

hence (a) follows from the following chain of implications:

$$\mathsf{P}\{\omega: \xi_n \nrightarrow \xi\} = 0 \Leftrightarrow \mathsf{P}\left(\bigcup_{\varepsilon>0} A^\varepsilon\right) = 0 \Leftrightarrow \mathsf{P}\left(\bigcup_{m=1}^\infty A^{1/m}\right) - 0$$

$$\Leftrightarrow \mathsf{P}(A^{1/m}) = 0, \; m \geq 1 \Leftrightarrow \mathsf{P}(A^\varepsilon) = 0, \; \varepsilon > 0$$

$$\Leftrightarrow \mathsf{P}\left(\bigcup_{k\geq n} A_k^\varepsilon\right) \to 0, \, n \to \infty, \, \varepsilon > 0$$

$$\Leftrightarrow \mathsf{P}\left(\sup_{k\geq n} |\xi_k - \xi| \geq \varepsilon\right) \to 0, \, n \to \infty, \, \varepsilon > 0.$$

(b) Let

$$B_{k,l}^\varepsilon = \{\omega: |\xi_k - \xi_l| \geq \varepsilon\}, \quad B^\varepsilon = \bigcap_{n=1}^\infty \bigcup_{\substack{k\geq n \\ l\geq n}} B_{k,l}^\varepsilon.$$

Then $\{\omega\colon \{\xi_n(\omega)\}_{n\geq 1}$ *is not fundamental*$\} = \bigcup_{\varepsilon>0} B^\varepsilon$, and it can be shown as in (a) that $\mathsf{P}\{\omega\colon \{\xi_n(\omega)\}_{n\geq 1}$ *is not fundamental*$\} = 0 \Leftrightarrow$ (6). The equivalence of (6) and (7) follows from the obvious inequalities

$$\sup_{k\geq 0} |\xi_{n+k} - \xi_n| \leq \sup_{\substack{k\geq 0 \\ l\geq 0}} |\xi_{n+k} - \xi_{n+l}| \leq 2 \sup_{k\geq 0} |\xi_{n+k} - \xi_n|.$$

This completes the proof of the theorem.

$\square$

Corollary. *Since*

$$\mathsf{P}\left\{\sup_{k\geq n} |\xi_k - \xi| \geq \varepsilon\right\} = \mathsf{P}\left\{\bigcup_{k\geq n} (|\xi_k - \xi| \geq \varepsilon)\right\} \leq \sum_{k\geq n} \mathsf{P}\{|\xi_k - \xi| \geq \varepsilon\},$$

a sufficient condition for $\xi_n \overset{\text{a. s.}}{\rightarrow} \xi$ is that

$$\sum_{k=1}^{\infty} \mathsf{P}\{|\xi_k - \xi| \geq \varepsilon\} < \infty \tag{8}$$

is satisfied for every $\varepsilon > 0$.

It is appropriate to observe at this point that the reasoning used in obtaining (8) lets us establish the following simple but important result which is essential in studying properties that are satisfied with probability 1.

Let $A_1, A_2, \ldots$ be a sequence of events in $\mathscr{F}$. Let (see Table 2.1 in Sect. 1) $\{A_n \text{ i.o.}\}$ denote the event $\limsup A_n$ that consists in the realization of infinitely many of $A_1, A_2, \ldots$

Borel–Cantelli Lemma.

(a) *If $\sum \mathsf{P}(A_n) < \infty$ then $\mathsf{P}\{A_n \text{ i.o.}\} = 0$.*
(b) *If $\sum \mathsf{P}(A_n) = \infty$ and $A_1, A_2, \ldots$ are independent, then $\mathsf{P}\{A_n \text{ i.o.}\} = 1$.*

PROOF. (a) By definition $\{A_n \text{ i.o.}\} = \limsup A_n = \bigcap_{n=1}^{\infty} \bigcup_{k\geq n} A_k$. Consequently

$$\mathsf{P}\{A_n \text{ i.o.}\} = \mathsf{P}\left(\bigcap_{n=1}^{\infty} \bigcup_{k\geq n} A_k\right) = \lim \mathsf{P}\left(\bigcup_{k\geq n} A_k\right) \leq \lim \sum_{k\geq n} \mathsf{P}(A_k),$$

and (a) follows.

(b) If $A_1, A_2, \ldots$ are independent, so are $\bar{A}_1, \bar{A}_2, \ldots$. Hence for $N \geq n$ we have

$$\mathsf{P}\left(\bigcap_{k=n}^{N} \bar{A}_k\right) = \prod_{k=n}^{N} \mathsf{P}(\bar{A}_k),$$

and it is then easy to deduce that

$$P\left(\bigcap_{k=n}^{\infty} \overline{A}_k\right) = \prod_{k=n}^{\infty} P(\overline{A}_k).\tag{9}$$

Since $\log(1 - x) \le -x$, $0 \le x < 1$,

$$\log \prod_{k=n}^{\infty}[1 - P(A_k)] = \sum_{k=n}^{\infty} \log[1 - P(A_k)] \le -\sum_{k=n}^{\infty} P(A_k) = -\infty.$$

Consequently

$$P\left(\bigcap_{k=n}^{\infty} \overline{A}_k\right) = 0$$

for all n, and therefore $P(A_n \text{ i.o.}) = 1$.

This completes the proof of the lemma.

$\square$

Corollary 1. *If $A_n^{\varepsilon} = \{\omega\colon |\xi_n - \xi| \ge \varepsilon\}$ then (8) means that $\sum_{n=1}^{\infty} P(A_n^{\varepsilon}) < \infty$, $\varepsilon > 0$, and then by the Borel–Cantelli lemma we have $P(A^{\varepsilon}) = 0$, $\varepsilon > 0$, where $A^{\varepsilon} = \limsup A_n^{\varepsilon}(= \{A_n^{\varepsilon} i.o.\})$. Therefore*

$$\sum_{k=1}^{\infty} P\{|\xi_k - \xi| \ge \varepsilon\} < \infty, \ \varepsilon > 0 \Rightarrow P(A^{\varepsilon}) = 0, \ \varepsilon > 0$$

$$\Leftrightarrow P\{\omega\colon \xi_n \nrightarrow \xi)\} = 0,$$

as we already observed above.

Corollary 2. *Let $(\varepsilon_n)_{n \ge 1}$ be a sequence of positive numbers such that $\varepsilon_n \downarrow 0$, $n \to \infty$. Then if ξ_n converges to ξ in probability sufficiently "fast" in the sense that*

$$\sum_{n=1}^{\infty} P\{|\xi_n - \xi| \ge \varepsilon_n\} < \infty,\tag{10}$$

then $\xi_n \xrightarrow{a.\,s.} \xi$.

In fact, let $A_n = \{|\xi_n - \xi| \ge \varepsilon_n\}$. Then $P(A_n \text{ i.o.}) = 0$ by the Borel–Cantelli lemma. This means that, for almost every $\omega \in \Omega$, there is an $N = N(\omega)$ such that $|\xi_n(\omega) - \xi(\omega)| \le \varepsilon_n$ for $n \ge N(\omega)$. But $\varepsilon_n \downarrow 0$, and therefore $\xi_n(\omega) \to \xi(\omega)$ for almost every $\omega \in \Omega$.

4. Theorem 2. *We have the following implications*:

$$\xi_n \xrightarrow{\text{a. s.}} \xi \Rightarrow \xi_n \xrightarrow{P} \xi, \tag{11}$$

$$\xi_n \xrightarrow{L^p} \xi \Rightarrow \xi_n \xrightarrow{P} \xi, \quad p > 0, \tag{12}$$

$$\xi_n \xrightarrow{P} \xi \Rightarrow \xi_n \xrightarrow{d} \xi. \tag{13}$$

PROOF. Statement (11) follows from comparing the definition of convergence in probability with (5), and (12) follows from Chebyshev's inequality.

To prove (13), let $f(x)$ be a continuous function, let $|f(x)| \leq c$, let $\varepsilon > 0$, and let N be such that $P(|\xi| > N) \leq \varepsilon/(4c)$. Take δ so that $|f(x) - f(y)| \leq \varepsilon/2$ for $|x| < N$ and $|x - y| \leq \delta$. Then (cf. the "probabilistic" proof of Weierstrass's theorem in Subsection 5, Sect. 5, Chap. 1)

$$
\begin{aligned}
\mathsf{E}\,|f(\xi_n) - f(\xi)| &= \mathsf{E}(|f(\xi_n) - f(\xi)|; |\xi_n - \xi| \leq \delta, |\xi| \leq N) \\
&\quad + \mathsf{E}(|f(\xi_n) - f(\xi)|; |\xi_n - \xi| \leq \delta, |\xi| > N) \\
&\quad + \mathsf{E}(|f(\xi_n) - f(\xi)|; |\xi_n - \xi| > \delta) \\
&\leq \varepsilon/2 + \varepsilon/2 + 2c\,\mathsf{P}\{|\xi_n - \xi| > \delta\} \\
&= \varepsilon + 2c\,\mathsf{P}\{|\xi_n - \xi| > \delta\}.
\end{aligned}
$$

But $\mathsf{P}\{|\xi_n - \xi| > \delta\} \to 0$, and hence $\mathsf{E}\,|f(\xi_n) - f(\xi)| \leq 2\varepsilon$ for sufficiently large n; since $\varepsilon > 0$ is arbitrary, this establishes (13).

□

We now present a number of examples which show, in particular, that the converses of (11) and (12) are false in general.

Example 1. $(\xi_n \xrightarrow{P} \xi \nRightarrow \xi_n \xrightarrow{\text{a. s.}} \xi;\ \xi_n \xrightarrow{L^p} \xi \nRightarrow \xi_n \xrightarrow{\text{a. s.}} \xi.)$ Let $\Omega = [0, 1]$, $\mathscr{F} = \mathscr{B}([0, 1])$, $\mathsf{P} = $ Lebesgue measure. Put

$$A_n^i = \left[\frac{i - 1}{n}, \frac{i}{n}\right], \quad \xi_n^i = I_{A_n^i}(\omega), \quad i = 1, 2, \ldots, n;\ n \geq 1.$$

Then the sequence

$$\{\xi_1^1; \xi_2^1, \xi_2^2; \xi_3^1, \xi_3^2, \xi_3^3; \ldots\}$$

of random variables converges both in probability and in the mean of order $p > 0$, but does not converge at any point $\omega \in [0, 1]$.

Example 2. $(\xi_n \xrightarrow{P} \xi \Leftarrow \xi_n \xrightarrow{\text{a. s.}} \xi \nRightarrow \xi_n \xrightarrow{L^p} \xi, p > 0.)$ Again let $\Omega = [0, 1]$, $\mathscr{F} = \mathscr{B}[0, 1]$, $\mathsf{P} = $ Lebesgue measure, and let

$$\xi_n(\omega) = \begin{cases} e^n, & 0 \leq \omega \leq 1/n, \\ 0, & \omega > 1/n. \end{cases}$$

Then $\{\xi_n\}$ converges with probability 1 (and therefore in probability) to zero, but

$$\mathsf{E}\,|\xi_n|^p = \frac{e^{np}}{n} \to \infty, \quad n \to \infty,$$

for every $p > 0$.

Example 3. ($\xi_n \xrightarrow{L^p} \xi \nRightarrow \xi_n \xrightarrow{\text{a. s.}} \xi$.) Let $\{\xi_n\}$ be a sequence of independent random variables with

$$P(\xi_n = 1) = p_n, \quad P(\xi_n = 0) = 1 - p_n.$$

Then it is easy to show that

$$\xi_n \xrightarrow{P} 0 \Leftrightarrow p_n \to 0, \quad n \to \infty, \tag{14}$$

$$\xi_n \xrightarrow{L^p} 0 \Leftrightarrow p_n \to 0, \quad n \to \infty, \tag{15}$$

$$\xi_n \xrightarrow{\text{a. s.}} 0 \Leftrightarrow \sum_{n=1}^{\infty} p_n < \infty. \tag{16}$$

In particular, if $p_n = 1/n$ then $\xi_n \xrightarrow{L^p} 0$ for every $p > 0$, but $\xi_n \xrightarrow{\text{a. s.}} 0$.

The following theorem singles out an interesting case when almost sure convergence implies convergence in L^1.

Theorem 3. Let (ξ_n) be a sequence of nonnegative random variables such that $\xi_n \xrightarrow{\text{a. s.}} \xi$ and $\mathsf{E}\,\xi_n \to \mathsf{E}\,\xi < \infty$. Then

$$\mathsf{E}\,|\xi_n - \xi| \to 0, \quad n \to \infty. \tag{17}$$

PROOF. We have $\mathsf{E}\,\xi_n < \infty$ for sufficiently large n, and therefore for such n we have

$$\begin{aligned}
\mathsf{E}\,|\xi - \xi_n| &= \mathsf{E}(\xi - \xi_n)I_{\{\xi \geq \xi_n\}} + \mathsf{E}(\xi_n - \xi)I_{\{\xi_n > \xi\}} \\
&= 2\,\mathsf{E}(\xi - \xi_n)I_{\{\xi \geq \xi_n\}} + \mathsf{E}(\xi_n - \xi).
\end{aligned}$$

But $0 \leq (\xi - \xi_n)I_{\{\xi \geq \xi_n\}} \leq \xi$. Therefore, by the dominated convergence theorem, $\lim_n \mathsf{E}(\xi - \xi_n)I_{\{\xi \geq \xi_n\}} = 0$, which together with $\mathsf{E}\,\xi_n \to \mathsf{E}\,\xi$ proves (17).
□

Remark. The dominated convergence theorem also holds when almost sure convergence is replaced by convergence in probability (see Problem 1). Hence in Theorem 3, we may replace "$\xi_n \xrightarrow{\text{a. s.}} \xi$" by "$\xi_n \xrightarrow{P} \xi$."

5. It is shown in analysis that every fundamental sequence (x_n), $x_n \in R$, is convergent (Cauchy criterion). Let us give similar results for the convergence of a sequence of random variables.

Theorem 4 (Cauchy Criterion for Almost Sure Convergence). *A necessary and sufficient condition for the sequence $(\xi_n)_{n \geq 1}$ of random variables to converge with probability 1 (to a random variable ξ) is that it is fundamental with probability 1.*

PROOF. If $\xi_n \xrightarrow{\text{a. s.}} \xi$ then

$$\sup_{\substack{k \geq n \\ l \geq n}} |\xi_k - \xi_l| \leq \sup_{k \geq n} |\xi_k - \xi| + \sup_{l \geq n} |\xi_l - \xi|,$$

whence the necessity follows (see Theorem 1).

Now let $(\xi_n)_{n \geq 1}$ be fundamental with probability 1. Let $\mathcal{N} = \{\omega: (\xi_n(\omega))$ *is not fundamental*$\}$. Then whenever $\omega \in \Omega \backslash \mathcal{N}$ the sequence of numbers $(\xi_n(\omega))_{n \geq 1}$ is fundamental and, by Cauchy's criterion for sequences of numbers, $\lim \xi_n(\omega)$ exists. Let

$$\xi(\omega) = \begin{cases} \lim \xi_n(\omega), & \omega \in \Omega \backslash \mathcal{N}, \\ 0, & \omega \in \mathcal{N}. \end{cases} \tag{18}$$

The function so defined is a random variable, and evidently $\xi_n \overset{\text{a. s.}}{\to} \xi$.

This completes the proof.

$\square$

Before considering the case of convergence in probability, let us establish the following useful result.

Theorem 5. *If the sequence (ξ_n) is fundamental (or convergent) in probability, it contains a subsequence (ξ_{n_k}) that is fundamental (or convergent) with probability* 1.

PROOF. Let (ξ_n) be fundamental in probability. By Theorem 4 it is enough to show that it contains a subsequence that converges almost surely.

Take $n_1 = 1$ and define n_k inductively as the smallest $n > n_{k-1}$ for which

$$\mathsf{P}\{|\xi_t - \xi_s| > 2^{-k}\} < 2^{-k}$$

for all $s \geq n$, $t \geq n$. Then

$$\sum_k \mathsf{P}\{|\xi_{n_{k+1}} - \xi_{n_k}| > 2^{-k}\} < \sum 2^{-k} < \infty$$

and by the Borel–Cantelli lemma

$$\mathsf{P}\{|\xi_{n_{k+1}} - \xi_{n_k}| > 2^{-k} \text{ i.o.}\} = 0.$$

Hence

$$\sum_{k=1}^{\infty} |\xi_{n_{k+1}} - \xi_{n_k}| < \infty$$

with probability 1.

Let $\mathcal{N} = \{\omega: \sum |\xi_{n_{k+1}} - \xi_{n_k}| = \infty\}$. Then if we put

$$\xi(\omega) = \begin{cases} \xi_{n_1}(\omega) + \sum_{k=1}^{\infty} (\xi_{n_{k+1}}(\omega) - \xi_{n_k}(\omega)), & \omega \in \Omega \backslash \mathcal{N}, \\ 0, & \omega \in \mathcal{N}, \end{cases}$$

we obtain $\xi_{n_k} \overset{\text{a. s.}}{\to} \xi$.

If the original sequence converges in probability, then it is fundamental in probability (see also (19) below), and consequently this case reduces to the one already considered.

This completes the proof of the theorem.

$\square$

Theorem 6 (Cauchy Criterion for Convergence in Probability). *A necessary and sufficient condition for a sequence $(\xi_n)_{n \geq 1}$ of random variables to converge in probability is that it is fundamental in probability.*

PROOF. If $\xi_n \overset{P}{\to} \xi$ then

$$P\{|\xi_n - \xi_m| \geq \varepsilon\} \leq P\{|\xi_n - \xi| \geq \varepsilon/2\} + P\{|\xi_m - \xi| \geq \varepsilon/2\} \tag{19}$$

and consequently (ξ_n) is fundamental in probability.

Conversely, if (ξ_n) is fundamental in probability, by Theorem 5 there are a subsequence (ξ_{n_k}) and a random variable ξ such that $\xi_{n_k} \overset{a.s.}{\to} \xi$. But then

$$P\{|\xi_n - \zeta| \geq \varepsilon\} \leq P\{|\xi_n - \xi_{n_k}| \geq \varepsilon/2\} + P\{|\xi_{n_k} - \xi| \geq \varepsilon/2\},$$

from which it is clear that $\xi_n \overset{P}{\to} \xi$. This completes the proof.

□

Before discussing convergence in the mean of order p, we make some observations about L^p spaces.

We denote by $L^p = L^p(\Omega, \mathscr{F}, P)$ the space of random variables $\xi = \xi(\omega)$ with $E|\xi|^p \equiv \int_\Omega |\xi|^p \, dP < \infty$. Suppose that $p \geq 1$ and put

$$\|\xi\|_p = (E|\xi|^p)^{1/p}.$$

It is clear that

$$\|\xi\|_p \geq 0, \tag{20}$$

$$\|c\xi\|_p = |c| \, \|\xi\|_p, \qquad c \text{ constant}, \tag{21}$$

and by Minkowski's inequality (31), Sect. 6,

$$\|\xi + \eta\|_p \leq \|\xi\|_p + \|\eta\|_p. \tag{22}$$

Hence, in accordance with the usual terminology of functional analysis, the function $\| \cdot \|_p$, defined on L^p and satisfying (20)–(22), is (for $p \geq 1$) a *semi-norm*.

For it to be a *norm*, it must also satisfy

$$\|\xi\|_p = 0 \Rightarrow \xi = 0. \tag{23}$$

This property is, of course, not satisfied, since according to Property **H** (Sect. 6) we can only say that $\xi = 0$ *almost surely*.

This fact leads to a somewhat different view of the space L^p. That is, we connect with every random variable $\xi \in L^p$ the class $[\xi]$ of random variables in L^p that are equivalent to it (ξ and η are *equivalent* if $\xi = \eta$ almost surely). It is easily verified that the property of equivalence is *reflexive, symmetric,* and *transitive,* and consequently the linear space L^p can be divided into disjoint equivalence *classes* of random variables. If we now think of $[L^p]$ as the collection of the classes $[\xi]$ of equivalent random variables $\xi \in L^p$, and define

$$[\xi] + [\eta] = [\xi + \eta].$$
$$a[\xi] = [a\xi], \quad \text{where } a \text{ is a constant,}$$
$$\|[\xi]\|_p = \|\xi\|_p,$$

then $[L^p]$ becomes a normed linear space.

In functional analysis, we ordinarily describe elements of a space $[L^p]$ not as *equivalence classes of functions*, but simply as *functions*. In the same way we do not actually use the notation $[L^p]$. From now on, we no longer think about sets of equivalence classes of functions, but simply about elements, functions, random variables, and so on.

It is a basic result of functional analysis that the spaces L^p, $p \geq 1$, are *complete*, i.e., that every fundamental sequence has a limit. Let us state and prove this in probabilistic language.

Theorem 7 (Cauchy Test for Convergence in the pth Mean). *A necessary and sufficient condition that a sequence $(\xi_n)_{n \geq 1}$ of random variables in L^p converges in the mean of order p to a random variable in L^p is that the sequence is fundamental in the mean of order p.*

PROOF. The necessity follows from Minkowski's inequality. Let (ξ_n) be fundamental ($\|\xi_n - \xi_m\|_p \to 0$, $n, m \to \infty$). As in the proof of Theorem 5, we select a subsequence (ξ_{n_k}) such that $\xi_{n_k} \overset{\text{a. s.}}{\to} \xi$, where ξ is a random variable with $\|\xi\|_p < \infty$.

Let $n_1 = 1$ and define n_k inductively as the smallest $n > n_{k-1}$ for which

$$\|\xi_t - \xi_s\|_p < 2^{-2k}$$

for all $s \geq n$, $t \geq n$. Let

$$A_k = \{\omega : |\xi_{n_{k+1}} - \xi_{n_k}| \geq 2^{-k}\}.$$

Then by Chebyshev's inequality

$$P(A_k) \leq \frac{E\,|\xi_{n_{k+1}} - \xi_{n_k}|^p}{2^{-kp}} \leq \frac{2^{-2kp}}{2^{-kp}} = 2^{-kp} \leq 2^{-k}.$$

As in Theorem 5, we deduce that there is a random variable ξ such that $\xi_{n_k} \overset{\text{a. s.}}{\to} \xi$.

We now deduce that $\|\xi_n - \xi\|_p \to 0$ as $n \to \infty$. To do this, we fix $\varepsilon > 0$ and choose $N = N(\varepsilon)$ so that $\|\xi_n - \xi_m\|_p^p < \varepsilon$ for all $n \geq N$, $m \geq N$. Then for any fixed $n \geq N$, by Fatou's lemma (Sect. 6)

$$E\,|\xi_n - \xi|^p = E\left\{\lim_{n_k \to \infty} |\xi_n - \xi_{n_k}|^p\right\} = E\left\{\liminf_{n_k \to \infty} |\xi_n - \xi_{n_k}|^p\right\}$$
$$\leq \liminf_{n_k \to \infty} E\,|\xi_n - \xi_{n_k}|^p = \liminf_{n_k \to \infty} \|\xi_n - \xi_{n_k}\|_p^p \leq \varepsilon.$$

Consequently $E |\xi_n - \xi|^p \to 0$, $n \to \infty$. It is also clear that since $\xi = (\xi - \xi_n) + \xi_n$ we have $E |\xi|^p < \infty$ by Minkowski's inequality.

This completes the proof of the theorem.

□

Remark 1. In the terminology of functional analysis a complete normed linear space is called a *Banach space*. Thus L^p, $p \geq 1$, is a Banach space.

Remark 2. If $0 < p < 1$, the function $\|\xi\|_p = (E |\xi|^p)^{1/p}$ does not satisfy the triangle inequality (22) and consequently is not a norm. Nevertheless the space (of equivalence classes) L^p, $0 < p < 1$, is complete in the metric $d(\xi, \eta) \equiv E |\xi - \eta|^p$.

Remark 3. Let $L^\infty = L^\infty(\Omega, \mathscr{F}, P)$ be the space (of equivalence classes of) random variables $\xi = \xi(\omega)$ for which $\|\xi\|_\infty < \infty$, where $\|\xi\|_\infty$, the *essential supremum* of ξ, is defined by

$$\|\xi\|_\infty \equiv \text{ess sup } |\xi| \equiv \inf\{0 \leq c \leq \infty : P(|\xi| > c) = 0\}.$$

The function $\| \cdot \|_\infty$ is a norm, and L^∞ is complete in this norm.

6. PROBLEMS

1. Use Theorem 5 to show that almost sure convergence can be replaced by convergence in probability in Theorems 3 and 4 of Sect. 6

2. Prove that L^∞ is complete.

3. Show that if $\xi_n \xrightarrow{P} \xi$ and also $\xi_n \xrightarrow{P} \eta$ then ξ and η are equivalent ($P(\xi \neq \eta) = 0$).

4. Let $\xi_n \xrightarrow{P} \xi$, $\eta_n \xrightarrow{P} \eta$, and let ξ and η be equivalent ($P\{\xi \neq \eta\} = 0$). Show that

$$P\{|\xi_n - \eta_n| \geq \varepsilon\} \to 0, \quad n \to \infty,$$

for every $\varepsilon > 0$.

5. Let $\xi_n \xrightarrow{P} \xi$, $\eta_n \xrightarrow{P} \eta$. Show that if $\varphi = \varphi(x, y)$ is a continuous function, then $\varphi(\xi_n, \eta_n) \xrightarrow{P} \varphi(\xi, \eta)$ (*Slutsky's lemma*).

6. Let $(\xi_n - \xi)^2 \xrightarrow{P} 0$. Show that $\xi_n^2 \xrightarrow{P} \xi^2$.

7. Show that if $\xi_n \xrightarrow{d} C$, where C is a constant, then this sequence converges in probability:

$$\xi_n \xrightarrow{d} C \Rightarrow \xi_n \xrightarrow{P} C.$$

8. Let $(\xi_n)_{n \geq 1}$ have the property that $\sum_{n=1}^\infty E |\xi_n|^p < \infty$ for some $p > 0$. Show that $\xi_n \to 0$ (P-a. s.).

9. Let $(\xi_n)_{n \geq 1}$ be a sequence of identically distributed random variables. Show that

$$E\,|\xi_1| < \infty \Leftrightarrow \sum_{n=1}^{\infty} P\{|\xi_1| > \varepsilon n\} < \infty,\ \varepsilon > 0$$

$$\Leftrightarrow \sum_{n=1}^{\infty} P\left\{\left|\frac{\xi_n}{n}\right| > \varepsilon\right\} < \infty,\ \varepsilon > 0 \Rightarrow \frac{\xi_n}{n} \to 0\ (\text{P -a. s.}).$$

10. Let $(\xi_n)_{n \geq 1}$ be a sequence of random variables. Suppose that there are a random variable ξ and a sequence $\{n_k\}$ such that $\xi_{n_k} \to \xi$ (P-a. s.) and $\max_{n_{k-1} < l \leq n_k} |\xi_l - \xi_{n_k-1}| \to 0$ (P-a. s.) as $k \to \infty$. Show that then $\xi_n \to \xi$ (P-a. s.).

11. Let the d-metric on the set of random variables be defined by

$$d(\xi, \eta) = E\,\frac{|\xi - \eta|}{1 + |\xi - \eta|}$$

and identify random variables that coincide almost surely. Show that $d = d(\xi, \eta)$ is a well defined metric and that convergence in probability is equivalent to convergence in the d-metric.

12. Show that there is no metric on the set of random variables such that convergence in that metric is equivalent to almost sure convergence.

13. Let $X_1 \leq X_2 \leq \ldots$ and $X_n \xrightarrow{P} X$. Show that $X_n \to X$ (P-a. s.).

14. Let $X_n \to X$ (P-a. s.). Then also $n^{-1} \sum_{k=1}^{n} X_k \to X$ (P-a. s.) (*Cesàro summation*). Show by an example that convergence P-a. s. here cannot be replaced by convergence in probability.

15. Let $(\Omega, \mathscr{F}, P)$ be a probability space and $X_n \xrightarrow{P} X$. Show that if the measure P is atomic, then $X_n \to X$ also with probability one. (A set $A \in \mathscr{F}$ is an P-*atom* if for any $B \in \mathscr{F}$ either $P(B \cap A) = P(A)$ or $P(B \cap A) = 0$. The measure P is atomic if there exists a countable family $\{A_n\}$ of disjoint P-atoms such that $P\left(\bigcup_{n=1}^{\infty} A_n\right) = 1$.)

16. By the (first) Borel–Cantelli lemma, if $\sum_{n=1}^{\infty} P(|\xi_n| > \varepsilon) < \infty$ for any $\varepsilon > 0$, then $\xi_n \to 0$ (P-a. s.). Show by an example that convergence $\xi_n \to 0$ (P-a. s.) may hold also under the condition $\sum_{n=1}^{\infty} P(|\xi_n| > \varepsilon) = \infty, \varepsilon > 0$.

17. (*To the second Borel–Cantelli lemma.*) Let $\Omega = (0, 1)$, $\mathscr{B} = \mathscr{B}((0, 1))$, and P Lebesgue measure. Consider the events $A_n = (0, 1/n)$. Show that $\sum P(A_n) = \infty$, but each $\omega \in (0, 1)$ can belong only to finitely many sets $A_1, \ldots, A_{[1/\omega]}$, i. e. $P\{A_n \text{ i. o.}\} = 0$.

18. Give an example of a sequence of random variables such that $\lim \sup \xi_n = \infty$ and $\lim \inf \xi_n = -\infty$ with probability one, but nevertheless there is a random variable η such that $\xi_n \xrightarrow{P} \eta$.

19. Let Ω be an at most countable set. Prove that $\xi_n \xrightarrow{P} \xi$ implies $\xi_n \to \xi$ (P-a. s.).

20. Let $A_1, A_2, \ldots$ be independent events and $\sum_{n=1}^{\infty} P(A_n) < \infty$. Prove that $S_n = \sum_{k=1}^{n} I(A_k)$ fulfills the following extension of the second Borel–Cantelli lemma:

$$\lim_{n} \frac{S_n}{E\,S_n} = 1 \quad (\text{P-a. s.}).$$

21. Let $(X_n)_{n\geq 1}$ and $(Y_n)_{n\geq 1}$ be two sequences of random variables having the same finite-dimensional distributions $(F_{X_1,...,X_n} = F_{Y_1,...,Y_n}, n \geq 1)$. Suppose $X_n \xrightarrow{P} X$. Prove that Y_n then converges in probability, $Y_n \xrightarrow{P} Y$, to a random variable Y with the same distribution as X.

22. Let $(X_n)_{n\geq 1}$ be a sequence of independent random variables such that $X_n \xrightarrow{P} X$ for some random variable X. Prove that X is a degenerate random variable.

23. Show that for any sequence of random variables $\xi_1, \xi_2, \ldots$ there is a sequence of *constants* $a_1, a_2, \ldots$ such that $\xi_n/a_n \to 0$ (P-a. s.).

24. Let $\xi_1, \xi_2, \ldots$ be a sequence of random variables and $S_n = \xi_1 + \cdots + \xi_n$, $n \geq 1$. Show that the set $\{S_n \to \}$, i. e. the set of $\omega \in \Omega$ such that the series $\sum_{k\geq 1} \xi_k(\omega)$ converges, can be represented as

$$\{S_n \to \} = \bigcap_{N\geq 1} \bigcup_{m\geq 1} \bigcap_{k\geq m} \left\{ \sup_{l\geq k} |S_l - S_k| \leq N^{-1} \right\}.$$

Accordingly, the set $\{S_n \nrightarrow\}$, where the series $\sum_{k\geq 1} \xi_k(\omega)$ diverges, is representable as

$$\{S_n \nrightarrow\} = \bigcup_{N\geq 1} \bigcap_{m\geq 1} \bigcup_{k\geq m} \left\{ \sup_{l\geq k} |S_l - S_k| > N^{-1} \right\}.$$

25. Prove the following version of the second Borel–Cantelli lemma: Let $A_1, A_2, \ldots$ be arbitrary (not necessarily independent) events such that

$$\sum_{n=1}^{\infty} P(A_n) = \infty \quad \text{and} \quad \liminf_n \frac{\sum_{i,k\leq n} P(A_i \cap A_k)}{\left(\sum_{1<k\leq n} P(A_k) \right)^2} \leq 1;$$

then $P(A_n \text{ i. o.}) = 1$.

26. Show that in the Borel–Cantelli lemma it suffices to assume only *pairwise* independence of the events $A_1, A_2, \ldots$ instead of their independence.

27. Prove the following version of the *zero–one law* (cf. zero–one laws in Sect. 1 of Chapter 4, Vol. 2): if the events $A_1, A_2, \ldots$ are pairwise independent, then

$$P\{A_n \text{ i. o.}\} = \begin{cases} 0, & \text{if } \sum P(A_n) < \infty, \\ 1, & \text{if } \sum P(A_n) = \infty. \end{cases}$$

28. Let $A_1, A_2, \ldots$ be an arbitrary sequence of events such that $\lim_n P(A_n) = 0$ and $\sum_n P(A_n \cap \bar{A}_{n+1}) < \infty$. Prove that then $P\{A_n \text{ i. o.}\} = 0$.

29. Prove that if $\sum_n P\{|\xi_n| > n\} < \infty$, then $\limsup_n(|\xi_n|/n) \leq 1$ (P-a. s.).

30. Let $\xi_n \downarrow \xi$ (P-a. s.), $E|\xi_n| < \infty$, $n \geq 1$, and $\inf_n E\xi_n > -\infty$. Show that then $\xi_n \xrightarrow{L^1} \xi$, i. e. $E|\xi_n - \xi| \to 0$.

31. In connection with the Borel–Cantelli lemma, show that $P\{A_n \text{ i. o.}\} = 1$ if and only if $\sum_n P(A \cap A_n) = \infty$ for any set A with $P(A) > 0$.

32. Let $A_1, A_2, \ldots$ be independent events with $P(A_n) < 1$ for all $n \geq 1$. Then $P\{A_n \text{ i.o.}\} = 1$ if and only if $P(\bigcup A_n) = 1$.

33. Let $X_1, X_2, \ldots$ be independent random variables with $P\{X_n = 0\} = 1/n$ and $P\{X_n = 1\} = 1 - 1/n$. Let $E_n = \{X_n = 0\}$. Show that $\sum_{n=1}^{\infty} P(E_n) = \infty$, $\sum_{n=1}^{\infty} P(\overline{E}_n) = \infty$. Conclude from these that $\lim_n X_n$ does not exist (P-a.s.).

34. Let $X_1, X_2, \ldots$ be a sequence of random variables. Show that $X_n \xrightarrow{P} 0$ if and only if

$$\mathsf{E}\, \frac{|X_n|^r}{1 + |X_n|^r} \to 0 \quad \text{for some } r > 0.$$

In particular, if $S_n = X_1 + \cdots + X_n$, then

$$\frac{S_n - \mathsf{E}\, S_n}{n} \xrightarrow{P} 0 \Leftrightarrow \mathsf{E}\, \frac{(S_n - \mathsf{E}\, S_n)^2}{n^2 + (S_n - \mathsf{E}\, S_n)^2} \to 0.$$

Show that

$$\max_{1 \leq k \leq n} |X_k| \xrightarrow{P} 0 \Rightarrow \frac{S_n}{n} \xrightarrow{P} 0$$

for any sequence $X_1, X_2, \ldots$.

35. Let $X_1, X_2, \ldots$ be independent identically distributed Bernoulli random variables with $P\{X_k = \pm 1\} = 1/2$. Let $U_n = \sum_{k=1}^{n} (X_k / 2^k)$, $n \geq 1$. Show that $U_n \to U$ (P-a.s.), where U is a random variable uniformly distributed on $(-1, +1)$.

11 The Hilbert Space of Random Variables with Finite Second Moment

1. An important role among the Banach spaces L^p, $p \geq 1$, is played by the space $L^2 = L^2(\Omega, \mathscr{F}, P)$, the space of (equivalence classes of) random variables with finite second moments.

If ξ and $\eta \in L^2$, we put

$$(\xi, \eta) \equiv \mathsf{E}\, \xi \eta. \tag{1}$$

It is clear that if $\xi, \eta, \zeta \in L^2$ then

$$(a\xi + b\eta, \zeta) = a(\xi, \zeta) + b(\eta, \zeta), \quad a, b \in R,$$
$$(\xi, \xi) \geq 0$$

and

$$(\xi, \xi) = 0 \Leftrightarrow \xi = 0.$$

Consequently (ξ, η) is a *scalar product*. The space L^2 is *complete* with respect to the norm

$$\|\xi\| = (\xi, \xi)^{1/2} \tag{2}$$

induced by this scalar product (as was shown in Sect. 10). In accordance with the terminology of functional analysis, a space with the scalar product (1) is a *Hilbert space*.

Hilbert space methods are extensively used in probability theory to study properties that depend only on the first two moments of random variables ("L^2-theory"). Here we shall introduce the basic concepts and facts that will be needed for an exposition of L^2-theory (Chapter 6, Vol. 2).

2. Two random variables ξ and η in L^2 are said to be orthogonal ($\xi \perp \eta$) if $(\xi, \eta) \equiv \mathsf{E}\,\xi\eta = 0$. According to Sect. 8, ξ and η are uncorrelated if $\mathrm{Cov}(\xi, \eta) = 0$, i.e., if

$$\mathsf{E}\,\xi\eta = \mathsf{E}\,\xi\,\mathsf{E}\,\eta.$$

It follows that the properties of being orthogonal and of being uncorrelated coincide for random variables with zero mean values.

A set $M \subseteq L^2$ is a *system of orthogonal random variables* if $\xi \perp \eta$ for every $\xi, \eta \in M$ ($\xi \neq \eta$).

If also $\|\xi\| = 1$ for every $\xi \in M$, then M is an *orthonormal system*.

3. Let $M = \{\eta_1, \ldots, \eta_n\}$ be an orthonormal system and ξ any random variable in L^2. Let us find, in the class of linear estimators $\sum_{i=1}^{n} a_i \eta_i$, the best mean-square estimator for ξ (cf. Subsection 2 of Sect. 8).

A simple computation shows that

$$\mathsf{E}\left|\xi - \sum_{i=1}^{n} a_i \eta_i\right|^2 \equiv \left\|\xi - \sum_{i=1}^{n} a_i \eta_i\right\|^2 = \left(\xi - \sum_{i=1}^{n} a_i \eta_i, \xi - \sum_{i=1}^{n} a_i \eta_i\right)$$

$$= \|\xi\|^2 - 2\sum_{i=1}^{n} a_i(\xi, \eta_i) + \left(\sum_{i=1}^{n} a_i \eta_i, \sum_{i=1}^{n} a_i \eta_i\right)$$

$$= \|\xi\|^2 - 2\sum_{i=1}^{n} a_i(\xi, \eta_i) + \sum_{i=1}^{n} a_i^2$$

$$= \|\xi\|^2 - \sum_{i=1}^{n} |(\xi, \eta_i)|^2 + \sum_{i=1}^{n} |a_i - (\xi, \eta_i)|^2$$

$$\geq \|\xi\|^2 - \sum_{i=1}^{n} |(\xi, \eta_i)|^2, \tag{3}$$

where we used the equation

$$a_i^2 - 2a_i(\xi, \eta_i) = |a_i - (\xi, \eta_i)|^2 - |(\xi, \eta_i)|^2.$$

It is now clear that the infimum of $\mathsf{E}\,|\xi - \sum_{i=1}^{n} a_i \eta_i|^2$ over all real $a_1, \ldots, a_n$ is attained for $a_i = (\xi, \eta_i)$, $i = 1, \ldots, n$.

Consequently the best (in the mean-square sense) linear estimator for ξ in terms of $\eta_1, \ldots, \eta_n$ is

$$\hat{\xi} = \sum_{i=1}^{n} (\xi, \eta_i) \eta_i. \tag{4}$$

Here

$$\Delta \equiv \inf \mathsf{E} \left| \xi - \sum_{i=1}^{n} a_i \eta_i \right|^2 = \mathsf{E} |\xi - \hat{\xi}|^2 = \|\xi\|^2 - \sum_{i=1}^{n} |(\xi, \eta_i)|^2 \tag{5}$$

(compare (17), Sect. 4, Chap. 1 and (13), Sect. 8).

Inequality (3) also implies *Bessel's inequality*: if $M = \{\eta_1, \eta_2, \ldots\}$ is an orthonormal system and $\xi \in L^2$, then

$$\sum_{i=1}^{\infty} |(\xi, \eta_i)|^2 \le \|\xi\|^2; \tag{6}$$

and equality is attained if and only if

$$\xi = \operatorname*{l.i.m.}_{n} \sum_{i=1}^{n} (\xi, \eta_i) \eta_i. \tag{7}$$

The *best linear estimator* of ξ is often denoted by $\hat{\mathsf{E}}(\xi \mid \eta_1, \ldots, \eta_n)$ and called the *conditional expectation* (of ξ with respect to $\eta_1, \ldots, \eta_n$) in the *wide sense*.

The reason for the terminology is as follows. If we consider all estimators $\varphi = \varphi(\eta_1, \ldots, \eta_n)$ of ξ in terms of $\eta_1, \ldots, \eta_n$ (where φ is a Borel function), the best estimator will be $\varphi^* = \mathsf{E}(\xi \mid \eta_1, \ldots, \eta_n)$, i.e., the conditional expectation of ξ with respect to $\eta_1, \ldots, \eta_n$ (cf. Theorem 1, Sect. 8). Hence the best linear estimator is, by analogy, denoted by $\hat{\mathsf{E}}(\xi \mid \eta_1, \ldots, \eta_n)$ and called the conditional expectation in the wide sense. We note that if $\eta_1, \ldots, \eta_n$ form a Gaussian system (see Sect. 13 below), then $\mathsf{E}(\xi \mid \eta_1, \ldots, \eta_n)$ and $\hat{\mathsf{E}}(\xi \mid \eta_1, \ldots, \eta_n)$ are the same.

Let us discuss the *geometric meaning* of $\hat{\xi} = \hat{\mathsf{E}}(\xi \mid \eta_1, \ldots, \eta_n)$.

Let $\mathscr{L} = \mathscr{L}\{\eta_1, \ldots, \eta_n\}$ denote the *linear manifold* spanned by the orthonormal system of random variables $\eta_1, \ldots, \eta_n$ (i.e., the set of random variables of the form $\sum_{i=1}^{n} a_i \eta_i$, $a_i \in R$).

Then it follows from the preceding discussion that ξ admits the "orthogonal decomposition"

$$\xi = \hat{\xi} + (\xi - \hat{\xi}), \tag{8}$$

where $\hat{\xi} \in \mathscr{L}$ and $\xi - \hat{\xi} \perp \mathscr{L}$ in the sense that $\xi - \hat{\xi} \perp \lambda$ for every $\lambda \in \mathscr{L}$. It is natural to call $\hat{\xi}$ the *projection* of ξ on $\mathscr{L}$ (the element of $\mathscr{L}$ "closest" to ξ), and to say that $\xi - \hat{\xi}$ is *perpendicular* to $\mathscr{L}$.

4. The concept of orthonormality of the random variables $\eta_1, \ldots, \eta_n$ makes it easy to find the best linear estimator (the projection) $\hat{\xi}$ of ξ in terms of $\eta_1, \ldots, \eta_n$. The situation becomes more complicated if we give up the hypothesis of orthonormality.

However, the case of arbitrary $\eta_1, \ldots, \eta_n$ can in a certain sense be reduced to the case of orthonormal random variables, as will be shown below. We shall suppose for the sake of simplicity that all our random variables have zero mean values.

We shall say that the random variables $\eta_1, \ldots, \eta_n$ are *linearly independent* if the equation

$$\sum_{i=1}^{n} a_i \eta_i = 0 \quad \text{(P-a. s.)}$$

is satisfied only when all a_i are zero.

Consider the covariance matrix

$$\mathbb{R} = \mathsf{E}\,\eta\eta^*$$

of the vector $\eta = (\eta_1, \ldots, \eta_n)^*$, where $*$ denotes the transpose. It is symmetric and positive semi-definite, and as noticed in Sect. 8, can be diagonalized by an orthogonal matrix $\mathcal{O}$:

$$\mathcal{O}^* \mathbb{R} \mathcal{O} = D,$$

where

$$D = \begin{pmatrix} d_1 & & 0 \\ & \ddots & \\ 0 & & d_n \end{pmatrix}$$

has nonnegative elements d_i, the eigenvalues of $\mathbb{R}$, i.e., the zeros of the characteristic equation $\det(\mathbb{R} - \lambda E) = 0$, where E is the identity matrix.

If $\eta_1, \ldots, \eta_n$ are linearly independent, the Gram determinant $(\det \mathbb{R})$ is not zero and therefore $d_i > 0$. Let

$$B = \begin{pmatrix} \sqrt{d_1} & & 0 \\ & \ddots & \\ 0 & & \sqrt{d_n} \end{pmatrix}$$

and

$$\beta = B^{-1}\mathcal{O}^*\eta. \tag{9}$$

Then the covariance matrix of β is

$$\mathsf{E}\,\beta\beta^* = B^{-1}\mathcal{O}^*\,\mathsf{E}\,\eta\eta^*\mathcal{O}B^{-1} = B^{-1}\mathcal{O}^*\mathbb{R}\mathcal{O}B^{-1} = E,$$

and therefore $\beta = (\beta_1, \ldots, \beta_n)$ consists of uncorrelated random variables. It is also clear that

$$\eta = (\mathcal{O}B)\beta. \tag{10}$$

Consequently if $\eta_1, \ldots, \eta_n$ are linearly independent there is an orthonormal system such that (9) and (10) hold. Here

$$\mathscr{L}\{\eta_1, \ldots, \eta_n\} = \mathscr{L}\{\beta_1, \ldots, \beta_n\}.$$

This method of constructing an orthonormal system $\beta_1, \ldots, \beta_n$ is frequently inconvenient. The reason is that if we think of η_i as the value of the random sequence $(\eta_1, \ldots, \eta_n)$ at the instant i, the value β_i constructed above depends not only on the "past," $(\eta_1, \ldots, \eta_i)$, but also on the "future," $(\eta_{i+1}, \ldots, \eta_n)$. The *Gram–Schmidt orthogonalization process*, described below, does not have this defect, and moreover has the advantage that it can be applied to an infinite sequence of *linearly independent* random variables (i.e., to a sequence in which every finite set of the variables are linearly independent).

Let $\eta_1, \eta_2, \ldots$ be a sequence of linearly independent random variables in L^2. We construct a sequence $\varepsilon_1, \varepsilon_2, \ldots$ as follows. Let $\varepsilon_1 = \eta_1/\|\eta_1\|$. If $\varepsilon_1, \ldots, \varepsilon_{n-1}$ have been selected so that they are orthonormal, then

$$\varepsilon_n = \frac{\eta_n - \hat{\eta}_n}{\|\eta_n - \hat{\eta}_n\|}, \tag{11}$$

where $\hat{\eta}_n$ is the projection of η_n on the linear manifold $\mathcal{L}(\varepsilon_1, \ldots, \varepsilon_{n-1})$ generated by $(\eta_1, \ldots, \eta_{n-1})$:

$$\hat{\eta}_n = \sum_{k=1}^{n-1} (\eta_n, \varepsilon_k)\varepsilon_k. \tag{12}$$

Since $\eta_1, \ldots, \eta_n$ are linearly independent and $\mathcal{L}\{\eta_1, \ldots, \eta_{n-1}\} = \mathcal{L}\{\varepsilon_1, \ldots, \varepsilon_{n-1}\}$, we have $\|\eta_n - \hat{\eta}_n\| > 0$ and consequently ε_n is well defined.

By construction, $\|\varepsilon_n\| = 1$ for $n \geq 1$, and it is clear that $(\varepsilon_n, \varepsilon_k) = 0$ for $k < n$. Hence the sequence $\varepsilon_1, \varepsilon_2, \ldots$ is orthonormal. Moreover, by (11),

$$\eta_n = \hat{\eta}_n + b_n\varepsilon_n,$$

where $b_n = \|\eta_n - \hat{\eta}_n\|$ and $\hat{\eta}_n$ is defined by (12).

Now let $\eta_1, \ldots, \eta_n$ be any set of random variables (not necessarily linearly independent). Let $\det \mathbb{R} = 0$, where $\mathbb{R} \equiv \|r_{ij}\|$ is the covariance matrix of $(\eta_1, \ldots, \eta_n)$, and let

$$\text{rank } \mathbb{R} = r < n.$$

Then, from linear algebra, the quadratic form

$$Q(a) = \sum_{i,j=1}^{n} r_{ij}a_i a_j, \quad a = (a_1, \ldots, a_n),$$

has the property that there are $n - r$ linearly independent vectors $a^{(1)}, \ldots, a^{(n-r)}$ such that $Q(a^{(i)}) = 0$, $i = 1, \ldots, n - r$.

But

$$Q(a) = \mathsf{E}\left(\sum_{k=1}^{n} a_k\eta_k\right)^2.$$

Consequently

$$\sum_{k=1}^{n} a_k^{(i)} \eta_k = 0, \quad i = 1, \dots, n - r,$$

with probability 1.

In other words, there are $n - r$ linear relations among the variables $\eta_1, \dots, \eta_n$. Therefore if, for example, $\eta_1, \dots, \eta_r$ are linearly independent, the other variables $\eta_{r+1}, \dots, \eta_n$ can be expressed linearly in terms of them, so that $\mathscr{L}\{\eta_1, \dots, \eta_n\} = \mathscr{L}\{\eta_1, \dots, \eta_r\}$. Hence it is clear that by means of the orthogonalization process we can find r orthonormal random variables $\varepsilon_1, \dots, \varepsilon_r$ such that $\eta_1, \dots, \eta_n$ can be expressed linearly in terms of them and $\mathscr{L}\{\eta_1, \dots, \eta_n\} = \mathscr{L}\{\varepsilon_1, \dots, \varepsilon_r\}$.

5. Let $\eta_1, \eta_2, \dots$ be a sequence of random variables in L^2. Let $\mathscr{L} = \mathscr{L}\{\eta_1, \eta_2 \dots\}$ be the *linear manifold* spanned by $\eta_1, \eta_2, \dots$, i.e., the set of random variables of the form $\sum_{i=1}^{n} a_i \eta_i$, $n \geq 1$, $a_i \in R$. Denote $\overline{\mathscr{L}} = \overline{\mathscr{L}}\{\eta_1, \eta_2, \dots\}$ the *closed linear manifold* spanned by $\eta_1, \eta_2, \dots$, i.e., the set of random variables in $\mathscr{L}$ together with their mean-square limits.

We say that a set $\eta_1, \eta_2, \dots$ is a *countable orthonormal basis* (or a *complete orthonormal system*) in L^2 if:

(a) $\eta_1, \eta_2, \dots$ is an orthonormal system,
(b) $\overline{\mathscr{L}}\{\eta_1, \eta_2, \dots\} = L^2$.

A Hilbert space with a *countable* orthonormal basis is said to be separable.

By (b), for every $\xi \in L^2$ and a given $\varepsilon > 0$ there are numbers $a_1, \dots, a_n$ such that

$$\left\| \xi - \sum_{i=1}^{n} a_i \eta_i \right\| \leq \varepsilon.$$

Then by (3)

$$\left\| \xi - \sum_{i=1}^{n} (\xi, \eta_i) \eta_i \right\| \leq \varepsilon.$$

Consequently every element of a separable Hilbert space L^2 can be represented as

$$\xi = \sum_{i=1}^{\infty} (\xi, \eta_i) \eta_i, \tag{13}$$

or more precisely as

$$\xi = \text{l.i.m.}_{n} \sum_{i=1}^{n} (\xi, \eta_i) \eta_i.$$

We infer from this and (3) that *Parseval's equation* holds:

$$\|\xi\|^2 = \sum_{i=1}^{\infty} |(\xi, \eta_i)|^2, \quad \xi \in L^2. \tag{14}$$

It is easy to show that the converse is also valid: if $\eta_1, \eta_2, \ldots$ is an orthonormal system and either (13) or (14) is satisfied, then the system is a *basis*.

We now give some examples of separable Hilbert spaces and their bases.

Example 1. Let $\Omega = R$, $\mathscr{F} = \mathscr{B}(R)$, and let P be the Gaussian measure,

$$\mathsf{P}(-\infty, a] = \int_{-\infty}^{a} \varphi(x)\,dx, \quad \varphi(x) = \frac{1}{\sqrt{2\pi}}e^{-x^2/2}.$$

Let $D = d/dx$ and

$$H_n(x) = \frac{(-1)^n D^n \varphi(x)}{\varphi(x)}, \quad n \geq 0. \tag{15}$$

We find easily that

$$\begin{aligned}
D\varphi(x) &= -x\varphi(x), \\
D^2\varphi(x) &= (x^2 - 1)\varphi(x), \\
D^3\varphi(x) &= (3x - x^3)\varphi(x), \\
&\ldots\ldots\ldots\ldots\ldots\ldots\ldots
\end{aligned} \tag{16}$$

It follows that $H_n(x)$ are polynomials (the *Hermite polynomials*). From (15) and (16) we find that

$$\begin{aligned}
H_0(x) &= 1, \\
H_1(x) &= x, \\
H_2(x) &= x^2 - 1, \\
H_3(x) &= x^3 - 3x, \\
&\ldots\ldots\ldots\ldots\ldots
\end{aligned}$$

A simple calculation shows that

$$\begin{aligned}
(H_m, H_n) &= \int_{-\infty}^{\infty} H_m(x) H_n(x)\,\mathsf{P}(dx) \\
&= \int_{-\infty}^{\infty} H_m(x) H_n(x) \varphi(x)\,dx = n!\delta_{mn},
\end{aligned}$$

where δ_{mn} is the Kronecker delta $(0$, if $m \neq n$, and 1 if $m = n)$. Hence if we put

$$h_n(x) = \frac{H_n(x)}{\sqrt{n!}},$$

the system of *normalized Hermite polynomials* $\{h_n(x)\}_{n \geq 0}$ will be an orthonormal system. We know from functional analysis (see, e.g., [52], Chapter VII, Sect. 3) that if

$$\lim_{c \downarrow 0} \int_{-\infty}^{\infty} e^{c|x|}\,\mathsf{P}(dx) < \infty, \tag{17}$$

the system $\{1, x, x^2, \ldots\}$ is complete in L^2, i.e., every function $\xi = \xi(x)$ in L^2 can be represented either as $\sum_{i=1}^{n} a_i \eta_i(x)$, where $\eta_i(x) = x^i$, or as a limit of these functions (in the mean-square sense). If we apply the Gram–Schmidt orthogonalization process to the sequence $\eta_1(x), \eta_2(x), \ldots$ with $\eta_i(x) = x^i$, the resulting orthonormal system will be precisely the system of normalized Hermite polynomials. In the present case, (17) is satisfied. Hence $\{h_n(x)\}_{n\geq 0}$ is a basis and therefore every random variable $\xi = \xi(x)$ on this probability space can be represented in the form

$$\xi(x) = \underset{n}{\text{l.i.m.}} \sum_{i=0}^{n} (\xi, h_i) h_i(x). \tag{18}$$

Example 2. Let $\Omega = \{0, 1, 2, \ldots\}$ and let $P = \{P_1, P_2, \ldots\}$ be the Poisson distribution

$$P_x = \frac{e^{-\lambda}\lambda^x}{x!}, \quad x = 0, 1, \ldots; \quad \lambda > 0.$$

Put $\Delta f(x) = f(x) - f(x-1)$ $(f(x) = 0, x < 0)$, and by analogy with (15) define the *Poisson–Charlier polynomials*

$$\Pi_n(x) = \frac{(-1)^n \Delta^n P_x}{P_x}, \quad n \geq 1, \quad \Pi_0 = 1. \tag{19}$$

Since

$$(\Pi_m, \Pi_n) = \sum_{x=0}^{\infty} \Pi_m(x)\Pi_n(x)P_x = c_n \delta_{mn},$$

where c_n are positive constants, the system of *normalized Poisson–Charlier polynomials* $\{\pi_n(x)\}_{n\geq 0}$, $\pi_n(x) = \Pi_n(x)/\sqrt{c_n}$, is an orthonormal system, which is a *basis* since it satisfies (17).

Example 3. In this example we describe the Rademacher and Haar systems, which are of interest in function theory as well as in probability theory.

Let $\Omega = [0, 1]$, $\mathscr{F} = \mathscr{B}([0, 1])$, and let P be Lebesgue measure. As we mentioned in Sect. 1, every $x \in [0, 1]$ has a unique binary expansion

$$x = \frac{x_1}{2} + \frac{x_2}{2^2} + \cdots,$$

where $x_i = 0$ or 1. (To ensure uniqueness of the expansion, we agree to consider only expansions containing an *infinite* number of zeros. Thus out of the two expansions

$$\frac{1}{2} = \frac{1}{2} + \frac{0}{2^2} + \frac{0}{2^3} + \cdots = \frac{0}{2} + \frac{1}{2^2} + \frac{1}{2^3} + \cdots$$

we choose the first one.)

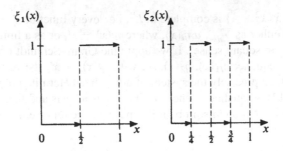

Fig. 30 Bernoulli variables

We define random variables $\xi_1(x)$, $\xi_2(x)$, ... by putting

$$\xi_n(x) = x_n.$$

Then for any numbers a_i, equal to 0 or 1,

$$\begin{aligned}
\mathsf{P}\{x\colon \xi_1 &= a_1, \ldots, \xi_n = a_n\} \\
&= \mathsf{P}\left\{x\colon \frac{a_1}{2} + \frac{a_2}{2^2} + \cdots + \frac{a_n}{2^n} \leq x < \frac{a_1}{2} + \frac{a_2}{2^2} + \cdots + \frac{a_n}{2^n} + \frac{1}{2^n}\right\} \\
&= \mathsf{P}\left\{x\colon x \in \left[\frac{a_1}{2} + \cdots + \frac{a_n}{2^n}, \frac{a_1}{2} + \cdots + \frac{a_n}{2^n} + \frac{1}{2^n}\right]\right\} = \frac{1}{2^n}.
\end{aligned}$$

It follows immediately that $\xi_1, \xi_2, \ldots$ form a *sequence of independent Bernoulli random variables* (Fig. 30 shows the construction of $\xi_1 = \xi_1(x)$ and $\xi_2 = \xi_2(x)$).

If we now set $R_n(x) = 1 - 2\xi_n(x)$, $n \geq 1$, it is easily verified that $\{R_n\}$ (the *Rademacher functions*, Fig. 31) are orthonormal:

$$\mathsf{E}\, R_n R_m = \int_0^1 R_n(x) R_m(x)\, dx - \delta_{nm}.$$

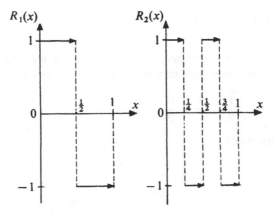

Fig. 31 Rademacher functions

Notice that $(1, R_n) \equiv \mathsf{E}\, R_n = 0$. It follows that this system is not complete.

However, the Rademacher system can be used to construct the *Haar system*, which also has a simple structure and is both *orthonormal* and *complete*.

Again let $\Omega = [0, 1)$ and $\mathscr{F} = \mathscr{B}([0, 1))$. Put

$$H_1(x) = 1,$$
$$H_2(x) = R_1(x)$$

. . . .

$$H_n(x) = \begin{cases} 2^{j/2} R_{j+1}(x) & \text{if } \frac{k-1}{2^j} \leq x < \frac{k}{2^j}, \ n = 2^j + k, \ 1 \leq k \leq 2^j, j \geq 1, \\ 0, & \text{otherwise.} \end{cases}$$

It is easy to see that $H_n(x), n \geq 3$, can also be written in the form

$$H_{2^m+1}(x) = \begin{cases} 2^{m/2}, & 0 \leq x < 2^{-(m+1)}, \\ -2^{m/2}, & 2^{-(m+1)} \leq x < 2^{-m}, \\ 0, & \text{otherwise,} \end{cases} \quad m = 1, 2, \ldots,$$

$$H_{2^m+j}(x) = H_{2^m+1}\left(x - \frac{j-1}{2^m}\right), \quad j = 1, \ldots, 2^m, \quad m = 1, 2, \ldots$$

Figure 32 shows graphs of the first eight functions, to give an idea of the structure of the Haar functions.

It is easy to see that the Haar system is orthonormal. Moreover, it is complete both in L^1 and in L^2, i.e., if $f = f(x) \in L^p$ for $p = 1$ or 2, then

$$\int_0^1 \left| f(x) - \sum_{k=1}^n (f, H_k) H_k(x) \right|^p dx \to 0, \quad n \to \infty.$$

The system also has the property that

$$\sum_{k=1}^n (f, H_k) H_k(x) \to f(x), \quad n \to \infty,$$

with probability 1 (with respect to Lebesgue measure).

In Sect. 4, Chap. 7, Vol. 2 we shall prove these facts by deriving them from general theorems on the convergence of martingales. This will, in particular, provide a good illustration of the application of martingale methods to the theory of functions.

6. If $\eta_1, \ldots, \eta_n$ is a finite orthonormal system then, as was shown above, for every random variable $\xi \in L^2$ there is a random variable $\hat{\xi}$ in the linear manifold $\mathscr{L} = \mathscr{L}\{\eta_1, \ldots, \eta_n\}$, namely the projection of ξ on $\mathscr{L}$, such that

$$\|\xi - \hat{\xi}\| = \inf\{\|\xi - \zeta\| : \zeta \in \mathscr{L}\{\eta_1, \ldots, \eta_n\}\}.$$

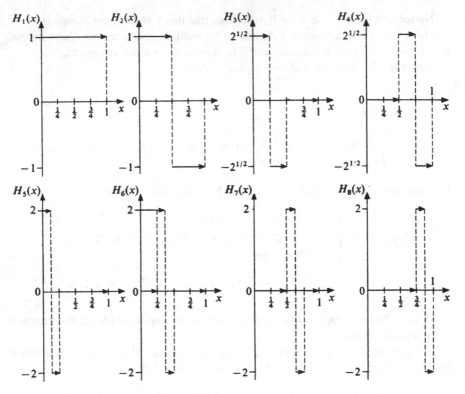

Fig. 32 The Haar functions $H_1(x), \ldots, H_8(x)$

Here $\hat{\xi} = \sum_{i=1}^{n}(\xi, \eta_i)\eta_i$. This result has a natural generalization to the case when $\eta_1, \eta_2, \ldots$ is a countable orthonormal system (not necessarily a basis). In fact, we have the following result.

Theorem. *Let $\eta_1, \eta_2, \ldots$ be an orthonormal system of random variables, and $\overline{\mathscr{L}} = \overline{\mathscr{L}}\{\eta_1, \eta_2, \ldots\}$ the closed linear manifold spanned by the system. Then there is a unique element $\hat{\xi} \in \overline{\mathscr{L}}$ such that*

$$\|\xi - \hat{\xi}\| = \inf\{\|\xi - \zeta\| : \zeta \in \overline{\mathscr{L}}\}. \tag{20}$$

Moreover,

$$\hat{\xi} = \underset{n}{\text{l.i.m.}} \sum_{i=1}^{n}(\xi, \eta_i)\eta_i \tag{21}$$

and $\xi - \hat{\xi} \perp \zeta$, $\zeta \in \overline{\mathscr{L}}$.

PROOF. Let $d = \inf\{\|\xi - \zeta\| : \zeta \in \overline{\mathscr{L}}\}$ and choose a sequence $\zeta_1, \zeta_2, \ldots$ such that $\|\xi - \zeta_n\| \to d$. Let us show that this sequence is fundamental. A simple calculation shows that

$$\|\zeta_n - \zeta_m\|^2 = 2\|\zeta_n - \xi\|^2 + 2\|\zeta_m - \xi\|^2 - 4\left\|\frac{\zeta_n + \zeta_m}{2} - \xi\right\|^2.$$

It is clear that $(\zeta_n + \zeta_m)/2 \in \overline{\mathscr{L}}$; consequently $\|[(\zeta_n + \zeta_m)/2] - \xi\|^2 \geq d^2$ and therefore $\|\zeta_n - \zeta_m\|^2 \to 0$, $n, m \to \infty$.

The space L^2 is complete (Theorem 7, Sect. 10). Hence there is an element $\hat{\xi}$ such that $\|\zeta_n - \hat{\xi}\| \to 0$. But $\overline{\mathscr{L}}$ is closed, so $\hat{\xi} \in \overline{\mathscr{L}}$. Moreover, $\|\zeta_n - \xi\| \to d$, and consequently $\|\xi - \hat{\xi}\| = d$, which establishes the existence of the required element.

Let us show that $\hat{\xi}$ is the only element of $\overline{\mathscr{L}}$ with the required property. Let $\tilde{\xi} \in \overline{\mathscr{L}}$ and let

$$\|\xi - \hat{\xi}\| = \|\xi - \tilde{\xi}\| = d.$$

Then (by Problem 3)

$$\|\hat{\xi} + \tilde{\xi} - 2\xi\|^2 + \|\hat{\xi} - \tilde{\xi}\|^2 = 2\|\hat{\xi} - \xi\|^2 + 2\|\tilde{\xi} - \xi\|^2 = 4d^2.$$

But

$$\|\hat{\xi} + \tilde{\xi} - 2\xi\|^2 = 4\|\tfrac{1}{2}(\hat{\xi} + \tilde{\xi}) - \xi\|^2 \geq 4d^2.$$

Consequently $\|\hat{\xi} - \tilde{\xi}\|^2 = 0$. This establishes the uniqueness of the element of $\overline{\mathscr{L}}$ that is closest to ξ.

Now let us show that $\xi - \hat{\xi} \perp \zeta, \zeta \in \overline{\mathscr{L}}$. By (20)

$$\|\xi - \hat{\xi} - c\zeta\| \geq \|\xi - \hat{\xi}\|$$

for every $c \in R$. But

$$\|\xi - \hat{\xi} - c\zeta\|^2 = \|\xi - \hat{\xi}\|^2 + c^2\|\zeta\|^2 - 2(\xi - \hat{\xi}, c\zeta).$$

Therefore

$$c^2\|\zeta\|^2 \geq 2(\xi - \hat{\xi}, c\zeta). \tag{22}$$

Take $c = \lambda(\xi - \hat{\xi}, \zeta)$, $\lambda \in R$. Then we find from (22) that

$$(\xi - \hat{\xi}, \zeta)^2[\lambda^2\|\zeta\|^2 - 2\lambda] \geq 0.$$

We have $\lambda^2\|\zeta\|^2 - 2\lambda < 0$ if λ is a sufficiently small positive number. Consequently $(\xi - \hat{\xi}, \zeta) = 0$, $\zeta \in \overline{\mathscr{L}}$.

It remains only to prove (21).

The set $\overline{\mathscr{L}} = \overline{\mathscr{L}}\{\eta_1, \eta_2, \ldots\}$ is a closed subspace of L^2 and therefore a Hilbert space (with the same scalar product). Now the system $\eta_1, \eta_2, \ldots$ is a basis for $\overline{\mathscr{L}}$ and consequently

$$\hat{\xi} = \text{l.i.m.}_n \sum_{k=1}^{n} (\hat{\xi}, \eta_k)\eta_k. \tag{23}$$

But $\xi - \hat{\xi} \perp \eta_k$, $k \geq 1$, and therefore $(\hat{\xi}, \eta_k) = (\xi, \eta_k)$, $k \geq 0$. This, with (23) establishes (21).

This completes the proof of the theorem.

$\square$

Remark. As in the finite-dimensional case, we say that $\hat{\xi}$ is the projection of ξ on $\mathscr{L} = \mathscr{L}\{\eta_1, \eta_2, \ldots\}$, that $\xi - \hat{\xi}$ is perpendicular to $\mathscr{L}$ and that the representation

$$\xi = \hat{\xi} + (\xi - \hat{\xi})$$

is the orthogonal decomposition of ξ.

We also denote $\hat{\xi}$ by $\hat{\mathsf{E}}(\xi \mid \eta_1, \eta_2, \ldots)$ (cf. $\hat{\mathsf{E}}(\xi \mid \eta_1, \ldots, \eta_n)$ in Subsection 3) and call it the *conditional expectation in the wide sense* (of ξ with respect to $\eta_1, \eta_2, \ldots$). From the point of view of estimating ξ in terms of $\eta_1, \eta_2, \ldots$, the variable $\hat{\xi}$ is the best linear estimator, with error

$$\Delta \equiv \mathsf{E}\,|\xi - \hat{\xi}|^2 \equiv \|\xi - \hat{\xi}\|^2 = \|\xi\|^2 - \sum_{i=1}^{\infty} |(\xi, \eta_i)|^2,$$

which follows from (5) and (23).

7. Problems

1. Show that if $\xi = \text{l.i.m.}\,\xi_n$ then $\|\xi_n\| \to \|\xi\|$.
2. Show that if $\xi = \text{l.i.m.}\,\xi_n$ and $\eta = \text{l.i.m.}\,\eta_n$ then $(\xi_n, \eta_n) \to (\xi, \eta)$.
3. Show that the norm $\|\cdot\|$ has the *parallelogram* property

$$\|\xi + \eta\|^2 + \|\xi - \eta\|^2 = 2(\|\xi\|^2 + \|\eta\|^2).$$

4. Let $(\xi_1, \ldots, \xi_n)$ be a family of orthogonal random variables. Show that they have the *Pythagorean property*,

$$\left\| \sum_{i=1}^{n} \xi_i \right\|^2 = \sum_{i=1}^{n} \|\xi_i\|^2.$$

5. Let $\xi_1, \xi_2, \ldots$ be a sequence of orthogonal random variables, $S_n = \xi_1 + \cdots + \xi_n$. Show that if $\sum_{n=1}^{\infty} \mathsf{E}\,\xi_n^2 < \infty$ then there is a random variable S with $\mathsf{E}\,S^2 < \infty$ such that $\text{l.i.m.}\,S_n = S$, i.e. $\|S_n - S\|^2 = \mathsf{E}\,|S_n - S|^2 \to 0, n \to \infty$.
6. Show that the Rademacher functions R_n can be defined as

$$R_n(x) = \text{sign}\,(\sin 2^n \pi x), \quad 0 \le x \le 1, \ n = 1, 2, \ldots$$

7. Prove that, for $\mathscr{G} \subseteq \mathscr{F}$,

$$\|\xi\| \ge \|\mathsf{E}(\xi \mid \mathscr{G})\| \quad \text{for } \xi \in L^2(\mathscr{F}),$$

where the equality holds if and only if $\xi = \mathsf{E}(\xi \mid \mathscr{G})$ a. s.
8. Prove that if $\xi, \eta \in L^2(\mathscr{F})$, $\mathsf{E}(\xi \mid \eta) = \eta$, and $\mathsf{E}(\eta \mid \xi) = \xi$, then $\xi = \eta$ a. s.

9. Suppose we are given three sequences $(\mathscr{G}_n^{(1)})$, $(\mathscr{G}_n^{(2)})$ and $(\mathscr{G}_n^{(3)})$ of σ-subalgebras of $\mathscr{F}$, and let ξ be a bounded random variable. Suppose we know that

$$\mathscr{G}_n^{(1)} \subseteq \mathscr{G}_n^{(2)} \subseteq \mathscr{G}_n^{(3)} \quad \text{for each } n,$$

$$\mathsf{E}(\xi \mid \mathscr{G}_n^{(1)}) \xrightarrow{\mathrm{P}} \eta, \quad \mathsf{E}(\xi \mid \mathscr{G}_n^{(3)}) \xrightarrow{\mathrm{P}} \eta.$$

Prove that $\mathsf{E}(\xi \mid \mathscr{G}_n^{(2)}) \xrightarrow{\mathrm{P}} \eta$.

12 Characteristic Functions

1. The method of characteristic functions is one of the main tools of the analytic theory of probability. This will appear very clearly in Chapter 3 in the proofs of limit theorems and, in particular, in the proof of the central limit theorem, which generalizes the de Moivre–Laplace theorem. In the present section we merely define characteristic functions and present their basic properties.

First we make some general remarks.

Besides random variables which take real values, the theory of characteristic functions requires random variables taking *complex values* (see Subsection 1 of Sect. 5).

Many definitions and properties involving random variables can easily be carried over to the complex case. For example, the expectation $\mathsf{E}\,\zeta$ of a complex random variable $\zeta = \xi + i\eta$ will exist if the expectations $\mathsf{E}\,\xi$ and $\mathsf{E}\,\eta$ exist. In this case we define $\mathsf{E}\,\zeta = \mathsf{E}\,\xi + i\,\mathsf{E}\,\eta$. It is easy to deduce from the Definition 6 (Sect. 5) of the independence of random elements that the complex random variables $\zeta_1 = \xi_1 + i\eta_1$ and $\zeta_2 = \xi_2 + i\eta_2$ are independent if and only if the pairs (ξ_1, η_1) and (ξ_2, η_2) are independent; or, equivalently, the σ-algebras $\mathscr{L}_{\xi_1, \eta_1}$ and $\mathscr{L}_{\xi_2, \eta_2}$ are independent.

Besides the space L^2 of real random variables with finite second moment, we shall consider the Hilbert space of complex random variables $\zeta = \xi + i\eta$ with $\mathsf{E}\,|\zeta|^2 < \infty$, where $|\zeta|^2 = \xi^2 + \eta^2$ and the scalar product (ζ_1, ζ_2) is defined by $\mathsf{E}\,\zeta_1 \bar{\zeta}_2$, where $\bar{\zeta}_2$ is the complex conjugate of ζ. The term "random variable" will now be used for both real and complex random variables, with a comment (when necessary) on which is intended.

Let us introduce some notation.

When a vector $a \in R^n$ is involved in algebraic operations, we consider it to be a *column vector*,

$$a = \begin{pmatrix} a_1 \\ \vdots \\ a_n \end{pmatrix},$$

and a^* to be a row vector, $a^* = (a_1, \ldots, a_n)$. If a and $b \in R^n$, their scalar product (a, b) is $\sum_{i=1}^n a_i b_i$. Clearly $(a, b) = a^* b$.

If $a \in R^n$ and $\mathbb{R} = \|r_{ij}\|$ is an n by n matrix,

$$(\mathbb{R}a, a) = a^* \mathbb{R}a = \sum_{i,j=1}^{n} r_{ij} a_i a_j. \tag{1}$$

2. **Definition 1.** Let $F = F(x)$ be an n-dimensional distribution function in $(R^n, \mathscr{B}(R^n))$, $x = (x_1, \ldots, x_n)^*$. Its *characteristic function* is

$$\varphi(t) = \int_{R^n} e^{i(t,x)} dF(x), \quad t \in R^n. \tag{2}$$

Definition 2. If $\xi = (\xi_1, \ldots, \xi_n)^*$ is a random vector defined on the probability space $(\Omega, \mathscr{F}, \mathsf{P})$ with values in R^n, its *characteristic function* is

$$\varphi_\xi(t) = \int_{R^n} e^{i(t,x)} dF_\xi(x), \quad t \in R^n, \tag{3}$$

where $F_\xi = F_\xi(x)$ is the distribution function of the vector $\xi = (\xi_1, \ldots, \xi_n)^*$, $x = (x_1, \ldots, x_n)^*$.

If $F(x)$ has a density $f = f(x)$ then

$$\varphi(t) = \int_{R^n} e^{i(t,x)} f(x)\, dx.$$

In other words, in this case the characteristic function is just the Fourier transform of $f(x)$.

It follows from (3) and Theorem 7 of Sect. 6 (on change of variable in a Lebesgue integral) that the characteristic function $\varphi_\xi(t)$ of a random vector can also be defined by

$$\varphi_\xi(t) = \mathsf{E}\, e^{i(t,\xi)}, \quad t \in R^n. \tag{4}$$

We now present some basic properties of characteristic functions, stated and proved for $n = 1$. Further important results for the general case will be given as problems.

Let $\xi = \xi(\omega)$ be a random variable, $F_\xi = F_\xi(x)$ its distribution function, and

$$\varphi_\xi(t) = \mathsf{E}\, e^{it\xi}$$

its characteristic function.

We see at once that if $\eta = a\xi + b$ then

$$\varphi_\eta(t) = \mathsf{E}\, e^{it\eta} = \mathsf{E}\, e^{it(a\xi+b)} = e^{itb}\, \mathsf{E}\, e^{iat\xi}.$$

Therefore

$$\varphi_\eta(t) = e^{itb} \varphi_\xi(at). \tag{5}$$

Moreover, if $\xi_1, \xi_2, \ldots, \xi_n$ are independent random variables and $S_n = \xi_1 + \cdots + \xi_n$, then

$$\varphi_{S_n}(t) = \prod_{j=1}^{n} \varphi_{\xi_j}(t). \qquad (6)$$

In fact,

$$\varphi_{S_n} = \mathsf{E}\, e^{it(\xi_1 + \cdots + \xi_n)} = \mathsf{E}\, e^{it\xi_1} \cdots e^{it\xi_n}$$

$$= \mathsf{E}\, e^{it\xi_1} \cdots \mathsf{E}\, e^{it\xi_n} = \prod_{j=1}^{n} \varphi_{\xi_j}(t),$$

where we have used the property that the expectation of a product of independent (bounded) random variables (either real or complex; see Theorem 6 of Sect. 6, and Problem 1) is equal to the product of their expectations.

Property (6) is the key to the proofs of limit theorems for sums of independent random variables by the method of characteristic functions (see Sect. 3, Chap. 3). In this connection we note that the distribution function F_{S_n} is expressed in terms of the distribution functions of the individual terms in a rather complicated way, namely $F_{S_n} = F_{\xi_1} * \cdots * F_{\xi_n}$, where $*$ denotes convolution (see Subsection 4 of Sect. 8).

Here are some examples of characteristic functions.

Example 1. The characteristic function of a Bernoulli random variable ξ with $\mathsf{P}(\xi = 1) = p$, $\mathsf{P}(\xi = 0) = q$, $p + q = 1$, $1 > p > 0$, is

$$\varphi_{\xi}(t) = pe^{it} + q.$$

If $\xi_1, \ldots, \xi_n$ are independent identically distributed random variables like ξ, then, writing $T_n = (S_n - np)/\sqrt{npq}$, we have

$$\varphi_{T_n}(t) = \mathsf{E}\, e^{iT_n t} = e^{-it\sqrt{np/q}}[pe^{it/\sqrt{npq}} + q]^n$$

$$= [pe^{it\sqrt{q/(np)}} + qe^{-it\sqrt{p/(nq)}}]^n. \qquad (7)$$

Notice that it follows that as $n \to \infty$

$$\varphi_{T_n}(t) \to e^{-t^2/2}, \quad T_n = \frac{S_n - np}{\sqrt{npq}}. \qquad (8)$$

Example 2. Let $\xi \sim \mathcal{N}(m, \sigma^2)$, $|m| < \infty$, $\sigma^2 > 0$. Let us show that

$$\varphi_{\xi}(t) = e^{itm - t^2 \sigma^2/2}. \qquad (9)$$

Let $\eta = (\xi - m)/\sigma$. Then $\eta \sim \mathcal{N}(0, 1)$ and, since

$$\varphi_{\xi}(t) = e^{itm}\varphi_{\eta}(\sigma t)$$

by (5), it is enough to show that

$$\varphi_\eta(t) = e^{-t^2/2}. \tag{10}$$

We have

$$\varphi_\eta(t) = \mathsf{E}\, e^{it\eta} = \frac{1}{\sqrt{2\pi}} \int_{-\infty}^{\infty} e^{itx} e^{-x^2/2}\, dx$$

$$= \frac{1}{\sqrt{2\pi}} \int_{-\infty}^{\infty} \sum_{n=0}^{\infty} \frac{(itx)^n}{n!} e^{-x^2/2}\, dx = \sum_{n=0}^{\infty} \frac{(it)^n}{n!} \frac{1}{\sqrt{2\pi}} \int_{-\infty}^{\infty} x^n e^{-x^2/2}\, dx$$

$$= \sum_{n=0}^{\infty} \frac{(it)^{2n}}{(2n)!} (2n-1)!! = \sum_{n=0}^{\infty} \frac{(it)^{2n}}{(2n)!} \frac{(2n)!}{2^n n!}$$

$$= \sum_{n=0}^{\infty} \left(-\frac{t^2}{2}\right)^n \cdot \frac{1}{n!} = e^{-t^2/2},$$

where we have used the formula (see Problem 7 in Sect. 8)

$$\frac{1}{\sqrt{2\pi}} \int_{-\infty}^{\infty} x^{2n} e^{-x^2/2}\, dx \equiv \mathsf{E}\, \eta^{2n} = (2n-1)!!.$$

Example 3. Let ξ be a Poisson random variable,

$$\mathsf{P}(\xi = k) = \frac{e^{-\lambda} \lambda^k}{k!}, \quad k = 0, 1, \ldots.$$

Then

$$\mathsf{E}\, e^{it\xi} = \sum_{k=0}^{\infty} e^{itk} \frac{e^{-\lambda} \lambda^k}{k!} = e^{-\lambda} \sum_{k=0}^{\infty} \frac{(\lambda e^{it})^k}{k!} = \exp\{\lambda(e^{it} - 1)\}. \tag{11}$$

3. As we observed in Subsection 1 of Sect. 9, with every distribution function in $(R, \mathscr{B}(R))$ we can associate a random variable of which it is the distribution function. Hence in discussing the properties of characteristic functions (in the sense either of Definition 1 or Definition 2), we may consider only characteristic functions $\varphi(t) = \varphi_\xi(t)$ of random variables $\xi = \xi(\omega)$.

Theorem 1. *Let ξ be a random variable with distribution function $F = F(x)$ and*

$$\varphi(t) = \mathsf{E}\, e^{it\xi}$$

its characteristic function. Then φ has the following properties:

(1) $|\varphi(t)| \leq \varphi(0) = 1$;
(2) $\varphi(t)$ *is uniformly continuous for $t \in R$;*
(3) $\varphi(t) = \overline{\varphi(-t)}$;
(4) $\varphi(t)$ *is real-valued if and only if F is symmetric ($\int_B dF(x) = \int_{-B} dF(x)$), $B \in \mathscr{B}(R)$, $-B = \{-x : x \in B\}$;*
(5) *if $\mathsf{E}\, |\xi|^n < \infty$ for some $n \geq 1$, then $\varphi^{(r)}(t)$ exists for every $r \leq n$, and*

$$\varphi^{(r)}(t) = \int_R (ix)^r e^{itx}\, dF(x), \tag{12}$$

$$\mathsf{E}\,\xi^r = \frac{\varphi^{(r)}(0)}{i^r}, \tag{13}$$

$$\varphi(t) = \sum_{r=0}^{n} \frac{(it)^r}{r!}\,\mathsf{E}\,\xi^r + \frac{(it)^n}{n!}\varepsilon_n(t), \tag{14}$$

where $|\varepsilon_n(t)| \le 3\,\mathsf{E}\,|\xi|^n$ and $\varepsilon_n(t) \to 0$, $t \to 0$;
(6) if $\varphi^{(2n)}(0)$ exists and is finite then $\mathsf{E}\,\xi^{2n} < \infty$;
(7) if $\mathsf{E}\,|\xi|^n < \infty$ for all $n \ge 1$ and

$$\limsup_n \frac{(\mathsf{E}\,|\xi|^n)^{1/n}}{n} = \frac{1}{T} < \infty,$$

then

$$\varphi(t) = \sum_{n=0}^{\infty} \frac{(it)^n}{n!}\,\mathsf{E}\,\xi^n \tag{15}$$

for all $|t| < T$.

PROOF. Properties (1) and (3) are evident. Property (2) follows from the inequality

$$|\varphi(t+h) - \varphi(t)| = |\,\mathsf{E}\,e^{it\xi}(e^{ih\xi} - 1)| \le \mathsf{E}\,|e^{ih\xi} - 1|$$

and the dominated convergence theorem, according to which $\mathsf{E}\,|e^{ih\xi} - 1| \to 0$ as $h \to 0$.

Property (4). Let F be symmetric. Then if $g(x)$ is a bounded odd Borel function, we have $\int_R g(x)\, dF(x) = 0$ (observe that for simple odd functions this follows directly from the definition of the symmetry of F). Consequently $\int_R \sin tx\, dF(x) = 0$ and therefore

$$\varphi(t) = \mathsf{E}\,\cos t\xi.$$

Conversely, let $\varphi_\xi(t)$ be a real function. Then by property (3)

$$\varphi_{-\xi}(t) = \varphi_\xi(-t) = \overline{\varphi_\xi(t)} = \varphi_\xi(t), \quad t \in R.$$

Hence (as will be shown below in Theorem 2) the distribution functions $F_{-\xi}$ and F_ξ of the random variables $-\xi$ and ξ are the same, and therefore (by Theorem 1 of Sect. 3)

$$P(\xi \in B) = P(-\xi \in B) = P(\xi \in -B)$$

for every $B \in \mathscr{B}(R)$.

Property (5). If $\mathsf{E}\,|\xi|^n < \infty$, we have $\mathsf{E}\,|\xi|^r < \infty$ for $r \le n$, by Lyapunov's inequality (28) (Sect. 6).

Consider the difference quotient

$$\frac{\varphi(t+h) - \varphi(t)}{h} = \mathsf{E}\, e^{it\xi} \left(\frac{e^{ih\xi} - 1}{h} \right).$$

Since

$$\left| \frac{e^{ihx} - 1}{h} \right| \le |x|,$$

and $\mathsf{E}\,|\xi| < \infty$, it follows from the dominated convergence theorem that the limit

$$\lim_{h \to 0} \mathsf{E}\, e^{it\xi} \left(\frac{e^{ih\xi} - 1}{h} \right)$$

exists and equals

$$\mathsf{E}\, e^{it\xi} \lim_{h \to 0} \left(\frac{e^{ih\xi} - 1}{h} \right) = i\,\mathsf{E}(\xi e^{it\xi}) = i \int_{-\infty}^{\infty} x e^{itx}\, dF(x). \tag{16}$$

Hence $\varphi'(t)$ exists and

$$\varphi'(t) = i(\mathsf{E}\,\xi e^{it\xi}) = i \int_{-\infty}^{\infty} x e^{itx}\, dF(x).$$

The existence of the derivatives $\varphi^{(r)}(t)$, $1 < r \le n$, and the validity of (12), follow by induction.

Formula (13) follows immediately from (12). Let us now establish (14).
Since

$$e^{iy} = \cos y + i \sin y = \sum_{k=0}^{n-1} \frac{(iy)^k}{k!} + \frac{(iy)^n}{n!} [\cos \theta_1 y + i \sin \theta_2 y]$$

for real y, with $|\theta_1| \le 1$ and $|\theta_2| \le 1$, we have

$$e^{it\xi} = \sum_{k=0}^{n-1} \frac{(it\xi)^k}{k!} + \frac{(it\xi)^n}{n!} [\cos \theta_1(\omega) t\xi + i \sin \theta_2(\omega) t\xi] \tag{17}$$

and

$$\mathsf{E}\, e^{it\xi} = \sum_{k=0}^{n-1} \frac{(it)^k}{k!} \mathsf{E}\,\xi^k + \frac{(it)^n}{n!} [\mathsf{E}\,\xi^n + \varepsilon_n(t)], \tag{18}$$

where

$$\varepsilon_n(t) = \mathsf{E}[\xi^n(\cos\theta_1(\omega)t\xi + i\sin\theta_2(\omega)t\xi - 1)].$$

It is clear that $|\varepsilon_n(t)| \le 3\,\mathsf{E}\,|\xi^n|$. The theorem on dominated convergence shows that $\varepsilon_n(t) \to 0$, $t \to 0$.

Property (6). We give a proof by induction. Suppose first that $\varphi''(0)$ exists and is finite. Let us show that in that case $\mathsf{E}\,\xi^2 < \infty$. By L'Hôpital's rule and Fatou's lemma,

$$\varphi''(0) = \lim_{h\to 0} \frac{1}{2}\left[\frac{\varphi'(2h) - \varphi'(0)}{2h} + \frac{\varphi'(0) - \varphi'(-2h)}{2h}\right]$$

$$= \lim_{h\to 0} \frac{2\varphi'(2h) - 2\varphi'(-2h)}{8h} = \lim_{h\to 0} \frac{1}{4h^2}[\varphi(2h) - 2\varphi(0) + \varphi(-2h)]$$

$$= \lim_{h\to 0} \int_{-\infty}^{\infty} \left(\frac{e^{ihx} - e^{-ihx}}{2h}\right)^2 dF(x)$$

$$= -\lim_{h\to 0} \int_{-\infty}^{\infty} \left(\frac{\sin hx}{hx}\right)^2 x^2\,dF(x) \le -\int_{-\infty}^{\infty} \lim_{h\to 0} \left(\frac{\sin hx}{hx}\right)^2 x^2\,dF(x)$$

$$= -\int_{-\infty}^{\infty} x^2\,dF(x).$$

Therefore,

$$\int_{-\infty}^{\infty} x^2\,dF(x) \le -\varphi''(0) < \infty.$$

Now let $\varphi^{(2k+2)}(0)$ exist, finite, and let $\int_{-\infty}^{+\infty} x^{2k}\,dF(x) < \infty$. If $\int_{-\infty}^{\infty} x^{2k}\,dF(x) = 0$, then $\int_{-\infty}^{\infty} x^{2k+2}\,dF(x) = 0$ also. Hence we may suppose that $\int_{-\infty}^{\infty} x^{2k}\,dF(x) > 0$. Then, by Property (5),

$$\varphi^{(2k)}(t) = \int_{-\infty}^{\infty} (ix)^{2k} e^{itx}\,dF(x)$$

and therefore,

$$(-1)^k \varphi^{(2k)}(t) = \int_{-\infty}^{\infty} e^{itx}\,dG(x),$$

where $G(x) = \int_{-\infty}^{x} u^{2k}\,dF(u)$.

Consequently the function $(-1)^k \varphi^{(2k)}(t)G^{-1}(\infty)$ is the characteristic function of the probability distribution $G(x) \cdot G^{-1}(\infty)$ and by what we have proved,

$$G^{-1}(\infty) \int_{-\infty}^{\infty} x^2\,dG(x) < \infty.$$

But $G^{-1}(\infty) > 0$, and therefore

$$\int_{-\infty}^{\infty} x^{2k+2}\,dF(x) = \int_{-\infty}^{\infty} x^2\,dG(x) < \infty.$$

Property (7). Let $0 < t_0 < T$. Then, by Stirling's formula (6) (Sect. 2 of Chap. 1) we find that

$$\limsup \frac{(\mathsf{E}\,|\xi|^n)^{1/n}}{n} < \frac{1}{t_0} \Rightarrow \limsup \frac{(\mathsf{E}\,|\xi|^n t_0^n)^{1/n}}{n} < 1$$

$$\Rightarrow \limsup \left(\frac{\mathsf{E}\,|\xi|^n t_0^n}{n!} \right)^{1/n} < 1.$$

Consequently the series $\sum [\mathsf{E}\,|\xi|^n t_0^n / n!]$ converges by Cauchy's test, and therefore the series $\sum_{r=0}^{\infty} [(it)^r / r!]\,\mathsf{E}\,\xi^r$ converges for $|t| \le t_0$. But by (14), for $n \ge 1$,

$$\varphi(t) = \sum_{r=0}^{n} \frac{(it)^r}{r!}\,\mathsf{E}\,\xi^r + R_n(t),$$

where $|R_n(t)| \le 3(|t|^n / n!)\,\mathsf{E}\,|\xi|^n$. Therefore

$$\varphi(t) = \sum_{r=0}^{\infty} \frac{(it)^r}{r!}\,\mathsf{E}\,\xi^r$$

for all $|t| < T$. This completes the proof of the theorem.
□

Remark 1. By a method similar to that used for (14), we can establish that if $\mathsf{E}\,|\xi|^n < \infty$ for some $n \ge 1$, then

$$\varphi(t) = \sum_{k=0}^{n} \frac{i^k (t-s)^k}{k!} \int_{-\infty}^{\infty} x^k e^{isx}\,dF(x) + \frac{i^n (t-s)^n}{n!}\,\varepsilon_n(t-s), \qquad (19)$$

where $|\varepsilon_n(t-s)| \le 3\,\mathsf{E}\,|\xi^n|$, and $\varepsilon_n(t-s) \to 0$ as $t - s \to 0$.

Remark 2. With reference to the condition that appears in Property (7), see also Subsection 9, below, on the "uniqueness of the solution of the moment problem."

4. The following theorem shows that the characteristic function is uniquely determined by the distribution function.

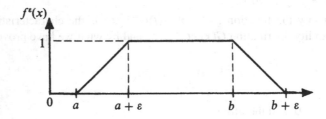

Fig. 33

Theorem 2 (Uniqueness). *Let F and G be distribution functions with the same characteristic function, i.e.*

$$\int_{-\infty}^{\infty} e^{itx}\, dF(x) = \int_{-\infty}^{\infty} e^{itx}\, dG(x) \tag{20}$$

for all $t \in R$. Then $F(x) \equiv G(x)$.

PROOF. Choose a and $b \in R$, and $\varepsilon > 0$, and consider the function $f^\varepsilon = f^\varepsilon(x)$ shown in Fig. 33. We show that

$$\int_{-\infty}^{\infty} f^\varepsilon(x)\, dF(x) = \int_{-\infty}^{\infty} f^\varepsilon(x)\, dG(x). \tag{21}$$

Let $n \geq 0$ be large enough so that $[a, b + \varepsilon] \subseteq [-n, n]$, and let the sequence $\{\delta_n\}$ be such that $1 \geq \delta_n \downarrow 0$, $n \to \infty$. Like every continuous function on $[-n, n]$ that has equal values at the endpoints, $f^\varepsilon = f^\varepsilon(x)$ can be uniformly approximated by trigonometric polynomials (the Weierstrass–Stone theorem, see [28]), i.e., there is a *finite* sum

$$f_n^\varepsilon(x) = \sum_k a_k \exp\left(i\pi x \frac{k}{n}\right) \tag{22}$$

such that

$$\sup_{-n \leq x \leq n} |f^\varepsilon(x) - f_n^\varepsilon(x)| \leq \delta_n. \tag{23}$$

Let us extend the periodic function $f_n^\varepsilon(x)$ to all of R, and observe that

$$\sup_x |f_n^\varepsilon(x)| \leq 2.$$

Then, since by (20) and (22)

$$\int_{-\infty}^{\infty} f_n^\varepsilon(x)\, dF(x) = \int_{-\infty}^{\infty} f_n^\varepsilon(x)\, dG(x),$$

we have

$$\left| \int_{-\infty}^{\infty} f^\varepsilon(x)\, dF(x) - \int_{-\infty}^{\infty} f^\varepsilon(x)\, dG(x) \right| = \left| \int_{-n}^{n} f^\varepsilon\, dF - \int_{-n}^{n} f^\varepsilon\, dG \right|$$

$$\leq \left| \int_{-n}^{n} f_n^\varepsilon\, dF - \int_{-n}^{n} f_n^\varepsilon\, dG \right| + 2\delta_n$$

$$\leq \left| \int_{-\infty}^{\infty} f_n^\varepsilon\, dF - \int_{-\infty}^{\infty} f_n^\varepsilon\, dG \right| + 2\delta_n$$

$$+ 2F(\overline{[-n, n]}) + 2G(\overline{[-n, n]}),$$

$$\tag{24}$$

where $F(A) = \int_A dF(x)$, $G(A) = \int_A dG(x)$. As $n \to \infty$, the right-hand side of (24) tends to zero, and this establishes (21).

As $\varepsilon \to 0$, we have $f^\varepsilon(x) \to I_{(a,b]}(x)$. It follows from (21) by the dominated convergence theorem that

$$\int_{-\infty}^{\infty} I_{(a,b]}(x)\, dF(x) = \int_{-\infty}^{\infty} I_{(a,b]}(x)\, dG(x),$$

i.e., $F(b) - F(a) = G(b) - G(a)$. Since a and b are arbitrary, it follows that $F(x) = G(x)$ for all $x \in R$.

This completes the proof of the theorem.

$\square$

5. The preceding theorem says that a distribution function $F = F(x)$ is uniquely determined by its characteristic function $\varphi = \varphi(t)$. The next theorem gives an explicit representation of F in terms of φ.

Theorem 3 (Inversion Formula). *Let $F = F(x)$ be a distribution function and*

$$\varphi(t) = \int_{-\infty}^{\infty} e^{itx}\, dF(x)$$

its characteristic function.

(a) *For any pair of points a and b ($a < b$) at which $F = F(x)$ is continuous,*

$$F(b) - F(a) = \lim_{c \to \infty} \frac{1}{2\pi} \int_{-c}^{c} \frac{e^{-ita} - e^{-itb}}{it}\, \varphi(t)\, dt. \tag{25}$$

(b) *If $\int_{-\infty}^{\infty} |\varphi(t)|\, dt < \infty$, the distribution function $F(x)$ has a density $f(x)$,*

$$F(x) = \int_{-\infty}^{x} f(y)\, dy \tag{26}$$

and

$$f(x) = \frac{1}{2\pi} \int_{-\infty}^{\infty} e^{-itx} \varphi(t)\, dt. \tag{27}$$

PROOF. We first observe that if $F(x)$ has density $f(x)$ then

$$\varphi(t) = \int_{-\infty}^{\infty} e^{itx} f(x)\, dx, \tag{28}$$

and (27) is just the Fourier transform of the (integrable) function $\varphi(t)$. Integrating both sides of (27) and applying Fubini's theorem, we obtain

$$F(b) - F(a) = \int_a^b f(x)\,dx = \frac{1}{2\pi} \int_a^b \left[\int_{-\infty}^{\infty} e^{-itx} \varphi(t)\,dt \right] dx$$

$$= \frac{1}{2\pi} \int_{-\infty}^{\infty} \varphi(t) \left[\int_a^b e^{-itx}\,dx \right] dt$$

$$= \frac{1}{2\pi} \int_{-\infty}^{\infty} \varphi(t) \frac{e^{-ita} - e^{-itb}}{it}\,dt.$$

After these remarks, which to some extent clarify (25), we turn to the proof.

(a) We have

$$\Phi_c \equiv \frac{1}{2\pi} \int_{-c}^{c} \frac{e^{-ita} - e^{-itb}}{it} \varphi(t)\,dt$$

$$= \frac{1}{2\pi} \int_{-c}^{c} \frac{e^{-ita} - e^{-itb}}{it} \left[\int_{-\infty}^{\infty} e^{itx}\,dF(x) \right] dt$$

$$= \frac{1}{2\pi} \int_{-\infty}^{\infty} \left[\int_{-c}^{c} \frac{e^{-ita} - e^{-itb}}{it} e^{itx}\,dt \right] dF(x)$$

$$= \int_{-\infty}^{\infty} \Psi_c(x)\,dF(x), \tag{29}$$

where we have put

$$\Psi_c(x) = \frac{1}{2\pi} \int_{-c}^{c} \frac{e^{-ita} - e^{-itb}}{it} e^{itx}\,dt$$

and applied Fubini's theorem, which is applicable in this case because

$$\left| \frac{e^{-ita} - e^{-itb}}{it} \cdot e^{itx} \right| = \left| \frac{e^{-ita} - e^{-itb}}{it} \right| = \left| \int_a^b e^{-itx}\,dx \right| \le b - a$$

and

$$\int_{-c}^{c} \int_{-\infty}^{\infty} (b - a)\,dF(x)\,dt \le 2c(b - a) < \infty.$$

In addition,

$$\Psi_c(x) = \frac{1}{2\pi} \int_{-c}^{c} \frac{\sin t(x - a) - \sin t(x - b)}{t}\,dt$$

$$= \frac{1}{2\pi} \int_{-c(x-a)}^{c(x-a)} \frac{\sin v}{v}\,dv - \frac{1}{2\pi} \int_{-c(x-b)}^{c(x-b)} \frac{\sin u}{u}\,du. \tag{30}$$

The function

$$g(s, t) = \int_s^t \frac{\sin v}{v}\,dv$$

is uniformly continuous in s and t, and

$$g(s, t) \to \pi \tag{31}$$

as $s \downarrow -\infty$ and $t \uparrow \infty$ (see [35], 3.721 or "Dirichlet integral" in Wikipedia). Hence there is a constant C such that $|\Psi_c(x)| < C < \infty$ for all c and x. Moreover, it follows from (30) and (31) that

$$\Psi_c(x) \to \Psi(x), \quad c \to \infty,$$

where

$$\Psi(x) = \begin{cases} 0, & x < a, x > b, \\ \frac{1}{2}, & x = a, x = b, \\ 1, & a < x < b. \end{cases}$$

Let μ be the measure on $(R, \mathscr{B}(R))$ such that $\mu(a, b] = F(b) - F(a)$. Recall that by assumption a and b are continuity points of $F(x)$, hence $F(a-) = F(a)$, $F(b-) = F(b)$ and $\mu(a) = \mu(b) = 0$. Then if we apply the dominated convergence theorem and use the formulas of Problem 1 of Sect. 3, we find that, as $c \to \infty$,

$$\Phi_c = \int_{-\infty}^{\infty} \Psi_c(x) \, dF(x) \to \int_{-\infty}^{\infty} \Psi(x) \, dF(x)$$

$$= \mu(a, b) + \tfrac{1}{2}\mu\{a\} + \tfrac{1}{2}\mu\{b\} = F(b) - F(a).$$

Hence (25) is established.

(b) Let $\int_{-\infty}^{\infty} |\varphi(t)| \, dt < \infty$. Write

$$f(x) = \frac{1}{2\pi} \int_{-\infty}^{\infty} e^{-itx} \varphi(t) \, dt.$$

It follows from the dominated convergence theorem that this is a continuous function of x and therefore is integrable on $[a, b]$. Consequently we find, applying Fubini's theorem again, that

$$\int_a^b f(x) \, dx = \int_a^b \frac{1}{2\pi} \left(\int_{-\infty}^{\infty} e^{-itx} \varphi(t) \, dt \right) dx$$

$$= \frac{1}{2\pi} \int_{-\infty}^{\infty} \varphi(t) \left[\int_a^b e^{-itx} \, dx \right] dt = \lim_{c \to \infty} \frac{1}{2\pi} \int_{-c}^c \varphi(t) \left[\int_a^b e^{-itx} dx \right] dt$$

$$= \lim_{c \to \infty} \frac{1}{2\pi} \int_{-c}^c \frac{e^{-ita} - e^{itb}}{it} \varphi(t) \, dt = F(b) - F(a)$$

for all points a and b of continuity of $F(x)$.

Hence it follows that

$$F(x) = \int_{-\infty}^{x} f(y)\, dy, \quad x \in R,$$

and since $f(x)$ is continuous and $F(x)$ is nondecreasing, $f(x)$ is the density of $F(x)$. This completes the proof of the theorem.

□

Remark. The inversion formula (25) provides a second proof of Theorem 2.

Theorem 4. *A necessary and sufficient condition for the components of the random vector* $\xi = (\xi_1, \ldots, \xi_n)^*$ *to be independent is that its characteristic function is the product of the characteristic functions of the components:*

$$\mathsf{E}\, e^{i(t_1\xi_1 + \cdots + t_n\xi_n)} = \prod_{k=1}^{n} \mathsf{E}\, e^{it_k\xi_k}, \quad (t_1, \ldots, t_n)^* \in R^n.$$

PROOF. The necessity follows from Problem 1. To prove the sufficiency we let $F(x_1, \ldots, x_n)$ be the distribution function of the vector $\xi = (\xi_1, \ldots, \xi_n)^*$ and $F_k(x)$, the distribution functions of the ξ_k, $1 \le k \le n$. Put $G = G(x_1, \ldots, x_n) = F_1(x_1) \cdots F_n(x_n)$. Then, by Fubini's theorem, for all $(t_1, \ldots, t_n)^* \in R^n$

$$\int_{R^n} e^{i(t_1 x_1 + \cdots + t_n x_n)}\, dG(x_1, \ldots, x_n) = \prod_{k=1}^{n} \int_{R} e^{it_k x_k}\, dF_k(x_k)$$

$$= \prod_{k=1}^{n} \mathsf{E}\, e^{it_k\xi_k} = \mathsf{E}\, e^{i(t_1\xi_1 + \cdots + t_n\xi_n)}$$

$$= \int_{R^n} e^{i(t_1 x_1 + \cdots + t_n x_n)}\, dF(x_1, \ldots, x_n).$$

Therefore by Theorem 2 (or rather, by its multidimensional analog; see Problem 3) we have $F = G$, and consequently, by the theorem of Sect. 5, the random variables $\xi_1, \ldots, \xi_n$ are independent.

□

6. Theorem 1 gives us necessary conditions for a function to be a characteristic function. Hence if $\varphi = \varphi(t)$ fails to satisfy, for example, one of the first three conclusions of the theorem, that function cannot be a characteristic function. We quote without proof some results in the same direction.

Bochner–Khinchin Theorem. *Let* $\varphi(t)$ *be continuous,* $t \in R$, *with* $\varphi(0) = 1$. *A necessary and sufficient condition that* $\varphi(t)$ *is a characteristic function is that it is positive semi-definite, i.e., that for all real* $t_1, \ldots, t_n$ *and all complex* $\lambda_1, \ldots, \lambda_n$, $n = 1, 2, \ldots$,

$$\sum_{i,j=1}^{n} \varphi(t_i - t_j)\lambda_i \bar{\lambda}_j \ge 0. \tag{32}$$

The necessity of (32) is evident since if $\varphi(t) = \int_{-\infty}^{\infty} e^{itx} \, dF(x)$ then

$$\sum_{i,j=1}^{n} \varphi(t_i - t_j)\lambda_i\bar{\lambda}_j = \int_{-\infty}^{\infty} \left| \sum_{k=1}^{n} \lambda_k e^{it_k x} \right|^2 dF(x) \geq 0.$$

The proof of the sufficiency of (32) is more difficult. (See [31], XIX.2.)

Pólya's Theorem. *Let a continuous even function $\varphi(t)$ satisfy $\varphi(t) \geq 0$, $\varphi(0) = 1$, $\varphi(t) \to 0$ as $t \to \infty$ and let $\varphi(t)$ be convex on $(-\infty, 0)$. (Hence also on $(0, \infty)$.) Then $\varphi(t)$ is a characteristic function ([31], XV.2).*

This theorem provides a very convenient method of constructing characteristic functions. Examples are

$$\varphi_1(t) = e^{-|t|},$$

$$\varphi_2(t) = \begin{cases} 1 - |t|, & |t| \leq 1, \\ 0, & |t| > 1. \end{cases}$$

Another is the function $\varphi_3(t)$ drawn in Fig. 34. On $[-a, a]$, the function $\varphi_3(t)$ coincides with $\varphi_2(t)$. However, the corresponding distribution functions F_2 and F_3 are evidently different. This example shows that in general two characteristic functions can be the same on a finite interval without their distribution functions being the same.

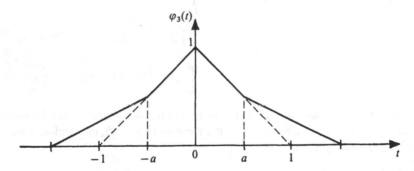

Fig. 34

Marcinkiewicz's Theorem. *If a characteristic function has the form $\exp \mathscr{P}(t)$, where $\mathscr{P}(t)$ is a polynomial, then this polynomial is of degree at most two ([65], 7.3).*

It follows, for example, that e^{-t^4} is not a characteristic function.

7. The following theorem shows that a property of the characteristic function of a random variable can lead to a nontrivial conclusion about the nature of the random variable.

Theorem 5. *Let* $\varphi_\xi(t)$ *be the characteristic function of the random variable* ξ.

(a) *If* $|\varphi_\xi(t_0)| = 1$ *for some* $t_0 \neq 0$, *then* ξ *is a lattice random variable concentrated at the points* $a + nh$, $h = 2\pi/|t_0|$, *that is,*

$$\sum_{n=-\infty}^{\infty} P\{\xi = a + nh\} = 1, \tag{33}$$

where a is a constant.

(b) *If* $|\varphi_\xi(t)| = |\varphi_\xi(\alpha t)| = 1$ *for two different points t and* αt, *where* α *is irrational, then* ξ *is degenerate, that is* $P\{\xi = a\} = 1$, *where a is some number.*

(c) *If* $|\varphi_\xi(t)| \equiv 1$, *then* ξ *is degenerate.*

PROOF. (a) If $|\varphi_\xi(t_0)| = 1$, $t_0 \neq 0$, there is a number a such that $\varphi(t_0) = e^{it_0 a}$. Then

$$e^{it_0 a} = \int_{-\infty}^{\infty} e^{it_0 x} \, dF(x) \Rightarrow 1 = \int_{-\infty}^{\infty} e^{it_0(x-a)} \, dF(x) \Rightarrow$$

$$1 = \int_{-\infty}^{\infty} \cos t_0(x-a) \, dF(x) \Rightarrow \int_{-\infty}^{\infty} [1 - \cos t_0(x-a)] \, dF(x) = 0.$$

Since $1 - \cos t_0(x - a) \geq 0$, it follows from property **H** (Subsection 2 of Sect. 6) that

$$1 = \cos t_0(\xi - a) \quad (\text{P-a. s.}),$$

which is equivalent to (33).

(b) It follows from $|\varphi_\xi(t)| = |\varphi_\xi(\alpha t)| = 1$ and from (33) that

$$\sum_{n=-\infty}^{\infty} P\left\{\xi = a + \frac{2\pi}{t}n\right\} = \sum_{m=-\infty}^{\infty} P\left\{\xi = b + \frac{2\pi}{\alpha t}m\right\} = 1.$$

If ξ is not degenerate, there must be at least two common points:

$$a + \frac{2\pi}{t}n_1 = b + \frac{2\pi}{\alpha t}m_1, \quad a + \frac{2\pi}{t}n_2 = b + \frac{2\pi}{\alpha t}m_2,$$

in the sets

$$\left\{a + \frac{2\pi}{t}n, \, n = 0, \pm 1, \ldots\right\} \quad \text{and} \quad \left\{b + \frac{2\pi}{\alpha t}m, \, m = 0, \pm 1, \ldots\right\},$$

whence

$$\frac{2\pi}{t}(n_1 - n_2) = \frac{2\pi}{\alpha t}(m_1 - m_2),$$

and this contradicts the assumption that α is irrational. Conclusion (c) follows from (b).

This completes the proof of the theorem.

□

8. Let $\xi = (\xi_1, \ldots, \xi_k)^*$ be a random vector and

$$\varphi_\xi(t) = \mathsf{E}\, e^{i(t,\xi)}, \quad t = (t_1, \ldots, t_k)^*,$$

its characteristic function. Let us suppose that $\mathsf{E}\,|\xi_i|^n < \infty$ for some $n \geq 1$, $i = 1, \ldots, k$. From Hölder's and Lyapunov's inequalities (Sect. 6, (29), and (27) respectively) it follows that the (mixed) moments $\mathsf{E}(\xi_1^{\nu_1} \cdots \xi_k^{\nu_k})$ exist for all nonnegative $\nu_1, \ldots, \nu_k$ such that $\nu_1 + \cdots + \nu_k \leq n$.

As in Theorem 1, this implies the existence and continuity of the partial derivatives

$$\frac{\partial^{\nu_1 + \cdots + \nu_k}}{\partial t_1^{\nu_1} \ldots \partial t_k^{\nu_k}} \varphi_\xi(t_1, \ldots, t_k)$$

for $\nu_1 + \cdots + \nu_k \leq n$. Then if we expand $\varphi_\xi(t_1, \ldots, t_k)$ in a Taylor series, we see that

$$\varphi_\xi(t_1, \ldots, t_k) = \sum_{\nu_1 + \cdots + \nu_k \leq n} \frac{i^{\nu_1 + \cdots + \nu_k}}{\nu_1! \cdots \nu_k!} m_\xi^{(\nu_1, \ldots, \nu_k)} t_1^{\nu_1} \cdots t_k^{\nu_k} + o(|t|^n), \qquad (34)$$

where $|t| = |t_1| + \cdots + |t_k|$ and

$$m_\xi^{(\nu_1, \ldots, \nu_k)} = \mathsf{E}\, \xi_1^{\nu_1} \cdots \xi_k^{\nu_k}$$

is the *mixed moment of order* $\nu = (\nu_1, \ldots, \nu_k)$.

Now $\varphi_\xi(t_1, \ldots, t_k)$ is continuous, $\varphi_\xi(0, \ldots, 0) = 1$, and consequently this function is different from zero in some neighborhood $|t| < \delta$ of zero. In this neighborhood the partial derivative

$$\frac{\partial^{\nu_1 + \cdots + \nu_k}}{\partial t_1^{\nu_1} \cdots \partial t_k^{\nu_k}} \log \varphi_\xi(t_1, \ldots, t_k)$$

exists and is continuous, where $\log z$ denotes the principal value of the logarithm (if $z = re^{i\theta}$, we take $\log z$ to be $\log r + i\theta$). Hence we can expand $\log \varphi_\xi(t_1, \ldots, t_k)$ by Taylor's formula,

$$\log \varphi_\xi(t_1, \ldots, t_k) = \sum_{\nu_1 + \cdots + \nu_k \leq n} \frac{i^{\nu_1 + \cdots + \nu_k}}{\nu_1! \cdots \nu_k!} s_\xi^{(\nu_1, \ldots, \nu_k)} t_1^{\nu_1} \cdots t_k^{\nu_k} + o(|t|^n), \qquad (35)$$

where the coefficients $s_\xi^{(\nu_1, \ldots, \nu_k)}$ are the (mixed) *semi-invariants* or *cumulants* of order $\nu = (\nu_1, \ldots, \nu_k)$ of $\xi = (\xi_1, \ldots, \xi_k)^*$.

---

* We write the integer-valued vector ν row-wise, since it is not subject to algebraic operations.

Observe that if ξ and η are independent, then

$$\log \varphi_{\xi+\eta}(t) = \log \varphi_\xi(t) + \log \varphi_\eta(t), \tag{36}$$

and therefore

$$s_{\xi+\eta}^{(\nu_1,\ldots,\nu_k)} = s_\xi^{(\nu_1,\ldots,\nu_k)} + s_\eta^{(\nu_1,\ldots,\nu_k)}. \tag{37}$$

(It is this property that gives rise to the term "semi-invariant" for $s_\xi^{(\nu_1,\ldots,\nu_k)}$.)

To simplify the formulas and make (34) and (35) look "one-dimensional," we introduce the following notation.

If $\nu = (\nu_1,\ldots,\nu_k)$ is a vector whose components are nonnegative integers, we put

$$\nu! = \nu_1! \cdots \nu_k!, \quad |\nu| = \nu_1 + \cdots + \nu_k, \quad t^\nu = t_1^{\nu_1} \cdots t_k^{\nu_k}.$$

We also put $s_\xi^{(\nu)} = s_\xi^{(\nu_1,\ldots,\nu_k)}$, $m_\xi^{(\nu)} = m_\xi^{(\nu_1,\ldots,\nu_k)}$.
Then (34) and (35) can be written

$$\varphi_\xi(t) = \sum_{|\nu|\le n} \frac{i^{|\nu|}}{\nu!} m_\xi^{(\nu)} t^\nu + o(|t|^n), \tag{38}$$

$$\log \varphi_\xi(t) = \sum_{|\nu|\le n} \frac{i^{|\nu|}}{\nu!} s_\xi^{(\nu)} t^\nu + o(|t|^n). \tag{39}$$

The following theorem and its corollaries give *formulas that connect moments and semi-invariants.*

Theorem 6. Let $\xi = (\xi_1,\ldots,\xi_k)^*$ be a random vector with $\mathsf{E}\,|\xi_i|^n < \infty$, $i = 1,\ldots,k$, $n \ge 1$. Then for $\nu = (\nu_1,\ldots,\nu_k)$ such that $|\nu| \le n$

$$m_\xi^{(\nu)} = \sum_{\lambda^{(1)}+\cdots+\lambda^{(q)}=\nu} \frac{1}{q!} \frac{\nu!}{\lambda^{(1)}!\cdots\lambda^{(q)}!} \prod_{p=1}^q s_\xi^{(\lambda^{(p)})}, \tag{40}$$

$$s_\xi^{(\nu)} = \sum_{\lambda^{(1)}+\cdots+\lambda^{(q)}=\nu} \frac{(-1)^{q-1}}{q} \frac{\nu!}{\lambda^{(1)}!\cdots\lambda^{(q)}!} \prod_{p=1}^q m_\xi^{(\lambda^{(p)})}, \tag{41}$$

where $\sum_{\lambda^{(1)}+\cdots+\lambda^{(q)}=\nu}$ indicates summation over all ordered sets of nonnegative integral vectors $\lambda^{(p)}$, $|\lambda^{(p)}| > 0$, whose sum is ν.

PROOF. Since

$$\varphi_\xi(t) = \exp(\log \varphi_\xi(t)),$$

if we expand the function exp by Taylor's formula and use (39), we obtain

$$\varphi_\xi(t) = 1 + \sum_{q=1}^n \frac{1}{q!} \left(\sum_{1\le|\lambda|\le n} \frac{i^{|\lambda|}}{\lambda!} s_\xi^{(\lambda)} t^\lambda \right)^q + o(|t|^n). \tag{42}$$

Comparing terms in t^λ on the right-hand sides of (38) and (42), and using $|\lambda^{(1)}| + \cdots + |\lambda^{(q)}| = |\lambda^{(1)} + \cdots + \lambda^{(q)}|$, we obtain (40).

Moreover,

$$\log \varphi_\xi(t) = \log \left[1 + \sum_{1 \le |\lambda| \le n} \frac{i^{|\lambda|}}{\lambda!} m_\xi^{(\lambda)} t^\lambda + o(|t|^n) \right]. \tag{43}$$

For small z we have the expansion

$$\log(1 + z) = \sum_{q=1}^{n} \frac{(-1)^{q-1}}{q} z^q + o(z^q).$$

Using this in (43) and then comparing the coefficients of t^λ with the corresponding coefficients on the right-hand side of (38), we obtain (41).

□

Corollary 1. *The following formulas connect moments and semi-invariants*:

$$m_\xi^{(\nu)} = \sum_{\{r_1 \lambda^{(1)} + \cdots + r_x \lambda^{(x)} = \nu\}} \frac{1}{r_1! \cdots r_x!} \frac{\nu!}{(\lambda^{(1)}!)^{r_1} \cdots (\lambda^{(x)}!)^{r_x}} \prod_{j=1}^{x} [s_\xi^{(\lambda^{(j)})}]^{r_j}, \tag{44}$$

$$s_\xi^{(\nu)} = \sum_{\{r_1 \lambda^{(1)} + \cdots + r_x \lambda^{(x)} = \nu\}} \frac{(-1)^{q-1}(q-1)!}{r_1! \cdots r_x!} \frac{\nu!}{(\lambda^{(1)}!)^{r_1} \cdots (\lambda^{(x)}!)^{r_x}} \prod_{j=1}^{x} [m_\xi^{(\lambda^{(j)})}]^{r_j}, \tag{45}$$

where $\sum_{\{r_1 \lambda^{(1)} + \cdots + r_x \lambda^{(x)} = \nu\}}$ *denotes summation over all unordered sets of different nonnegative integral vectors* $\lambda^{(j)}$, $|\lambda^{(j)}| > 0$, *and over all ordered sets of positive integral numbers* r_j *such that* $r_1 \lambda^{(1)} + \cdots + r_x \lambda^{(x)} = \nu$.

To establish (44) we suppose that among all the vectors $\lambda^{(1)}, \ldots, \lambda^{(q)}$ that occur in (40), there are r_1 equal to $\lambda^{(i_1)}, \ldots, r_x$ equal to $\lambda^{(i_x)}$ ($r_j > 0$, $r_1 + \cdots + r_x = q$), where all the $\lambda^{(i_s)}$ are different. There are $q!/(r_1! \ldots, r_x!)$ different sets of vectors, corresponding (except for order) with the set $\{\lambda^{(1)}, \ldots, \lambda^{(q)}\}$. But if two sets, say, $\{\lambda^{(1)}, \ldots, \lambda^{(q)}\}$ and $\{\bar{\lambda}^{(1)}, \ldots, \bar{\lambda}^{(q)}\}$ differ only in order, then $\prod_{p=1}^{q} s_\xi^{(\lambda^{(p)})} = \prod_{p=1}^{q} s_\xi^{(\bar{\lambda}^{(p)})}$. Hence if we identify sets that differ only in order, we obtain (44) from (40).

Formula (45) can be deduced from (41) in a similar way.

Corollary 2. *Let us consider the special case when* $\nu = (1, \ldots, 1)$. *In this case the moments* $m_\xi^{(\nu)} \equiv \mathsf{E}\, \xi_1 \cdots \xi_k$, *and the corresponding semi-invariants are called simple.*

Formulas connecting simple moments and simple semi-invariants can be read off from the formulas given above. However, it is useful to have them written in a different way.

For this purpose, we introduce the following notation.

Let $\xi = (\xi_1, \ldots, \xi_k)^*$ be a vector, and $I_\xi = \{1, 2, \ldots, k\}$ its set of indices. If $I \subseteq I_\xi$, let ξ_I denote the vector consisting of the components of ξ whose indices belong to I. Let $\chi(I)$ be the vector $\{\chi_1, \ldots, \chi_k\}$ for which $\chi_i = 1$ if $i \in I$, and $\chi_i = 0$ if $i \notin I$. These vectors are in one-to-one correspondence with the sets $I \subseteq I_\xi$. Hence we can write

$$m_\xi(I) = m_\xi^{(\chi(I))}, \quad s_\xi(I) = s_\xi^{(\chi(I))}.$$

In other words, $m_\xi(I)$ and $s_\xi(I)$ are simple moments and semi-invariants of the subvector ξ_I of ξ.

In accordance with the definition given in Subsection 3 of Sect. 1, Chap. 1, a *decomposition* of a set I is an unordered collection of disjoint nonempty sets I_p such that $\sum_p I_p = I$.

In terms of these definitions, we have the formulas

$$m_\xi(I) = \sum_{\sum_{p=1}^q I_p = I} \prod_{p=1}^q s_\xi(I_p), \tag{46}$$

$$s_\xi(I) = \sum_{\sum_{p=1}^q I_p = I} (-1)^{q-1}(q-1)! \prod_{p=1}^q m_\xi(I_p), \tag{47}$$

where $\sum_{\sum_{p=1}^q I_p = I}$ denotes summation over all decompositions of I, $1 \leq q \leq N(I)$, with $N(I)$ being the number of elements of the set I.

We shall derive (46) from (44). If $\nu = \chi(I)$ and $\lambda^{(1)} + \cdots + \lambda^{(q)} = \nu$, then $\lambda^{(p)} = \chi(I_p)$, $I_p \subseteq I$, where the $\lambda^{(p)}$ are all different, $\lambda^{(p)}! = \nu! = 1$, and every unordered set $\{\chi(I_1), \ldots, \chi(I_q)\}$ is in one-to-one correspondence with the decomposition $I = \sum_{p=1}^q I_p$. Consequently (46) follows from (44).

In a similar way, (47) follows from (45).

Example 4. Let ξ be a random variable $(k = 1)$ and $m_n = m_\xi^{(n)} = \mathsf{E}\,\xi^n$, $s_n = s_\xi^{(n)}$. Then (40) and (41) imply the following formulas:

$$\begin{aligned}
m_1 &= s_1, \\
m_2 &= s_2 + s_1^2, \\
m_3 &= s_3 + 3s_1 s_2 + s_1^3, \\
m_4 &= s_4 + 3s_2^2 + 4s_1 s_3 + 6s_1^2 s_2 + s_1^4,
\end{aligned} \tag{48}$$

$$\cdots\cdots\cdots\cdots\cdots\cdots\cdots\cdots\cdots\cdots\cdots\cdots\cdots$$

and

$$\begin{aligned}
s_1 &= m_1 = \mathsf{E}\,\xi, \\
s_2 &= m_2 - m_1^2 = \operatorname{Var}\xi, \\
s_3 &= m_3 - 3m_1 m_2 + 2m_1^3, \\
s_4 &= m_4 - 3m_2^2 - 4m_1 m_3 + 12m_1^2 m_2 - 6m_1^4,
\end{aligned} \tag{49}$$

$$\cdots\cdots\cdots\cdots\cdots\cdots\cdots\cdots\cdots\cdots\cdots\cdots\cdots$$

Example 5. Let $\xi \sim \mathcal{N}(m, \sigma^2)$. Since, by (9),

$$\log \varphi_\xi(t) = itm - \frac{t^2 \sigma^2}{2},$$

we have $s_1 = m$, $s_2 = \sigma^2$ by (39), and all the semi-invariants, from the third on, are zero: $s_n = 0$, $n \geq 3$.

We may observe that by Marcinkiewicz's theorem a function $\exp \mathscr{P}(t)$, where $\mathscr{P}$ is a polynomial, can be a characteristic function only when the degree of that polynomial is at most 2. It follows, in particular, that the Gaussian distribution is the only distribution with the property that all its semi-invariants s_n are zero from a certain index onward.

Example 6. If ξ is a Poisson random variable with parameter $\lambda > 0$, then by (11)

$$\log \varphi_\xi(t) = \lambda(e^{it} - 1).$$

It follows that

$$s_n = \lambda \tag{50}$$

for all $n \geq 1$.

Example 7. Let $\xi = (\xi_1, \ldots, \xi_n)^*$ be a random vector. Then

$$
\begin{aligned}
m_\xi(1) &= s_\xi(1), \\
m_\xi(1,2) &= s_\xi(1,2) + s_\xi(1)s_\xi(2), \\
m_\xi(1,2,3) &= s_\xi(1,2,3) + s_\xi(1,2)s_\xi(3) + s_\xi(1,3)s_\xi(2) \\
&\quad + s_\xi(2,3)s_\xi(1) + s_\xi(1)s_\xi(2)s_\xi(3),
\end{aligned} \tag{51}
$$

$$\ldots \ldots \ldots \ldots \ldots \ldots \ldots \ldots \ldots \ldots \ldots \ldots \ldots \ldots \ldots \ldots \ldots \ldots \ldots$$

These formulas show that the simple moments can be expressed in terms of the simple semi-invariants in a very *symmetric* way. If we put $\xi_1 \equiv \xi_2 \equiv \cdots \equiv \xi_k$, we then, of course, obtain (48).

The group-theoretical origin of the coefficients in (48) becomes clear from (51). It also follows from (51) that

$$s_\xi(1,2) = m_\xi(1,2) - m_\xi(1)m_\xi(2) = \mathsf{E}\,\xi_1\xi_2 - \mathsf{E}\,\xi_1\,\mathsf{E}\,\xi_2, \tag{52}$$

i.e., $s_\xi(1,2)$ is just the *covariance* of ξ_1 and ξ_2.

9. Let ξ be a random variable with distribution function $F = F(x)$ and characteristic function $\varphi(t)$. Let us suppose that all the moments $m_n = \mathsf{E}\,\xi^n$, $n \geq 1$ exist.

It follows from Theorem 2 that a characteristic function uniquely determines a probability distribution. Let us now ask the following question (uniqueness for the moments problem): Do the moments $\{m_n\}_{n \geq 1}$ determine the *probability distribution uniquely*?

More precisely, let F and G be distribution functions with the same moments, i.e.

$$\int_{-\infty}^{\infty} x^n \, dF(x) = \int_{-\infty}^{\infty} x^n \, dG(x) \tag{53}$$

for all integers $n \geq 0$. The question is whether F and G must be the same.

In general, the answer is "no." To see this, consider the distribution F with density

$$f(x) = \begin{cases} ke^{-\alpha x^\lambda}, & x > 0, \\ 0, & x < 0, \end{cases}$$

where $\alpha > 0$, $0 < \lambda < \frac{1}{2}$, and k is determined by the condition $\int_0^\infty f(x)dx = 1$. Write $\beta = \alpha \tan \lambda \pi$ and let $g(x) = 0$ for $x \leq 0$ and

$$g(x) = ke^{-\alpha x^\lambda}[1 + \varepsilon \sin(\beta x^\lambda)], \qquad |\varepsilon| < 1, \ x > 0.$$

It is evident that $g(x) \geq 0$. Let us show that

$$\int_0^\infty x^n e^{-\alpha x^\lambda} \sin \beta x^\lambda \, dx = 0 \tag{54}$$

for all integers $n \geq 0$.

For $p > 0$ and complex q with $\operatorname{Re} q > 0$, we have (see [1], Chapter 6, formula 6.1.1)

$$\int_0^\infty t^{p-1} e^{-qt} \, dt = \frac{\Gamma(p)}{q^p}.$$

Take $p = (n+1)/\lambda$, $q = \alpha + i\beta$, $t = x^\lambda$. Then

$$\int_0^\infty x^{\lambda\{[(n+1)/\lambda]-1\}} e^{-(\alpha+i\beta)x^\lambda} \lambda x^{\lambda-1} dx = \lambda \int_0^\infty x^n e^{-(\alpha+i\beta)x^\lambda} \, dx$$

$$= \lambda \int_0^\infty x^n e^{-\alpha x^\lambda} \cos \beta x^\lambda \, dx - i\lambda \int_0^\infty x^n e^{-\alpha x^\lambda} \sin \beta x^\lambda \, dx$$

$$= \frac{\Gamma(\frac{n+1}{\lambda})}{\alpha^{(n+1)/\lambda}(1 + i\tan \lambda \pi)^{(n+1)/\lambda}}. \tag{55}$$

But

$$(1 + i\tan \lambda \pi)^{(n+1)/\lambda} = (\cos \lambda \pi + i \sin \lambda \pi)^{(n+1)/\lambda}(\cos \lambda \pi)^{-(n+1)/\lambda}$$
$$= e^{i\pi(n+1)}(\cos \lambda \pi)^{-(n+1)/\lambda}$$
$$= \cos \pi(n+1) \cdot \cos(\lambda \pi)^{-(n+1)/\lambda},$$

since $\sin \pi(n+1) = 0$.

Hence the right-hand side of (55) is real and therefore (54) is valid for all integral $n \geq 0$. Now let $G(x)$ be the distribution function with density $g(x)$. It follows from (54) that the distribution functions F and G have equal moments, i.e., (53) holds for all integers $n \geq 0$.

We now give some conditions that guarantee the uniqueness of the solution of the moment problem.

Theorem 7. *Let $F = F(x)$ be a distribution function and $\mu_n = \int_{-\infty}^{\infty} |x|^n dF(x)$. If*

$$\limsup_{n\to\infty} \frac{\mu_n^{1/n}}{n} < \infty, \tag{56}$$

the moments $\{m_n\}_{n\geq 1}$, where $m_n = \int_{-\infty}^{\infty} x^n \, dF(x)$, determine the distribution function $F = F(x)$ uniquely.

PROOF. It follows from (56) and conclusion (7) of Theorem 1 that there is a $t_0 > 0$ such that, for all $|t| \leq t_0$, the characteristic function

$$\varphi(t) = \int_{-\infty}^{\infty} e^{itx} \, dF(x)$$

can be represented in the form

$$\varphi(t) = \sum_{k=0}^{\infty} \frac{(it)^k}{k!} m_k$$

and consequently the moments $\{m_n\}_{n\geq 1}$ uniquely determine the characteristic function $\varphi(t)$ for $|t| \leq t_0$.

Take a point s with $|s| \leq t_0/2$. Then, as in the proof of (15), we deduce from (56) that

$$\varphi(t) = \sum_{k=0}^{\infty} \frac{(t-s)^k}{k!} \varphi^{(k)}(s)$$

for $|t - s| \leq t_0$, where

$$\varphi^{(k)}(s) = i^k \int_{-\infty}^{\infty} x^k e^{isx} \, dF(x)$$

is uniquely determined by the moments $\{m_n\}_{n\geq 1}$. Consequently the moments determine $\varphi(t)$ uniquely for $|t| \leq \frac{3}{2}t_0$. Continuing this process, we see that $\{m_n\}_{n\geq 1}$ determines $\varphi(t)$ uniquely for all t, and therefore also determines $F(x)$.

This completes the proof of the theorem.

□

Corollary 3. *The moments uniquely determine the probability distribution if it is concentrated on a finite interval.*

Corollary 4. *A sufficient condition for the moment problem to have a unique solution is that*

$$\limsup_{n\to\infty} \frac{(m_{2n})^{1/2n}}{2n} < \infty. \tag{57}$$

For the proof it is enough to observe that the odd moments can be estimated in terms of the even ones, and then use (56).

Example. Let $F(x)$ be the normal distribution function,

$$F(x) = \frac{1}{\sqrt{2\pi\sigma^2}} \int_{-\infty}^{x} e^{-t^2/2\sigma^2} \, dt.$$

Then $m_{2n+1} = 0$, $m_{2n} = [(2n)!/2^n n!]\sigma^{2n}$, and it follows from (57) that these are the moments only of the normal distribution.

Finally we state, without proof:
Carleman's test for the uniqueness of the moments problem. ([31], VII.3)

(a) *Let $\{m_n\}_{n\geq 1}$ be the moments of a probability distribution, and let*

$$\sum_{n=0}^{\infty} \frac{1}{(m_{2n})^{1/2n}} = \infty.$$

Then they determine the probability distribution uniquely.
(b) *If $\{m_n\}_{n\geq 1}$ are the moments of a distribution that is concentrated on $[0,\infty)$, then the solution will be unique if we require only that*

$$\sum_{n=0}^{\infty} \frac{1}{(m_n)^{1/2n}} = \infty.$$

10. Let $F = F(x)$ and $G = G(x)$ be distribution functions with characteristic functions $f = f(t)$ and $g = g(t)$, respectively. The following theorem, which we give without proof, makes it possible to estimate how close F and G are to each other (in the uniform metric) in terms of the closeness of f and g.

Theorem (Esseen's Inequality). *Let $G(x)$ have derivative $G'(x)$ with $\sup_x |G'(x)| \leq C$. Then for every $T > 0$*

$$\sup_x |F(x) - G(x)| \leq \frac{2}{\pi} \int_0^T \left| \frac{f(t) - g(t)}{t} \right| dt + \frac{24}{\pi T} \sup_x |G'(x)|. \tag{58}$$

11. We present two tables of characteristic functions $\varphi(t)$ of some frequently used probability distributions (see Tables 2.2 and 2.3 of distributions and their parameters in Subsection 1 of Sect. 3).

12. PROBLEMS

1. Let ξ and η be independent random variables, $f(x) = f_1(x) + if_2(x)$, $g(x) = g_1(x) + ig_2(x)$, where $f_k(x)$ and $g_k(x)$ are Borel functions, $k = 1, 2$. Show that if $\mathsf{E}|f(\xi)| < \infty$ and $\mathsf{E}|g(\eta)| < \infty$, then

$$\mathsf{E}|f(\xi)g(\eta)| < \infty$$

Table 2.4

Discrete distributions	Characteristic functions
Discrete uniform	$\frac{1}{N}\frac{e^{it}}{1-e^{it}}\left(1-e^{itN}\right)$
Bernoulli	$q+pe^{it}$
Binomial	$[q+pe^{it}]^n$
Poisson	$\exp\{\lambda(e^{it}-1)\}$
Geometric	$\frac{p}{1-qe^{it}}$
Negative binomial	$\left[\frac{p}{1-qe^{it}}\right]^r$

Table 2.5

Distributions having a density	Characteristic functions
Uniform on $[a,b]$	$\frac{e^{itb}-e^{ita}}{it(b-a)}$
Normal, or Gaussian	$\exp\left\{itm-\frac{\sigma^2 t^2}{2}\right\}$
Gamma	$(1-it\beta)^{-\alpha}$
Beta	$\frac{\Gamma(\alpha+\beta)}{\Gamma(\alpha)}\sum\limits_{k=0}^{\infty}\frac{(it)^k\Gamma(\alpha+k)}{k!\,\Gamma(\alpha+\beta+k)\,\Gamma(1+k)}$
Exponential	$\frac{\lambda}{\lambda-it}$
Two-sided exponential	$\frac{\lambda^2 e^{it\alpha}}{t^2+\lambda^2}$
Chi-square	$(1-2it)^{-n/2}$
Student, or t-distribution	$\frac{\sqrt{\pi}\,\Gamma((n+1)/2)}{\Gamma(n/2)}\frac{\exp\{-\sqrt{n}\}|t|}{2^{2(m-1)}(m-1)!}\sum\limits_{k=0}^{m-1}(2k)!\,C_{n-1+k}^{2k}(2\sqrt{n}|t|)^{m-1-k}$, for integer $m=\frac{n+1}{2}$
Cauchy	$e^{-\theta|t|}$

and

$$\mathsf{E}f(\xi)g(\eta)=\mathsf{E}f(\xi)\cdot\mathsf{E}g(\eta).$$

2. Let $\xi=(\xi_1,\ldots,\xi_n)^*$ and $\mathsf{E}\|\xi\|^n<\infty$, where $\|\xi\|=+\sqrt{\sum\xi_i^2}$. Show that

$$\varphi_\xi(t)=\sum_{k=0}^n\frac{i^k}{k!}\mathsf{E}(t,\xi)^k+\varepsilon_n(t)\|t\|^n,$$

where $t=(t_1,\ldots,t_n)^*$ and $\varepsilon_n(t)\to0$, $t\to0$.

3. Prove Theorem 2 for n-dimensional distribution functions $F=F_n(x_1,\ldots,x_n)$ and $G=G_n(x_1,\ldots,x_n)$.

4. Let $F = F(x_1, \ldots, x_n)$ be an n-dimensional distribution function and $\varphi = \varphi(t_1, \ldots, t_n)$ its characteristic function. Using the notation (12) of Sect. 3, establish the *inversion formula*

$$P(a, b] = \lim_{c \to \infty} \frac{1}{(2\pi)^n} \int_{-c}^{c} \cdots \int_{-c}^{c} \prod_{k=1}^{n} \frac{e^{-it_k a_k} - e^{-it_k b_k}}{it_k} \, \varphi(t_1, \ldots, t_n) \, dt_1 \cdots dt_n.$$

(We are to suppose that $(a, b]$ is an interval of continuity of $P(a, b]$, i.e., for $k = 1, \ldots, n$ the points a_k, b_k are points of continuity of the marginal distribution functions $F_k(x_k)$ which are obtained from $F(x_1, \ldots, x_n)$ by taking all the variables except x_k equal to $+\infty$.)

5. Let $\varphi_k(t)$, $k \geq 1$, be characteristic functions, and let the nonnegative numbers λ_k, $k \geq 1$, satisfy $\sum \lambda_k = 1$. Show that $\sum \lambda_k \varphi_k(t)$ is a characteristic function.

6. If $\varphi(t)$ is a characteristic function, are $\operatorname{Re} \varphi(t)$ and $\operatorname{Im} \varphi(t)$ characteristic functions?

7. Let φ_1, φ_2 and φ_3 be characteristic functions, and $\varphi_1 \varphi_2 = \varphi_1 \varphi_3$. Does it follow that $\varphi_2 = \varphi_3$?

8. Prove the formulas for characteristic functions given in Tables 2.4 and 2.5.

9. Let ξ be an integral-valued random variable and $\varphi_\xi(t)$ its characteristic function. Show that

$$P(\xi = k) = \frac{1}{2\pi} \int_{-\pi}^{\pi} e^{-ikt} \varphi_\xi(t) \, dt, \quad k = 0, \pm 1, \pm 2 \ldots.$$

10. Show that the system of functions $\left\{ \frac{1}{\sqrt{2\pi}} e^{i\lambda n}, n = 0, \pm 1, \ldots \right\}$ forms an orthonormal basis in the space $L^2 = L^2([-\pi, \pi], \mathscr{B}([-\pi, \pi]))$ with Lebesgue measure μ.

11. In the Bochner–Khinchin theorem the function $\varphi(t)$ under consideration is assumed to be *continuous*. Prove the following result (due to Riesz) showing to what extent we can get rid of the continuity assumption.
    Let $\varphi = \varphi(t)$ be a complex-valued Lebesgue measurable function such that $\varphi(0) = 1$. Then $\varphi = \varphi(t)$ is positive semi-definite if and only if it equals (Lebesgue almost everywhere on the real line) some characteristic function.

12. Which of the following functions

$$\varphi(t) = e^{-|t|^k}, \quad 0 \leq k \leq 2, \qquad \varphi(t) = e^{-|t|^k}, \quad k > 2,$$

$$\varphi(t) = (1 + |t|)^{-1}, \qquad\qquad\qquad \varphi(t) = (1 + t^4)^{-1},$$

$$\varphi(t) = \begin{cases} 1 - |t|^3, & |t| \leq 1, \\ 0, & |t| > 1, \end{cases} \qquad \varphi(t) = \begin{cases} 1 - |t|, & |t| \leq 1/2, \\ 1/(4|t|), & |t| > 1/2, \end{cases}$$

are characteristic functions?

13. Let $\varphi(t)$ be the characteristic function of a distribution $F = F(x)$. Let $\{x_n\}$ be the set of discontinuity points of F $(\Delta F(x_n) \equiv F(x_n) - F(x_n-) > 0)$. Show that

$$\lim_{T \to \infty} \frac{1}{T} \int_{-T}^{T} |\varphi(t)|^2 \, dt = \sum_{n \geq 1} (\Delta F(x_n))^2.$$

14. The *concentration function* of a random variable X is

$$Q(X; l) = \sup_{x \in R} P\{x \leq X \leq x + l\}.$$

Show that:

(a) If X and Y are independent random variables, then

$$Q(X + Y; l) \leq \min(Q(x; l), Q(Y; l)) \quad \text{for all } l \geq 0;$$

(b) There exists x_l' such that $Q(X; l) = P\{x_l' \leq X \leq x_l' + l\}$ and the distribution function of X is continuous if and only if $Q(X; 0) = 0$.

15. Let $(m_n)_{n \geq 1}$ be the sequence of moments of a random variable X with distribution function $F = F(x)$ $\left(m_k = \int_{-\infty}^{\infty} x^k \, dF(x)\right)$. Show that $(m_n)_{n \geq 1}$ determines $F = F(x)$ uniquely whenever the series $\sum_{k=1}^{\infty} \frac{m_k}{k!} s^k$ absolutely converges for some $s > 0$.

16. Let $\varphi(t) = \int_{-\infty}^{\infty} e^{itx} \, dF(x)$ be the characteristic function of $F = F(x)$. Show that

$$\lim_{c \to \infty} \frac{1}{2c} \int_{-c}^{c} e^{-itx} \varphi(t) \, dt = F(x) - F(x-),$$

$$\lim_{c \to \infty} \frac{1}{2c} \int_{-c}^{c} |\varphi(t)|^2 \, dt = \sum_{x \in R} [F(x) - F(x-)]^2.$$

17. Show that every characteristic function $\varphi(t)$ satisfies the inequaltiy $1 - \operatorname{Re} \varphi(2t) \leq 4[1 - \operatorname{Re} \varphi(t)]$.

18. Suppose a characteristic function $\varphi(t)$ is such that $\varphi(t) = 1 + f(t) + o(t^2)$, $t \to 0$, where $f(t) = -f(-t)$. Show that then $\varphi(t) \equiv 1$.

19. Show that the functions

$$\varphi_n(t) = \frac{e^{it} - \sum_{k=0}^{n-1} (it)^k / k!}{(it)^n / n!}$$

are characteristic functions for any $n \geq 1$.

20. Prove that

$$\frac{2}{\pi} \int_{-\infty}^{\infty} \frac{1 - \operatorname{Re} \varphi(t)}{t^2} \, dt = \int_{-\infty}^{\infty} |x| \, dF(x).$$

21. Let a characteristic function $\varphi(t)$ be such that $\varphi(t) = 1 + O(|t|^\alpha)$, $t \to 0$, where $\alpha \in (0, 2]$. Show that the random variable ξ with characteristic function $\varphi(t)$ has then the following property:

$$P\{|\xi| > x\} = O(x^{-\alpha}), \quad x \to \infty.$$

22. If $\varphi(t)$ is a characteristic function, then $|\varphi(t)|^2$ is also a characteristic function.
23. Let X and Y be independent identically distributed random variables with zero mean and unit variance. Prove using characteristic functions that if the distribution F of $(X + Y)/\sqrt{2}$ is the same as that of X and Y, then F is the normal distribution.
24. If φ is a characteristic function, then so is $e^{\lambda(\varphi-1)}$ for any $\lambda \geq 0$.
25. The *Laplace transform* of a nonnegative random variable X with distribution function F is the function $\widehat{F} = \widehat{F}(\lambda)$ defined by

$$\widehat{F}(\lambda) = E\,e^{-\lambda X} = \int_{[0,\infty)} e^{-\lambda x}\, dF(x), \quad \lambda \geq 0.$$

Prove the following criterion (S. N. Bernstein): a function $f = f(\lambda)$ on $(0, \infty)$ is the Laplace transform of a distribution function $F = F(x)$ on $[0, \infty)$ if and only if this function is completely monotone (i.e., the derivatives $f^{(n)}(\lambda)$ of any order $n \geq 0$ exist and $(-1)^n f^{(n)}(\lambda) \geq 0$).

26. Let $\varphi(t)$ be a characteristic function. Show that so are

$$\int_0^1 \varphi(ut)\, du \quad \text{and} \quad \int_0^\infty e^{-u}\varphi(ut)\, du.$$

13 Gaussian Systems

1. Gaussian, or normal, distributions, random variables, processes, and systems play an extremely important role in probability theory and in mathematical statistics. This is explained in the first instance by the central limit theorem (Sect. 4 of Chap. 3), of which the De Moivre–Laplace limit theorem is a special case (Sect. 6 of Chap. 1). According to this theorem, the normal distribution is universal in the sense that the distribution of the sum of a large number of random variables or random vectors, subject to some not very restrictive conditions, is closely approximated by this distribution.

This is what provides a theoretical explanation of the "law of errors" of applied statistics, which says that errors of measurement that result from large numbers of independent "elementary" errors obey the normal distribution.

A multidimensional Gaussian distribution is specified by a small number of parameters; this is a definite advantage in using it in the construction of simple probabilistic models. Gaussian random variables have finite second moments, and consequently they can be studied by Hilbert space methods. Here it is important that in the Gaussian case "uncorrelated" is equivalent to "independent," so that the results of L^2-theory can be significantly strengthened.

2. Let us recall that (see Sect. 8) a random variable $\xi = \xi(\omega)$ is Gaussian, or normally distributed, with parameters m and σ^2 ($\xi \sim \mathcal{N}(m, \sigma^2)$), $|m| < \infty$, $\sigma^2 > 0$, if its density $f_\xi(x)$ has the form

$$f_\xi(x) = \frac{1}{\sqrt{2\pi}\sigma} e^{-(x-m)^2/2\sigma^2}, \tag{1}$$

where $\sigma = +\sqrt{\sigma^2}$. (This quantity σ is called the *standard deviation* of ξ from its mean value $\mathsf{E}\,\xi$, cf. Definition 5 in Sect. 4 of Chap. 1.)

As $\sigma \downarrow 0$, the density $f_\xi(x)$ "converges to the δ-function supported at $x = m$." It is natural to say that ξ is normally distributed with mean m and $\sigma^2 = 0$ ($\xi \sim \mathcal{N}(m, 0)$) if ξ has the property that $\mathsf{P}(\xi = m) = 1$.

We can, however, give a definition that applies both to the *nondegenerate* ($\sigma^2 > 0$) and the *degenerate* ($\sigma^2 = 0$) cases. Let us consider the characteristic function $\varphi_\xi(t) \equiv \mathsf{E}\,e^{it\xi}$, $t \in R$.

If $\mathsf{P}(\xi = m) = 1$, then evidently

$$\varphi_\xi(t) = e^{itm}, \tag{2}$$

whereas if $\xi \sim \mathcal{N}(m, \sigma^2)$, $\sigma^2 > 0$,

$$\varphi_\xi(t) = e^{itm - (1/2)t^2\sigma^2}. \tag{3}$$

It is obvious that when $\sigma^2 = 0$ the right-hand sides of (2) and (3) are the same. It follows, by Theorem 1 of Sect. 12, that the Gaussian random variable with parameters m and σ^2 ($|m| < \infty$, $\sigma^2 \geq 0$) must be the same as the random variable whose characteristic function is given by (3). The approach based on characteristic functions is especially useful in the multidimensional case.

Let $\xi = (\xi_1, \ldots, \xi_n)^*$ be a random vector and

$$\varphi_\xi(t) = \mathsf{E}\,e^{i(t,\xi)}, \quad t = (t_1, \ldots, t_n)^* \in R^n, \tag{4}$$

its characteristic function (see Definition 2, Sect. 12).

Definition 1. A random vector $\xi = (\xi_1, \ldots, \xi_n)^*$ is *Gaussian*, or *normally distributed*, if its characteristic function has the form

$$\varphi_\xi(t) = e^{i(t,m) - (1/2)(\mathbb{R}t, t)}, \tag{5}$$

where $m = (m_1, \ldots, m_n)^*$, $|m_k| < \infty$, and $\mathbb{R} = \|r_{kl}\|$ is a symmetric positive semi-definite $n \times n$ matrix; we use the abbreviation $\xi \sim \mathcal{N}(m, \mathbb{R})$.

This definition immediately makes us ask whether (5) is in fact a characteristic function. Let us show that it is.

First suppose that $\mathbb{R}$ is nonsingular. Then we can define the inverse $A = \mathbb{R}^{-1}$ and the function

$$f(x) = \frac{|A|^{1/2}}{(2\pi)^{n/2}} \exp\{-\tfrac{1}{2}(A(x - m), (x - m))\}, \tag{6}$$

where $x = (x_1, \ldots, x_n)^*$ and $|A| = \det A$. This function is nonnegative. Let us show that

$$\int_{R^n} e^{i(t,x)} f(x)\, dx = e^{i(t,m) - (1/2)(\mathbb{R}t, t)},$$

or equivalently that

$$I_n \equiv \int_{R^n} e^{i(t, x - m)} \frac{|A|^{1/2}}{(2\pi)^{n/2}} e^{-(1/2)(A(x-m), (x-m))} \, dx = e^{-(1/2)(\mathbb{R}t, t)}. \tag{7}$$

Let us make the change of variable

$$x - m = \mathcal{O}u, \quad t = \mathcal{O}v,$$

where $\mathcal{O}$ is an orthogonal matrix such that

$$\mathcal{O}^* \mathbb{R} \mathcal{O} = D,$$

and

$$D = \begin{pmatrix} d_1 & & 0 \\ & \ddots & \\ 0 & & d_n \end{pmatrix}$$

is a diagonal matrix with $d_i \geq 0$ (see the proof of the lemma in Sect. 8). Since $|\mathbb{R}| = \det \mathbb{R} \neq 0$, we have $d_i > 0$, $i = 1, \ldots, n$. Therefore

$$|A| = |\mathbb{R}^{-1}| = d_1^{-1} \cdots d_n^{-1}. \tag{8}$$

Moreover (for notation, see Subsection 1, Sect. 12)

$$
\begin{aligned}
i(t, x - m) - \tfrac{1}{2}(A(x - m), x - m) &= i(\mathcal{O}v, \mathcal{O}u) - \tfrac{1}{2}(A\mathcal{O}u, \mathcal{O}u) \\
&= i(\mathcal{O}v)^* \mathcal{O}u - \tfrac{1}{2}(\mathcal{O}u)^* A(\mathcal{O}u) \\
&= iv^* u - \tfrac{1}{2} u^* \mathcal{O}^* A \mathcal{O}u \\
&= iv^* u - \tfrac{1}{2} u^* D^{-1} u.
\end{aligned}
$$

Together with (9) of Sect. 12 and (8), this yields

$$I_n = (2\pi)^{-n/2}(d_1 \cdots d_n)^{-1/2} \int_{R^n} \exp(iv^*u - \tfrac{1}{2}u^*D^{-1}u)\, du$$

$$= \prod_{k=1}^{n}(2\pi d_k)^{-1/2} \int_{-\infty}^{\infty} \exp\left(iv_k u_k - \frac{u_k^2}{2d_k}\right) du_k = \prod_{k=1}^{n} \exp(-\tfrac{1}{2}v_k^2 d_k)$$

$$= \exp(-\tfrac{1}{2}v^*Dv) = \exp(-\tfrac{1}{2}v^*\mathcal{O}^*\mathbb{R}\mathcal{O}v) = \exp(-\tfrac{1}{2}t^*\mathbb{R}t) = \exp(-\tfrac{1}{2}(\mathbb{R}t, t)).$$

It also follows from (6) that

$$\int_{R^n} f(x)\, dx = 1. \tag{9}$$

Therefore (5) is the characteristic function of a nondegenerate n-dimensional Gaussian distribution (see Subsection 3, Sect. 3).

Now let $\mathbb{R}$ be singular. Take $\varepsilon > 0$ and consider the positive definite symmetric matrix $\mathbb{R}^\varepsilon \equiv \mathbb{R} + \varepsilon E$, where E is the identity matrix. Then by what has been proved,

$$\varphi^\varepsilon(t) = \exp\{i(t, m) - \tfrac{1}{2}(\mathbb{R}^\varepsilon t, t)\}$$

is a characteristic function:

$$\varphi^\varepsilon(t) = \int_{R^n} e^{i(t,x)}\, dF_\varepsilon(x),$$

where $F_\varepsilon(x) = F_\varepsilon(x_1, \ldots, x_n)$ is an n-dimensional distribution function.
As $\varepsilon \to 0$,

$$\varphi^\varepsilon(t) \to \varphi(t) = \exp\{i(t, m) - \tfrac{1}{2}(\mathbb{R}t, t)\}.$$

The limit function $\varphi(t)$ is continuous at $(0, \ldots, 0)$. Hence, by Theorem 1 and Problem 1 of Sect. 3, Chap. 3, it is a characteristic function.

Thus we have shown that Definition 1 is correct.

3. Let us now discuss the meaning of the vector m and the matrix $\mathbb{R} = \|r_{kl}\|$ that appear in (5). Since

$$\log \varphi_\xi(t) = i(t, m) - \tfrac{1}{2}(\mathbb{R}t, t) = i \sum_{k=1}^{n} t_k m_k - \frac{1}{2} \sum_{k,l=1}^{n} r_{kl} t_k t_l, \tag{10}$$

we find from (35) of Sect. 12 and the formulas that connect the moments and the semi-invariants that

$$m_1 = s_\xi^{(1,0,\ldots,0)} = \mathsf{E}\,\xi_1, \ldots, m_n = s_\xi^{(0,\ldots,0,1)} = \mathsf{E}\,\xi_n.$$

Similarly

$$r_{11} = s_\xi^{(2,0,\ldots,0)} = \mathrm{Var}\,\xi_1, \quad r_{12} = s_\xi^{(1,1,0,\ldots)} = \mathrm{Cov}(\xi_1, \xi_2),$$

and generally
$$r_{kl} = \mathrm{Cov}(\xi_k, \xi_l).$$

Consequently m is the *mean-value vector* of ξ and $\mathbb{R}$ is its *covariance matrix*.

If $\mathbb{R}$ is nonsingular, we can obtain this result in a different way. In fact, in this case ξ has a density $f(x)$ given by (6). A direct calculation then shows that

$$\mathsf{E}\,\xi_k \equiv \int x_k f(x)\,dx = m_k, \tag{11}$$

$$\mathrm{Cov}(\xi_k, \xi_l) = \int (x_k - m_k)(x_l - m_l)f(x)\,dx = r_{kl}.$$

4. Let us discuss some properties of Gaussian vectors.

Theorem 1

(a) *The components of a Gaussian vector are uncorrelated if and only if they are independent.*

(b) *A vector $\xi = (\xi_1, \ldots, \xi_n)^*$ is Gaussian if and only if, for every vector $\lambda = (\lambda_1, \ldots, \lambda_n)^* \in R^n$ the random variable $(\xi, \lambda) = \lambda_1 \xi_1 + \cdots + \lambda_n \xi_n$ has a Gaussian distribution.*

PROOF. (a) If the components of $\xi = (\xi_1, \ldots, \xi_n)^*$ are uncorrelated, it follows from the form of the characteristic function $\varphi_\xi(t)$ that it is a product of characteristic functions:

$$\varphi_\xi(t) = \prod_{k=1}^n \varphi_{\xi_k}(t_k).$$

Therefore, by Theorem 4 of Sect. 12, the components are independent.

The converse is evident, since independence always implies lack of correlation.

(b) If ξ is a Gaussian vector, it follows from (5) that

$$\mathsf{E}\exp\{it(\xi_1\lambda_1 + \cdots + \xi_n\lambda_n)\} = \exp\left\{it\left(\sum \lambda_k m_k\right) - \frac{t^2}{2}\left(\sum r_{kl}\lambda_k\lambda_l\right)\right\}, \quad t \in R,$$

and consequently

$$(\xi, \lambda) \sim \mathcal{N}\left(\sum \lambda_k m_k, \sum r_{kl}\lambda_k\lambda_l\right).$$

Conversely, to say that the random variable $(\xi, \lambda) = \xi_1\lambda_1 + \cdots + \xi_n\lambda_n$ is Gaussian means, in particular, that

$$\mathsf{E}\,e^{i(\xi,\lambda)} = \exp\left\{i\,\mathsf{E}(\xi,\lambda) - \tfrac{1}{2}\mathrm{Var}(\xi,\lambda)\right\} = \exp\left\{i\sum \lambda_k\,\mathsf{E}\,\xi_k - \tfrac{1}{2}\sum \lambda_k\lambda_l\,\mathrm{Cov}(\xi_k, \xi_l)\right\}.$$

Since $\lambda_1, \ldots, \lambda_n$ are arbitrary it follows from Definition 1 that the vector $\xi = (\xi_1, \ldots, \xi_n)$ is Gaussian.

This completes the proof of the theorem.

□

Remark. Let $\binom{\theta}{\xi}$ be a Gaussian vector with $\theta = (\theta_1, \ldots, \theta_k)^*$ and $\xi = (\xi_1, \ldots, \xi_l)^*$. If θ and ξ are uncorrelated, i.e., $\mathrm{Cov}(\theta_i, \xi_j) = 0$, $i = 1, \ldots, k; j = 1, \ldots, l$, they are independent.

The proof is the same as for conclusion (a) of the theorem.

Let $\xi = (\xi_1, \ldots, \xi_n)^*$ be a Gaussian vector; let us suppose, for simplicity, that its mean-value vector is zero. If $\mathrm{rank}\, \mathbb{R} = r < n$, then (as was shown in Sect. 11), there are $n - r$ linear relations connecting $\xi_1, \ldots, \xi_n$. We may then suppose that, say, $\xi_1, \ldots, \xi_r$ are linearly independent, and the others can be expressed linearly in terms of them. Hence all the basic properties of the vector $\xi = (\xi_1, \ldots, \xi_n)^*$ are determined by the first r components $(\xi_1, \ldots, \xi_r)$ for which the corresponding covariance matrix is already known to be nonsingular.

Thus we may suppose that the original vector $\xi = (\xi_1, \ldots, \xi_n)^*$ had linearly independent components and therefore that $|\mathbb{R}| > 0$.

Let $\mathcal{O}$ be an orthogonal matrix that diagonalizes $\mathbb{R}$,

$$\mathcal{O}^* \mathbb{R} \mathcal{O} = D.$$

As was pointed out in Subsection 3, the diagonal elements of D are positive and therefore determine the inverse matrix. Put $B^2 = D$ and

$$\beta = B^{-1} \mathcal{O}^* \xi.$$

Then it is easily verified that

$$\mathsf{E}\, e^{i(t, \beta)} = \mathsf{E}\, e^{i\beta^* t} = e^{-(1/2)(Et, t)},$$

i.e., the vector $\beta = (\beta_1, \ldots, \beta_n)^*$ is a Gaussian vector with components that are uncorrelated and therefore (by Theorem 1) independent. Then if we write $A = \mathcal{O}B$ we find that the original Gaussian vector $\xi = (\xi_1, \ldots, \xi_n)^*$ can be represented as

$$\xi = A\beta, \tag{12}$$

where $\beta = (\beta_1, \ldots, \beta_n)^*$ is a Gaussian vector with independent components, $\beta_k \sim \mathcal{N}(0, 1)$. Hence we have the following result. Let $\xi = (\xi_1, \ldots, \xi_n)^*$ be a vector with linearly independent components such that $\mathsf{E}\, \xi_k = 0$, $k = 1, \ldots, n$. This vector is Gaussian if and only if there is a Gaussian vector $\beta = (\beta_1, \ldots, \beta_n)^*$ with independent components $\beta_1, \ldots, \beta_n$, $\beta_k \sim \mathcal{N}(0, 1)$, and a nonsingular matrix A of order n such that $\xi = A\beta$. Here $\mathbb{R} = AA^*$ is the covariance matrix of ξ.

If $|\mathbb{R}| \neq 0$, then by the Gram–Schmidt method (see Sect. 11)

$$\xi_k = \hat{\xi}_k + b_k \varepsilon_k, \quad k = 1, \ldots, n, \tag{13}$$

where, since $\varepsilon = (\varepsilon_1, \ldots, \varepsilon_k)^* \sim \mathcal{N}(0, E)$ is a Gaussian vector,

$$\hat{\xi}_k = \sum_{l=1}^{k-1} (\xi_k, \varepsilon_l)\varepsilon_l, \tag{14}$$

$$b_k = \|\xi_k - \hat{\xi}_k\| \tag{15}$$

and

$$\mathscr{L}\{\xi_1, \ldots, \xi_k\} = \mathscr{L}\{\varepsilon_1, \ldots, \varepsilon_k\}. \tag{16}$$

We see immediately from the orthogonal decomposition (13) that

$$\hat{\xi}_k = \mathsf{E}(\xi_k \mid \xi_{k-1}, \ldots, \xi_1). \tag{17}$$

From this, with (16) and (14), it follows that in the Gaussian case the conditional expectation $\mathsf{E}(\xi_k \mid \xi_{k-1}, \ldots, \xi_1)$ is a linear function of $(\xi_1, \ldots, \xi_{k-1})$:

$$\mathsf{E}(\xi_k \mid \xi_{k-1}, \ldots, \xi_1) = \sum_{i=1}^{k-1} a_i \xi_i. \tag{18}$$

(This was proved in Sect. 8 for the case $k = 2$.)

Since, according to a remark to Theorem 1 of Sect. 8, $\mathsf{E}(\xi_k \mid \xi_{k-1}, \ldots, \xi_1)$ is an optimal estimator (in the mean-square sense) for ξ_k in terms of $\xi_1, \ldots, \xi_{k-1}$, it follows from (18) that in the Gaussian case the optimal estimator is *linear*.

We shall use these results in looking for optimal estimators of $\theta = (\theta_1, \ldots, \theta_k)^*$ in terms of $\xi = (\xi_1, \ldots, \xi_l)^*$ under the hypothesis that $(\theta^*, \xi^*)^*$ is Gaussian. Let

$$m_\theta = \mathsf{E}\,\theta, \quad m_\xi = \mathsf{E}\,\xi$$

be the column-vectors of mean values and

$$\begin{aligned}
\mathbb{R}_{\theta\theta} &\equiv \mathrm{Cov}(\theta, \theta) \equiv \|\mathrm{Cov}(\theta_i, \theta_j)\|, \quad 1 \le i, j \le k, \\
\mathbb{R}_{\theta\xi} &\equiv \mathrm{Cov}(\theta, \xi) \equiv \|\mathrm{Cov}(\theta_i, \xi_j)\|, \quad 1 \le i \le k, 1 < j < l, \\
\mathbb{R}_{\xi\xi} &\equiv \mathrm{Cov}(\xi, \xi) \equiv \|\mathrm{Cov}(\xi_i, \xi_j)\|, \quad 1 \le i, j \le l,
\end{aligned}$$

the covariance matrices. Let us suppose that $\mathbb{R}_{\xi\xi}$ has an inverse. Then we have the following theorem (cf. Theorem 2 in Sect. 8).

Theorem 2 (Theorem on Normal Correlation). *For a Gaussian vector* $(\theta^*, \xi^*)^*$, *the optimal estimator* $\mathsf{E}(\theta \mid \xi)$ *of* θ *in terms of* ξ, *and its error matrix*

$$\Delta = \mathsf{E}[\theta - \mathsf{E}(\theta \mid \xi)][\theta - \mathsf{E}(\theta \mid \xi)]^*$$

are given by the formulas

$$\mathsf{E}(\theta \mid \xi) = m_\theta + \mathbb{R}_{\theta\xi}\mathbb{R}_{\xi\xi}^{-1}(\xi - m_\xi), \tag{19}$$

$$\Delta = \mathbb{R}_{\theta\theta} - \mathbb{R}_{\theta\xi}\mathbb{R}_{\xi\xi}^{-1}(\mathbb{R}_{\theta\xi})^*. \tag{20}$$

PROOF. Form the vector

$$\eta = (\theta - m_\theta) - \mathbb{R}_{\theta\xi}\mathbb{R}_{\xi\xi}^{-1}(\xi - m_\xi). \tag{21}$$

We can verify at once that $\mathsf{E}\,\eta(\xi - m_\xi)^* = 0$, i.e., η is not correlated with $(\xi - m_\xi)$. But since $(\theta^*, \xi^*)^*$ is Gaussian, the vector $(\eta^*, \xi^*)^*$ is also Gaussian. Hence by the remark on Theorem 1, η and $\xi - m_\xi$ are independent. Therefore η and ξ are independent, and consequently $\mathsf{E}(\eta \mid \xi) = \mathsf{E}\,\eta = 0$. Therefore

$$\mathsf{E}[\theta - m_\theta \mid \xi] - \mathbb{R}_{\theta\xi}\mathbb{R}_{\xi\xi}^{-1}(\xi - m_\xi) = 0,$$

which establishes (19).

To establish (20) we consider the conditional covariance

$$\mathrm{Cov}(\theta, \theta \mid \xi) \equiv \mathsf{E}[(\theta - \mathsf{E}(\theta \mid \xi))(\theta - \mathsf{E}(\theta \mid \xi))^* \mid \xi]. \tag{22}$$

Since $\theta - \mathsf{E}(\theta \mid \xi) = \eta$, and η and ξ are independent, we find that

$$\begin{aligned}
\mathrm{Cov}(\theta, \theta \mid \xi) &= \mathsf{E}(\eta\eta^* \mid \xi) = \mathsf{E}\,\eta\eta^* \\
&= \mathbb{R}_{\theta\theta} + \mathbb{R}_{\theta\xi}\mathbb{R}_{\xi\xi}^{-1}\mathbb{R}_{\xi\xi}\mathbb{R}_{\xi\xi}^{-1}\mathbb{R}_{\theta\xi}^* - 2\mathbb{R}_{\theta\xi}\mathbb{R}_{\xi\xi}^{-1}\mathbb{R}_{\xi\xi}\mathbb{R}_{\xi\xi}^{-1}\mathbb{R}_{\theta\xi}^* \\
&= \mathbb{R}_{\theta\theta} - \mathbb{R}_{\theta\xi}\mathbb{R}_{\xi\xi}^{-1}\mathbb{R}_{\theta\xi}^*.
\end{aligned}$$

Since $\mathrm{Cov}(\theta, \theta \mid \xi)$ does not depend on "chance," we have

$$\Delta = \mathsf{E}\,\mathrm{Cov}(\theta, \theta \mid \xi) = \mathrm{Cov}(\theta, \theta \mid \xi),$$

and this establishes (20).

□

Corollary. *Let* $(\theta, \xi_1, \ldots, \xi_n)^*$ *be an* $(n + 1)$-*dimensional Gaussian vector, with* $\xi_1, \ldots, \xi_n$ *independent. Then*

$$\mathsf{E}(\theta \mid \xi_1, \ldots, \xi_n) = \mathsf{E}\,\theta + \sum_{i=1}^{n} \frac{\mathrm{Cov}(\theta, \xi_i)}{\mathrm{Var}\,\xi_i}(\xi_i - \mathsf{E}\,\xi_i),$$

$$\Delta = \mathrm{Var}\,\theta - \sum_{i=1}^{n} \frac{\mathrm{Cov}^2(\theta, \xi_i)}{\mathrm{Var}\,\xi_i}$$

(cf. (12) and (13) in Sect. 8).

5. Let $\xi_1, \xi_2, \ldots$ be a sequence of Gaussian random vectors that converge in probability to ξ. Let us show that ξ is also Gaussian.

In accordance with (a) of Theorem 1, it is enough to establish this only for random variables.

Let $m_n = \mathsf{E}\,\xi_n, \sigma_n^2 = \mathrm{Var}\,\xi_n$. Then by Lebesgue's dominated convergence theorem

$$\lim_{n\to\infty} e^{itm_n - (1/2)\sigma_n^2 t^2} = \lim_{n\to\infty} \mathsf{E}\,e^{it\xi_n} = \mathsf{E}\,e^{it\xi}.$$

It follows from the existence of the limit on the left-hand side that there are numbers m and σ^2 such that

$$m = \lim_{n \to \infty} m_n, \quad \sigma^2 = \lim_{n \to \infty} \sigma_n^2.$$

Consequently

$$\mathsf{E}\, e^{it\xi} = e^{itm - (1/2)\sigma^2 t^2}$$

i.e., $\xi \sim \mathcal{N}(m, \sigma^2)$.

It follows, in particular, that the closed linear manifold $\mathscr{L}(\xi_1, \xi_2, \ldots)$ generated by the Gaussian variables $\xi_1, \xi_2, \ldots$ (see Subsection 5, Sect. 11) consists of Gaussian variables.

6. We now turn to the concept of Gaussian systems in general.

Definition 2. A collection of random variables $\xi = (\xi_\alpha)$, where α belongs to some index set $\mathfrak{A}$, is a *Gaussian system* if the random vector $(\xi_{\alpha_1}, \ldots, \xi_{\alpha_n})^*$ is Gaussian for every $n \geq 1$ and all indices $\alpha_1, \ldots, \alpha_n$ chosen from $\mathfrak{A}$.

Let us notice some properties of Gaussian systems.

(a) If $\xi = (\xi_\alpha)$, $\alpha \in \mathfrak{A}$, is a Gaussian system, then every subsystem $\xi' = (\xi_{\alpha'})$, $\alpha' \in \mathfrak{A}' \subseteq \mathfrak{A}$, is also Gaussian.

(b) If ξ_α, $\alpha \in \mathfrak{A}$, are independent Gaussian variables, then the system $\xi = (\xi_\alpha)$, $\alpha \in \mathfrak{A}$, is Gaussian.

(c) If $\xi = (\xi_\alpha)$, $\alpha \in \mathfrak{A}$, is a Gaussian system, the closed linear manifold $\mathscr{L}(\xi)$, consisting of all variables of the form $\sum_{i=1}^n c_{\alpha_i} \xi_{\alpha_i}$, together with their mean-square limits, forms a Gaussian system.

Let us observe that the converse of (a) is false in general. For example, let ξ_1 and η_1 be independent and $\xi_1 \sim \mathcal{N}(0, 1)$, $\eta_1 \sim \mathcal{N}(0, 1)$. Define the system

$$(\xi, \eta) = \begin{cases} (\xi_1, |\eta_1|) & \text{if } \xi_1 \geq 0, \\ (\xi_1, -|\eta_1|) & \text{if } \xi_1 < 0. \end{cases} \tag{23}$$

Then it is easily verified that ξ and η are both Gaussian, but (ξ, η) is not.

Let $\xi = (\xi_\alpha)_{\alpha \in \mathfrak{A}}$ be a Gaussian system with mean-value "vector" $m = (m_\alpha)$, $\alpha \in \mathfrak{A}$, and covariance "matrix" $\mathbb{R} = (r_{\alpha\beta})_{\alpha, \beta \in \mathfrak{A}}$, where $m_\alpha = \mathsf{E}\, \xi_\alpha$. Then $\mathbb{R}$ is evidently symmetric ($r_{\alpha\beta} = r_{\beta\alpha}$) and positive semi-definite in the sense that for every vector $c = (c_\alpha)_{\alpha \in \mathfrak{A}}$ with values in $R^{\mathfrak{A}}$, and only a *finite* number of nonzero coordinates c_α

$$(\mathbb{R}c, c) \equiv \sum_{\alpha, \beta} r_{\alpha\beta} c_\alpha c_\beta \geq 0. \tag{24}$$

We now ask the converse question. Suppose that we are given a parameter set $\mathfrak{A} = \{\alpha\}$, a "vector" $m = (m_\alpha)_{\alpha \in \mathfrak{A}}$ and a symmetric positive semi-definite "matrix" $\mathbb{R} = (r_{\alpha\beta})_{\alpha, \beta \in \mathfrak{A}}$. Do there exist a probability space $(\Omega, \mathscr{F}, \mathsf{P})$ and a Gaussian system of random variables $\xi = (\xi_\alpha)_{\alpha \in \mathfrak{A}}$ on it, such that

$$\mathsf{E}\,\xi_\alpha = m_\alpha,$$
$$\operatorname{Cov}(\xi_\alpha, \xi_\beta) = r_{\alpha,\beta}, \quad \alpha,\,\beta \in \mathfrak{A}?$$

If we take a finite set $\alpha_1, \ldots, \alpha_n$, then for the vector $\overline{m} = (m_{\alpha_1}, \ldots, m_{\alpha_n})^*$ and the matrix $\overline{\mathbb{R}} = (r_{\alpha\beta})$, $\alpha,\,\beta = \alpha_1, \ldots, \alpha_n$, we can construct in R^n the Gaussian distribution $F_{\alpha_1, \ldots, \alpha_n}(x_1, \ldots, x_n)$ with characteristic function

$$\varphi(t) = \exp\{i(t, \overline{m}) - \frac{1}{2}(\overline{\mathbb{R}}t,\, t)\}, \quad t = (t_{\alpha_1}, \ldots, t_{\alpha_n})^*.$$

It is easily verified that the family

$$\{F_{\alpha_1 \ldots, \alpha_n}(x_1, \ldots, x_n); \alpha_i \in \mathfrak{A}\}$$

is consistent. Consequently by Kolmogorov's theorem (Theorem 1, Sect. 9, and Remark 2 on this) the answer to our question is positive.

7. If $\mathfrak{A} = \{1, 2, \ldots\}$, then in accordance with the terminology of Sect. 5 the system of random variables $\xi = (\xi_\alpha)_{\alpha \in \mathfrak{A}}$ is a *random sequence* and is denoted by $\xi = (\xi_1, \xi_2, \ldots)$. A Gaussian sequence is completely described by its mean-value vector $m = (m_1, m_2, \ldots)$ and covariance matrix $\mathbb{R} = \|r_{ij}\|$, $r_{ij} = \operatorname{Cov}(\xi_i,\,\xi_j)$. In particular, if $r_{ij} = \sigma_i^2 \delta_{ij}$, then $\xi = (\xi_1, \xi_2, \ldots)$ is a Gaussian sequence of independent random variables with $\xi_i \sim \mathcal{N}(m_i, \sigma_i^2)$, $i \geq 1$.

When $\mathfrak{A} = [0, 1]$, $[0, \infty)$, $(-\infty, \infty)$, $\ldots$, the system $\xi = (\xi_t)$, $t \in \mathfrak{A}$, is a *random process with continuous time*.

Let us mention some examples of Gaussian random processes. If we take their mean values to be zero, their probabilistic properties are completely described by the covariance matrices $\|r_{st}\|$. We write $r(s, t)$ instead of r_{st} and call it the *covariance function*.

Example 1. If $\mathfrak{A} = [0, \infty)$ and

$$r(s,\, t) = \min(s,\, t), \tag{25}$$

the Gaussian process $B = (B_t)_{t \geq 0}$ with this covariance function (see Problem 2) and $B_0 \equiv 0$ is a *Brownian motion* or *Wiener process*.

Observe that this process has *independent increments*; that is, for arbitrary $t_1 < t_2 < \cdots < t_n$ the random variables

$$B_{t_2} - B_{t_1}, \ldots, B_{t_n} - B_{t_{n-1}}$$

are independent. In fact, because the process is Gaussian it is enough to verify only that the increments are uncorrelated. But if $s < t < u < v$ then

$$\mathsf{E}[B_t - B_s][B_v - B_u] = [r(t, v) - r(t, u)] - [r(s, v) - r(s, u)]$$
$$= (t - t) - (s - s) = 0.$$

Remark. The example of the renewal process built constructively (Subsection 4, Sect. 9) based on a sequence of independent identically distributed random variables $\sigma_1, \sigma_2, \ldots$, suggests that it may be possible to construct a version of the Brownian motion in a similar manner.

In fact, there are such constructions using a sequence $\xi_1, \xi_2, \ldots$ of independent identically distributed standard Gaussian random variables $\xi_i \sim \mathcal{N}(0, 1)$.

For example, form the variables

$$B_t = \frac{\sqrt{2}}{\pi} \sum_{n=1}^{\infty} \frac{\xi_n}{n + 1/2} \sin((n + 1/2)\pi t), \quad t \in [0, 1]. \tag{26}$$

The "two series" theorem stated below (Theorem 2 in Sect. 3, Chap. 4, Vol. 2) implies that the series specifying B_t *converges* (P-a. s.) for each $t \in [0, 1]$. A more detailed analysis shows that this series converges (P-a. s.) *uniformly* and therefore $B = (B_t)_{0 \le t \le 1}$ has (P-a. s.) continuous paths. This process has Gaussian finite-dimensional distributions, which follows from Theorem 1 (b) and the statement in Subsection 5 on preserving the Gaussian distribution under taking the limit in probability for Gaussian random variables. It is not hard to see that the covariance function of this process is $r(s, t) = \mathsf{E}\, B_s B_t = \min(s, t)$.

Thus the process $B = (B_t)_{0 \le t \le 1}$ satisfies all the requirements specifying the Brownian motion process, but, what is more, it has (P-a. s.) *continuous* paths. As a rule continuity of paths (a desirable property justified by physical applications) is included in the definition of the Brownian motion. As we see, the process with this property does exist.

Let us describe one more well-known way of constructing the Brownian motion based on the *Haar functions* $H_n(x)$, $x \in [0, 1]$, $n = 1, 2, \ldots$, introduced in Subsection 5 of Sect. 11.

Using them, we construct the *Schauder functions* $S_n(t)$, $t \in [0, 1]$, $n - 1, 2, \ldots$:

$$S_n(t) = \int_0^t H_n(x)\, dx. \tag{27}$$

Then if $\xi_0, \xi_1, \xi_2, \ldots$ is a sequence of independent identically distributed random variables with standard normal distribution, $\xi_i \sim \mathcal{N}(0, 1)$, then the series

$$B_t = \sum_{n=1}^{\infty} \xi_n S_n(t) \tag{28}$$

converges uniformly in $t \in [0, 1]$ with probability one. The process $B = (B_t)_{0 \le t \le 1}$ is the Brownian motion.

Example 2. The process $B^0 = (B_t^0)$, $t \in \mathfrak{A}$, with $\mathfrak{A} = [0, 1]$, $B_0^0 \equiv 0$ and

$$r(s, t) = \min(s, t) - st \tag{29}$$

is a *conditional Wiener process* or a *Brownian bridge* (observe that since $r(1, 1) = 0$ we have $\mathsf{P}(B_1^0 = 0) = 1$).

Example 3. The process $X = (X_t)$, $t \in \mathfrak{A}$, with $\mathfrak{A} = (-\infty, \infty)$ and

$$r(s, t) = e^{-|t-s|} \tag{30}$$

is *a Gauss–Markov process.*

8. We state now another interesting property of the Brownian motion, whose proof illustrates very well an application of the Borel–Cantelli lemma (Sect. 10), or rather of Corollary 1 to it.

Theorem 3. *Let $B = (B_t)_{t \geq 0}$ be the standard Brownian motion. Then, with probability one,*

$$\lim_{n \to \infty} \sum_{k=1}^{2^n T} [B_{k2^{-n}} - B_{(k-1)2^{-n}}]^2 = T \tag{31}$$

for any $T > 0$.

PROOF. Without loss of generality we can take $T = 1$. Let

$$A_n^\varepsilon = \left\{ \omega : \left| \sum_{k=1}^{2^n} (B_{k2^{-n}} - B_{(k-1)2^{-n}})^2 - 1 \right| \geq \varepsilon \right\}.$$

Since the random variables $B_{k2^{-n}} - B_{(k-1)2^{-n}}$ are Gaussian with zero mean and variance 2^{-n}, we have

$$\mathrm{Var}\left(\sum_{k=1}^{2^n} (B_{k2^{-n}} - B_{(k-1)2^{-n}})^2 \right) = 2^{-n+1}.$$

Hence, by Chebyshev's inequality, $\mathsf{P}(A_n^\varepsilon) \leq \varepsilon^{-2} 2^{-n+1}$, and therefore

$$\sum_{n=1}^{\infty} \mathsf{P}(A_n^\varepsilon) \leq \varepsilon^{-2} \sum_{n=1}^{\infty} 2^{-n+1} = 2\varepsilon^{-2} < \infty. \tag{32}$$

The required statement (31) follows from this bound and Corollary 1 to the Borel–Cantelli lemma (Sect. 10). □

9. PROBLEMS

1. Let ξ_1, ξ_2, ξ_3 be independent Gaussian random variables, $\xi_i \sim \mathcal{N}(0, 1)$. Show that

$$\frac{\xi_1 + \xi_2 \xi_3}{\sqrt{1 + \xi_3^2}} \sim \mathcal{N}(0, 1).$$

(In this case we encounter the interesting problem of describing the *nonlinear* transformations of independent Gaussian variables $\xi_1, \ldots, \xi_n$ whose distributions are still Gaussian.)

2. Show that the "matrices" $\mathbb{R} = (r(s, t))_{s,t \in \mathfrak{A}}$ specified by the functions $r(s, t)$ in (25), (29), and (30) are positive semi-definite.

3. Let A be an $m \times n$ matrix. An $n \times m$ matrix $A^{\otimes}$ is a *pseudoinverse* of A if there are matrices U and V such that

$$AA^{\otimes}A = A, \quad A^{\otimes} = UA^* = A^*V.$$

Show that $A^{\otimes}$ exists and is unique.

4. Show that (19) and (20) in the theorem on normal correlation remain valid when $\mathbb{R}_{\xi\xi}$ is *singular* provided that $\mathbb{R}_{\xi\xi}^{-1}$ is replaced by $\mathbb{R}_{\xi\xi}^{\otimes}$.

5. Let $(\theta, \xi) = (\theta_1, \ldots, \theta_k; \xi_1, \ldots, \xi_l)^*$ be a Gaussian vector with nonsingular matrix $\Delta \equiv \mathbb{R}_{\theta\theta} - \mathbb{R}_{\xi\xi}^{\otimes} \mathbb{R}_{\theta\xi}^*$. Show that the distribution function

$$P(\theta \leq a \mid \xi) = P(\theta_1 \leq a_1, \ldots, \theta_k \leq a_k \mid \xi)$$

has (P-a. s.) the density $p(a_1, \ldots, a_k \mid \xi)$ defined by

$$\frac{|\Delta^{-1/2}|}{(2\pi)^{k/2}} \exp\left\{ -\tfrac{1}{2}(a - \mathsf{E}(\theta \mid \xi))^* \Delta^{-1} (a - \mathsf{E}(\theta \mid \xi)) \right\}.$$

6. (S. N. Bernstein). Let ξ and η be independent identically distributed random variables with finite variances. Show that if $\xi + \eta$ and $\xi - \eta$ are independent, then ξ and η are Gaussian.

7. *Mercer's theorem.* Let $r = r(s,t)$ be a continuous covariance function on $[a,b] \times [a,b]$, where $-\infty < a < b < \infty$. Prove that the equation

$$\lambda \int_a^b r(s,t)u(t)\,dt = u(s), \quad a \leq s \leq b,$$

has infinitely many values $\lambda_k > 0$ and the corresponding system of continuous solutions $\{u_k, k \geq 1\}$, which form a complete orthonormal system in $L^2(a,b)$, such that

$$r(s,t) = \sum_{k=1}^{\infty} \frac{u_k(s)u_k(t)}{\lambda_k},$$

where the series converges absolutely and uniformly on $[a,b] \times [a,b]$.

8. Let $X = \{X_t, t \geq 0\}$ be a Gaussian process with $\mathsf{E}\,X_t = 0$ and covariance function $r(s,t) = e^{-|t-s|}$, $s,t \geq 0$. Let $0 < t_1 < \cdots < t_n$ and let $f_{t_1,\ldots,t_n}(x_1, \ldots, x_n)$ be the density of $X_{t_1}, \ldots, X_{t_n}$. Prove that

$$f_{t_1,\ldots,t_n}(x_1, \ldots, x_n) = \left[(2\pi)^n \prod_{i=2}^{n} \left(1 - e^{2(t_{i-1}-t_i)} \right) \right]^{-1/2}$$

$$\times \exp\left\{ -\frac{x_1^2}{2} - \frac{1}{2} \sum_{i=2}^{n} \frac{(x_i - e^{(t_{i-1}-t_i)}x_{i-1})^2}{1 - e^{2(t_{i-1}-t_i)}} \right\}.$$

9. Let $f = \{f_n, n \geq 1\} \subset L^2(0,1)$ be a complete orthonormal system and (ξ_n) independent identically distributed $\mathcal{N}(0,1)$-random variables. Show that $B_t = \sum_{n\geq 1} \xi_n \int_0^t f_n(u)\, du$ is the Brownian motion process.

10. Prove that for Gaussian systems $(\xi, \eta_1, \ldots, \eta_n)$ the conditional expectations $\mathsf{E}(\xi \mid \eta_1, \ldots, \eta_n)$ are the same as the conditional expectations $\hat{\mathsf{E}}(\xi \mid \eta_1, \ldots, \eta_n)$ in the wide sense.

11. Let $(\xi, \eta_1, \ldots, \eta_k)$ be a Gaussian system. Determine the structure of the conditional moments $\mathsf{E}(\xi^n \mid \eta_1, \ldots, \eta_k)$, $n \geq 1$ (as functions of $\eta_1, \ldots, \eta_k$).

12. Let $X = (X_k)_{1\leq k\leq n}$ and $Y = (Y_k)_{1\leq k\leq n}$ be Gaussian random sequences with $\mathsf{E}X_k = \mathsf{E}Y_k$, $\operatorname{Var} X_k = \operatorname{Var} Y_k$, $1 \leq k \leq n$, and

$$\operatorname{Cov}(X_k, X_l) \leq \operatorname{Cov}(Y_k, Y_l), \quad 1 \leq k, l \leq n.$$

Prove *Slepyan's inequality*:

$$\mathsf{P}\left\{ \sup_{1\leq k\leq n} X_k < x \right\} \leq \mathsf{P}\left\{ \sup_{1\leq k\leq n} Y_k < x \right\}, \quad x \in R.$$

13. Prove that if $B^\circ = (B_t^\circ)_{0\leq t\leq 1}$ is a Brownian bridge, then the process $B = (B_t)_{t\geq 0}$ with $B_t = (1+t)B_{t/(1+t)}^\circ$ is a Brownian motion.

14. Verfy that if $B = (B_t)_{t\geq 0}$ is a Brownian motion, then so are the following processes:

$$B_t^{(1)} = -B_t;$$
$$B_t^{(2)} = tB_{1/t}, \ t > 0, \quad \text{and} \quad B_0^{(2)} = 0;$$
$$B_t^{(3)} = B_{t+s} - B_s, \quad s > 0;$$
$$B_t^{(4)} = B_T - B_{T-t} \quad \text{for } 0 \leq t \leq T, \ T > 0;$$
$$B_t^{(5)} = \frac{1}{a}B_{a^2 t}, \quad a > 0 \text{ (automodelling property)}.$$

15. For a Gaussian sequence $X = (X_k)_{1\leq k\leq n}$ denote $m = \max_{1\leq k\leq n} \mathsf{E}X_k$ and $\sigma^2 = \max_{1\leq k\leq n} \operatorname{Var} X_k$, and let

$$\mathsf{P}\left\{ \max_{1\leq k\leq n} (X_k - \mathsf{E}X_k) \geq a \right\} \leq 1/2 \quad \text{for some } a.$$

Then the following *Borel's inequality* holds:

$$\mathsf{P}\left\{ \max_{1\leq k\leq n} X_k > x \right\} \leq 2\Psi\left(\frac{x - m - a}{\sigma} \right),$$

where $\Psi(x) = (2\pi)^{-1/2} \int_x^\infty e^{-y^2/2}\, dy$.

16. Let (X, Y) be a two-variate Gaussian random variable with $\mathsf{E} X = \mathsf{E} Y = 0$, $\mathsf{E} X^2 > 0$, $\mathsf{E} Y^2 > 0$ and correlation coefficient $\rho = \frac{\mathsf{E} XY}{\sqrt{\mathsf{E} X^2 \, \mathsf{E} Y^2}}$. Show that

$$\mathsf{P}\{XY < 0\} = 1 - 2\,\mathsf{P}\{X > 0, Y > 0\} = \pi^{-1} \arccos \rho.$$

17. Let $Z = XY$, where $X \sim \mathcal{N}(0, 1)$ and $\mathsf{P}\{Y = 1\} = \mathsf{P}\{Y = -1\} = \frac{1}{2}$. Find the distributions of pairs (X, Z) and (Y, Z) and the distribution of $X + Z$. Show that $Z \sim \mathcal{N}(0, 1)$ and that X and Z are uncorrelated but dependent.

18. Give a detailed proof that the processes $(B_t)_{0 \le t \le 1}$ defined by (26) and (28) are Brownian motions.

19. Let $B^\mu = (B_t + \mu t)_{t \ge 0}$ be a Brownian motion with a drift.
    (a) Find the distribution of $B_{t_1}^\mu + B_{t_2}^\mu$, $t_1 < t_2$.
    (b) Find $\mathsf{E} B_{t_0}^\mu B_{t_1}^\mu$ and $\mathsf{E} B_{t_0}^\mu B_{t_1}^\mu B_{t_2}^\mu$ for $t_0 < t_1 < t_2$.

20. For the process B^μ as in the previous problem, find the conditional distributions

$$\mathsf{P}(B_{t_2}^\mu \in \cdot \mid B_{t_1}^\mu)$$

for $t_1 < t_2$ and $t_1 > t_2$ and

$$\mathsf{P}(B_{t_2}^\mu \in \cdot \mid B_{t_0}^\mu, B_{t_1}^\mu)$$

for $t_0 < t_1 < t_2$.

Chapter 3
Convergence of Probability Measures.
Central Limit Theorem

In the formal construction of a course in the theory of probability, limit theorems appear as a kind of superstructure over elementary chapters, in which all problems have finite, purely arithmetical character. In reality, however, the epistemological value of the theory of probability is revealed only by limit theorems. Moreover, without limit theorems it is impossible to understand the real content of the primary concept of all our sciences – the concept of probability.

B. V. Gnedenko and A. N. Kolmogorov,

"Limit Distributions for Sums of Independent Random Variables" [34]

1 Weak Convergence of Probability Measures and Distributions

1. Many of the fundamental results in probability theory are formulated as *limit theorems*. Bernoulli's law of large numbers was formulated as a limit theorem; so was the de Moivre–Laplace theorem, which can fairly be called the origin of a genuine theory of probability and, in particular, which led the way to numerous investigations that clarified the conditions for the validity of the central limit theorem. Poisson's theorem on the approximation of the binomial distribution by the "Poisson" distribution in the case of rare events was formulated as a limit theorem. After the example of these propositions, and of results on the rate of convergence in the de Moivre–Laplace and Poisson theorems, it became clear that in probability it is necessary to deal with various kinds of convergence of distributions, and to establish the rate of convergence requires the introduction of various "natural" measures of the distance between distributions.

In the present chapter we shall discuss some general features of the *convergence* of probability distributions and of the *distance* between them. In this section we take up questions in the general theory of *weak convergence of probability measures in metric spaces*. (This is the area to which J. Bernoulli's law of large numbers, as well as the de Moivre–Laplace theorem, the progenitor of the central limit theorem,

© Springer Science+Business Media New York 2016
A.N. Shiryaev, *Probability-1*, Graduate Texts
in Mathematics 95, DOI 10.1007/978-0-387-72206-1_3

belong.) From Sect. 3, it will become clear that the method of characteristic functions is one of the most powerful means for proving limit theorems on the weak convergence of probability distributions in R^n. In Sect. 7, we consider questions of metrizability of weak convergence. Then, in Sect. 9, we turn our attention to a different kind of convergence of distributions (stronger than weak convergence), namely *convergence in variation*. Proofs of the simplest results on the rate of convergence in the central limit theorem and Poisson's theorem will be given in Sects. 11 and 12. In Sect. 13 the results on weak convergence of Sects. 1 and 2 are applied to certain (conceptually important) problems of mathematical statistics.

2. We begin by recalling the statement of the *law of large numbers* (Sect. 5 of Chap. 1) *for the Bernoulli scheme.*

Let $\xi_1, \xi_2, \ldots$ be a sequence of independent identically distributed random variables with $P(\xi_i = 1) = p$, $P(\xi_i = 0) = q$, $p + q = 1$. In terms of the concept of convergence in probability (Sect. 10, Chap. 2), Bernoulli's law of large numbers can be stated as follows:

$$\frac{S_n}{n} \xrightarrow{P} p, \quad n \to \infty, \tag{1}$$

where $S_n = \xi_1 + \cdots + \xi_n$. (It will be shown in Chapter 4, Vol. 2 that in fact we have convergence with probability 1.)

We put

$$F_n(x) = P\left\{\frac{S_n}{n} \leq x\right\},$$
$$F(x) = \begin{cases} 1, & x \geq p, \\ 0, & x < p, \end{cases} \tag{2}$$

where $F(x)$ is the distribution function of the degenerate random variable $\xi \equiv p$. Also let P_n and P be the probability measures on $(R, \mathscr{B}(R))$ corresponding to the distributions F_n and F.

In accordance with Theorem 2 of Sect. 10, Chap. 2, convergence in probability, $S_n/n \xrightarrow{P} p$, implies convergence *in distribution*, $S_n/n \xrightarrow{d} p$, which means that

$$E f\left(\frac{S_n}{n}\right) \to E f(p), \quad n \to \infty, \tag{3}$$

for every function $f = f(x)$ belonging to the class C of *bounded continuous* functions on R.

Since

$$E f\left(\frac{S_n}{n}\right) = \int_R f(x) P_n(dx), \quad E f(p) = \int_R f(x) P(dx),$$

(3) can be written in the form

$$\int_R f(x) P_n(dx) \to \int_R f(x) P(dx), \quad f \in C(R), \tag{4}$$

or (in accordance with Sect. 6 of Chap. 2) in the form

$$\int_R f(x)\,dF_n(x) \to \int_R f(x)\,dF(x), \quad f \in C. \tag{5}$$

In analysis, (4) is called *weak convergence* (of P_n to P, $n \to \infty$) and written $P_n \xrightarrow{w} P$ (cf. Definition 2 below). It is also natural to call (5) weak convergence of F_n to F and denote it by $F_n \xrightarrow{w} F$.

Thus we may say that in a Bernoulli scheme

$$\frac{S_n}{n} \xrightarrow{P} p \Rightarrow F_n \xrightarrow{w} F. \tag{6}$$

It is also easy to see from (1) that, for the distribution functions defined in (2),

$$F_n(x) \to F(x), \quad n \to \infty,$$

for all points $x \in R$ *except for the single point* $x = p$, where $F(x)$ has a discontinuity.

This shows that weak convergence $F_n \to F$ *does not imply* pointwise convergence of $F_n(x)$ to $F(x)$ as $n \to \infty$, for *all* points $x \in R$. However, it turns out that, both for Bernoulli schemes and for arbitrary distribution functions, weak convergence is equivalent (see Theorem 2 below) to "convergence in general" in the sense of the following definition.

Definition 1. A sequence of distribution functions $\{F_n\}$, defined on the real line, converges *in general* to the distribution function F (notation: $F_n \Rightarrow F$) if, as $n \to \infty$,

$$F_n(x) \to F(x), \quad x \in \mathbb{C}(F),$$

where $\mathbb{C}(F)$ is the set of points of continuity of $F = F(x)$.

For Bernoulli schemes, $F = F(x)$ is degenerate, and it is easy to see (see Problem 7 of Sect. 10, Chap. 2) that

$$(F_n \Rightarrow F) \Rightarrow \left(\frac{S_n}{n} \xrightarrow{P} p\right).$$

Therefore, taking account of Theorem 2 below,

$$\left(\frac{S_n}{n} \xrightarrow{P} p\right) \Rightarrow (F_n \xrightarrow{w} F) \Leftrightarrow (F_n \Rightarrow F) \Rightarrow \left(\frac{S_n}{n} \xrightarrow{P} p\right) \tag{7}$$

and consequently the law of large numbers can be considered as a theorem on the *weak convergence of the distribution functions* defined in (2).

Let us write

$$F_n(x) = \mathsf{P}\left\{ \frac{S_n - np}{\sqrt{npq}} \le x \right\},$$

$$F(x) = \frac{1}{\sqrt{2\pi}} \int_{-\infty}^{x} e^{-u^2/2} du. \tag{8}$$

The de Moivre–Laplace theorem (Sect. 6, Chap. 1) states that $F_n(x) \to F(x)$ for all $x \in R$, and consequently $F_n \Rightarrow F$. Since, as we have observed, weak convergence $F_n \overset{w}{\to} F$ and convergence in general, $F_n \Rightarrow F$, are equivalent, we may therefore say that the de Moivre–Laplace theorem is also a theorem on the weak convergence of the distribution functions defined by (8).

These examples justify the concept of weak convergence of probability measures that will be introduced below in Definition 2. Although, on the real line, weak convergence is equivalent to convergence in general of the corresponding distribution functions, it is preferable to use weak convergence from the beginning. This is because in the first place it is easier to work with, and in the second place it remains useful in more general spaces than the real line, and in particular for metric spaces, including the especially important spaces R^n, R^∞, C, and D (see Sect. 3 of Chapter 2).

3. Let $(E, \mathscr{E}, \rho)$ be a metric space with metric $\rho = \rho(x, y)$ and σ-algebra $\mathscr{E}$ of Borel subsets generated by the open sets, and let $\mathsf{P}, \mathsf{P}_1, \mathsf{P}_2, \ldots$ be probability measures on $(E, \mathscr{E}, \rho)$.

Definition 2. A sequence of probability measures $\{\mathsf{P}_n\}$ *converges weakly* to the probability measure P (notation: $\mathsf{P}_n \overset{w}{\to} \mathsf{P}$) if

$$\int_E f(x)\, \mathsf{P}_n(dx) \to \int_E f(x)\, \mathsf{P}(dx) \tag{9}$$

for every function $f = f(x)$ in the class $C(E)$ of continuous bounded functions on E.

Definition 3. A sequence of probability measures $\{\mathsf{P}_n\}$ *converges in general* to the probability measure P (notation: $\mathsf{P}_n \Rightarrow \mathsf{P}$) if

$$\mathsf{P}_n(A) \to \mathsf{P}(A) \tag{10}$$

for every set A of $\mathscr{E}$ for which

$$\mathsf{P}(\partial A) = 0. \tag{11}$$

(Here ∂A denotes the boundary of A:

$$\partial A = [A] \cap [\overline{A}],$$

where $[A]$ is the closure of A.)

The following fundamental theorem shows the equivalence of the concepts of weak convergence and convergence in general for probability measures, and contains still other equivalent statements.

Theorem 1. *The following statements are equivalent.*

(I) $P_n \xrightarrow{w} P$,
(II) $\limsup P_n(A) \le P(A)$, *A closed*,
(III) $\liminf P_n(A) \ge P(A)$, *A open*,
(IV) $P_n \Rightarrow P$.

PROOF. (I)$\Rightarrow$(II). Let A be closed and

$$f_A^\varepsilon(x) = \left[1 - \frac{\rho(x,A)}{\varepsilon}\right]^+, \quad \varepsilon > 0,$$

where

$$\rho(x,A) = \inf\{\rho(x,y): y \in A\}, \quad [x]^+ = \max[0,x].$$

Let us also put

$$A^\varepsilon = \{x: \rho(x,A) < \varepsilon\}$$

and observe that $A^\varepsilon \downarrow A$ as $\varepsilon \downarrow 0$.

Since $f_A^\varepsilon(x)$ is bounded, continuous, and satisfies

$$P_n(A) = \int_E I_A(x)\, P_n(dx) \le \int_E f_A^\varepsilon(x)\, P_n(dx),$$

we have

$$\limsup_n P_n(A) \le \limsup_n \int_E f_A^\varepsilon(x)\, P_n(dx)$$

$$= \int_E f_A^\varepsilon(x)\, P(dx) \le P(A^\varepsilon) \downarrow P(A), \quad \varepsilon \downarrow 0,$$

which establishes the required implication.

The implications (II) $\Rightarrow$ (III) and (III) $\Rightarrow$ (II) become obvious if we take the complements of the sets concerned.

(III) $\Rightarrow$ (IV). Let $A^0 = A \backslash \partial A$ be the interior, and $[A]$ the closure, of A. Then from (II), (III), and the hypothesis $P(\partial A) = 0$, we have

$$\limsup_n P_n(A) \le \limsup_n P_n([A]) \le P([A]) = P(A),$$

$$\liminf_n P_n(A) \ge \liminf_n P_n(A^0) \ge P(A^0) = P(A),$$

and therefore $P_n(A) \to P(A)$ for every A such that $P(\partial A) = 0$.

(IV) $\to$ (I). Let $f = f(x)$ be a bounded continuous function with $|f(x)| \le M$. We put

$$D = \{t \in R: P\{x: f(x) = t\} \ne 0\}$$

and consider a decomposition $T_k = (t_0, t_1, \ldots, t_k)$ of $[-M, M]$:

$$-M = t_0 < t_1 < \cdots < t_k = M, \quad k \geq 1,$$

with $t_i \notin D$, $i = 0, 1, \ldots, k$. (Observe that D is at most countable since the sets $f^{-1}\{t\}$ are disjoint and P is finite.)

Let $B_i = \{x : t_i \leq f(x) < t_{i+1}\}$. Since $f(x)$ is continuous and therefore the set $f^{-1}(t_i, t_{i+1})$ is open, we have $\partial B_i \subseteq f^{-1}\{t_i\} \cup f^{-1}\{t_{i+1}\}$. The points $t_i, t_{i+1} \notin D$; therefore $\mathsf{P}(\partial B_i) = 0$ and, by (IV),

$$\sum_{i=0}^{k-1} t_i \, \mathsf{P}_n(B_i) \to \sum_{t=0}^{k-1} t_i \, \mathsf{P}(B_i). \tag{12}$$

But

$$\left| \int_E f(x) \, \mathsf{P}_n(dx) - \int_E f(x) \, \mathsf{P}(dx) \right| \leq \left| \int_E f(x) \, \mathsf{P}_n(dx) - \sum_{i=0}^{k-1} t_i \, \mathsf{P}_n(B_i) \right|$$

$$+ \left| \sum_{i=0}^{k-1} t_i \, \mathsf{P}_n(B_i) - \sum_{i=0}^{k-1} t_i \, \mathsf{P}(B_i) \right|$$

$$+ \left| \sum_{i=0}^{k-1} t_i \, \mathsf{P}(B_i) - \int_E f(x) \, \mathsf{P}(dx) \right|$$

$$\leq 2 \max_{0 \leq i \leq k-1} (t_{i+1} - t_i)$$

$$+ \left| \sum_{i=0}^{k-1} t_i \, \mathsf{P}_n(B_i) - \sum_{i=0}^{k-1} t_i \, \mathsf{P}(B_i) \right|,$$

whence, by (12), since the T_k $(k \geq 1)$ are arbitrary,

$$\lim_n \int_E f(x) \, \mathsf{P}_n(dx) = \int_E f(x) \, \mathsf{P}(dx).$$

This completes the proof of the theorem. $\qquad\qquad\square$

Remark 1. The functions $f(x) = I_A(x)$ and $f_A^\varepsilon(x)$ that appear in the proof that (I) $\Rightarrow$ (II) are respectively *upper semicontinuous* and *uniformly continuous*. Hence it is easy to show that each of the conditions of the theorem is equivalent to one of the following:

(V) $\int_E f(x) \, \mathsf{P}_n(dx) \to \int_E f(x) \, \mathsf{P}(dx)$ *for all bounded uniformly continuous* $f(x)$;

(VI) $\int_E f(x) \, \mathsf{P}_n(dx) \to \int_E f(x) \, \mathsf{P}(dx)$ *for all bounded functions satisfying the Lip-schitz condition* (see Lemma 2 in Sect. 7);

(VII) $\limsup \int_E f(x) \, \mathsf{P}_n(dx) \leq \int_E f(x) \, \mathsf{P}(dx)$ *for all bounded* $f(x)$ *that are upper semicontinuous* ($\limsup f(x_n) \leq f(x)$, $x_n \to x$);

(VIII) $\liminf \int_E f(x) \, \mathsf{P}_n(dx) \geq \int_E f(x) \, \mathsf{P}(dx)$ *for all bounded* $f(x)$ *that are lower semicontinuous* ($\liminf_n f(x_n) \geq f(x)$, $x_n \to x$).

Remark 2. Theorem 1 admits a natural generalization to the case when the probability measures P and P_n defined on $(E, \mathscr{E}, \rho)$ are replaced by *arbitrary* (not necessarily probability) *finite measures* μ and μ_n. For such measures we can introduce weak convergence $\mu_n \xrightarrow{w} \mu$ and convergence in general $\mu_n \Rightarrow \mu$ and, just as in Theorem 1, we can establish the equivalence of the following conditions:

(I*) $\mu_n \xrightarrow{w} \mu$;
(II*) $\limsup \mu_n(A) \leq \mu(A)$, *where A is closed and $\mu_n(E) \to \mu(E)$*;
(III*) $\liminf \mu_n(A) \geq \mu(A)$, *where A is open and $\mu_n(E) \to \mu(E)$*;
(IV*) $\mu_n \Rightarrow \mu$.

Each of these is equivalent to any of (V*)–(VIII*), which are (V)–(VIII) with P_n and P replaced by μ_n and μ.

4. Let $(R, \mathscr{B}(R))$ be the real line with the system $\mathscr{B}(R)$ of Borel sets generated by the Euclidean metric $\rho(x, y) = |x - y|$ (compare Remark 2 in Sect. 2, Chapter 2). Let P and P_n, $n \geq 1$, be probability measures on $(R, \mathscr{B}(R))$ and let F and F_n, $n \geq 1$, be the corresponding distribution functions.

Theorem 2. *The following conditions are equivalent*:

(1) $P_n \xrightarrow{w} P$,
(2) $P_n \Rightarrow P$,
(3) $F_n \xrightarrow{w} F$,
(4) $F_n \Rightarrow F$.

PROOF. Since (2) $\Leftrightarrow$(1)$\Leftrightarrow$(3), it is enough to show that (2) $\Leftrightarrow$(4).
If $P_n \Rightarrow P$, then in particular

$$P_n(-\infty, x] \to P(-\infty, x]$$

for all $x \in R$ such that $P\{x\} = 0$. But this means that $F_n \Rightarrow F$.
Now let $F_n \Rightarrow F$. To prove that $P_n \Rightarrow P$ it is enough (by Theorem 1) to show that $\liminf_n P_n(A) > P(A)$ for every open set A.
If A is open, there is a countable collection of disjoint open intervals $I_1, I_2, \dots$ (of the form (a, b)) such that $A = \sum_{k=1}^{\infty} I_k$. Choose $\varepsilon > 0$ and in each interval $I_k = (a_k, b_k)$ select a subinterval $I_k' = (a_k', b_k']$ such that $a_k', b_k' \in \mathbb{C}(F)$ and $P(I_k) \leq P(I_k') + \varepsilon \cdot 2^{-k}$. (Since $F(x)$ has at most countably many discontinuities, such intervals I_k', $k \geq 1$, certainly exist.) By Fatou's lemma,

$$\liminf_n P_n(A) = \liminf_n \sum_{k=1}^{\infty} P_n(I_k)$$

$$\geq \sum_{k=1}^{\infty} \liminf_n P_n(I_k) \geq \sum_{k=1}^{\infty} \liminf_n P_n(I_k').$$

But

$$P_n(I_k') = F_n(b_k') - F_n(a_k') \to F(b_k') - F(a_k') = P(I_k').$$

Therefore

$$\liminf_n P_n(A) \geq \sum_{k=1}^{\infty} P(I_k') \geq \sum_{k=1}^{\infty} (P(I_k) - \varepsilon \cdot 2^{-k}) = P(A) - \varepsilon.$$

Since $\varepsilon > 0$ is arbitrary, this shows that $\liminf_n P_n(A) \geq P(A)$ if A is open.
This completes the proof of the theorem.   □

5. Let $(E, \mathscr{E})$ be a measurable space. A collection $\mathscr{K}_0(E) \subseteq \mathscr{E}$ of subsets is a *determining class* if whenever two probability measures P and Q on $(E, \mathscr{E})$ satisfy

$$P(A) = Q(A) \quad \text{for all } A \in \mathscr{K}_0(E)$$

it follows that the measures are identical, i.e.,

$$P(A) = Q(A) \quad \text{for all } A \in \mathscr{E}.$$

If $(E, \mathscr{E}, \rho)$ is a metric space, a collection $\mathscr{K}_1(E) \subseteq \mathscr{E}$ is a *convergence-determining class* if whenever probability measures $P, P_1, P_2, \ldots$ satisfy

$$P_n(A) \to P(A) \quad \text{for all } A \in \mathscr{K}_1(E) \text{ with } P(\partial A) = 0$$

it follows that

$$P_n(A) \to P(A) \quad \text{for all } A \in E \text{ with } P(\partial A) = 0.$$

When $(E, \mathscr{E}) = (R, \mathscr{B}(R))$, we can take a determining class $\mathscr{K}_0(R)$ to be the class of "elementary" sets $\mathscr{K} = \{(-\infty, x], \ x \in R\}$ (Theorem 1 in Sect. 3, Chap. 2). It follows from the equivalence of (2) and (4) of Theorem 2 that this class $\mathscr{K}$ is also a convergence-determining class.

It is natural to ask about such determining classes in more general spaces.

For R^n, $n \geq 2$, the class $\mathscr{K}$ of "elementary" sets of the form $(-\infty, x] = (-\infty, x_1] \times \cdots \times (-\infty, x_n]$, where $x = (x_1, \ldots, x_n) \in R^n$, is both a determining class (Theorem 2 in Sect. 3, Chap. 2) and a convergence-determining class (Problem 2).

For R^∞ the *cylindrical* sets are the "elementary" sets whose probabilities uniquely determine the probabilities of the Borel sets (Theorem 3 in Sect. 3, Chap. 2). It turns out that in this case the class of cylindrical sets is also the class of convergence-determining sets (Problem 3).

We might expect that the cylindrical sets would still constitute determining classes in more general spaces. However, this is, in general, not the case.

For example, consider the space $(C, \mathscr{B}(C), \rho)$ with the uniform metric ρ (see Subsection 6 in Sect. 2, Chap. 2). Let P be the probability measure concentrated on the element $x = x(t) \equiv 0, 0 \leq t \leq 1$, and let $P_n, n \geq 1$, be the probability measures each of which is concentrated on the element $x = x_n(t)$ shown in Fig. 35. It is easy

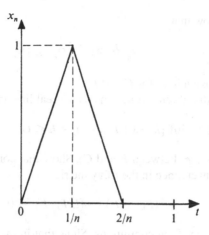

Fig. 35

to see that $P_n(A) \to P(A)$ for all cylindrical sets A with $P(\partial A) = 0$. But if we consider, for example, the set

$$A = \{\alpha \in C: |\alpha(t)| \leq \tfrac{1}{2}, \ 0 \leq t \leq 1\} \in \mathscr{B}_0(C),$$

then $P(\partial A) = 0$, $P_n(A) = 0$, $P(A) = 1$ and consequently $P_n \not\Rightarrow P$. ($\mathscr{B}_0(C)$ is the σ-algebra generated by open sets, see Subsection 6 in Sect. 2 of Chap. 2.)

Therefore the class of cylindrical sets in this case is a determining, but not a convergence-determining class.

6. PROBLEMS

1. Let us say that a function $F = F(x)$, defined on R^m, is *continuous at $x \in R^m$* provided that, for every $\varepsilon > 0$, there is a $\delta > 0$ such that $|F(x) - F(y)| < \varepsilon$ for all $y \in R^m$ that satisfy

$$x - \delta e < y < x + \delta e,$$

where $e = (1, \ldots, 1) \in R^m$. Let us say that a sequence of distribution functions $\{F_n\}$ *converges in general* to the distribution function F ($F_n \Rightarrow F$) if $F_n(x) \to F(x)$ for all points $x \in R^m$, where $F = F(x)$ is continuous. Show that the conclusion of Theorem 2 remains valid for R^m, $m > 1$. (See the Remark 1 on Theorem 1.)

2. Show that the class $\mathscr{K}$ of "elementary" sets in R^n is a convergence-determining class.

3. Let E be one of the spaces R^∞, C, or D. Let us say that a sequence $\{P_n\}$ of probability measures (defined on the σ-algebra $\mathscr{E}$ of Borel sets generated by the open sets) *converges in general in the sense of finite-dimensional distributions* to the probability measure P (notation: $P_n \overset{f}{\Rightarrow} P$) if $P_n(A) \to P(A)$, $n \to \infty$, for all *cylindrical* sets A with $P(\partial A) = 0$.

For R^∞, show that

$$(P_n \overset{f}{\Rightarrow} P) \Leftrightarrow (P_n \Rightarrow P).$$

Does this conclusion hold for C and D?

4. Let F and G be distribution functions on the real line and let

$$L(F, G) = \inf \{h > 0 : F(x - h) - h \le G(x) \le F(x + h) + h\}$$

be the *Lévy distance* (between F and G). Show that convergence in general is equivalent to convergence in the Lévy metric:

$$(F_n \Rightarrow F) \Leftrightarrow (L(F_n, F) \to 0).$$

5. Let $F_n \Rightarrow F$ and let F be continuous. Show that in this case $F_n(x)$ converges *uniformly* to $F(x)$:

$$\sup_x |F_n(x) - F(x)| \to 0, \quad n \to \infty.$$

6. Prove the statement in Remark 1 on Theorem 1.
7. Establish the equivalence of (I^*)–(IV^*) as stated in Remark 2 on Theorem 1.
8. Show that $P_n \overset{w}{\to} P$ if and only if every subsequence $\{P_{n'}\}$ of $\{P_n\}$ contains a subsequence $\{P_{n''}\}$ such that $P_{n''} \overset{w}{\to} P$.
9. Give an example of probability measures P, P_n on $(R, \mathscr{B}(R))$, $n \ge 1$, such that $P_n \overset{w}{\to} P$, but convergence $P_n(B) \to P(B)$ need not hold for *all* Borel sets $B \in \mathscr{B}(R)$.
10. Give an example of distribution functions $F = F(x)$, $F_n = F_n(x)$, $n \ge 1$, such that $F_n \overset{w}{\to} F$, but $\sup_x |F_n(x) - F(x)| \nrightarrow 0$, $n \to \infty$.
11. In many handbooks on probability theory the statement $(4) \Rightarrow (3)$ of Theorem 2 on convergence of distribution functions F_n, $n \ge 1$, to a distribution function F is related to the names of Helly and Bray. In this connection we propose to prove the following statements:
    (a) *Helly–Bray Lemma.* If $F_n \Rightarrow F$ (see Definition 1), then

$$\lim_n \int_a^b g(x) \, dF_n(x) = \int_a^b g(x) \, dF(x),$$

where a and b are continuity points of the distribution function $F = F(x)$ and $g = g(x)$ is a continuous function on $[a, b]$.
    (b) *Helly–Bray Theorem.* If $F_n \Rightarrow F$ and $g = g(x)$ is a continuous function on R, then

$$\lim_n \int_{-\infty}^\infty g(x) \, dF_n(x) = \int_{-\infty}^\infty g(x) \, dF(x).$$

12. Show that if $F_n \Rightarrow F$ and the sequence $\left(\int |x|^b \, dF_n(x) \right)_{n \geq 1}$ is bounded for some $b > 0$, then

$$\lim_n \int |x|^a \, dF_n(x) = \int |x|^a \, dF(x), \quad 0 \leq a \leq b,$$

$$\lim_n \int x^k \, dF_n(x) = \int x^k \, dF(x) \quad \text{for any } k = 1, 2, \ldots, [b], \ k \neq b.$$

13. Let $F_n \Rightarrow F$ and $m = \text{med}(F)$, $m_n = \text{med}(F_n)$ be the medians of F and F_n respectively (see Problem 5 in Sect. 4, Chap. 1). Suppose that m and m_n are uniquely defined for all $n \geq 1$. Prove that $m_n \to m$.

14. Let F be a distribution function that is uniquely determined by its moments $a_k = \int_{-\infty}^{\infty} x^k \, dF(x)$, $k = 1, 2, \ldots$ Let $(F_n)_{n \geq 1}$ be a sequence of distribution functions whose moments converge to those of F,

$$a_{n,k} = \int_{-\infty}^{\infty} x^k \, dF_n(x) \to a_k = \int_{-\infty}^{\infty} x^k \, dF(x), \quad k = 1, 2, \ldots$$

Show that then $F_n \Rightarrow F$.

15. Prove the following version of the *law of large numbers* (due to Khinchin): Let $X_1, X_2, \ldots$ be pairwise independent identically distributed random variables with a finite expectation $\mathsf{E} X_1 = m$, and let $S_n = X_1 + \cdots + X_n$. Then $S_n / n \xrightarrow{\mathsf{P}} m$.

2 Relative Compactness and Tightness of Families of Probability Distributions

1. If we are given a sequence of probability measures, then before we can consider the question of its (weak) convergence to some probability measure, we have of course to establish whether the sequence converges at all to some measure, or has at least one convergent subsequence.

For example, the sequence $\{\mathsf{P}_n\}$, where $\mathsf{P}_{2n} = \mathsf{P}$, $\mathsf{P}_{2n+1} = \mathsf{Q}$, and P and Q are different probability measures, is evidently not convergent, but has the two convergent subsequences $\{\mathsf{P}_{2n}\}$ and $\{\mathsf{P}_{2n+1}\}$.

It is easy to construct a sequence $\{\mathsf{P}_n\}$ of probability measures P_n, $n \geq 1$, that not only fails to converge, but contains no convergent subsequences at all. All that we have to do is to take P_n, $n \geq 1$, to be concentrated at $\{n\}$ (that is, $\mathsf{P}_n\{n\} = 1$). In fact, since $\lim_n \mathsf{P}_n(a, b] = 0$ whenever $a < b$, a limit measure would have to be identically zero, contradicting the fact that $1 = \mathsf{P}_n(R) \nrightarrow 0$, $n \to \infty$. It is interesting to observe that in this example the corresponding sequence $\{F_n\}$ of distribution functions,

$$F_n(x) = \begin{cases} 1, & x \geq n, \\ 0, & x < n, \end{cases}$$

is evidently convergent: for every $x \in R$,

$$F_n(x) \to G(x) \equiv 0.$$

However, the limit function $G = G(x)$ is not a distribution function (in the sense of Definition 1 of Sect. 3, Chap. 2).

This instructive example shows that the space of distribution functions is not *compact*. It also shows that if a sequence of distribution functions is to converge to a limit that is also a distribution function, we must have some conditions that will prevent mass from "escaping to infinity." (See in this connection Problem 3 in Sect. 3.)

After these introductory remarks, which illustrate the kinds of difficulty that can arise, we turn to the basic definitions.

2. Let us suppose that all measures are defined on the metric space $(E, \mathcal{E}, \rho)$.

Definition 1. A family of probability measures $\mathcal{P} = \{P_\alpha; \alpha \in \mathfrak{A}\}$ is *relatively compact* if every sequence of measures from $\mathcal{P}$ contains a subsequence which converges weakly to a probability measure.

We emphasize that in this definition the limit measure is to be a *probability measure*, although it need not belong to the original class $\mathcal{P}$. (This is why the word "relatively" appears in the definition.)

It is often far from simple to verify that a given family of probability measures is relatively compact. Consequently it is desirable to have simple and useable tests for this property. We need the following definitions.

Definition 2. A family of probability measures $\mathcal{P} = \{P_\alpha; \alpha \in \mathfrak{A}\}$ is *tight* if, for every $\varepsilon > 0$, there is a compact set $K \subseteq E$ such that

$$\sup_{\alpha \in \mathfrak{A}} P_\alpha(E \backslash K) \leq \varepsilon. \tag{1}$$

Definition 3. A family of distribution functions $F = \{F_\alpha; \alpha \in \mathfrak{A}\}$ defined on R^n, $n \geq 1$, is *relatively compact* (or *tight*) if the same property is possessed by the family $\mathcal{P} = \{P_\alpha; \alpha \in \mathfrak{A}\}$ of probability measures, where P_α is the measure constructed from F_α.

3. The following result is fundamental for the study of weak convergence of probability measures.

Theorem 1 (Prokhorov's Theorem). *Let $\mathcal{P} = \{P_\alpha; \alpha \in \mathfrak{A}\}$ be a family of probability measures defined on a complete separable metric space $(E, \mathcal{E}, \rho)$. Then $\mathcal{P}$ is relatively compact if and only if it is tight.*

PROOF. We shall give the proof only when the space is the real line. (The proof can be carried over (see [9], [76]), almost unchanged, to arbitrary Euclidean spaces R^n, $n \geq 2$. Then the theorem can be extended successively to R^∞, to σ-compact spaces; and finally to general complete separable metric spaces, by reducing each case to the preceding one.)

Necessity. Let the family $\mathscr{P} = \{P_\alpha; \alpha \in \mathfrak{A}\}$ of probability measures defined on $(R, \mathscr{B}(R))$ be relatively compact but not tight. Then there is an $\varepsilon > 0$ such that for every compact set $K \subseteq R$

$$\sup_\alpha P_\alpha(R \backslash K) > \varepsilon,$$

and therefore, for each interval $I = (a, b)$,

$$\sup_\alpha P_\alpha(R \backslash I) > \varepsilon.$$

It follows that for every interval $I_n = (-n, n)$, $n \geq 1$, there is a measure P_{α_n} such that

$$P_{\alpha_n}(R \backslash I_n) > \varepsilon.$$

Since the original family $\mathscr{P}$ is relatively compact, we can select from $\{P_{\alpha_n}\}_{n \geq 1}$ a subsequence $\{P_{\alpha_{n_k}}\}$ such that $P_{\alpha_{n_k}} \xrightarrow{w} Q$, where Q is a probability measure.

Then, by the equivalence of conditions (I) and (II) in Theorem 1 of Sect. 1, we have

$$\limsup_{k \to \infty} P_{\alpha_{n_k}}(R \backslash I_n) \leq Q(R \backslash I_n) \tag{2}$$

for every $n \geq 1$. But $Q(R \backslash I_n) \downarrow 0$, $n \to \infty$, and the left side of (2) exceeds $\varepsilon > 0$. This contradiction shows that relatively compact families are tight.

To prove the sufficiency we need a general result (*Helly's theorem*) on the *sequential compactness* of families of generalized distribution functions (Subsection 2 of Sect. 3, Chap. 2).

Let $\mathscr{I} = \{G\}$ be the collection of generalized distribution functions $G = G(x)$ that satisfy:

(1) $G(x)$ is *nondecreasing*;
(2) $0 \leq G(-\infty)$, $G(+\infty) \leq 1$;
(3) $G(x)$ is *continuous on the right*.

Then $\mathscr{I}$ clearly contains the class of distribution functions $\mathscr{F} = \{F\}$ for which $F(-\infty) = 0$ and $F(+\infty) = 1$.

Theorem 2 (Helly's Theorem). *The class* $\mathscr{I} = \{G\}$ *of generalized distribution functions is sequentially compact, i.e., for every sequence* $\{G_n\}$ *of functions from* $\mathscr{I}$ *we can find a function* $G \in \mathscr{I}$ *and a subsequence* $\{n_k\} \subseteq \{n\}$ *such that*

$$G_{n_k}(x) \to G(x), \quad k \to \infty,$$

for every point x of the set $\mathbb{C}(G)$ of points of continuity of $G = G(x)$.

PROOF. Let $T = \{x_1, x_2, \ldots\}$ be a countable dense subset of R. Since the sequence of numbers $\{G_n(x_1)\}$ is *bounded*, there is a subsequence $N_1 = \{n_1^{(1)}, n_2^{(1)}, \ldots\}$ such that $G_{n_i^{(1)}}(x_1)$ approaches a limit g_1 as $i \to \infty$. Then we extract from N_1 a subsequence $N_2 = \{n_1^{(2)}, n_2^{(2)}, \ldots\}$ such that $G_{n_i^{(2)}}(x_2)$ approaches a limit g_2 as $i \to \infty$; and so on.

Define a function $G_T(x)$ on the set $T \subseteq R$ by

$$G_T(x_i) = g_i, \quad x_i \in T,$$

and consider the "Cantor" diagonal sequence $N = \{n_1^{(1)}, n_2^{(2)}, \ldots\}$. Then, for each $x_i \in T$, as $m \to \infty$, we have

$$G_{n_m^{(m)}}(x_i) \to G_T(x_i).$$

Finally, let us define $G = G(x)$ for all $x \in R$ by putting

$$G(x) = \inf\{G_T(y): y \in T, \ y > x\}. \tag{3}$$

We claim that $G = G(x)$ is the required function and $G_{n_m^{(m)}}(x) \to G(x)$ at all points x of continuity of G.

Since all the functions G_n under consideration are nondecreasing, we have $G_{n_m^{(m)}}(x) \leq G_{n_m^{(m)}}(y)$ for all x and y that belong to T and satisfy the inequality $x \leq y$. Hence $G_T(x) \leq G_T(y)$ for such x and y. It follows from this and (3) that $G = G(x)$ is nondecreasing.

Now let us show that it is continuous on the right. Let $x_k \downarrow x$ and $d = \lim_k G(x_k)$. Clearly $G(x) \leq d$, and we have to show that actually $G(x) = d$. Suppose the contrary, that is, let $G(x) < d$. It follows from (3) that there is a $y \in T$, $x < y$, such that $G_T(y) < d$. But $x < x_k < y$ for sufficiently large k, and therefore $G(x_k) \leq G_T(y) < d$ and $\lim_k G(x_k) < d$, which contradicts $d = \lim_k G(x_k)$. Thus we have constructed a function G that belongs to $\mathscr{I}$.

We now establish that $G_{n_m^{(m)}}(x^0) \to G(x^0)$ for every $x^0 \in \mathbb{C}(G)$.

If $x^0 < y \in T$, then

$$\limsup_m G_{n_m^{(m)}}(x^0) \leq \limsup_m G_{n_m^{(m)}}(y) = G_T(y),$$

whence

$$\limsup_m G_{n_m^{(m)}}(x^0) \leq \inf\{G_T(y): y > x^0, \ y \in T\} = G(x^0). \tag{4}$$

On the other hand, let $x^1 < y < x^0$, $y \in T$. Then

$$G(x^1) \leq G_T(y) = \lim_m G_{n_m^{(m)}}(y) = \liminf_m G_{n_m^{(m)}}(y) \leq \liminf_m G_{n_m^{(m)}}(x^0).$$

Hence if we let $x^1 \uparrow x^0$ we find that

$$G(x^0-) \le \liminf_m G_{n_m^{(m)}}(x^0). \tag{5}$$

But if $G(x^0-) = G(x^0)$ then (4) and (5) imply that $G_{n_m^{(m)}}(x^0) \to G(x^0)$, $m \to \infty$.

This completes the proof of the theorem.

$\square$

We can now complete the proof of Theorem 1.

Sufficiency. Let the family $\mathscr{P}$ be tight and let $\{P_n\}$ be a sequence of probability measures from $\mathscr{P}$. Let $\{F_n\}$ be the corresponding sequence of distribution functions.

By Helly's theorem, there are a subsequence $\{F_{n_k}\} \subseteq \{F_n\}$ and a generalized distribution function $G \in \mathscr{I}$ such that $F_{n_k}(x) \to G(x)$ for $x \in \mathbb{C}(G)$. Let us show that because $\mathscr{P}$ was assumed tight, the function $G = G(x)$ is in fact a genuine distribution function $(G(-\infty) = 0,\ G(+\infty) = 1)$.

Take $\varepsilon > 0$, and let $I = (a, b]$ be the interval for which

$$\sup_n P_n(R\backslash I) < \varepsilon,$$

or, equivalently,

$$1 - \varepsilon \le P_n(a, b], \quad n \ge 1.$$

Choose points $a', b' \in \mathbb{C}(G)$ such that $a' < a$, $b' > b$. Then $1 - \varepsilon \le P_{n_k}(a, b] \le P_{n_k}(a', b'] = F_{n_k}(b') - F_{n_k}(a') \to G(b') - G(a')$. It follows that $G(+\infty) - G(-\infty) = 1$, and since $0 \le G(-\infty) \le G(+\infty) \le 1$, we have $G(-\infty) = 0$ and $G(+\infty) = 1$.

Therefore the limit function $G = G(x)$ is a distribution function and $F_{n_k} \Rightarrow G$. Together with Theorem 2 of Sect. 1 this shows that $P_{n_k} \overset{w}{\to} Q$, where Q is the probability measure corresponding to the distribution function G.

This completes the proof of Theorem 1. $\square$

4. PROBLEMS

1. Carry out the proofs of Theorems 1 and 2 for R^n, $n \ge 2$.
2. Let P_α be a Gaussian measure on the real line, with parameters m_α and σ_α^2, $\alpha \in \mathfrak{A}$. Show that the family $\mathscr{P} = \{P_\alpha; \alpha \in \mathfrak{A}\}$ is tight if and only if

$$|m_\alpha| \le a, \quad \sigma_\alpha^2 \le b, \quad \alpha \in \mathfrak{A}.$$

3. Construct examples of tight and nontight families $\mathscr{P} = \{P_\alpha; \alpha \in \mathfrak{A}\}$ of probability measures defined on $(R^\infty, \mathscr{B}(R^\infty))$.
4. Let P be a probability measure on a metric space $(E, \mathscr{E}, \rho)$. The measure P is said to be *tight* (cf. Definition 2), if for any $\varepsilon > 0$ there is a compact set $K \subseteq E$ such that $P(K) \ge 1 - \varepsilon$. Prove the following assertion ("Ulam's theorem"): *Each probability measure on a Polish space* (i.e., complete separable metric space) *is tight*.

5. Let $X = \{X_\alpha; \alpha \in \mathfrak{A}\}$ be a family of random vectors ($X_\alpha \in R^d$, $\alpha \in \mathfrak{A}$) such that $\sup_\alpha \mathsf{E} \|X_\alpha\|^r < \infty$ for some $r > 0$. Show that the family $\mathscr{P} = \{P_\alpha; \alpha \in \mathfrak{A}\}$ of distributions $P_\alpha = \text{Law}(X_\alpha)$ is tight.

3 Proof of Limit Theorems by the Method of Characteristic Functions

1. The proofs of the first limit theorems of probability theory—the law of large numbers, and the de Moivre-Laplace and Poisson theorems for Bernoulli schemes—were based on *direct analysis* of the distribution functions F_n, which are expressed rather simply in terms of binomial probabilities. (In the Bernoulli scheme, we are adding random variables that take only two values, so that in principle we can find F_n explicitly.) However, it is practically impossible to apply a similar direct method to the study of more complicated random variables.

The first step in proving limit theorems for sums of arbitrarily distributed random variables was taken by Chebyshev. The inequality that he discovered, and which is now known as Chebyshev's inequality, not only makes it possible to give an elementary proof of James Bernoulli's law of large numbers, but also lets us establish very general conditions for this law to hold, when stated in the form

$$\mathsf{P}\left\{ \left| \frac{S_n}{n} - \frac{\mathsf{E} S_n}{n} \right| \geq \varepsilon \right\} \to 0, \quad n \to \infty, \quad \text{every } \varepsilon > 0, \tag{1}$$

for sums $S_n = \xi_1 + \cdots + \xi_n$, $n \geq 1$, of independent random variables. (See Problem 2.)

Furthermore, Chebyshev created (and Markov perfected) the "method of moments" which made it possible to show that the conclusion of the de Moivre-Laplace theorem, written in the form

$$\mathsf{P}\left\{ \frac{S_n - \mathsf{E} S_n}{\sqrt{\text{Var} \, S_n}} \leq x \right\} \to \frac{1}{\sqrt{2\pi}} \int_{-\infty}^{x} e^{-u^2/2} du \ (= \Phi(x)), \tag{2}$$

is "universal," in the sense that it is valid under very general hypotheses concerning the nature of the random variables. For this reason it is known as the *Central Limit Theorem* of probability theory.

Somewhat later Lyapunov proposed a different method for proving the central limit theorem, based on the idea (which goes back to Laplace) of the characteristic function of a probability distribution. Subsequent developments have shown that Lyapunov's method of characteristic functions is extremely effective for proving the most diverse limit theorems. Consequently it has been extensively developed and widely applied.

In essence, the method is as follows.

2. We already know (Chap. 2, Sect. 12) that there is a one-to-one correspondence between distribution functions and characteristic functions. Hence we can study the properties of distribution functions by using the corresponding characteristic functions. It is a fortunate circumstance that weak convergence $F_n \xrightarrow{w} F$ of distributions is equivalent to pointwise convergence $\varphi_n \to \varphi$ of the corresponding characteristic functions. Namely, we have the following result, which provides the basic method of proving theorems on weak convergence for distributions on the real line.

Theorem 1 (Continuity Theorem). *Let $\{F_n\}$ be a sequence of distribution functions $F_n = F_n(x)$, $x \in R$, and let $\{\varphi_n\}$ be the corresponding sequence of characteristic functions,*

$$\varphi_n(t) = \int_{-\infty}^{\infty} e^{itx} \, dF_n(x), \quad t \in R.$$

(1) *If $F_n \xrightarrow{w} F$, where $F = F(x)$ is a distribution function, then $\varphi_n(t) \to \varphi(t)$, $t \in R$, where $\varphi(t)$ is the characteristic function of $F = F(x)$.*

(2) *If $\lim_n \varphi_n(t)$ exists for each $t \in R$ and $\varphi(t) = \lim_n \varphi_n(t)$ is continuous at $t = 0$, then $\varphi(t)$ is the characteristic function of a probability distribution $F = F(x)$, and*

$$F_n \xrightarrow{w} F.$$

The proof of conclusion (1) is an immediate consequence of the definition of weak convergence, applied to the functions $\text{Re } e^{itX}$ and $\text{Im } e^{itx}$.

The proof of (2) requires some preliminary propositions.

Lemma 1. *Let $\{\mathsf{P}_n\}$ be a tight family of probability measures. Suppose that every weakly convergent subsequence $\{\mathsf{P}_{n'}\}$ of $\{\mathsf{P}_n\}$ converges to the same probability measure P. Then the whole sequence $\{\mathsf{P}_n\}$ converges to P.*

PROOF. Suppose that $\mathsf{P}_n \nrightarrow \mathsf{P}$. Then there is a bounded continuous function $f = f(x)$ such that

$$\int_R f(x) \, \mathsf{P}_n(dx) \nrightarrow \int_R f(x) \, \mathsf{P}(dx).$$

It follows that there exist $\varepsilon > 0$ and an infinite sequence $\{n'\} \subseteq \{n\}$ such that

$$\left| \int_R f(x) \, \mathsf{P}_{n'}(dx) - \int_R f(x) \, \mathsf{P}(dx) \right| \geq \varepsilon > 0. \tag{3}$$

By Prokhorov's theorem (Sect. 2) we can select a subsequence $\{\mathsf{P}_{n''}\}$ of $\{\mathsf{P}_{n'}\}$ such that $\mathsf{P}_{n''} \xrightarrow{w} \mathsf{Q}$, where Q is a probability measure.

By the hypotheses of the lemma, $Q = P$, and therefore

$$\int_R f(x) \, P_{n''}(dx) \to \int_R f(x) \, P(dx),$$

which leads to a contradiction with (3). This completes the proof of the lemma.

□

Lemma 2. *Let $\{P_n\}$ be a tight family of probability measures on $(R, \mathscr{B}(R))$. A necessary and sufficient condition for the sequence $\{P_n\}$ to converge weakly to a probability measure is that for each $t \in R$ the limit $\lim_n \varphi_n(t)$ exists, where $\varphi_n(t)$ is the characteristic function of P_n:*

$$\varphi_n(t) = \int_R e^{itx} \, P_n(dx).$$

PROOF. If $\{P_n\}$ is tight, by Prohorov's theorem there is a subsequence $\{P_{n'}\}$ and a probability measure P such that $P_{n'} \xrightarrow{w} P$. Suppose that the whole sequence $\{P_n\}$ does not converge to P ($P_n \xrightarrow{w}\!\!\!\!\!/ \ P$). Then, by Lemma 1, there is a subsequence $\{P_{n''}\}$ and a probability measure Q such that $P_{n''} \xrightarrow{w} Q$, and $P \neq Q$.

Now we use the existence of $\lim_n \varphi_n(t)$ for each $t \in R$. Then

$$\lim_{n'} \int_R e^{itx} \, P_{n'}(dx) = \lim_{n''} \int_R e^{itx} \, P_{n''}(dx)$$

and therefore

$$\int_R e^{itx} \, P(dx) = \int_R e^{itx} \, Q(dx), \quad t \in R.$$

But the characteristic function determines the distribution uniquely (Theorem 2, Sect. 12, Chap. 2). Hence $P = Q$, which contradicts the assumption that $P_n \xrightarrow{w}\!\!\!\!\!/ \ P$.

The converse part of the lemma follows immediately from the definition of weak convergence.

□

The following lemma estimates the "tails" of a distribution function in terms of the behavior of its characteristic function in a neighborhood of zero.

Lemma 3. *Let $F = F(x)$ be a distribution function on the real line and let $\varphi = \varphi(t)$ be its characteristic function. Then there is a constant $K > 0$ such that for every $a > 0$*

$$\int_{|x| \geq 1/a} dF(x) \leq \frac{K}{a} \int_0^a [1 - \operatorname{Re} \varphi(t)] \, dt. \tag{4}$$

PROOF. Since $\operatorname{Re}\varphi(t) = \int_{-\infty}^{\infty} \cos tx\, dF(x)$, we find by Fubini's theorem that

$$
\frac{1}{a}\int_0^a [1 - \operatorname{Re}\varphi(t)]\, dt = \frac{1}{a}\int_0^a \left[\int_{-\infty}^{\infty}(1 - \cos tx)\, dF(x)\right] dt
$$

$$
= \int_{-\infty}^{\infty}\left[\frac{1}{a}\int_0^a (1 - \cos tx)\, dt\right] dF(x)
$$

$$
= \int_{-\infty}^{\infty}\left(1 - \frac{\sin ax}{ax}\right) dF(x)
$$

$$
\geq \inf_{|y|\geq 1}\left(1 - \frac{\sin y}{y}\right)\cdot \int_{|ax|\geq 1} dF(x)
$$

$$
= \frac{1}{K}\int_{|x|\geq 1/a} dF(x),
$$

where

$$
\frac{1}{K} = \inf_{|y|\geq 1}\left(1 - \frac{\sin y}{y}\right) = 1 - \sin 1 \geq \tfrac{1}{7},
$$

so that (4) holds with $K = 7$. This establishes the lemma.
□

PROOF OF CONCLUSION (2) OF THEOREM 1. Let $\varphi_n(t) \to \varphi(t)$, $n \to \infty$, where $\varphi(t)$ is continuous at 0. Let us show that it follows that the family of probability measures $\{P_n\}$ is tight, where P_n is the measure corresponding to F_n.

By (4) and the dominated convergence theorem,

$$
P_n\left\{R\setminus\left(-\frac{1}{a}, \frac{1}{a}\right)\right\} = \int_{|x|\geq 1/a} dF_n(x) \leq \frac{K}{a}\int_0^a [1 - \operatorname{Re}\varphi_n(t)]\, dt
$$

$$
\to \frac{K}{a}\int_0^a [1 - \operatorname{Re}\varphi(t)]\, dt
$$

as $n \to \infty$.

Since, by hypothesis, $\varphi(t)$ is continuous at 0 and $\varphi(0) = 1$, for every $\varepsilon > 0$ there is an $a > 0$ such that

$$
P_n\left\{R\setminus\left(-\frac{1}{a}, \frac{1}{a}\right)\right\} \leq \varepsilon
$$

for all $n \geq 1$. Consequently $\{P_n\}$ is tight, and by Lemma 2 there is a probability measure P such that

$$
P_n \overset{w}{\to} P.
$$

Hence

$$
\varphi_n(t) = \int_{-\infty}^{\infty} e^{itx}\, P_n(dx) \to \int_{-\infty}^{\infty} e^{itx}\, P(dx),
$$

but also $\varphi_n(t) \to \varphi(t)$. Therefore $\varphi(t)$ is the characteristic function of P.
This completes the proof of the theorem. □

Corollary 1. *Let $\{F_n\}$ be a sequence of distribution functions and $\{\varphi_n\}$ the corresponding sequence of characteristic functions. Also let F be a distribution function and φ its characteristic function. Then $F_n \xrightarrow{w} F$ if and only if $\varphi_n(t) \to \varphi(t)$ for all $t \in R$.*

Remark 1. Let $\eta, \eta_1, \eta_2, \dots$ be random variables and $F_{\eta_n} \xrightarrow{w} F_\eta$. In accordance with the definition 4 of Sect. 10, Chap. 2, we then say that *the random variables $\eta_1, \eta_2, \dots$ converge to η in distribution, and write $\eta_n \xrightarrow{d} \eta$.*

Since this notation is self-explanatory, we shall frequently use it instead of $F_{\eta_n} \xrightarrow{w} F_\eta$ when stating limit theorems.

3. In the next section, Theorem 1 will be applied to prove the central limit theorem for independent but not identically distributed random variables. We will prove it assuming the condition known as *Lindeberg's condition*. Then we will show that *Lyapounov's condition* implies *Lindeberg's condition*. In the present section we shall merely apply the method of characteristic functions to prove some simple limit theorems.

Theorem 2 (Khinchin's Law of Large Numbers). *Let $\xi_1, \xi_2, \dots$ be a sequence of independent identically distributed random variables with $\mathsf{E}\,|\xi_1| < \infty$, $S_n = \xi_1 + \dots + \xi_n$ and $\mathsf{E}\,\xi_1 = m$. Then $S_n/n \xrightarrow{P} m$, that is, for every $\varepsilon > 0$*

$$\mathsf{P}\left\{\left|\frac{S_n}{n} - m\right| \geq \varepsilon\right\} \to 0, \quad n \to \infty.$$

PROOF. Let $\varphi(t) = \mathsf{E}\,e^{it\xi_1}$ and $\varphi_{S_n/n}(t) = \mathsf{E}\,e^{itS_n/n}$. Since the random variables are independent, we have

$$\varphi_{S_n/n}(t) = \left[\varphi\left(\frac{t}{n}\right)\right]^n$$

by (6) of Sect. 12, Chap. 2. But according to (14) of Sect. 12, Chap. 2

$$\varphi(t) = 1 + itm + o(t), \quad t \to 0.$$

Therefore for each given $t \in R$

$$\varphi\left(\frac{t}{n}\right) = 1 + i\frac{t}{n}m + o\left(\frac{1}{n}\right), \quad n \to \infty,$$

and therefore

$$\varphi_{S_n/n}(t) = \left[1 + i\frac{t}{n}m + o\left(\frac{1}{n}\right)\right]^n \to e^{itm}.$$

The function $\varphi(t) = e^{itm}$ is continuous at 0 and is the characteristic function of the degenerate probability distribution that is concentrated at m. Therefore

$$\frac{S_n}{n} \xrightarrow{d} m,$$

and consequently (see Problem 7 in Sect. 10, Chap. 2)

$$\frac{S_n}{n} \xrightarrow{\mathsf{P}} m.$$

This completes the proof of the theorem.

□

Theorem 3 (Central Limit Theorem for Independent Identically Distributed Random Variables). *Let* $\xi_1, \xi_2, \ldots$ *be a sequence of independent identically distributed (nondegenerate) random variables with* $\mathsf{E}\,\xi_1^2 < \infty$ *and* $S_n = \xi_1 + \cdots + \xi_n$. *Then as* $n \to \infty$

$$\mathsf{P}\left\{\frac{S_n - \mathsf{E}\,S_n}{\sqrt{\mathrm{Var}\,S_n}} \le x\right\} \to \Phi(x), \quad x \in R, \tag{5}$$

where

$$\Phi(x) = \frac{1}{\sqrt{2\pi}} \int_{-\infty}^{x} e^{-u^2/2}\, du.$$

PROOF. Let $\mathsf{E}\,\xi_1 = m$, $\mathrm{Var}\,\xi_1 = \sigma^2$ and

$$\varphi(t) = \mathsf{E}\,e^{it(\xi_1 - m)}.$$

Then if we put

$$\varphi_n(t) = \mathsf{E}\exp\left\{it\frac{S_n - \mathsf{E}\,S_n}{\sqrt{\mathrm{Var}\,S_n}}\right\},$$

we find that

$$\varphi_n(t) = \left[\varphi\left(\frac{t}{\sigma\sqrt{n}}\right)\right]^n.$$

But by (14) of Sect. 12, Chap. 2

$$\varphi(t) = 1 - \frac{\sigma^2 t^2}{2} + o(t^2), \quad t \to 0.$$

Therefore

$$\varphi_n(t) = \left[1 - \frac{\sigma^2 t^2}{2\sigma^2 n} + o\left(\frac{1}{n}\right)\right]^n \to e^{-t^2/2},$$

as $n \to \infty$ for fixed t.

The function $e^{-t^2/2}$ is the characteristic function of a normally distributed random variable with mean zero and unit variance (denoted by $\mathcal{N}(0,1)$). This, by Theorem 1, establishes (5). In accordance with Remark 1 this can also be written in the form

$$\frac{S_n - \mathsf{E}\,S_n}{\sqrt{\mathrm{Var}\,S_n}} \xrightarrow{d} \mathcal{N}(0,1). \tag{6}$$

This completes the proof of the theorem.

□

The preceding two theorems have dealt with the behavior of the probabilities of (normalized and centered) sums of independent and identically distributed random variables. However, in order to state Poisson's theorem (Sect. 6, Chap. 1) we have to use a more general model.

Let us suppose that for each $n \geq 1$ we are given independent random variables $\xi_{n1}, \ldots, \xi_{nn}$. In other words, let there be given a triangular array

$$\begin{pmatrix} \xi_{11} \\ \xi_{21}, \ \xi_{22} \\ \xi_{31}, \ \xi_{32}, \ \xi_{33} \\ \cdots\cdots\cdots \end{pmatrix}$$

of random variables, those in each row being independent. Put $S_n = \xi_{n1} + \cdots + \xi_{nn}$.

Theorem 4 (Poisson's Theorem). *For each $n \geq 1$ let the independent random variables $\xi_{n1}, \ldots, \xi_{nn}$ be such that*

$$\mathsf{P}(\xi_{nk} = 1) = p_{nk}, \quad \mathsf{P}(\xi_{nk} = 0) = q_{nk}$$

with $p_{nk} + q_{nk} = 1$. Suppose that

$$\max_{1 \leq k \leq n} p_{nk} \to 0, \quad \sum_{k=1}^{n} p_{nk} \to \lambda > 0, \quad n \to \infty.$$

Then, for each $m = 0, 1, \ldots,$

$$\mathsf{P}(S_n = m) \to \frac{e^{-\lambda}\lambda^m}{m!}, \quad n \to \infty. \tag{7}$$

PROOF. Since

$$\mathsf{E}\, e^{it\xi_{nk}} = p_{nk}e^{it} + q_{nk}$$

for $1 \leq k \leq n$, we have

$$\varphi_{S_n}(t) = \mathsf{E}\, e^{itS_n} = \prod_{k=1}^{n}(p_{nk}e^{it} + q_{nk})$$

$$= \prod_{k=1}^{n}(1 + p_{nk}(e^{it} - 1)) \to \exp\{\lambda(e^{it} - 1)\}, \quad n \to \infty.$$

The function $\varphi(t) = \exp\{\lambda(e^{it} - 1)\}$ is the characteristic function of the Poisson distribution (Example 3 in Sect. 12, Chap. 2), so that (7) is established. This completes the proof of the theorem.

□

If $\pi(\lambda)$ denotes a Poisson random variable with parameter λ, then (7) can be written like (6), in the form

$$S_n \xrightarrow{d} \pi(\lambda).$$

4. PROBLEMS

1. Prove Theorem 1 for R^n, $n \geq 2$.
2. Let $\xi_1, \xi_2, \ldots$ be a sequence of independent random variables with finite means $\mathsf{E}\,|\xi_n|$ and variances $\mathrm{Var}\,\xi_n$ such that $\mathrm{Var}\,\xi_n \leq K < \infty$, where K is a constant. Use Chebyshev's inequality to prove the law of large numbers (1).
3. In Corollary 1, show that the family $\{\varphi_n\}$ is *uniformly continuous* and that $\varphi_n \to \varphi$ uniformly on every finite interval.
4. Let ξ_n, $n \geq 1$, be random variables with characteristic functions $\varphi_{\xi_n}(t)$, $n \geq 1$. Show that $\xi_n \xrightarrow{d} 0$ if and only if $\varphi_{\xi_n}(t) \to 1$, $n \to \infty$, in some neighborhood of $t = 0$.
5. Let $X_1, X_2, \ldots$ be a sequence of independent identically distributed random vectors (with values in R^k) with mean zero and (finite) covariance matrix Γ. Show that
$$\frac{X_1 + \cdots + X_n}{\sqrt{n}} \xrightarrow{d} \mathcal{N}(0, \Gamma).$$
(Compare with Theorem 3.)
6. Let $\xi_1, \xi_2, \ldots$ and $\eta_1, \eta_2, \ldots$ be sequences of random variables such that ξ_n and η_n are independent for each n. Suppose that $\xi_n \xrightarrow{d} \xi$, $\eta_n \xrightarrow{d} \eta$ as $n \to \infty$, where ξ and η are independent. Prove that the sequence of two-dimensional random variables (ξ_n, η_n) converges in distribution to (ξ, η).

   Let $f = f(x, y)$ be a continuous function. Verify that the sequence $f(\xi_n, \eta_n)$ converges in distribution to $f(\xi, \eta)$.
7. Show by an example that in the statement (2) of Theorem 1 the condition of continuity at zero of the "limiting" characteristic function $\varphi(t) = \lim_n \varphi_n(t)$ cannot be, in general, relaxed. (In other words, if the characteristic function $\varphi(t)$ of F is not continuous at zero, then it is possible that $\varphi_n(t) \to \varphi(t)$, but $F_n \xrightarrow{w} F$.) Establish by an example that without continuity of $\varphi(t)$ at zero the tightness property of the family of distributions P_n with characteristic functions $\varphi_n(t)$, $n \geq 1$, may fail.

4 Central Limit Theorem for Sums of Independent Random Variables: I—Lindeberg's Condition

In this section we prove the Central Limit Theorem for (normalized and centered) sums S_n of independent random variables $\xi_1, \xi_2, \ldots, \xi_n$, $n \geq 1$, under the *classical Lindeberg condition*. In the next section we will consider a more general set-up: first, the Central Limit Theorem will be stated for a "triangle array" of random variables, and, secondly, it will be proved under the so-called *non-classical conditions*.

Theorem 1. *Let $\xi_1, \xi_2, \ldots$ be a sequence of independent random variables with finite second moments. Let $m_k = \mathsf{E}\,\xi_k$, $\sigma_k^2 = \mathrm{Var}\,\xi_k > 0$, $S_n = \xi_1 + \cdots + \xi_n$, $D_n^2 = \sum_{k=1}^n \sigma_k^2$ and let $F_k = F_k(x)$ be the distribution function of ξ_k.*
Suppose that the following "Lindeberg's condition" holds: for any $\varepsilon > 0$

$$(L) \qquad \frac{1}{D_n^2} \sum_{k=1}^{n} \int_{\{x:\, |x-m_k| \geq \varepsilon D_n\}} (x - m_k)^2 \, dF_k(x) \to 0, \qquad n \to \infty. \qquad (1)$$

Then

$$\frac{S_n - \mathsf{E}\, S_n}{\sqrt{\mathrm{Var}\, S_n}} \xrightarrow{d} \mathcal{N}(0, 1). \qquad (2)$$

PROOF. Without loss of generality we may assume that $m_k = 0$, $k \geq 1$. Denote $\varphi_k(t) = \mathsf{E}\, e^{it\xi_k}$, $T_n = \frac{S_n}{\sqrt{\mathrm{Var}\, S_n}} = \frac{S_n}{D_n}$, $\varphi_{S_n}(t) = \mathsf{E}\, e^{itS_n}$, $\varphi_{T_n}(t) = \mathsf{E}\, e^{itT_n}$.

Then

$$\varphi_{T_n}(t) = \mathsf{E}\, e^{itT_n} = \mathsf{E}\, e^{itS_n/D_n} = \varphi_{S_n}\!\left(\frac{t}{D_n}\right) = \prod_{k=1}^{n} \varphi_k\!\left(\frac{t}{D_n}\right) \qquad (3)$$

and by Theorem 1 of Sect. 3 for the proof of (2) it suffices to show that for any $t \in R$

$$\varphi_{T_n}(t) \to e^{-t^2/2}, \qquad n \to \infty. \qquad (4)$$

Take some $t \in R$, which will be fixed throughout the proof. Using the expansions

$$e^{iy} = 1 + iy + \frac{\theta_1 y^2}{2},$$

$$e^{iy} = 1 + iy - \frac{y^2}{2} + \frac{\theta_2 |y|^3}{3!},$$

which hold for any real y with some $\theta_1 = \theta_1(y)$, $\theta_2 = \theta_2(y)$ such that $|\theta_1| \leq 1$, $|\theta_2| \leq 1$, we find that

$$\varphi_k(t) = \mathsf{E}\, e^{it\xi_k} = \int_{-\infty}^{\infty} e^{itx} \, dF_k(x) = \int_{|x| \geq \varepsilon D_n} \left(1 + itx + \frac{\theta_1 (tx)^2}{2}\right) dF_k(x)$$

$$+ \int_{|x| < \varepsilon D_n} \left(1 + itx - \frac{t^2 x^2}{2} + \frac{\theta_2 |tx|^3}{6}\right) dF_k(x)$$

$$= 1 + \frac{t^2}{2} \int_{|x| \geq \varepsilon D_n} \theta_1 x^2 \, dF_k(x) - \frac{t^2}{2} \int_{|x| < \varepsilon D_n} x^2 \, dF_k(x) + \frac{|t|^3}{6} \int_{|x| < \varepsilon D_n} \theta_2 |x|^3 \, dF_k(x)$$

(we have used here that $m_k = \int_{-\infty}^{\infty} x \, dF_k(x) = 0$ by assumption).

Therefore

$$\varphi_k\!\left(\frac{t}{D_n}\right) = 1 - \frac{t^2}{2D_n^2} \int_{|x| < \varepsilon D_n} x^2 \, dF_k(x) + \frac{t^2}{2D_n^2} \int_{|x| \geq \varepsilon D_n} \theta_1 x^2 \, dF_k(x)$$

$$+ \frac{|t|^3}{6D_n^3} \int_{|x| < \varepsilon D_n} \theta_2 |x|^3 \, dF_k(x). \qquad (5)$$

Since

$$\left| \frac{1}{2} \int_{|x| \geq \varepsilon D_n} \theta_1 x^2 \, dF_k(x) \right| \leq \frac{1}{2} \int_{|x| \geq \varepsilon D_n} x^2 \, dF_k(x),$$

we have

$$\frac{1}{2} \int_{|x| \geq \varepsilon D_n} \theta_1 x^2 \, dF_k(x) = \tilde{\theta}_1 \int_{|x| \geq \varepsilon D_n} x^2 dF_k(x), \qquad (6)$$

where $\tilde{\theta}_1 = \tilde{\theta}_1(t, k, n)$ and $|\tilde{\theta}_1| \leq 1/2$.

In the same way

$$\left| \frac{1}{6} \int_{|x| < \varepsilon D_n} \theta_2 |x|^3 \, dF_k(x) \right| \leq \frac{1}{6} \int_{|x| < \varepsilon D_n} \frac{\varepsilon D_n}{|x|} |x|^3 \, dF_k(x) \leq \frac{1}{6} \int_{|x| < \varepsilon D_n} \varepsilon D_n x^2 \, dF_k(x)$$

and therefore

$$\frac{1}{6} \int_{|x| < \varepsilon D_n} \theta_2 |x|^3 \, dF_k(x) = \tilde{\theta}_2 \int_{|x| < \varepsilon D_n} \varepsilon D_n x^2 \, dF_k(x), \qquad (7)$$

where $\tilde{\theta}_2 = \tilde{\theta}_2(t, k, n)$ and $|\tilde{\theta}_2| \leq 1/6$.

Now let

$$A_{kn} = \frac{1}{D_n^2} \int_{|x| < \varepsilon D_n} x^2 \, dF_k(x), \qquad B_{kn} = \frac{1}{D_n^2} \int_{|x| \geq \varepsilon D_n} x^2 \, dF_k(x).$$

Then by (5)–(7)

$$\varphi_k \left(\frac{t}{D_n} \right) = 1 - \frac{t^2 A_{kn}}{2} + t^2 \tilde{\theta}_1 B_{kn} + |t|^3 \varepsilon \tilde{\theta}_2 A_{kn} \ (= 1 + C_{kn}). \qquad (8)$$

Note that

$$\sum_{k=1}^{n} (A_{kn} + B_{kn}) = 1 \qquad (9)$$

and according to Condition (1)

$$\sum_{k=1}^{n} B_{kn} \to 0, \quad n \to \infty. \qquad (10)$$

Therefore for all sufficiently large n

$$\max_{1 \leq k \leq n} |C_{kn}| \leq t^2 \varepsilon^2 + \varepsilon |t|^3 \qquad (11)$$

and

$$\sum_{k=1}^{n} |C_{kn}| \leq t^2 + \varepsilon |t|^3. \qquad (12)$$

Now we use that for any complex number z with $|z| \leq 1/2$

$$\log(1 + z) = z + \theta |z|^2,$$

where $\theta = \theta(z)$ with $|\theta| \leq 1$ and log is the *principal value* of the logarithm ($\log z = \log |z| + i \arg z$, $-\pi < \arg z \leq \pi$). Then (8) and (11) imply that for small enough $\varepsilon > 0$ and sufficiently large n

$$\log \varphi_k \left(\frac{t}{D_n} \right) = \log(1 + C_{kn}) = C_{kn} + \theta_{kn} |C_{kn}|^2,$$

where $|\theta_{kn}| \leq 1$. Therefore we obtain from (3) that

$$\frac{t^2}{2} + \log \varphi_{T_n}(t) = \frac{t^2}{2} + \sum_{k=1}^{n} \log \varphi_k \left(\frac{t}{D_n} \right) = \frac{t^2}{2} + \sum_{k=1}^{n} C_{kn} + \sum_{k=1}^{n} \theta_{kn} |C_{kn}|^2.$$

But

$$\frac{t^2}{2} + \sum_{k=1}^{n} C_{kn} = \frac{t^2}{2} \left(1 - \sum_{k=1}^{n} A_{kn} \right) + t^2 \sum_{k=1}^{n} \tilde{\theta}_1(t, k, n) B_{kn} +$$

$$+ \varepsilon |t|^3 \sum_{k=1}^{n} \tilde{\theta}_2(t, k, n) A_{kn},$$

and in view of (9), (10) for any $\delta > 0$ we can find a large enough n_0 and an $\varepsilon > 0$ such that for all $n \geq n_0$

$$\left| \frac{t^2}{2} + \sum_{k=1}^{n} C_{kn} \right| \leq \frac{\delta}{2}.$$

Next, by (11) and (12)

$$\left| \sum_{k=1}^{n} \theta_{kn} |C_{kn}|^2 \right| \leq \max_{1 \leq k \leq n} |C_{kn}| \cdot \sum_{k=1}^{n} |C_{kn}| \leq (t^2 \varepsilon^2 + \varepsilon |t|^3)(t^2 + \varepsilon |t|^3).$$

Therefore for sufficiently large n we can choose $\varepsilon > 0$ to satisfy the inequality

$$\left| \sum_{k=1}^{n} \theta_{kn} |C_{kn}|^2 \right| \leq \frac{\delta}{2},$$

so that

$$\left| \frac{t^2}{2} + \log \varphi_{T_n}(t) \right| \leq \delta.$$

Thus for any real t

$$\varphi_{T_n}(t)e^{t^2/2} \to 1, \quad n \to \infty,$$

hence

$$\varphi_{T_n}(t) \to e^{-t^2/2}, \quad n \to \infty.$$

□

2. Consider some particular cases where *Lindeberg's condition* (1) is fulfilled, so that the Central Limit Theorem is valid.

(a) Suppose that *Lyapunov's condition* holds, i.e., for some $\delta > 0$

$$\frac{1}{D_n^{2+\delta}} \sum_{k=1}^{n} \mathsf{E}\,|\xi_k - m_k|^{2+\delta} \to 0, \quad n \to \infty. \tag{13}$$

Take an $\varepsilon > 0$, then

$$\mathsf{E}\,|\xi_k - m_k|^{2+\delta} = \int_{-\infty}^{\infty} |x - m_k|^{2+\delta}\,dF_k(x)$$

$$\geq \int_{\{x:\,|x-m_k|\geq \varepsilon D_n\}} |x - m_k|^{2+\delta}\,dF_k(x)$$

$$\geq \varepsilon^\delta D_n^\delta \int_{\{x:\,|x-m_k|\geq \varepsilon D_n\}} (x - m_k)^2\,dF_k(x),$$

consequently

$$\frac{1}{D_n^2} \sum_{k=1}^{n} \int_{\{x:\,|x-m_k|\geq \varepsilon D_n\}} (x - m_k)^2\,dF_k(x) \leq \frac{1}{\varepsilon^\delta} \cdot \frac{1}{D_n^{2+\delta}} \sum_{k=1}^{n} \mathsf{E}\,|\xi_k - m_k|^{2+\delta}.$$

Therefore *Lyapunov's condition* implies *Lindeberg's condition*.

(b) Let $\xi_1, \xi_2, \ldots$ be independent *identically distributed* random variables with $m = \mathsf{E}\,\xi_1$ and variance $0 < \sigma^2 = \operatorname{Var}\xi_1 < \infty$. Then

$$\frac{1}{D_n^2} \sum_{k=1}^{n} \int_{\{x:\,|x-m|\geq \varepsilon D_n\}} |x-m|^2\,dF_k(x) = \frac{n}{n\sigma^2} \int_{\{x:\,|x-m|\geq \varepsilon\sigma^2\sqrt{n}\}} |x-m|^2\,dF_1(x) \to 0,$$

since $\{x:\,|x - m| \geq \varepsilon\sigma^2\sqrt{n}\} \downarrow \varnothing$, $n \to \infty$ and $\sigma^2 = \mathsf{E}\,|\xi_1 - m|^2 < \infty$.

Thus *Lindeberg's condition* is fulfilled and hence Theorem 3 of Sect. 3 follows from Theorem 1 just proved.

(c) Let $\xi_1, \xi_2, \ldots$ be independent random variables such that for all $n \geq 1$

$$|\xi_n| \leq K < \infty,$$

where K is a constant and $D_n \to \infty$ as $n \to \infty$.

Then by Chebyshev's inequality

$$\int_{\{x:\ |x-m_k|\geq \varepsilon D_n\}} |x-m_k|^2\, dF_k(x) = \mathsf{E}[(\xi_k - m_k)^2\, I(|\xi_k - m_k| \geq \varepsilon D_n)]$$

$$\leq (2K)^2\, \mathsf{P}\{|\xi_k - m_k| \geq \varepsilon D_n\} \leq (2K)^2\, \frac{\sigma_k^2}{\varepsilon^2 D_n^2}.$$

Hence

$$\frac{1}{D_n^2}\sum_{k=1}^{n} \int_{\{x:\ |x-m_k|\geq \varepsilon D_n\}} |x-m_k|^2 dF_k(x) \leq \frac{(2K)^2}{\varepsilon^2 D_n^2} \to 0, \quad n \to \infty.$$

Therefore *Lindeberg's condition* is again fulfilled, so that the Central Limit Theorem holds.

3. Remark 1. Let $T_n = \frac{S_n - \mathsf{E}\, S_n}{D_n}$ and $F_{T_n}(x) = \mathsf{P}(T_n \leq x)$. Then the statement (2) means that for any $x \in R$

$$F_{T_n}(x) \to \Phi(x), \quad n \to \infty.$$

Since $\Phi(x)$ is continuous, convergence here is uniform (Problem 5 in Sect. 1), i.e.,

$$\sup_{x\in R} |F_{T_n}(x) - \Phi(x)| \to 0, \quad n \to \infty. \tag{14}$$

In particular, this implies that

$$\mathsf{P}\{S_n \leq x\} - \Phi\left(\frac{x - \mathsf{E}\, S_n}{D_n}\right) \to 0, \quad n \to \infty.$$

This fact is often stated by saying that S_n for sufficiently large n is approximately normally distributed with mean $\mathsf{E}\, S_n$ and variance $D_n^2 = \operatorname{Var} S_n$.

Remark 2. Since by the above remark convergence $F_{T_n}(x) \to \Phi(x), n \to \infty$, is *uniform* in x, it is natural to ask about the *rate of convergence* in (14). When $\xi_1, \xi_2, \ldots$ are *independent identically distributed* random variables with $\mathsf{E}\, |\xi_1|^3 < \infty$, the answer to this question is given by the *Berry–Esseen theorem (inequality)*:

$$\sup_x |F_{T_n}(x) - \Phi(x)| \leq C\, \frac{\mathsf{E}\, |\xi_1 - \mathsf{E}\, \xi_1|^3}{\sigma^3 \sqrt{n}}, \tag{15}$$

where C is a universal constant whose exact value is unknown so far. At present the following inequalities for this constant are known:

$$0.4097 = \frac{\sqrt{10} + 3}{6\sqrt{2\pi}} \leq C \leq 0.469$$

(the lower bound was obtained by Esseen [29], for the upper bound see [87]).
The proof of (15) is given in Sect. 11 below.

Remark 3. Now we state Lindeberg's condition in a somewhat different (and even more compact) form, which is especially appropriate in the case of a "triangle array" of random variables.

Let $\xi_1, \xi_2, \ldots$ be a sequence of independent random variables, let $m_k = \mathsf{E}\,\xi_k$, $\sigma_k^2 = \mathrm{Var}\,\xi_k$, $D_n^2 = \sum_{k=1}^n \sigma_k^2 > 0$, $n \geq 1$, and $\xi_{nk} = \frac{\xi_k - m_k}{D_n}$. With this notation Condition (1) takes the form

$$(L) \qquad \sum_{k=1}^n \mathsf{E}[\xi_{nk}^2 I(|\xi_{nk}| \geq \varepsilon)] \to 0, \quad n \to \infty. \qquad (16)$$

If $S_n = \xi_{n1} + \cdots + \xi_{nn}$, then $\mathrm{Var}\,S_n = 1$ and Theorem 1 says that under Condition (16)

$$S_n \xrightarrow{d} \mathcal{N}(0,1).$$

In this form the Central Limit Theorem is true without assuming that ξ_{nk}'s have the special form $\frac{\xi_k - m_k}{D_n}$. In fact, the following result holds, which can be proved by repeating word by word the proof of Theorem 1.

Theorem 2. *Let for each $n \geq 1$*

$$\xi_{n1}, \xi_{n2}, \ldots, \xi_{nn}$$

be independent random variables such that $\mathsf{E}\,\xi_{nk} = 0$ and $\mathrm{Var}\,S_n = 1$, where $S_n = \xi_{n1} + \cdots + \xi_{nn}$.

Then Lindeberg's condition (16) is sufficient for convergence $S_n \xrightarrow{d} \mathcal{N}(0,1)$.

4. Since

$$\max_{1 < k \leq n} \mathsf{E}\,\xi_{nk}^2 \leq \varepsilon^2 + \sum_{k=1}^n \mathsf{E}[\xi_{nk}^2 I(|\xi_{nk}| \geq \varepsilon)],$$

it is clear that Lindeberg's condition (16) implies that

$$\max_{1 \leq k \leq n} \mathsf{E}\,\xi_{kn}^2 \to 0, \quad n \to \infty. \qquad (17)$$

Remarkably, subject to this condition the validity of the Central Limit Theorem automatically implies Lindeberg's condition.

Theorem 3. *Let for each $n \geq 1$*

$$\xi_{n1}, \xi_{n2}, \ldots, \xi_{nn}$$

be independent random variables such that $\mathsf{E}\,\xi_{nk} = 0$ and $\mathrm{Var}\,S_n = 1$, where $S_n = \xi_{n1} + \cdots + \xi_{nn}$. Suppose that (17) is fulfilled. Then Lindeberg's condition is necessary and sufficient for the Central Limit Theorem, $S_n \xrightarrow{d} \mathcal{N}(0,1)$, to hold.

Sufficiency follows from Theorem 2. For the proof of the necessity part we will need the following lemma (cf. Lemma 3 in Sect. 3).

Lemma. *Let ξ be a random variable with distribution function $F = F(x)$, $\mathsf{E}\,\xi = 0$, $\mathrm{Var}\,\xi = \gamma > 0$. Then for any $a > 0$*

$$\int_{|x| \geq 1/a} x^2 \, dF(x) \leq \frac{1}{a^2} \left[\mathrm{Re} f(\sqrt{6}a) - 1 + 3\gamma a^2 \right], \tag{18}$$

where $f(t) = \mathsf{E}\,e^{it\xi}$ is the characteristic function of ξ.

PROOF. We have

$$\mathrm{Re} f(t) - 1 + \frac{1}{2} \gamma t^2 = \frac{1}{2} \gamma t^2 - \int_{-\infty}^{\infty} [1 - \cos tx] \, dF(x)$$

$$= \frac{1}{2} \gamma t^2 - \int_{|x| < 1/a} [1 - \cos tx] \, dF(x) - \int_{|x| \geq 1/a} [1 - \cos tx] \, dF(x)$$

$$\geq \frac{1}{2} \gamma t^2 - \frac{1}{2} t^2 \int_{|x| < 1/a} x^2 \, dF(x) - 2a^2 \int_{|x| \geq 1/a} x^2 \, dF(x)$$

$$= \left(\frac{1}{2} t^2 - 2a^2 \right) \int_{|x| \geq 1/a} x^2 \, dF(x).$$

Letting $t = \sqrt{6}a$ we obtain (18).

□

Now we turn to the proof of necessity in Theorem 3. Let

$$F_{nk}(x) = \mathsf{P}\{\xi_{nk} \leq x\}, \quad f_{nk}(t) = \mathsf{E}\,e^{it\xi_{nk}},$$
$$\mathsf{E}\,\xi_{nk} = 0, \quad \mathrm{Var}\,\xi_{nk} = \gamma_{nk} > 0, \tag{19}$$
$$\sum_{k=1}^{n} \gamma_{nk} = 1, \quad \max_{1 \leq k \leq n} \gamma_{nk} \to 0, \ n \to \infty.$$

Let $\log z$ denote the *principal value* of the logarithm of a complex number z (i.e., $\log z = \log |z| + i \arg z$, $-\pi < \arg z \leq \pi$). Then

$$\log \prod_{k=1}^{n} f_{nk}(t) = \sum_{k=1}^{n} \log f_{nk}(t) + 2\pi i m,$$

where $m = m(n, t)$ is an integer. Hence

$$\mathrm{Re} \log \prod_{k=1}^{n} f_{nk}(t) = \mathrm{Re} \sum_{k=1}^{n} \log f_{nk}(t). \tag{20}$$

Since

$$\prod_{k=1}^{n} f_{nk}(t) \to e^{-\frac{1}{2} t^2},$$

we have

$$\left| \prod_{k=1}^{n} f_{nk}(t) \right| \to e^{-\frac{1}{2}t^2}.$$

Hence

$$\mathrm{Re} \log \prod_{k=1}^{n} f_{nk}(t) = \mathrm{Re} \log \left| \prod_{k=1}^{n} f_{nk}(t) \right| \to -\frac{1}{2}t^2. \tag{21}$$

For $|z| < 1$

$$\log(1 + z) = z - \frac{z^2}{2} + \frac{z^3}{3} - \cdots, \tag{22}$$

and for $|z| < 1/2$

$$|\log(1 + z) - z| \le |z|^2. \tag{23}$$

By (19), for any fixed t and all sufficiently large n we have

$$|f_{nk}(t) - 1| \le \frac{1}{2}\gamma_{nk}t^2 \le \frac{1}{2}, \quad k = 1, 2, \ldots, n. \tag{24}$$

Hence we obtain from (23), (24)

$$\left| \sum_{k=1}^{n} \{\log[1 + (f_{nk}(t) - 1)] - (f_{nk}(t) - 1)\} \right| \le \sum_{k=1}^{n} |f_{nk}(t) - 1|^2$$

$$\le \frac{t^4}{4} \max_{1 \le k \le n} \gamma_{nk} \cdot \sum_{k=1}^{n} \gamma_{nk} = \frac{t^4}{4} \max_{1 \le k \le n} \gamma_{nk} \to 0, \quad n \to \infty,$$

therefore

$$\left| \mathrm{Re} \sum_{k=1}^{n} \log f_{nk}(t) - \mathrm{Re} \sum_{k=1}^{n} (f_{nk}(t) - 1) \right| \to 0, \quad n \to \infty. \tag{25}$$

Now (20), (21) and (25) imply that

$$\mathrm{Re} \sum_{k=1}^{n} (f_{nk}(t) - 1) + \frac{1}{2}t^2 = \sum_{k=1}^{n} \left[\mathrm{Re} f_{nk}(t) - 1 + \frac{1}{2}t^2\gamma_{nk} \right] \to 0, \quad n \to \infty.$$

Letting $t = \sqrt{6}a$ we find that for any $a > 0$

$$\sum_{k=1}^{n} [\mathrm{Re} f_{nk}(\sqrt{6}a) - 1 + 3a^2\gamma_{nk}] \to 0, \quad n \to \infty. \tag{26}$$

Finally, we obtain from (26) and (18) with $a = 1/\varepsilon$ that

$$\sum_{k=1}^{n} \mathsf{E}[\xi_{nk} I(|\xi_{nk}| \geq \varepsilon)] = \sum_{k=1}^{n} \int_{|x| \geq \varepsilon} x^2 \, dF_{nk}(x)$$

$$\leq \varepsilon^2 \sum_{k=1}^{n} [\mathrm{Re} f_{nk}(\sqrt{6}a) - 1 + 3a^2 \gamma_{nk}] \to 0, \quad n \to \infty,$$

which proves Lindeberg's condition.

□

5. PROBLEMS

1. Let $\xi_1, \xi_2, \ldots$ be a sequence of independent identically distributed random variables with $\mathsf{E}\,\xi^2 < \infty$. Show that

$$\max\left(\frac{|\xi_1|}{\sqrt{n}}, \ldots, \frac{|\xi_n|}{\sqrt{n}}\right) \overset{d}{\to} 0, \quad n \to \infty.$$

2. Give a direct proof of the fact that in the Bernoulli scheme $\sup_x |F_{T_n}(x) - \Phi(x)|$ is of order $1/\sqrt{n}$ as $n \to \infty$.

3. Let $X_1, X_2, \ldots$ be a sequence of exchangeable random variables (see Problem 4 in Sect. 5 of Chap. 2) with $\mathsf{E}\,X_1 = 0$, $\mathsf{E}\,X_1^2 = 1$ and

$$\mathrm{Cov}(X_1, X_2) = \mathrm{Cov}(X_1^2, X_2^2). \tag{27}$$

Prove that they obey the Central Limit Theorem,

$$\frac{1}{\sqrt{n}} \sum_{i=1}^{n} X_i \overset{d}{\to} \mathcal{N}(0,1). \tag{28}$$

Conversely, if $\mathsf{E}\,X_n^2 < \infty$ and (28) holds, then (27) is fulfilled.

4. *Local Central Limit Theorem.* Let $X_1, X_2, \ldots$ be independent identically distributed random variables with $\mathsf{E}\,X_1 = 0$, $\mathsf{E}\,X_1^2 = 1$. Assume that their characteristic function $\varphi(t) = \mathsf{E}\,e^{itX_1}$ satisfies the condition

$$\int_{-\infty}^{\infty} |\varphi(t)|^r \, dt < \infty \quad \text{for some } r \geq 1.$$

Show that the random variables $S_n/\sqrt{n}$ have densities $f_n(x)$ such that

$$f_n(x) \to (2\pi)^{-1/2} e^{-x^2/2}, \quad n \to \infty, \quad \text{uniformly in } x \in R.$$

What is the corresponding result for lattice random variables?

5. Let $X_1, X_2, \ldots$ be independent identically distributed random variables with $\mathsf{E}\,X_1 = 0$, $\mathsf{E}\,X_1^2 = 1$. Let $d_1^2, d_2^2, \ldots$ be nonnegative constants such that

$d_n = o(D_n)$, where $D_n^2 = \sum_{k=1}^n d_k^2$. Show that the sequence of the weighted random variables $d_1 X_1, d_2 X_2, \ldots$ fulfills the Central Limit Theorem:

$$\frac{1}{D_n} \sum_{k=1}^n d_k X_k \xrightarrow{d} \mathcal{N}(0,1).$$

6. Let $\xi_1, \xi_2, \ldots$ be independent identically distributed random variables with $\mathsf{E}\,\xi_1 = 0$, $\mathsf{E}\,\xi_1^2 = 1$. Let $(\tau_n)_{n \geq 1}$ be a sequence of random variables taking values $1, 2, \ldots$ such that $\tau_n/n \xrightarrow{\mathsf{P}} c$, where $c > 0$ is a constant. Show that

$$\mathrm{Law}(\tau_n^{-1/2} S_{\tau_n}) \to \Phi, \quad \text{where } S_n = \xi_1 + \cdots + \xi_n$$

(i.e., $\tau_n^{-1/2} S_{\tau_n} \xrightarrow{d} \mathcal{N}(0,1)$). (Note that it is not assumed that the sequences $(\tau_n)_{n \geq 1}$ and $(\xi_n)_{n \geq 1}$ are independent.)

7. Let $\xi_1, \xi_2, \ldots$ be independent identically distributed random variables with $\mathsf{E}\,\xi_1 = 0$, $\mathsf{E}\,\xi_1^2 = 1$. Prove that

$$\mathrm{Law}\left(n^{-1/2} \max_{1 \leq m \leq n} S_m\right) \to \mathrm{Law}(|\xi|), \quad \text{where } \xi \sim \mathcal{N}(0,1).$$

In other words, for $x > 0$,

$$\mathsf{P}\left\{n^{-1/2} \max_{1 \leq m \leq n} S_m \leq x\right\} \to \sqrt{\frac{2}{\pi}} \int_0^x e^{-y^2/2}\, dy \quad \left(= \frac{1}{\sqrt{2}}\,\mathrm{erf}(x)\right).$$

Hint: Establish first that this statement holds for symmetric *Bernoulli* random variables $\xi_1, \xi_2, \ldots$, i.e., such that $\mathsf{P}(\xi_n = \pm 1) = 1/2$, and then prove that the limiting distribution will be the same for any sequence $\xi_1, \xi_2, \ldots$ satisfying the conditions of the problem. (The property that the limiting distribution does not depend on the specific choice of independent identically distributed random variables $\xi_1, \xi_2, \ldots$ with $\mathsf{E}\,\xi_1 = 0$, $\mathsf{E}\,\xi_1^2 = 1$ is known as the "invariance principle," cf. Sect. 7.)

8. Under the conditions of the previous problem prove that

$$\mathsf{P}\left\{n^{-1/2} \max_{1 \leq m \leq n} |S_m| \leq x\right\} \to H(x), \quad x > 0,$$

where

$$H(x) = \frac{4}{\pi} \sum_{k=0}^\infty \frac{(-1)^k}{2k+1} \exp\left\{-\frac{(2k+1)^2 \pi^2}{8x^2}\right\}.$$

9. Let $X_1, X_2, \ldots$ be a sequence of independent random variables with

$$\mathsf{P}\{X_n = \pm n^\alpha\} = \frac{1}{2n^\beta}, \quad \mathsf{P}\{X_n = 0\} = 1 - \frac{1}{n^\beta}, \quad \text{where } 2\alpha > \beta - 1.$$

Show that Lindeberg's condition is fulfilled if and only if $0 \leq \beta < 1$.

10. Let $X_1, X_2, \ldots$ be a sequence of independent random variables such that $|X_n| \leq C_n$ (P-a.s.) and $C_n = o(D_n)$, where

$$D_n^2 = \sum_{k=1}^{n} \mathsf{E}(X_k - \mathsf{E} X_k)^2 \to \infty.$$

Show that

$$\frac{S_n - \mathsf{E} S_n}{D_n} \to \mathscr{N}(0,1), \quad \text{where } S_n = X_1 + \cdots + X_n.$$

11. Let $X_1, X_2, \ldots$ be a sequence of independent random variables with $\mathsf{E} X_n = 0$, $\mathsf{E} X_n^2 = \sigma_n^2$. Assume that they obey the Central Limit Theorem and

$$\mathsf{E}\left(D_n^{-1/2} \sum_{i=1}^{n} X_i\right)^k \to \frac{(2k)!}{2^k k!} \quad \text{for some } k \geq 1.$$

Show that in this case Lindeberg's condition of order k holds, i.e.,

$$\sum_{j=1}^{n} \int_{\{|x|>\varepsilon\}} |x|^k \, dF_j(x) = o(D_n^k), \quad \varepsilon > 0.$$

(The ordinary Lindeberg's condition corresponds to $k = 2$, see (1).)

12. Let $X = X(\lambda)$ and $Y = Y(\mu)$ be independent random variables having the Poisson distributions with parameters λ and μ respectively. Show that

$$\frac{(X(\lambda) - \lambda) - (Y(\mu) - \mu)}{\sqrt{X(\lambda) + Y(\mu)}} \to \mathscr{N}(0,1) \quad \text{as } \lambda \to \infty, \mu \to \infty.$$

13. Let $X_1^{(n)}, \ldots, X_{n+1}^{(n)}$ for any $n \geq 1$ be an $(n+1)$-dimensional random vector uniformly distributed on the *unit sphere*. Prove the following "Poincaré's theorem":

$$\lim_{n \to \infty} \mathsf{P}\{\sqrt{n} X_{n+1}^{(n)} \leq x\} = \frac{1}{\sqrt{2\pi}} \int_{-\infty}^{x} e^{-u^2/2} \, du.$$

5 Central Limit Theorem for Sums of Independent Random Variables: II—Nonclassical Conditions

1. It was shown in Sect. 4 that the Lindeberg condition (16) of Sect. 4 implies that the condition

$$\max_{1 \leq k \leq n} \mathsf{E} \xi_{nk}^2 \to 0$$

is satisfied. In turn, this implies the so-called condition of *asymptotic negligibility*, that is, the condition that for every $\varepsilon > 0$,

$$\max_{1 \leq k \leq n} \mathsf{P}\{|\xi_{nk}| \geq \varepsilon\} \to 0, \qquad n \to \infty.$$

Consequently, we may say that Theorems 1 and 2 of Sect. 4 provide a condition of validity of the central limit theorem for sums of independent random variables under the condition of asymptotic negligibility. Limit theorems in which the condition of asymptotic negligibility is imposed on individual terms are usually called theorems with a classical formulation. It is easy, however, to give examples of nondegenerate random variables for which neither the Lindeberg condition nor the asymptotic negligibility condition is satisfied, but nevertheless the central limit theorem is satisfied. Here is the simplest example.

Let $\xi_1, \xi_2, \ldots$ be a sequence of independent normally distributed random variables with $\mathsf{E}\, \xi_n = 0$, $\operatorname{Var} \xi_1 = 1$, $\operatorname{Var} \xi_k = 2^{k-2}$, $k \geq 2$. Let $S_n = \xi_{n1} + \cdots + \xi_{nn}$ with

$$\xi_{nk} = \xi_k \Bigg/ \sqrt{\sum_{k=1}^{n} \operatorname{Var} \xi_i}\,.$$

It is easily verified that here neither the Lindeberg condition nor the asymptotic negligibility condition is satisfied, although the validity of the central limit theorem is evident, since S_n is normally distributed with $\mathsf{E}\, S_n = 0$ and $\operatorname{Var} S_n = 1$.

Theorem 1 (below) provides a sufficient (and necessary) condition for the central limit theorem without assuming the "classical" condition of asymptotic negligibility. In this sense, condition (Λ), presented below, is an example of "nonclassical" conditions which reflect the title of this section.

2. We shall suppose that we are given a "triangle array" of random variables, i.e., for each $n \geq 1$ we have n independent random variables

$$\xi_{n1}, \xi_{n2}, \ldots, \xi_{nn}$$

with $\mathsf{E}\, \xi_{nk} = 0$, $\operatorname{Var} \xi_{nk} = \sigma_{nk}^2 > 0$, $\sum_{k=1}^{n} \sigma_{nk}^2 = 1$. Let $S_n = \xi_{n1} + \cdots + \xi_{nn}$,

$$F_{nk}(x) = \mathsf{P}\{\xi_{nk} \leq x\}, \quad \Phi(x) = (2\pi)^{-1/2} \int_{-\infty}^{x} e^{-y^2/2} dy, \quad \Phi_{nk}(x) = \Phi\left(\frac{x}{\sigma_{nk}}\right).$$

Theorem 1. *To have*

$$S_n \xrightarrow{d} \mathcal{N}(0, 1), \tag{1}$$

it is sufficient (and necessary) that for every $\varepsilon > 0$ the condition

$$(\Lambda) \qquad \sum_{k=1}^{n} \int_{|x| > \varepsilon} |x| |F_{nk}(x) - \Phi_{nk}(x)| \, dx \to 0, \qquad n \to \infty, \tag{2}$$

is satisfied.

The following theorem clarifies the connection between condition (Λ) and the classical Lindeberg condition

$$\text{(L)} \qquad \sum_{k=1}^{n} \int_{|x|>\varepsilon} x^2 \, dF_{nk}(x) \to 0, \quad n \to \infty. \tag{3}$$

Theorem 2. 1. *The Lindeberg condition implies that condition (Λ) is satisfied*:

$$\text{(L)} \Rightarrow (\Lambda).$$

2. *If* $\max_{1\le k\le n} \mathsf{E}\,\xi_{nk}^2 \to 0$ *as* $n \to \infty$, *the condition* (Λ) *implies the Lindeberg condition* (L):

$$(\Lambda) \Rightarrow \text{(L)}.$$

PROOF OF THEOREM 1. The proof of the necessity of condition (Λ) is rather complicated (see [63, 82, 99]). Here we only prove the sufficiency.

Let

$$f_{nk}(t) = \mathsf{E}\,e^{it\xi_{nk}}, \qquad f_n(t) = \mathsf{E}\,e^{itS_n},$$

$$\varphi_{nk}(t) = \int_{-\infty}^{\infty} e^{itx} \, d\Phi_{nk}(x), \qquad \varphi(t) = \int_{-\infty}^{\infty} e^{itx} \, d\Phi(x).$$

It follows from Sect. 12 of Chap. 2 that

$$\varphi_{nk}(t) = e^{-(t^2\sigma_{nk}^2)/2}, \quad \varphi(t) = e^{-t^2/2}.$$

By the corollary of Theorem 1 of Sect. 3 , we have $S_n \xrightarrow{d} \mathcal{N}(0,1)$ if and only if $f_n(t) \to \varphi(t)$ as $n \to \infty$, for every real t.

We have

$$f_n(t) - \varphi(t) = \prod_{k=1}^{n} f_{nk}(t) - \prod_{k=1}^{n} \varphi_{nk}(t).$$

Since $|f_{nk}(t)| \le 1$ and $|\varphi_{nk}(t)| \le 1$, we have

$$|f_n(t) - \varphi(t)| = \left| \prod_{k=1}^{n} f_{nk}(t) - \prod_{k=1}^{n} \varphi_{nk}(t) \right|$$

$$\le \sum_{k=1}^{n} |f_{nk}(t) - \varphi_{nk}(t)| = \sum_{k=1}^{n} \left| \int_{-\infty}^{\infty} e^{itx} d(F_{nk} - \Phi_{nk}) \right|$$

$$= \sum_{k=1}^{n} \left| \int_{-\infty}^{\infty} \left(e^{itx} - itx + \frac{t^2x^2}{2} \right) d(F_{nk} - \Phi_{nk}) \right|, \tag{4}$$

where we have used the fact that

$$\int_{-\infty}^{\infty} x^k dF_{nk} = \int_{-\infty}^{\infty} x^k d\Phi_{nk} \qquad \text{for} \quad k = 1, 2.$$

If we apply the formula for integration by parts (Theorem 11 in Sect. 6, Chap. 2) to the integral

$$\int_a^b \left(e^{itx} - itx + \frac{t^2 x^2}{2} \right) d(F_{nk} - \Phi_{nk}),$$

we obtain (taking account of the limits $x^2[1 - F_{nk}(x) + F_{nk}(-x)] \to 0$, and $x^2[1 - \Phi_{nk}(x) + \Phi_{nk}(-x)] \to 0$, $x \to \infty$)

$$\int_{-\infty}^{\infty} \left(e^{itx} - itx + \frac{t^2 x^2}{2} \right) d(F_{nk} - \Phi_{nk})$$

$$= it \int_{-\infty}^{\infty} (e^{itx} - 1 - itx)(F_{nk}(x) - \Phi_{nk}(x)) \, dx. \tag{5}$$

From (4) and (5), we obtain

$$|f_n(t) - \varphi(t)| \leq \sum_{k=1}^n \left| t \int_{-\infty}^{\infty} (e^{itx} - 1 - itx)(F_{nk}(x) - \Phi_{nk}(x)) \right| dx$$

$$\leq \frac{|t|^3}{2} \varepsilon \sum_{k=1}^n \int_{|x| \leq \varepsilon} |x| \, |F_{nk}(x) - \Phi_{nk}(x)| \, dx$$

$$+ 2t^2 \sum_{k=1}^n \int_{|x| > \varepsilon} |x| \, |F_{nk}(x) - \Phi_{nk}(x)| \, dx$$

$$\leq \varepsilon |t|^3 \sum_{k=1}^n \sigma_{nk}^2 + 2t^2 \sum_{k=1}^n \int_{|x| > \varepsilon} |x| |F_{nk}(x) - \Phi_{nk}(x)| \, dx, \tag{6}$$

where we have used the inequality

$$\int_{|x| \leq \varepsilon} |x| |F_{nk}(x) - \Phi_{nk}(x)| \, dx \leq 2\sigma_{nk}^2, \tag{7}$$

which is easily established by using the formula (71) in Sect. 6, Chap. 2.

It follows from (6) that $f_n(t) \to \varphi(t)$ as $n \to \infty$, because ε is an arbitrary positive number and condition (Λ) is satisfied.

This completes the proof of the theorem.

$\square$

PROOF OF THEOREM 2. 1. According to Sect. 4, *Lindeberg's condition* (L) implies that $\max_{1 \leq k \leq n} \sigma_{nk}^2 \to 0$. Hence, if we use the fact that $\sum_{k=1}^n \sigma_{nk}^2 = 1$, we obtain

$$\sum_{k=1}^n \int_{|x| > \varepsilon} x^2 \, d\Phi_{nk}(x) \leq \int_{|x| > \varepsilon / \sqrt{\max_{1 \leq k \leq n} \sigma_{nk}^2}} x^2 \, d\Phi(x) \to 0, \qquad n \to \infty. \tag{8}$$

Together with Condition (L), this shows that, for every $\varepsilon > 0$,

$$\sum_{k=1}^{n} \int_{|x|>\varepsilon} x^2 \, d[F_{nk}(x) + \Phi_{nk}(x)] \to 0, \qquad n \to \infty. \tag{9}$$

Let us fix $\varepsilon > 0$. Then there is a continuous differentiable even function $h = h(x)$ for which $|h(x)| \le x^2$, $|h'(x)| \le 4x$, and

$$h(x) = \begin{cases} x^2, & |x| > 2\varepsilon, \\ 0, & |x| \le \varepsilon. \end{cases}$$

For $h(x)$, we have by (9)

$$\sum_{k=1}^{n} \int_{|x|>\varepsilon} h(x) \, d[F_{nk}(x) + \Phi_{nk}(x)] \to 0, \qquad n \to \infty. \tag{10}$$

By integrating by parts in (10), we obtain

$$\sum_{k=1}^{n} \int_{x \ge \varepsilon} h'(x)[(1 - F_{nk}(x)) + (1 - \Phi_{nk}(x))] \, dx$$

$$= \sum_{k=1}^{n} \int_{x \ge \varepsilon} h(x) \, d[F_{nk} + \Phi_{nk}] \to 0,$$

$$\sum_{k=1}^{n} \int_{x \le -\varepsilon} h'(x)[F_{nk}(x) + \Phi_{nk}(x)] \, dx = \sum_{k=1}^{n} \int_{x \le -\varepsilon} h(x) \, d[F_{nk} + \Phi_{nk}] \to 0.$$

Since $h'(x) = 2x$ for $|x| \ge 2\varepsilon$, we obtain

$$\sum_{k=1}^{n} \int_{|x| \ge 2\varepsilon} |x| \, |F_{nk}(x) - \Phi_{nk}(x)| \, dx \to 0, \qquad n \to \infty.$$

Therefore, since ε is an arbitrary positive number, we find that $(L) \Rightarrow (\Lambda)$.

2. For the function $h = h(x)$ introduced above, we find by (8) and the condition $\max_{1 \le k \le n} \sigma_{nk}^2 \to 0$ that

$$\sum_{k=1}^{n} \int_{|x|>\varepsilon} h(x) \, d\Phi_{nk}(x) \le \sum_{k=1}^{n} \int_{|x|>\varepsilon} x^2 \, d\Phi_{nk}(x) \to 0, \qquad n \to \infty. \tag{11}$$

If we integrate by parts, we obtain

$$
\left| \sum_{k=1}^{n} \int_{|x| \geq \varepsilon} h(x)\, d[F_{nk} - \Phi_{nk}] \right| \leq \left| \sum_{k=1}^{n} \int_{x \geq \varepsilon} h(x)\, d[(1 - F_{nk}) - (1 - \Phi_{nk})] \right|
$$

$$
+ \left| \sum_{k=1}^{n} \int_{x \leq -\varepsilon} h(x)\, d[F_{nk} - \Phi_{nk}] \right|
$$

$$
\leq \sum_{k=1}^{n} \int_{x \geq \varepsilon} |h'(x)| |[(1 - F_{nk}) - (1 - \Phi_{nk})]|\, dx
$$

$$
+ \sum_{k=1}^{n} \int_{x \leq -\varepsilon} |h'(x)| |F_{nk} - \Phi_{nk}|\, dx
$$

$$
\leq 4 \sum_{k=1}^{n} \int_{|x| \geq \varepsilon} |x| |F_{nk}(x) - \Phi_{nk}(x)|\, dx. \tag{12}
$$

It follows from (11) and (12) that

$$
\sum_{k=1}^{n} \int_{|x| \geq 2\varepsilon} x^2\, dF_{nk}(x) \leq \sum_{k=1}^{n} \int_{|x| \geq \varepsilon} h(x)\, dF_{nk}(x) \to 0, \qquad n \to \infty,
$$

i.e., the Lindeberg condition (L) is satisfied.

This completes the proof of the theorem.   □

3. PROBLEMS

1. Establish formula (5).
2. Verify relations (10) and (12).
3. Let $N = (N_t)_{t \geq 0}$ be a renewal process introduced in Subsection 4 of Sect. 9, Chap. 2 ($N_t = \sum_{n=1}^{\infty} I(T_n \leq t)$, $T_n = \sigma_1 + \cdots + \sigma_n$, where $\sigma_1, \sigma_2, \ldots$ is a sequence of independent identically distributed random variables). Assuming that $\mu = \mathsf{E}\sigma_1 < \infty$, $0 < \mathrm{Var}\,\sigma_1 < \infty$ prove the following central limit theorem for N:

$$
\frac{N_t - t\mu^{-1}}{\sqrt{t\mu^{-3}\,\mathrm{Var}\,\sigma_1}} \xrightarrow{d} \mathcal{N}(0, 1),
$$

where $\mathcal{N}(0, 1)$ is a normal random variable with zero mean and unit variance.

6 Infinitely Divisible and Stable Distributions

1. In stating Poisson's theorem in Sect. 3 we found it necessary to use a triangular array, supposing that for each $n \geq 1$ there was a sequence of independent random variables $\{\xi_{n,k}\}, 1 \leq k \leq n$.

Put

$$T_n = \xi_{n,1} + \cdots + \xi_{n,n}, \quad n \geq 1. \tag{1}$$

The idea of an infinitely divisible distribution arises in the following problem: how can we determine all the distributions that can be expressed as limits of sequences of distributions of random variables T_n, $n \geq 1$?

Generally speaking, the problem of limit distributions is indeterminate in such great generality. Indeed, if ξ is a random variable and $\xi_{n,1} = \xi$, $\xi_{n,k} = 0$, $1' < k \leq n$, then $T_n \equiv \xi$ and consequently the limit distribution is the distribution of ξ, which can be arbitrary.

In order to have a more meaningful problem, we shall suppose in the present section that the variables $\xi_{n,1}, \ldots, \xi_{n,n}$ are, for each $n \geq 1$, not only independent, but also *identically* distributed.

Recall that this was the situation in Poisson's theorem (Theorem 4 of Sect. 3). The same framework also includes the central limit theorem (Theorem 3 of Sect. 3) for sums $S_k = \xi_1 + \cdots + \xi_n$, $n \geq 1$, of independent identically distributed random variables $\xi_1, \xi_2, \ldots$. In fact, if we put

$$\xi_{n,k} = \frac{\xi_k - \mathsf{E}\,\xi_k}{\mathrm{Var}_n}, \quad D_n^2 = \mathrm{Var}\,S_n,$$

then

$$T_n = \sum_{k=1}^{n} \xi_{n,k} = \frac{S_n - \mathsf{E}\,S_n}{D_n}.$$

Consequently both the normal and the Poisson distributions can be presented as limits in a triangular array. If $T_n \overset{d}{\to} T$, it is intuitively clear that since T_n is a sum of independent identically distributed random variables, the limit variable T must also be a sum of independent identically distributed random variables. With this in mind, we introduce the following definition.

Definition 1. A random variable T, its distribution F_T, and its characteristic function φ_T are said to be *infinitely divisible* if, for each $n \geq 1$, there are independent identically distributed random variables $\eta_1, \ldots, \eta_n$ such that* $T \overset{d}{=} \eta_1 + \cdots + \eta_n$ (or, equivalently, $F_T = F_{\eta_1} * \cdots * F_{\eta_n}$, or $\varphi_T = (\varphi_{\eta_1})^n$).

Remark 1. If the basic probability space on which T is defined is "poor" enough, it may happen that the distribution function F_T and its characteristic function φ_T admit the representations $F_T = F^{(n)} * \cdots * F^{(n)}$ (n times) and $\varphi_T = (\varphi^{(n)})^n$ with some distribution functions $F^{(n)}$ and their characteristic functions $\varphi^{(n)}$, whereas the representation $T \overset{d}{=} \eta_1 + \cdots + \eta_n$ is impossible. J.L. Doob (see [34]) gave an example of such a "poor" probability space on which a random variable T is defined having the Poisson distribution with parameter $\lambda = 1$ (which is infinitely divisible because $F_T = F^{(n)} * \cdots * F^{(n)}$ with distribution functions $F^{(n)}$ of the Poisson distribution

---

* The notation $\xi \overset{d}{=} \eta$ means that the random variables ξ and η agree *in distribution*, i.e., $F_\xi(x) = F_\eta(x)$, $x \in R$.

with parameter $\lambda = 1/n$) but there are no random variables η_1 and η_2 having the Poisson distribution with parameter $\lambda = 1/2$.

Having all this in mind, we stress that the above Definition 1 actually tacitely assumes that the probability space $(\Omega, \mathscr{F}, \mathsf{P})$ is sufficiently "rich" to avoid the effects pointed out by J.L. Doob (Problem 11).

Theorem 1. *A random variable T can be a limit in distribution of sums $T_n = \sum_{k=1}^{n} \xi_{n,k}$ if and only if T is infinitely divisible.*

PROOF. If T is infinitely divisible, for each $n \geq 1$ there are independent identically distributed random variables $\xi_{n,1}, \ldots, \xi_{n,k}$ such that $T \overset{d}{=} \xi_{n,1} + \cdots + \xi_{n,k}$, and this means that $T \overset{d}{=} T_n$, $n \geq 1$.

Conversely, let $T_n \overset{d}{\to} T$. Let us show that T is infinitely divisible, i.e., for each k there are independent identically distributed random variables $\eta_1, \ldots, \eta_k$ such that $T \overset{d}{=} \eta_1 + \cdots + \eta_k$.

Choose a $k \geq 1$ and represent $T_{nk} = \sum_{i=1}^{nk} \xi_{nk,i}$ in the form $\zeta_n^{(1)} + \cdots + \zeta_n^{(k)}$, where

$$\zeta_n^{(1)} = \xi_{nk,1} + \cdots + \xi_{nk,n}, \ldots, \zeta_n^{(k)} = \xi_{nk,n(k-1)+1} + \cdots + \xi_{nk,nk}.$$

Since $T_{nk} \overset{d}{\to} T$, $n \to \infty$, the sequence of distribution functions corresponding to the random variables T_{nk}, $n \geq 1$, is relatively compact and therefore, by Prohorov's theorem, is *tight* (Sect. 2). Moreover,

$$[\mathsf{P}(\zeta_n^{(1)} > z)]^k = \mathsf{P}(\zeta_n^{(1)} > z, \ldots, \zeta_n^{(k)} > z) \leq \mathsf{P}(T_{nk} > kz)$$

and

$$[\mathsf{P}(\zeta_n^{(1)} < -z)]^k = \mathsf{P}(\zeta_n^{(1)} < -z, \ldots, \zeta_n^{(k)} < -z) \leq \mathsf{P}(T_{nk} < -kz).$$

The family of distributions for $\zeta_n^{(1)}$, $n \geq 1$, is tight because of the preceding two inequalities and because the family of distributions for T_{nk}, $n \geq 1$, is tight. Therefore there is a subsequence $\{n_i\} \subset \{n\}$ and a random vector $(\eta_1, \ldots, \eta_k)$, which without loss of generality may be assumed to be defined on our "rich" probability space, such that

$$(\zeta_{n_j}^{(1)}, \ldots, \zeta_{n_j}^{(k)}) \overset{d}{\to} (\eta_1, \ldots, \eta_k),$$

or, equivalently, that

$$\mathsf{E}\, e^{i(\lambda_1 \zeta_{n_j}^{(1)} + \cdots + \lambda_k \zeta_{n_j}^{(k)})} \to \mathsf{E}\, e^{i(\lambda_1 \eta_1 + \cdots + \lambda_k \eta_k)}$$

for any $\lambda_1, \ldots, \lambda_k \in R$. Since the variables $\zeta_{n_j}^{(1)}, \ldots, \zeta_{n_j}^{(k)}$ are independent,

$$\mathsf{E}\, e^{i(\lambda_1 \zeta_{n_j}^{(1)} + \cdots + \lambda_k \zeta_{n_j}^{(k)})} = \mathsf{E}\, e^{i\lambda_1 \zeta_{n_j}^{(1)}} \cdots \mathsf{E}\, e^{i\lambda_k \zeta_{n_j}^{(k)}} \to \mathsf{E}\, e^{i\lambda_1 \eta_1} \cdots \mathsf{E}\, e^{i\lambda_k \eta_k}.$$

Therefore

$$\mathsf{E}\,e^{i(\lambda_1\eta_1+\cdots+\lambda_k\eta_k)} = \mathsf{E}\,e^{i\lambda_1\eta_1}\cdots\mathsf{E}\,e^{i\lambda_k\eta_k}$$

and by Theorem 4 of Sect. 12, Chap. 2, $\eta_1,\ldots,\eta_k$ are independent. Clearly, they are identically distributed.

Thus we have

$$T_{njk} = \zeta_{n_j}^{(1)} + \cdots + \zeta_{n_j}^{(k)} \xrightarrow{d} \eta_1 + \cdots + \eta_k$$

and moreover $T_{njk} \xrightarrow{d} T$. Consequently (Problem 1)

$$T \overset{d}{=} \eta_1 + \cdots + \eta_k.$$

This completes the proof of the theorem.

$\square$

Remark 2. The conclusion of the theorem remains valid if we replace the hypothesis that $\xi_{n,1},\ldots,\xi_{n,n}$ are *identically distributed* for each $n \geq 1$ by the condition of their *asymptotic negligibility* $\max_{1\leq k\leq n}\mathsf{P}\{|\xi_{nk}| \geq \varepsilon\} \to 0$.

2. To test whether a given random variable T is infinitely divisible, it is simplest to begin with its characteristic function $\varphi(t)$. If we can find characteristic functions $\varphi_n(t)$ such that $\varphi(t) = [\varphi_n(t)]^n$ for every $n \geq 1$, then T is infinitely divisible.

In the Gaussian case,

$$\varphi(t) = e^{itm}e^{-(1/2)t^2\sigma^2},$$

and if we put

$$\varphi_n(t) = e^{itm/n}e^{-(1/2)t^2\sigma^2/n},$$

we see at once that $\varphi(t) = [\varphi_n(t)]^n$.

In the Poisson case,

$$\varphi(t) = \exp\{\lambda(e^{it} - 1)\},$$

and if we put $\varphi_n(t) = \exp\{(\lambda/n)(e^{it} - 1)\}$ then $\varphi(t) = [\varphi_n(t)]^n$.

If a random variable T has a Γ-distribution with density

$$f(x) = \begin{cases} \dfrac{x^{\alpha-1}e^{-x/\beta}}{\Gamma(\alpha)\beta^\alpha}, & x \geq 0, \\[2mm] 0, & x < 0, \end{cases}$$

then (see Table 2.5, Sect. 12, Chap. 2) its characteristic function is

$$\varphi(t) = \frac{1}{(1 - i\beta t)^\alpha}.$$

Consequently $\varphi(t) = [\varphi_n(t)]^n$ where

$$\varphi_n(t) = \frac{1}{(1 - i\beta t)^{\alpha/n}},$$

and therefore T is infinitely divisible.

We quote without proof the following result on the general form of the characteristic functions of infinitely divisible distributions.

Theorem 2 (Kolmogorov–Lévy–Khinchin Representation). *A random variable T is infinitely divisible if and only if its characteristic function has the form $\varphi(t) = \exp \psi(t)$ with*

$$\psi(t) = it\beta - \frac{t^2\sigma^2}{2} + \int_{-\infty}^{\infty} \left(e^{itx} - 1 - \frac{itx}{1 + x^2} \right) \frac{1 + x^2}{x^2} \, d\lambda(x), \qquad (2)$$

where $\beta \in R$, $\sigma^2 \geq 0$ and λ is a measure on $(R, \mathscr{B}(R))$ with $\lambda\{0\} = 0$.

3. Let $\xi_1, \xi_2, \ldots$ be a sequence of independent identically distributed random variables and $S_n = \xi_1 + \cdots + \xi_n$. Suppose that there are constants b_n and $a_n > 0$, and a random variable T, such that

$$\frac{S_n - b_n}{a_n} \xrightarrow{d} T. \qquad (3)$$

We ask for a description of the distributions (random variables T) that can be obtained as limit distributions in (3).

If the random variables $\xi_1, \xi_2, \ldots$ satisfy $0 < \sigma^2 \equiv \operatorname{Var} \xi_1 < \infty$, then if we put $b_n = n \, \mathsf{E} \, \xi_1$ and $a_n = \sigma\sqrt{n}$, we find by Sect. 4 that T has the normal distribution $\mathscr{N}(0, 1)$.

If $f(x) = \theta/\pi(x^2 + \theta^2)$ is the Cauchy density (with parameter $\theta > 0$) and $\xi_1, \xi_2, \ldots$ are independent random variables with density $f(x)$, the characteristic function $\varphi_{\xi_1}(t)$ is equal to $e^{-\theta|t|}$ and therefore $\varphi_{S_n/n}(t) = (e^{-\theta|t|/n})^n = e^{-\theta|t|}$, i.e., S_n/n also has the Cauchy distribution (with the same parameter θ).

Consequently there are other limit distributions besides the normal, for example the Cauchy distribution.

If we put $\xi_{nk} = (\xi_k/a_n) - (b_n/na_n)$, $1 \leq k \leq n$, we find that

$$\frac{S_n - b_n}{a_n} = \sum_{k=1}^{n} \xi_{n,k} \; (= T_n).$$

Therefore all conceivable distributions for T that can appear as limits in (3) are necessarily (in agreement with Theorem 1) infinitely divisible. However, the specific characteristics of the variable $T_n = (S_n - b_n)/a_n$ may make it possible to obtain further information on the structure of the limit distributions that arise.

For this reason we introduce the following definition.

Definition 2. A random variable T, its distribution function $F(x)$, and its characteristic function $\varphi(t)$ are *stable* if, for every $n \geq 1$, there are constants $a_n > 0$, b_n, and independent random variables $\xi_1, \ldots, \xi_n$, distributed like T, such that

$$a_n T + b_n \stackrel{d}{=} \xi_1 + \cdots + \xi_n \tag{4}$$

or, equivalently, $F((x - b_n)/a_n) = F * \cdots * F(x)$, or

$$[\varphi(t)]^n = [\varphi(a_n t)]e^{ib_n t}. \tag{5}$$

Theorem 3. *A necessary and sufficient condition for the random variable T to be a limit in distribution of random variables $(S_n - b_n)/a_n$, $a_n > 0$, is that T is stable.*

PROOF. If T is stable, then by (4)

$$T \stackrel{d}{=} \frac{S_n - b_n}{a_n},$$

where $S_n = \xi_1 + \cdots + \xi_n$, and consequently $(S_n - b_n)/a_n \stackrel{d}{\to} T$.

Conversely, let $\xi_1, \xi_2, \ldots$ be a sequence of independent identically distributed random variables, $S_n = \xi_1 + \cdots + \xi_n$ and $(S_n - b_n)/a_n \stackrel{d}{\to} T$, $a_n > 0$. Let us show that T is a stable random variable.

If T is degenerate, it is evidently stable. Let us suppose that T is nondegenerate. Choose $k \geq 1$ and write

$$S_n^{(1)} = \xi_1 + \cdots + \xi_n, \quad \ldots, \quad S_n^{(k)} = \xi_{(k-1)n+1} + \cdots + \xi_{kn},$$

$$T_n^{(1)} = \frac{S_n^{(1)} - b_n}{a_n}, \quad \ldots, \quad T_n^{(k)} = \frac{S_n^{(k)} - b_n}{a_n}.$$

It is clear that all the variables $T_n^{(1)}, \ldots, T_n^{(k)}$ have the same distribution and

$$T_n^{(i)} \stackrel{d}{\to} T, \quad n \to \infty, \, i = 1, \ldots, k.$$

Write

$$U_n^{(k)} = T_n^{(1)} + \cdots + T_n^{(k)}.$$

Then we obtain as in the proof of Theorem 1 that

$$U_n^{(k)} \stackrel{d}{\to} T^{(1)} + \cdots + T^{(k)},$$

where $T^{(i)}$, $1 \leq i \leq k$, are independent and $T^{(1)} \stackrel{d}{=} \ldots \stackrel{d}{=} T^{(k)} \stackrel{d}{=} T$.

On the other hand,

$$U_n^{(k)} = \frac{\xi_1 + \cdots + \xi_{kn} - kb_n}{a_n}$$

$$= \frac{a_{kn}}{a_n}\left(\frac{\xi_1 + \cdots + \xi_{kn} - b_{kn}}{a_{kn}}\right) + \frac{b_{kn} - kb_n}{a_n}$$

$$= \alpha_n^{(k)} V_{kn} + \beta_n^{(k)}, \tag{6}$$

where

$$\alpha_n^{(k)} = \frac{a_{kn}}{a_n}, \qquad \beta_n^{(k)} = \frac{b_{kn} - kb_n}{a_n}$$

and

$$V_{kn} = \frac{\xi_1 + \cdots + \xi_{kn} - b_{kn}}{a_{kn}}.$$

It is clear from (6) that

$$V_{kn} = \frac{U_n^{(k)} - \beta_n^{(k)}}{\alpha_n^{(k)}},$$

where $V_{kn} \xrightarrow{d} T$, $U_n^{(k)} \xrightarrow{d} T^{(1)} + \cdots + T^{(k)}$, $n \to \infty$.

It follows from the lemma established below that there are constants $\alpha^{(k)} > 0$ and $\beta^{(k)}$ such that $\alpha_n^{(k)} \to \alpha^{(k)}$ and $\beta_n^{(k)} \to \beta^{(k)}$ as $n \to \infty$. Therefore

$$T \stackrel{d}{=} \frac{T^{(1)} + \cdots + T^{(k)} - \beta^{(k)}}{\alpha^{(k)}},$$

which shows that T is a stable random variable.

This completes the proof of the theorem.

□

We now state and prove the lemma that we used above.

Lemma. *Let $\xi_n \xrightarrow{d} \xi$ and let there be constants $a_n > 0$ and b_n such that*

$$a_n \xi_n + b_n \xrightarrow{d} \tilde{\xi},$$

where the random variables ξ and $\tilde{\xi}$ are not degenerate. Then there are constants $a > 0$ and b such that $\lim a_n = a$, $\lim b_n = b$, and

$$\tilde{\xi} \stackrel{d}{=} a\xi + b.$$

PROOF. Let φ_n, φ and $\tilde{\varphi}$ be the characteristic functions of ξ_n, ξ and $\tilde{\xi}$ respectively. Then $\varphi_{a_n\xi_n+b_n}(t)$, the characteristic function of $a_n\xi_n + b_n$, is equal to $e^{itb_n}\varphi_n(a_nt)$ and, by Corollary of Theorem 1 and Problem 3 of Sect. 3,

$$e^{itb_n}\varphi_n(a_nt) \to \tilde{\varphi}(t), \tag{7}$$

$$\varphi_n(t) \to \varphi(t) \tag{8}$$

uniformly in t on every finite interval.

Let $\{n_i\}$ be a subsequence of $\{n\}$ such that $a_{n_i} \to a$. Let us first show that $a < \infty$. Suppose that $a = \infty$. By (7),

$$\sup_{|t| \le c} \left\| \varphi_n(a_n t)| - |\tilde{\varphi}(t) \right\| \to 0, \quad n \to \infty,$$

for every $c > 0$. We replace t by t_0/a_{n_i}. Then, since $a_{n_i} \to \infty$, we have

$$\left| \varphi_{n_i} \left(a_{n_i} \frac{t_0}{a_{n_i}} \right) \right| - \left| \tilde{\varphi} \left(\frac{t_0}{a_{n_i}} \right) \right| \to 0$$

and therefore

$$|\varphi_{n_i}(t_0)| \to |\tilde{\varphi}(0)| = 1.$$

But $|\varphi_{n_i}(t_0)| \to |\varphi(t_0)|$. Therefore $|\varphi(t_0)| = 1$ for every $t_0 \in R$, and consequently, by Theorem 5 of Sect. 12, Chap. 2, the random variable ξ must be degenerate, which contradicts the hypotheses of the lemma.

Thus $a < \infty$. Now suppose that there are two subsequences $\{n_i\}$ and $\{n_i'\}$ such that $a_{n_i} \to a$, $a_{n_i'} \to a'$, where $a \ne a'$; suppose for definiteness that $0 \le a' < a$. Then by (7) and (8),

$$|\varphi_{n_i}(a_{n_i} t)| \to |\varphi(at)|, \quad |\varphi_{n_i}(a_{n_i} t)| \to |\tilde{\varphi}(t)|$$

and

$$|\varphi_{n_i'}(a_{n_i'} t)| \to |\varphi(a't)|, \quad |\varphi_{n_i'}(a_{n_i'} t)| \to |\tilde{\varphi}(t)|.$$

Consequently

$$|\varphi(at)| = |\varphi(a't)|,$$

and therefore, for all $t \in R$,

$$|\varphi(t)| = \left| \varphi \left(\frac{a'}{a} t \right) \right| = \cdots = \left| \varphi \left(\left(\frac{a'}{a} \right)^n t \right) \right| \to 1, \quad n \to \infty.$$

Therefore $|\varphi(t)| \equiv 1$ and, by Theorem 5 of Sect. 12, Chap. 2, it follows that ξ is a degenerate random variable. This contradiction shows that $a = a'$ and therefore that there is a finite limit $\lim a_n = a$, with $a \ge 0$.

Let us now show that there is a limit $\lim b_n = b$, and that $a > 0$. Since (8) is satisfied uniformly on each finite interval, we have

$$\varphi_n(a_n t) \to \varphi(at),$$

and therefore, by (7), the limit $\lim_{n \to \infty} e^{itb_n}$ exists for all t such that $\varphi(at) \ne 0$. Let $\delta > 0$ be such that $\varphi(at) \ne 0$ for all $|t| < \delta$. For such t, $\lim e^{itb_n}$ exists. Hence we can deduce (Problem 9) that $\limsup |b_n| < \infty$.

Let there be two sequences $\{n_i\}$ and $\{n_i'\}$ such that $\lim b_{n_i} = b$ and $\lim b_{n_i'} = b'$. Then

$$e^{itb} = e^{itb'},$$

for $|t| < \delta$, and consequently $b = b'$. Thus there is a finite limit $b = \lim b_n$ and, by (7),

$$\tilde{\varphi}(t) = e^{itb}\varphi(at),$$

which means that $\tilde{\xi} \overset{d}{=} a\xi + b$. Since $\tilde{\xi}$ is not degenerate, we have $a > 0$.

This completes the proof of the lemma.

□

4. We quote without proof a theorem on the general form of the characteristic functions of *stable* distributions.

Theorem 4 (Lévy–Khinchin Representation). *A random variable T is stable if and only if its characteristic function $\varphi(t)$ has the form $\varphi(t) = \exp \psi(t)$,*

$$\psi(t) = it\beta - d|t|^\alpha \left(1 + i\theta\frac{t}{|t|}G(t,\alpha)\right), \tag{9}$$

where $0 < \alpha < 2$, $\beta \in R$, $d \geq 0$, $|\theta| \leq 1$, $t/|t| = 0$ for $t = 0$, and

$$G(t, \alpha) = \begin{cases} \tan \frac{1}{2}\pi\alpha & \text{if } \alpha \neq 1, \\ (2/\pi)\log|t| & \text{if } \alpha = 1. \end{cases} \tag{10}$$

Observe that it is easy to exhibit characteristic functions of *symmetric* stable distributions:

$$\varphi(t) = e^{-d|t|^\alpha}, \tag{11}$$

where $0 < \alpha \leq 2$, $d \geq 0$.

5. PROBLEMS

1. Show that if $\xi_n \overset{d}{\to} \xi$ and $\xi_n \overset{d}{\to} \eta$ then $\xi \overset{d}{=} \eta$.
2. Show that if φ_1 and φ_2 are infinitely divisible characteristic functions, then so is $\varphi_1 \cdot \varphi_2$.
3. Let φ_n be infinitely divisible characteristic functions and let $\varphi_n(t) \to \varphi(t)$ for every $t \in R$, where $\varphi(t)$ is a characteristic function. Show that $\varphi(t)$ is infinitely divisible.
4. Show that the characteristic function of an infinitely divisible distribution cannot take the value 0.
5. Give an example of a random variable that is infinitely divisible but not stable.
6. Show that a stable random variable ξ satisfies the inequality $E|\xi|^r < \infty$ for all $r \in (0, \alpha), 0 < \alpha < 2$.
7. Show that if ξ is a stable random variable with parameter $0 < \alpha \leq 1$, then $\varphi(t)$ is not differentiable at $t = 0$.
8. Prove that $e^{-d|t|^\alpha}$ is a characteristic function provided that $d \geq 0, 0 < \alpha \leq 2$.
9. Let $(b_n)_{n\geq 1}$ be a sequence of numbers such that $\lim_n e^{itb_n}$ exists for all $|t| < \delta$, $\delta > 0$. Show that $\limsup |b_n| < \infty$.
10. Show that the binomial and uniform distributions are *not* infinitely divisible.

11. Let F and φ be a distribution function and its characteristic function that are representable as $F = F^{(n)} * \cdots * F^{(n)}$ (n times) and $\varphi = [\varphi^{(n)}]^n$ with some distribution functions $F^{(n)}$ and their characteristic functions $\varphi^{(n)}$, $n \geq 1$. Show that there are ("rich" enough) probability space $(\Omega, \mathscr{F}, \mathsf{P})$ and random variables T and $(\eta_k^n)_{k \leq n}$ defined on it (such that $T \sim F$ and $\eta_1^{(n)}, \ldots, \eta_n^{(n)}$ are independent and identically distributed with distribution $F^{(n)}$) such that $T \overset{d}{=} \eta_1^{(n)} + \cdots + \eta_n^{(n)}$, $n \geq 1$.

12. Give an example of a random variable which is not infinitely divisible, but whose characteristic function nevertheless does not vanish.

7 Metrizability of Weak Convergence

1. Let $(E, \mathscr{E}, \rho)$ be a metric space and $\mathscr{P}(E) = \{P\}$, a family of probability measures on $(E, \mathscr{E})$. It is natural to raise the question of whether it is possible to "metrize" the weak convergence $P_n \overset{w}{\to} P$ that was introduced in Sect. 1, that is, whether it is possible to introduce a distance $\mu(P, \tilde{P})$ between any two measures P and $\tilde{P}$ in $\mathscr{P}(E)$ in such a way that the limit $\mu(P_n, P) \to 0$ is equivalent to the limit $P_n \overset{w}{\to} P$.

In connection with this formulation of the problem, it is useful to recall that *convergence* of random variables *in probability*, $\xi_n \overset{\mathsf{P}}{\to} \xi$, can be metrized by using, for example, the distance $d_{\mathsf{P}}(\xi, \eta) = \inf\{\varepsilon > 0 \colon \mathsf{P}(|\xi - \eta| \geq \varepsilon) \leq \varepsilon\}$ or the distances $d(\xi, \eta) = \mathsf{E}(|\xi - \eta|/(1 + |\xi - \eta|))$, $d(\xi, \eta) = \mathsf{E}\min(1, |\xi - \eta|)$. (More generally, we can set $d(\xi, \eta) = \mathsf{E}\,g(|\xi - \eta|)$, where the function $g = g(x)$, $x \geq 0$, can be chosen as any nonnegative increasing Borel function that is continuous at zero and has the properties $g(x + y) \leq g(x) + g(y)$ for $x \geq 0$, $y \geq 0$, $g(0) = 0$, and $g(x) > 0$ for $x > 0$.) However, at the same time there is, in the space of random variables over $(\Omega, \mathscr{F}, \mathsf{P})$, no distance $d(\zeta, \eta)$ such that $d(\xi_n, \xi) \to 0$ if and only if ξ_n converges to ξ *with probability one*. (In this connection, it is easy to find a sequence of random variables ξ_n, $n \geq 1$, that converges to ξ in probability but does not converge with probability one.) In other words, *convergence with probability one is not metrizable*. (See Problems 11 and 12 in Sect. 10, Chap. 2.)

The aim of this section is to obtain concrete instances of two metrics, $L(P, \tilde{P})$ and $\|P - \tilde{P}\|_{BL}^*$ in the space $\mathscr{P}(E)$ of measures, that metrize weak convergence:

$$P_n \overset{w}{\to} P \Leftrightarrow L(P_n, P) \to 0 \Leftrightarrow \|P_n - P\|_{BL}^* \to 0. \tag{1}$$

2. The Lévy–Prokhorov metric $L(P, \tilde{P})$. Let

$$\rho(x, A) = \inf\{\rho(x, y) \colon y \in A\},$$
$$A^{\varepsilon} = \{x \in E \colon \rho(x, A) < \varepsilon\}, \quad A \in \mathscr{E}.$$

For any two measures P and $\tilde{P} \in \mathcal{P}(E)$, we set

$$\sigma(P, \tilde{P}) = \inf\{\varepsilon > 0 : P(F) \leq \tilde{P}(F^\varepsilon) + \varepsilon \text{ for all closed sets } F \in \mathcal{E}\} \quad (2)$$

and

$$L(P, \tilde{P}) = \max[\sigma(P, \tilde{P}), \sigma(\tilde{P}, P)]. \quad (3)$$

The following lemma shows that the function $L(P, \tilde{P})$, $P, \tilde{P} \in \mathcal{P}(\mathcal{E})$, which is defined in this way, and is called the *Lévy-Prokhorov metric*, actually defines a metric.

Lemma 1. *The function* $L(P, \tilde{P})$ *has the following properties:*

(a) $L(P, \tilde{P}) = L(\tilde{P}, P)(= \sigma(P, \tilde{P}) = \sigma(\tilde{P}, P))$,
(b) $L(P, \tilde{P}) \leq L(P, \hat{P}) + L(\hat{P}, \tilde{P})$,
(c) $L(P, \tilde{P}) = 0$ *if and only if* $\tilde{P} = P$.

PROOF. (a) It is sufficient to show that (with $\alpha > 0$ and $\beta > 0$)

$$\text{"} P(F) \leq \tilde{P}(F^\alpha) + \beta \quad \text{for all closed sets } F \in \mathcal{E} \text{"} \quad (4)$$

if and only if

$$\text{"} \tilde{P}(F) \leq P(F^\alpha) + \beta \quad \text{for all closed sets } F \in \mathcal{E}. \text{"} \quad (5)$$

Let T be a closed set in $\mathcal{E}$. Then the set T^α is open and it is easy to verify that $T \subseteq E\backslash(E\backslash T^\alpha)^\alpha$. If (4) is satisfied, then, in particular,

$$P(E\backslash T^\alpha) \leq \tilde{P}((E\backslash T^\alpha)^\alpha) + \beta$$

and therefore,

$$\tilde{P}(T) \leq \tilde{P}(E\backslash(E\backslash T^\alpha)^\alpha) \leq P(T^\alpha) + \beta,$$

which establishes the equivalence of (4) and (5). Hence, it follows that

$$\sigma(P, \tilde{P}) = \sigma(P, P) \quad (6)$$

and therefore,

$$L(P, \tilde{P}) = \sigma(P, \tilde{P}) = \sigma(\tilde{P}, P) = L(\tilde{P}, P). \quad (7)$$

(b) Let $L(P, \hat{P}) < \delta_1$ and $L(\hat{P}, \tilde{P}) < \delta_2$. Then for each closed set $F \in \mathcal{E}$

$$\tilde{P}(F) \leq \hat{P}(F^{\delta_2}) + \delta_2 \leq P((F^{\delta_2})^{\delta_1}) + \delta_1 + \delta_2 \leq P(F^{\delta_1 + \delta_2}) + \delta_1 + \delta_2$$

and therefore, $L(P, \tilde{P}) \leq \delta_1 + \delta_2$. Hence, it follows that

$$L(P, \tilde{P}) \leq L(P, \hat{P}) + L(\hat{P}, \tilde{P}).$$

(c) If $L(P, \tilde{P}) = 0$, then for every closed set $F \in \mathcal{E}$ and every $\alpha > 0$

$$P(F) \leq \tilde{P}(F^\alpha) + \alpha. \quad (8)$$

Since $F^\alpha \downarrow F$, $\alpha \downarrow 0$, we find, by taking the limit in (8) as $\alpha \downarrow 0$, that $P(F) \leq \tilde{P}(F)$ and by symmetry $\tilde{P}(F) \leq P(F)$. Hence, $P(F) = \tilde{P}(F)$ for all closed sets $F \in \mathscr{E}$. For each Borel set $A \in \mathscr{E}$ and every $\varepsilon > 0$, there is an open set $G_\varepsilon \supseteq A$ and a closed set $F_\varepsilon \subseteq A$ such that $P(G_\varepsilon \backslash F_\varepsilon) \leq \varepsilon$. Hence, it follows that every probability measure P on a metric space $(E, \mathscr{E}, \rho)$ is completely determined by its values on closed sets. Consequently, it follows from the condition $\tilde{P}(F) = P(F)$ for all closed sets $F \in \mathscr{E}$ that $\tilde{P}(A) = P(A)$ for all Borel sets $A \in \mathscr{E}$.

□

Theorem 1. *The Lévy-Prokhorov metric $L(P, \tilde{P})$ metrizes weak convergence:*

$$L(P_n, P) \to 0 \Leftrightarrow P_n \overset{w}{\to} P. \tag{9}$$

PROOF. $(\Rightarrow)$ Let $L(P_n, P) \to 0$, $n \to \infty$. Then for every specified closed set $F \in \mathscr{E}$ and every $\varepsilon > 0$, we have, by (2) and equation (a) of Lemma 1,

$$\limsup_n P_n(F) \leq P(F^\varepsilon) + \varepsilon. \tag{10}$$

If we then let $\varepsilon \downarrow 0$, we find that

$$\limsup_n P_n(F) \leq P(F).$$

According to Theorem 1 of Sect. 1, it follows that

$$P_n \overset{w}{\to} P. \tag{11}$$

The proof of the implication $(\Leftarrow)$ will be based on a series of deep and powerful facts that illuminate the content of the concept of weak convergence and the method of establishing it, as well as methods of studying rates of convergence.

Thus, let $P_n \overset{w}{\to} P$. This means that for every *bounded continuous* function $f = f(x)$

$$\int_E f(x) P_n(dx) \to \int_E f(x) P(dx). \tag{12}$$

Now suppose that $\mathscr{G}$ is a class of equicontinuous functions $g = g(x)$ (for every $\varepsilon > 0$ there is a $\delta > 0$ such that $|g(y) - g(x)| < \varepsilon$ if $\rho(x, y) < \delta$ for all $g \in \mathscr{G}$) and $|g(x)| \leq C$ for the same constant $C > 0$ (for all $x \in E$ and $g \in \mathscr{G}$). By Theorem 3 of Sect. 8, the following condition, stronger than (12), is valid for $\mathscr{G}$:

$$P_n \overset{w}{\to} P \Rightarrow \sup_{g \in \mathscr{G}} \left| \int_E g(x) P_n(dx) - \int_E g(x) P(dx) \right| \to 0. \tag{13}$$

For each $A \in \mathscr{E}$ and $\varepsilon > 0$, we set (as in Theorem 1, Sect. 1)

$$f_A^\varepsilon(x) = \left[1 - \frac{\rho(x, A)}{\varepsilon} \right]^+. \tag{14}$$

It is clear that

$$I_A(x) \le f_A^\varepsilon(x) \le I_{A^\varepsilon}(x) \tag{15}$$

and

$$|f_A^\varepsilon(x) - f_A^\varepsilon(y)| \le \varepsilon^{-1}|\rho(x, A) - \rho(y, A)| \le \varepsilon^{-1}\rho(x, y).$$

Therefore, we have (13) for the class $\mathscr{G}^\varepsilon = \{f_A^\varepsilon(x), \, A \in \mathscr{E}\}$, i.e.,

$$\Delta_n \equiv \sup_{A \in \mathscr{E}} \left| \int_E f_A^\varepsilon(x) P_n(dx) - \int_E f_A^\varepsilon(x) P(dx) \right| \to 0, \quad n \to \infty. \tag{16}$$

From this and (15) we conclude that, for every closed set $A \in \mathscr{E}$ and $\varepsilon > 0$,

$$P(A^\varepsilon) \ge \int_E f_A^\varepsilon(x) \, dP \ge \int_E f_A^\varepsilon(x) \, dP_n - \Delta_n \ge P_n(A) - \Delta_n. \tag{17}$$

We choose $n(\varepsilon)$ so that $\Delta_n \le \varepsilon$ for $n \ge n(\varepsilon)$. Then, by (17), for $n \ge n(\varepsilon)$

$$P(A^\varepsilon) \ge P_n(A) - \varepsilon. \tag{18}$$

Hence, it follows from definitions (2) and (3) that $L(P_n, P) \le \varepsilon$ as soon as $n \ge n(\varepsilon)$. Consequently,

$$P_n \xrightarrow{w} P \Rightarrow \Delta_n \to 0 \Rightarrow L(P_n, P) \to 0.$$

The theorem is now proved (up to (13)).

□

3. The metric $\|P - \tilde{P}\|_{BL}^*$. We denote by BL the set of bounded continuous functions $f = f(x)$, $x \in E$ (with $\|f\|_\infty = \sup_x |f(x)| < \infty$) that also satisfy the Lipschitz condition

$$\|f\|_L = \sup_{x \ne y} \frac{|f(x) - f(y)|}{\rho(x, y)} < \infty.$$

We set $\|f\|_{BL} = \|f\|_\infty + \|f\|_L$. The space BL with the norm $\| \cdot \|_{BL}$ is a Banach space.

We define the metric $\|P - \tilde{P}\|_{BL}^*$ by setting

$$\|P - \tilde{P}\|_{BL}^* = \sup_{f \in BL} \left\{ \left| \int f \, d(P - \tilde{P}) \right| : \|f\|_{BL} \le 1 \right\}. \tag{19}$$

(We can verify that $\|P - \tilde{P}\|_{BL}^*$ actually satisfies the conditions for a metric; Problem 2.)

Theorem 2. *The metric $\|P - \tilde{P}\|_{BL}^*$ metrizes weak convergence:*

$$\|P_n - P\|_{BL}^* \to 0 \Leftrightarrow P_n \xrightarrow{w} P.$$

PROOF. The implication ($\Leftarrow$) follows directly from (13). To prove ($\Rightarrow$), it is enough to show that in the definition of weak convergence $P_n \overset{w}{\rightarrow} P$ as given by (12) for every continuous bounded function $f = f(x)$, it is enough to restrict consideration to the class of bounded functions that satisfy a Lipschitz condition. In other words, the implication ($\Rightarrow$) will be proved if we establish the following result.

Lemma 2. *Weak convergence $P_n \overset{w}{\rightarrow} P$ occurs if and only if property* (12) *is satisfied for every function $f = f(x)$ of class BL.*

PROOF. The proof is obvious in one direction. Let us now consider the functions $f_A^\varepsilon = f_A^\varepsilon(x)$ defined in (14). As was established above in the proof of Theorem 1, for each $\varepsilon > 0$ the class $\mathscr{G}^\varepsilon = \{f_A^\varepsilon(x), A \in \mathscr{G}\} \subseteq BL$. If we now analyze the proof of the implication (I) $\Rightarrow$ (II) in Theorem 1 of Sect. 1, we can observe that it actually uses property (12) *not for all* bounded continuous functions but only for functions of class $\mathscr{G}^\varepsilon$, $\varepsilon > 0$. Since $\mathscr{G}^\varepsilon \subseteq BL$, $\varepsilon > 0$, it is evidently true that the satisfaction of (12) for functions of class BL implies proposition (II) of Theorem 1, Sect. 1, which is equivalent (by the same Theorem 1, Sect. 1) to the weak convergence $P_n \overset{w}{\rightarrow} P$.

$\square$

Remark. The conclusion of Theorem 2 can be derived from Theorem 1 (and conversely) if we use the following inequalities between the metrics $L(P, \tilde{P})$ and $\|P - \tilde{P}\|_{BL}^*$, which are valid for the separable metric spaces $(E, \mathscr{E}, \rho)$:

$$\|P - \tilde{P}\|_{BL}^* \leq 2L(P, \tilde{P}), \tag{20}$$

$$\varphi(L(P, \tilde{P})) \leq \|P - \tilde{P}\|_{BL}^*, \tag{21}$$

where $\varphi(x) = 2x^2/(2 + x)$.

Taking into account that, for $x \geq 0$, we have $0 \leq \varphi \leq 2/3$ if and only if $x \leq 1$, and $(2/3)x^2 \leq \varphi(x)$ for $0 \leq x \leq 1$, we deduce from (20) and (21) that if $L(P, \tilde{P}) \leq 1$ or $\|P - \tilde{P}\|_{BL}^* \leq 2/3$, then

$$\tfrac{2}{3}L^2(P, \tilde{P}) \leq \|P - \tilde{P}\|_{BL}^* \leq 2L(P, \tilde{P}). \tag{22}$$

4. PROBLEMS

1. Show that in case $E = R$ the Lévy–Prokhorov distance between the probability distributions P and $\tilde{P}$ is *no less* than the *Lévy distance* $L(F, \tilde{F})$ between the distribution functions F and $\tilde{F}$ that correspond to P and $\tilde{P}$ (see Problem 4 in Sect. 1). Give an example of a strict inequality between these distances.

2. Show that formula (19) defines a metric on the space BL.

3. Establish the inequalities (20), (21), and (22).

4. Let $F = F(x)$ and $G = G(x)$ be two distribution functions and P_c and Q_c be the points of their intersection of the line $x + y = c$. Show that the Lévy distance between F and G (see Problem 4 in Sect. 1) equals

$$L(F, G) = \sup_c \frac{\overline{P_c Q_c}}{\sqrt{2}},$$

where $\overline{P_c Q_c}$ is the length of the interval between the points P_c and Q_c.

5. Show that the set of all distribution functions endowed with Lévy distance is a complete space.

8 On the Connection of Weak Convergence of Measures with Almost Sure Convergence of Random Elements ("Method of a Single Probability Space")

1. Let us suppose that on the probability space $(\Omega, \mathscr{F}, \mathsf{P})$ there are given random elements $X = X(\omega)$, $X_n = X_n(\omega)$, $n \geq 1$, taking values in the metric space $(E, \mathscr{E}, \rho)$; see Sect. 5, Chap. 2. We denote by P and P_n the probability distributions of X and X_n, i.e., let

$$P(A) = \mathsf{P}\{\omega : X(\omega) \in A\}, \quad P_n(A) = \mathsf{P}\{\omega : X_n(\omega) \in A\}, \quad A \in \mathscr{E}.$$

Generalizing the concept of convergence in distribution of random variables (see Sect. 10, Chap. 2), we introduce the following definition.

Definition 1. *A sequence of random elements* X_n, $n \geq 1$, *is said to converge in distribution, or in law (notation:* $X_n \xrightarrow{\mathscr{D}} X$, *or* $X_n \xrightarrow{\mathscr{L}} X$), *if* $P_n \xrightarrow{w} P$.

By analogy with the definitions of convergence of random variables in probability or with probability one (Sect. 10, Chap. 2), it is natural to introduce the following definitions.

Definition 2. A sequence of random elements X_n, $n \geq 1$, is said to converge *in probability* to X if

$$\mathsf{P}\{\omega : \rho(X_n(\omega), X(\omega)) \geq \varepsilon\} \to 0, \quad n \to \infty. \tag{1}$$

Definition 3. A sequence of random elements X_n, $n \geq 1$, is said to converge to X *with probability one* (*almost surely, almost everywhere*) if $\rho(X_n(\omega), X(\omega)) \xrightarrow{\text{a.s.}} 0$ as $n \to \infty$.

Remark 1. Both of the preceding definitions make sense, of course, provided that $\rho(X_n(\omega), X(\omega))$ are, as functions of $\omega \in \Omega$, random variables, i.e., $\mathscr{F}$-measurable functions. This will certainly be the case if the space $(E, \mathscr{E}, \rho)$ is separable (Problem 1).

Remark 2. In connection with Definition 2 note that convergence in probability introduced therein is metrized by the following Ky Fan distance (see [55]) between random elements X and Y (defined on $(\Omega, \mathscr{F}, \mathsf{P})$ and ranging in E; Problem 2):

$$d_\mathsf{P}(X, Y) = \inf\{\varepsilon > 0 \colon \mathsf{P}\{\rho(X(\omega), Y(\omega)) \geq \varepsilon\} \leq \varepsilon\}. \tag{2}$$

Remark 3. Whereas the definitions of convergence in probability and with probability one presume that all the random elements are defined on the same probability space, the definition $X_n \overset{\mathscr{D}}{\to} X$ of convergence in distribution is connected only with the convergence of distributions, and consequently, we may suppose that $X(\omega), X_1(\omega), X_2(\omega), \ldots$ have values in the same space E, but may be defined on "their own" probability spaces $(\Omega, \mathscr{F}, \mathsf{P}), (\Omega_1, \mathscr{F}_1, \mathsf{P}_1), (\Omega_2, \mathscr{F}_2, \mathsf{P}_2), \ldots$. However, without loss of generality we may always suppose that they are defined on the same probability space, taken as the direct product of the underlying spaces and with the definitions $X(\omega, \omega_1, \omega_2, \ldots) = X(\omega), X_1(\omega, \omega_1, \omega_2, \ldots) = X_1(\omega_1), \ldots$.

2. By Definition 1 and the theorem on change of variables under the Lebesgue integral sign (Theorem 7 of Sect. 6, Chap. 2)

$$X_n \overset{\mathscr{D}}{\to} X \Leftrightarrow \mathsf{E}f(X_n) \to \mathsf{E}f(X) \tag{3}$$

for every bounded continuous function $f = f(x)$, $x \in E$.

From (3) it is clear that, by Lebesgue's theorem on dominated convergence (Theorem 3 of Sect. 6, Chap. 2), the limit $X_n \overset{\text{a.s.}}{\to} X$ immediately implies the limit $X_n \overset{\mathscr{D}}{\to} X$, which is hardly surprising if we think of the situation when X and X_n are random variables (Theorem 2 of Sect. 10, Chap. 2). More unexpectedly, in a certain sense there is a converse result, the precise formulation and application we now turn to.

Preliminarily, we introduce a definition.

Definition 4. Random elements $X = X(\omega')$ and $Y = Y(\omega'')$, defined on probability spaces $(\Omega', \mathscr{F}', \mathsf{P}')$ and $(\Omega'', \mathscr{F}'', \mathsf{P}'')$ and with values in the same space E, are said to be *equivalent in distribution* (notation: $X \overset{\mathscr{D}}{=} Y$), if they have the same probability distribution.

Theorem 1. *Let* $(E, \mathscr{E}, \rho)$ *be a separable metric space.*

1. *Let random elements* X, X_n, $n \geq 1$, *defined on a probability space* $(\Omega, \mathscr{F}, \mathsf{P})$, *and with values in* E, *have the property that* $X_n \overset{\mathscr{D}}{\to} X$. *Then we can find a probability space* $(\Omega^*, \mathscr{F}^*, \mathsf{P}^*)$ *and random elements* X^*, X_n^*, $n \geq 1$, *defined on it, with values in* E, *such that*

$$X_n^* \overset{\text{a.s.}}{\to} X^*$$

and

$$X^* \overset{\mathscr{D}}{=} X, \quad X_n^* \overset{\mathscr{D}}{=} X_n, \quad n \geq 1.$$

2. *Let* P, P_n, $n \geq 1$, *be probability measures on* $(E, \mathscr{E}, \rho)$ *such that* $P_n \overset{w}{\to} P$. *Then there is a probability space* $(\Omega^*, \mathscr{F}^*, \mathsf{P}^*)$ *and random elements* X^*, X_n^*, $n \geq 1$, *defined on it, with values in* E, *such that*

$$X_n^* \overset{\text{a.s.}}{\to} X^*$$

and

$$P^* = P, \quad P_n^* = P_n, \quad n \geq 1,$$

where P^* and P_n^* are the probability distributions of X^* and X_n^*.

Before turning to the proof, we first notice that it is enough to prove only the second conclusion, since the first follows from it if we take P and P_n to be the distributions of X and X_n. Similarly, the second conclusion follows from the first. Second, we notice that the proof of the theorem in full generality is technically rather complicated. For this reason, here we give a proof only of the case $E = R$. This proof is rather transparent and moreover, provides a simple, clear construction of the required objectives. (Unfortunately, this construction does not work in the general case, even for $E = R^2$.)

PROOF OF THE THEOREM IN THE CASE $E = R$. Let $F = F(x)$ and $F_n = F_n(x)$ be the distribution functions corresponding to the measures P and P_n on $(R, \mathcal{B}(R))$. We associate with a function $F = F(x)$ its corresponding *quantile function* $Q = Q(u)$, uniquely defined by the formula

$$Q(u) = \inf\{x \colon F(x) \geq u\}, \quad 0 < u < 1. \tag{4}$$

It is easily verified that

$$F(x) \geq u \Leftrightarrow Q(u) \leq x. \tag{5}$$

We now take $\Omega^* = (0,1)$, $\mathscr{F}^* = \mathscr{B}(0,1)$, P^* to be Lebesgue measure, $P^*(d\omega^*) = d\omega^*$. We also take $X^*(\omega^*) = Q(\omega^*)$ for $\omega^* \in \Omega^*$. Then

$$P^*\{\omega^* \colon X^*(\omega^*) \leq x\} = P^*\{\omega^* \colon Q(\omega^*) \leq x\} = P^*\{\omega^* \colon \omega^* \leq F(x)\} = F(x),$$

i.e., the distribution of the random variable $X^*(\omega^*) = Q(\omega^*)$ coincides exactly with P. Similarly, the distribution of $X_n^*(\omega^*) = Q_n(\omega^*)$ coincides with P_n.

In addition, it is not difficult to show that the convergence of $F_n(x)$ to $F(x)$ at each point of continuity of the limit function $F = F(x)$ (equivalent, if $E = R$, to the convergence $P_n \overset{w}{\to} P$; see Theorem 1 in Sect. 1) implies that the sequence of quantiles $Q_n(u)$, $n \geq 1$, also converges to $Q(u)$ at every point of continuity of the limit $Q = Q(u)$. Since the set of points of discontinuity of $Q = Q(u)$, $u \in (0,1)$, is at most countable, its Lebesgue measure P^* is zero and therefore,

$$X_n^*(\omega^*) = Q_n(\omega^*) \overset{\text{a.s.}}{\to} X^*(\omega^*) = Q(\omega^*).$$

The theorem is established in the case of $E = R$.

$\square$

This construction in Theorem 1 of a passage from given random elements X and X_n to new elements X^* and X_n^*, defined on the same probability space, explains the announcement in the heading of this section of the *method of a single probability space*.

We now turn to a number of propositions that are established very simply by using this method.

3. Let us assume that the random elements X and X_n, $n \geq 1$, are defined, for example, on a probability space $(\Omega, \mathscr{F}, \mathsf{P})$ with values in a separable metric space $(E, \mathscr{E}, \rho)$, so that $X_n \overset{\mathscr{D}}{\to} X$. Also let $h = h(x)$, $x \in E$, be a measurable mapping of $(E, \mathscr{E}, \rho)$ into another separable metric space $(E', \mathscr{E}', \rho')$. In probability and mathematical statistics it is often necessary to deal with the search for conditions under which we can say of $h = h(x)$ that the limit $X_n \overset{\mathscr{D}}{\to} X$ implies the limit $h(X_n) \overset{\mathscr{D}}{\to} h(X)$.

For example, let ξ_1, ξ_2, ... be independent identically distributed random variables with $\mathsf{E}\,\xi_1 = m$, $\mathrm{Var}\,\xi_1 = \sigma^2 > 0$. Let $\overline{X}_n = (\xi_1 + \cdots + \xi_n)/n$. The central limit theorem shows that

$$\frac{\sqrt{n}(\overline{X}_n - m)}{\sigma} \overset{d}{\to} \mathscr{N}(0,1).$$

Let us ask, for what functions $h = h(x)$ can we guarantee that

$$h\left(\frac{\sqrt{n}(\overline{X}_n - m)}{\sigma}\right) \overset{d}{\to} h(\mathscr{N}(0,1))?$$

(The well-known *Mann–Wald theorem* states with respect to this question that this is certainly true for continuous functions $h = h(x)$, hence it immediately follows, e.g., that $n(\overline{X} - m)^2/\sigma^2 \overset{d}{\to} \chi_1^2$, where χ_1^2 is a random variable with a chi-squared distribution with one degree of freedom; see Table 2.3 in Sect. 3, Chap. 2.)

A second example. If $X = X(t, \omega), X_n = X_n(t, \omega)$, $t \in T$, are random processes (see Sect. 5, Chap. 2) and $h(X) = \sup_{t \in T} |X(t, \omega)|$, $h(X_n) = \sup_{t \in T} |X_n(t, \omega)|$, our problem amounts to asking under what conditions the convergence in distribution of the processes $X_n \overset{\mathscr{D}}{\to} X$ will imply the convergence in distribution of their suprema, $h(X_n) \overset{\mathscr{D}}{\to} h(X)$.

A simple condition that guarantees the validity of the implication

$$X_n \overset{\mathscr{D}}{\to} X \Rightarrow h(X_n) \overset{\mathscr{D}}{\to} h(X),$$

is that the mapping $h = h(x)$ is continuous. In fact, if $f = f(x')$ *is a bounded continuous* function on E', the function $f(h(x))$ will also be a bounded continuous function on E. Consequently,

$$X_n \overset{\mathscr{D}}{\to} X \Rightarrow \mathsf{E}f(h(X_n)) \to \mathsf{E}f(h(X)).$$

The theorem given below shows that in fact the requirement of continuity of the function $h = h(x)$ can be somewhat weakened by using the properties of the limit random element X.

We denote by Δ_h the set $\{x \in E \colon h(x)$ is not ρ-continuous at $x\}$; i.e., let Δ_h be the set of points of discontinuity of the function $h = h(x)$. We note that $\Delta_h \in \mathscr{E}$ (Problem 4).

Theorem 2. 1. *Let* $(E, \mathscr{E}, \rho)$ *and* $(E', \mathscr{E}', \rho')$ *be separable metric spaces, and let* $X_n \overset{\mathscr{D}}{\to} X$. *Let the mapping* $h = h(x)$, $x \in E$, *have the property that*

$$\mathsf{P}\{\omega \colon X(\omega) \in \Delta_h\} = 0. \tag{6}$$

Then $h(X_n) \overset{\mathscr{D}}{\to} h(X)$.

2. *Let* P, P_n, $n \geq 1$, *be probability distributions on the separable metric space* $(E, \mathscr{E}, \rho)$ *and* $h = h(x)$ *a measurable mapping of* $(E, \mathscr{E}, \rho)$ *on a separable metric space* $(E', \mathscr{E}', \rho')$. *Let*

$$P\{x \colon x \in \Delta_h\} = 0.$$

Then $P_n^h \overset{w}{\to} P^h$, *where* $P_n^h(A) = P_n\{h(x) \in A\}$, $P^h(A) = P\{h(x) \in A\}$, $A \in \mathscr{E}'$.

PROOF. As in Theorem 1, it is enough to prove the validity of, for example, the first proposition.

Let X^* and X_n^*, $n \geq 1$, be random elements constructed by the "method of a single probability space," so that $X^* \overset{\mathscr{D}}{=} X$, $X_n^* \overset{\mathscr{D}}{=} X_n$, $n \geq 1$, and $X_n^* \overset{a.\,s.}{\to} X^*$. Let $A^* = \{\omega^* \colon \rho(X_n^*, X^*) \nrightarrow 0\}$, $B^* = \{\omega^* \colon X^*(\omega^*) \in \Delta_n\}$. Then $\mathsf{P}^*(A^* \cup B^*) = 0$, and for $\omega^* \notin A^* \cup B^*$

$$h(X_n^*(\omega^*)) \to h(X^*(\omega^*)),$$

which implies that $h(X_n^*) \overset{a.s.}{\to} h(X^*)$. As we noticed in Subsection 1, it follows that $h(X_n^*) \overset{\mathscr{D}}{\to} h(X^*)$. But $h(X_n^*) \overset{\mathscr{D}}{=} h(X_n)$ and $h(X^*) \overset{\mathscr{D}}{=} h(X)$. Therefore, $h(X_n^*) \overset{\mathscr{D}}{\to} h(X)$.

This completes the proof of the theorem.

$\square$

4. In Sect. 7, in the proof of the implication ($\Leftarrow$) in Theorem 1, we used (13). We now give a proof that again relies on the "method of a single probability space."

Let $(E, \mathscr{E}, \rho)$ be a separable metric space, and $\mathscr{G}$ a class of equicontinuous functions $g = g(x)$ for which also $|g(x)| \leq C$, $C > 0$ for all $x \in E$ and $g \in \mathscr{G}$.

Theorem 3. *Let* P *and* P_n, $n \geq 1$, *be probability measures on* $(E, \mathscr{E}, \rho)$ *for which* $P_n \overset{w}{\to} P$. *Then*

$$\sup_{g \in \mathscr{G}} \left| \int_E g(x)\, P_n(dx) - \int_E g(x)\, P(dx) \right| \to 0, \quad n \to \infty. \tag{7}$$

PROOF. Let (7) not occur. Then there are an $a > 0$ and functions $g_1, g_2, \ldots$ from $\mathscr{G}$ such that

$$\left| \int_E g_n(x)\, P_n(dx) - \int_E g_n(x)\, P(dx) \right| \geq a > 0 \tag{8}$$

for *infinitely many* values of n. Turning by the "method of a single probability space" to random elements X^* and X_n^* (see Theorem 1), we transform (8) to the form

$$|\mathsf{E}^* g_n(X_n^*) - \mathsf{E}^* g_n(X^*)| \geq a > 0 \qquad (9)$$

for infinitely many values of n. But, by the properties of $\mathscr{G}$, for every $\varepsilon > 0$ there is a $\delta > 0$ for which $|g(y) - g(x)| < \varepsilon$ for all $g \in \mathscr{G}$, if $\rho(x, y) < \delta$. In addition, $|g(x)| \leq C$ for all $x \in E$ and $g \in \mathscr{G}$. Therefore,

$$
\begin{aligned}
|\mathsf{E}^* g_n(X_n^*) - \mathsf{E}^* g_n(X^*)| &\leq \mathsf{E}^*\{|g_n(X_n^*) - g_n(X^*)|; \rho(X_n^*, X^*) > \delta\} \\
&\quad + \mathsf{E}^*\{|g_n(X_n^*) - g_n(X^*)|; \rho(X_n^*, X^*) \leq \delta\} \\
&\leq 2C\,\mathsf{P}^*\{\rho(X_n^*, X^*) > \delta\} + \varepsilon.
\end{aligned}
$$

Since $X_n^* \overset{\text{a.s.}}{\to} X^*$, we have $\mathsf{P}^*\{\rho(X_n^*, X^*) > \delta\} \to 0$ as $n \to \infty$. Consequently, since $\varepsilon > 0$ is arbitrary,

$$\limsup_n |\mathsf{E}^* g_n(X_n^*) - \mathsf{E}^* g_n(X^*)| = 0,$$

which contradicts (9).

This completes the proof of the theorem. □

5. In this subsection the idea of the "method of a single probability space" used in Theorem 1 will be applied to estimating upper bounds of the Lévy–Prokhorov metric $L(P, \tilde{P})$ between two probability distributions on a separable space $(E, \mathscr{E}, \rho)$.

Theorem 4. *For each pair $P, \tilde{P}$ of measures we can find a probability space $(\Omega^*, \mathscr{F}^*, \mathsf{P}^*)$ and random elements X and $\tilde{X}$ on it with values in E such that their distributions coincide respectively with P and $\tilde{P}$ and*

$$L(P, \tilde{P}) \leq d_{\mathsf{P}*}(X, \tilde{X}) = \inf\{\varepsilon > 0 \colon \mathsf{P}^*(\rho(X, \tilde{X}) \geq \varepsilon) \leq \varepsilon\}. \qquad (10)$$

PROOF. By Theorem 1, we can find a probability space $(\Omega^*, \mathscr{F}^*, \mathsf{P}^*)$ and random elements X and $\tilde{X}$ such that $\mathsf{P}^*(X \in A) = P(A)$ and $\mathsf{P}^*(\tilde{X} \in A) = \tilde{P}(A)$, $A \in \mathscr{E}$.

Let $\varepsilon > 0$ have the property that

$$\mathsf{P}^*(\rho(X, \tilde{X}) \geq \varepsilon) \leq \varepsilon. \qquad (11)$$

Then for every $A \in \mathscr{E}$ we have, denoting $A^\varepsilon = \{x \in E : \rho(x, A) < \varepsilon\}$,

$$
\begin{aligned}
\tilde{P}(A) &= \mathsf{P}^*(\tilde{X} \in A) = \mathsf{P}^*(\tilde{X} \in A, X \in A^\varepsilon) + \mathsf{P}^*(\tilde{X} \in A, X \notin A^\varepsilon) \\
&\leq \mathsf{P}^*(X \in A^\varepsilon) + \mathsf{P}^*(\rho(X, \tilde{X}) \geq \varepsilon) \leq P(A^\varepsilon) + \varepsilon.
\end{aligned}
$$

Hence, by the definition of the Lévy–Prokhorov metric (Subsection 2, Sect. 7)

$$L(P, \tilde{P}) \leq \varepsilon. \qquad (12)$$

From (11) and (12), if we take the infimum for $\varepsilon > 0$ we obtain the required assertion (10). □

Corollary. *Let X and X̃ be random elements defined on a probability space* $(\Omega, \mathscr{F}, \mathsf{P})$ *with values in E. Let P_X and $P_{\tilde{X}}$ be their probability distributions. Then*

$$L(P_X, P_{\tilde{X}}) \leq d_{\mathsf{P}}(X, \tilde{X}).$$

Remark 4. The preceding proof shows that in fact (10) is valid whenever we can exhibit on *any* probability space $(\Omega^*, \mathscr{F}^*, \mathsf{P}^*)$ random elements X and $\tilde{X}$ with values in E whose distributions coincide with P and $\tilde{P}$ and for which the set $\{\omega^*: \rho(X(\omega^*), \tilde{X}(\omega^*)) \geq \varepsilon\} \in \mathscr{F}^*, \varepsilon > 0$. Hence, the property of (10) depends in an essential way on how well, with respect to the measures P and $\tilde{P}$, the objects $(\Omega^*, \mathscr{F}^*, \mathsf{P}^*)$ and $X, \tilde{X}$ are constructed. (The procedure for constructing $\Omega^*, \mathscr{F}^*, \mathsf{P}^*$ and $X, \tilde{X}$ for given $P, \tilde{P}$, is called *coupling*.) We could, for example, choose P^* equal to the direct product of the measures P and $\tilde{P}$, but this choice would, as a rule, not lead to a good estimate (10).

Remark 5. It is natural to raise the question of when there is equality in (10). In this connection we state the following result without proof: *Let P and P̃ be two probability measures on a separable metric space* $(E, \mathscr{E}, \rho)$; *then there are* $(\Omega^*, \mathscr{F}^*, \mathsf{P}^*)$ *and $X, \tilde{X}$ such that*

$$L(P, \tilde{P}) = d_{\mathsf{P}*}(X, \tilde{X}) = \inf\{\varepsilon > 0: \mathsf{P}^*(\rho(X, \tilde{X}) \geq \varepsilon) \leq \varepsilon\}.$$

5. PROBLEMS

1. Prove that in the case of separable metric spaces the real-valued function $\rho(X(\omega), Y(\omega))$ is a random variable for all random elements $X(\omega)$ and $Y(\omega)$ defined on a probability space $(\Omega, \mathscr{F}, \mathsf{P})$.
2. Prove that the function $d_{\mathsf{P}}(X, Y)$ defined in (2) is a metric in the space of random elements with values in E.
3. Establish (5).
4. Prove that the set $\Delta_h = \{x \in E: h(x) \text{ is not } \rho\text{-continuous at } x\} \in \mathscr{E}$.

9 The Distance in Variation Between Probability Measures: Kakutani–Hellinger Distance and Hellinger Integrals—Application to Absolute Continuity and Singularity of Measures

1. Let $(\Omega, \mathscr{F})$ be a measurable space and $\mathscr{P} = \{P\}$ a family of probability measures on it.

Definition 1. The *distance in variation* between measures P and $\tilde{P}$ in $\mathscr{P}$ (notation: $\|P - \tilde{P}\|$) is the total variation of $P - \tilde{P}$, i.e.,

$$\|P - \tilde{P}\| = \mathrm{var}\,(P - \tilde{P}) \equiv \sup \left| \int_{\Omega} \varphi(\omega)\, d(P - \tilde{P}) \right|, \tag{1}$$

where the sup is over the class of all $\mathscr{F}$-measurable functions that satisfy the condition that $|\varphi(\omega)| \leq 1$.

Lemma 1. *The distance in variation is given by*

$$\|P - \tilde{P}\| = 2 \sup_{A \in \mathscr{F}} |P(A) - \tilde{P}(A)|. \tag{2}$$

PROOF. Since, for all $A \in \mathscr{F}$,

$$P(A) - \tilde{P}(A) = \tilde{P}(\overline{A}) - P(\overline{A}),$$

we have

$$2|P(A) - \tilde{P}(A)| = |P(A) - \tilde{P}(A)| + |P(\overline{A}) - \tilde{P}(\overline{A})| \leq \|P - \tilde{P}\|,$$

where the last inequality follows from (1).

For the proof of the converse inequality we turn to the Hahn decomposition (see, for example, [52], § 5, Chapter VI, or [39], p. 121) of the *signed measure* $\mu \equiv P - \tilde{P}$. In this decomposition the measure μ is represented in the form $\mu = \mu_+ - \mu_-$, where the nonnegative measures μ_+ and μ_- (the upper and lower variations of μ) are of the form

$$\mu_+(A) = \int_{A \cap M} d\mu, \quad \mu_-(A) = -\int_{A \cap \overline{M}} d\mu, \quad A \in \mathscr{F},$$

where M is a set in $\mathscr{F}$. Here

$$\operatorname{var} \mu = \operatorname{var} \mu_+ + \operatorname{var} \mu_- = \mu_+(\Omega) + \mu_-(\Omega).$$

Since

$$\mu_+(\Omega) = P(M) - \tilde{P}(M), \quad \mu_-(\Omega) = \tilde{P}(\overline{M}) - P(\overline{M}),$$

we have

$$\|P - \tilde{P}\| = (P(M) - \tilde{P}(M)) + (\tilde{P}(\overline{M}) - P(\overline{M})) \leq 2 \sup_{A \in \mathscr{F}} |P(A) - \tilde{P}(A)|.$$

This completes the proof of the lemma.

$\square$

Definition 2. A sequence of probability measures P_n, $n \geq 1$, is said to be *convergent in variation* to the measure P (denoted $P_n \xrightarrow{\text{var}} P$), if

$$\|P_n - P\| \to 0, \quad n \to \infty. \tag{3}$$

From this definition and Theorem 1 of Sect. 1 it is easily seen that convergence in variation of probability measures defined on a metric space $(\Omega, \mathscr{F}, \rho)$ implies their weak convergence.

The proximity in variation of distributions is, perhaps, the strongest form of closeness of probability distributions, since if two distributions are close in variation, then in practice, in specific situations, they can be considered indistinguishable. In this connection, the impression may be created that the study of distance in variation is not of much probabilistic interest. However, for example, in Poisson's theorem (Sect. 6, Chap. 1) the convergence of the binomial to the Poisson distribution takes place in the sense of convergence to zero of the distance in variation between these distributions. (Later, in Sect. 12, we shall obtain an upper bound for this distance.)

We also provide an example from the field of mathematical statistics, where the necessity of determining the distance in variation between measures P and $\tilde{P}$ arises in a natural way in connection with the problem of discrimination (based on observed data) between two statistical hypotheses H (the true distribution is P) and $\tilde{H}$ (the true distribution is $\tilde{P}$) in order to decide which probabilistic model $(\Omega, \mathscr{F}, P)$ or $(\Omega, \mathscr{F}, \tilde{P})$ better fits the statistical data. If $\omega \in \Omega$ is treated as the result of an observation, by a *test* (for discrimination between the hypotheses H and $\tilde{H}$) we understand any $\mathscr{F}$-measurable function $\varphi = \varphi(\omega)$ with values in $[0, 1]$, the statistical meaning of which is that $\varphi(\omega)$ is "the probability with which hypothesis $\tilde{H}$ is accepted if the result of the observation is ω."

We shall characterize the performance of this rule for discrimination between H and $\tilde{H}$ by the *probabilities of errors of the first and second kind*:

$$
\begin{aligned}
\alpha(\varphi) &= E\varphi(\omega) \quad (= \text{Prob}\,(accepting\ \tilde{H}\,|\,H\ is\ true)), \\
\beta(\varphi) &= \tilde{E}(1 - \varphi(\omega)) \quad (= \text{Prob}\,(accepting\ H\,|\,\tilde{H}\ is\ true)).
\end{aligned}
$$

In the case when hypotheses H and $\tilde{H}$ are *equally significant* to us, it is natural to consider the test $\varphi^* = \varphi^*(\omega)$ (if there is such a test) that minimizes the sum $\alpha(\varphi) + \beta(\varphi)$ of the errors as the *optimal* one.

We set

$$
\mathscr{E}r(P, \tilde{P}) = \inf_{\varphi}[\alpha(\varphi) + \beta(\varphi)]. \tag{4}
$$

Let $Q = (P + \tilde{P})/2$ and $z = dP/dQ$, $\tilde{z} = d\tilde{P}/dQ$. Then

$$
\begin{aligned}
\mathscr{E}r(P, \tilde{P}) &= \inf_{\varphi}[E\varphi + \tilde{E}(1 - \varphi)] \\
&= \inf_{\varphi} E_Q[z\varphi + \tilde{z}(1 - \varphi)] = 1 + \inf_{\varphi} E_Q[\varphi(z - \tilde{z})],
\end{aligned}
$$

where E_Q is the expectation with respect to the measure Q.

It is easy to see that the inf is attained by the function

$$
\varphi^*(\omega) = I\{\tilde{z} < z\}
$$

and, since $E_Q(z - \tilde{z}) = 0$, that

$$
\mathscr{E}r(P, \tilde{P}) = 1 - \tfrac{1}{2}E_Q|z - \tilde{z}| = 1 - \tfrac{1}{2}\|P - \tilde{P}\|, \tag{5}
$$

where the last equation will follow from Lemma 2, below. Thus it is seen from (5) that the performance $\mathscr{E}r(P, \tilde{P})$ of the optimal test for discrimination between the two hypotheses depends on the *total variation distance* between P and $\tilde{P}$.

Lemma 2. *Let Q be a σ-finite measure such that $P \ll Q$, $\tilde{P} \ll Q$ and let $z = dP/dQ$, $\tilde{z} = d\tilde{P}/dQ$ be the Radon–Nikodym derivatives of P and $\tilde{P}$ with respect to Q. Then*

$$\|P - \tilde{P}\| = E_Q|z - \tilde{z}| \tag{6}$$

and if $Q = (P + \tilde{P})/2$, we have

$$\|P - \tilde{P}\| = E_Q|z - \tilde{z}| = 2E_Q|1 - z| = 2E_Q|1 - \tilde{z}|. \tag{7}$$

PROOF. For all $\mathscr{F}$-measurable functions $\psi = \psi(\omega)$ with $|\psi(\omega)| \leq 1$, we see from the definitions of z and $\tilde{z}$ that

$$|E\psi - \tilde{E}\psi| = |E_Q\psi(z - \tilde{z})| \leq E_Q|\psi||z - \tilde{z}| \leq E_Q|z - \tilde{z}|. \tag{8}$$

Therefore,

$$\|P - \tilde{P}\| \leq E_Q|z - \tilde{z}|. \tag{9}$$

However, for the function

$$\psi = \text{sign}(\tilde{z} - z) = \begin{cases} 1, & \tilde{z} \geq z, \\ -1, & \tilde{z} < z, \end{cases}$$

we have

$$|E\psi - \tilde{E}\psi| = E_Q|z - \tilde{z}|. \tag{10}$$

We obtain the required equation (6) from (9) and (10). Then (7) follows from (6) because $z + \tilde{z} = 2$ (Q-a. s.).

□

Corollary 1. *Let P and $\tilde{P}$ be two probability distributions on $(R, \mathscr{B}(R))$ with probability densities (with respect to Lebesgue measure dx) $p(x)$ and $\tilde{p}(x)$, $x \in R$. Then*

$$\|P - \tilde{P}\| = \int_{-\infty}^{\infty} |p(x) - \tilde{p}(x)|\, dx. \tag{11}$$

(As the measure Q, we are to take Lebesgue measure on $(R, \mathscr{B}(R))$.)

Corollary 2. *Let P and $\tilde{P}$ be two discrete measures, $P = (p_1, p_2, \ldots)$, $\tilde{P} = (\tilde{p}_1, \tilde{p}_2, \ldots)$, concentrated on a countable set of points $x_1, x_2, \ldots$. Then*

$$\|P - \tilde{P}\| = \sum_{i=1}^{\infty} |p_i - \tilde{p}_i|. \tag{12}$$

(As the measure Q, we are to take the counting measure, i.e., that with $Q(\{x_i\}) = 1$, $i = 1, 2, \ldots$.)

2. We now turn to still another measure of the proximity of two probability measures, closely related (as will follow later) to the proximity measure in variation.

Let P and $\tilde{P}$ be probability measures on $(\Omega, \mathscr{F})$ and Q, the third probability measure, *dominating* P and $\tilde{P}$, i.e., such that $P \ll Q$ and $\tilde{P} \ll Q$. We again use the notation

$$z = \frac{dP}{dQ}, \quad \tilde{z} = \frac{d\tilde{P}}{dQ}.$$

Definition 3. The *Kakutani–Hellinger distance* between the measures P and $\tilde{P}$ is the nonnegative number $\rho(P, \tilde{P})$ such that

$$\rho^2(P, \tilde{P}) = \tfrac{1}{2}E_Q[\sqrt{z} - \sqrt{\tilde{z}}]^2. \tag{13}$$

Since

$$E_Q[\sqrt{z} - \sqrt{\tilde{z}}]^2 = \int_\Omega \left[\sqrt{\frac{dP}{dQ}} - \sqrt{\frac{d\tilde{P}}{dQ}} \right]^2 dQ, \tag{14}$$

it is natural to write $\rho^2(P, \tilde{P})$ symbolically in the form

$$\rho^2(P, \tilde{P}) = \tfrac{1}{2}\int_\Omega [\sqrt{dP} - \sqrt{d\tilde{P}}]^2. \tag{15}$$

If we set

$$H(P, \tilde{P}) = E_Q\sqrt{z\tilde{z}}, \tag{16}$$

then, by analogy with (15), we may write symbolically

$$H(P, \tilde{P}) = \int_\Omega \sqrt{dP\, d\tilde{P}}. \tag{17}$$

From (13) and (16), as well as from (15) and (17), it is clear that

$$\rho^2(P, \tilde{P}) = 1 - H(P, \tilde{P}). \tag{18}$$

The number $H(P, \tilde{P})$ is called the *Hellinger integral* of the measures P and $\tilde{P}$. It turns out to be convenient, for many purposes, to consider the *Hellinger integrals* $H(\alpha; P, \tilde{P})$ *of order* $\alpha \in (0, 1)$, defined by the formula

$$H(\alpha; P, \tilde{P}) = E_Q z^\alpha \tilde{z}^{1-\alpha}, \tag{19}$$

or, symbolically,

$$H(\alpha; P, \tilde{P}) = \int_\Omega (dP)^\alpha (d\tilde{P})^{1-\alpha}. \tag{20}$$

It is clear that $H(1/2; P, \tilde{P}) = H(P, \tilde{P})$.

For Definition 3 to be reasonable, we need to show that the number $\rho^2(P, \tilde{P})$ is independent of the choice of the dominating measure and that in fact $\rho(P, \tilde{P})$ satisfies the requirements of the concept of "distance."

Lemma 3. 1. *The Hellinger integral of order $\alpha \in (0, 1)$ (and consequently also $\rho(P, \tilde{P})$) is independent of the choice of the dominating measure Q.*

2. *The function ρ defined in (13) is a metric on the set of probability measures.*

PROOF. 1. If the measure Q' dominates P and $\tilde{P}$, Q' also dominates $Q = (P + \tilde{P})/2$. Hence, it is enough to show that if $Q \ll Q'$, we have

$$E_Q(z^\alpha \tilde{z}^{1-\alpha}) = E_{Q'}(z')^\alpha (\tilde{z}')^{1-\alpha},$$

where $z' = dP/dQ'$ and $\tilde{z}' = d\tilde{P}/dQ'$.

Let us set $v = dQ/dQ'$. Then $z' = zv, \tilde{z}' = \tilde{z}v$, and

$$E_Q(z^\alpha \tilde{z}^{1-\alpha}) = E_{Q'}(vz^\alpha \tilde{z}^{1-\alpha}) = E_{Q'}(z')^\alpha (\tilde{z}')^{1-\alpha},$$

which establishes the first assertion.

2. If $\rho(P, \tilde{P}) = 0$ we have $z = \tilde{z}$ (Q-a.e.) and hence, $P = \tilde{P}$. By symmetry, we evidently have $\rho(P, \tilde{P}) = \rho(\tilde{P}, P)$. Finally, let P, P', and P'' be three measures, $P \ll Q$, $P' \ll Q$, and $P'' \ll Q$, with $z = dP/dQ$, $z' = dP'/dQ$, and $z'' = dP''/dQ$. By using the validity of the triangle inequality for the norm in $L^2(\Omega, \mathscr{F}, Q)$, we obtain

$$[E_Q(\sqrt{z} - \sqrt{z''})^2]^{1/2} \le [E_Q(\sqrt{z} - \sqrt{z'})^2]^{1/2} + [E_Q(\sqrt{z'} - \sqrt{z''})^2]^{1/2},$$

i.e.,

$$\rho(P, P'') \le \rho(P, P') + \rho(P', P'').$$

This completes the proof of the lemma.

□

By Definition (19) and Fubini's theorem (Sect. 6, Chap. 2), it follows immediately that in the case when the measures P and $\tilde{P}$ are *direct products* of measures, $P = P_1 \times \cdots \times P_n, \tilde{P} = \tilde{P}_1 \times \cdots \times \tilde{P}_n$ (see Subsection 10 in Sect. 6, Chap. 2), the Hellinger integral between the measures P and $\tilde{P}$ is equal to the product of the corresponding Hellinger integrals:

$$H(\alpha; P, \tilde{P}) = \prod_{i=1}^{n} H(\alpha; P_i, \tilde{P}_i).$$

The following theorem shows the connection between distance in variation and Kakutani–Hellinger distance (or, equivalently, the Hellinger integral). In particular, it shows that these distances define *the same topology* in the space of probability measures on $(\Omega, \mathscr{F})$.

Theorem 1. *We have the following inequalities*:

$$2[1 - H(P, \tilde{P})] \leq \|P - \tilde{P}\| \leq \sqrt{8[1 - H(P, \tilde{P})]}, \tag{21}$$

$$\|P - \tilde{P}\| \leq 2\sqrt{1 - H^2(P, \tilde{P})}. \tag{22}$$

In particular,

$$2\rho^2(P, \tilde{P}) \leq \|P - \tilde{P}\| \leq \sqrt{8}\rho(P, \tilde{P}). \tag{23}$$

PROOF. Since $H(P, \tilde{P}) \leq 1$ and $1 - x^2 \leq 2(1 - x)$ for $0 \leq x \leq 1$, the right-hand inequality in (21) follows from (22), the proof of which is provided by the following chain of inequalities (where $Q = (1/2)(P + \tilde{P})$):

$$\tfrac{1}{2}\|P - \tilde{P}\| = E_Q|1 - z| \leq \sqrt{E_Q|1 - z|^2} = \sqrt{1 - E_Q z(2 - z)}$$

$$= \sqrt{1 - E_Q z\tilde{z}} = \sqrt{1 - E_Q(\sqrt{z\tilde{z}})^2} \leq \sqrt{1 - (E_Q\sqrt{z\tilde{z}})^2}$$

$$= \sqrt{1 - H^2(P, \tilde{P})}.$$

Finally, the first inequality in (21) follows from the fact that by the inequality

$$\tfrac{1}{2}[\sqrt{z} - \sqrt{2 - z}]^2 \leq |z - 1|, \quad z \in [0, 2],$$

we have (again, $Q = (1/2)(P + \tilde{P})$)

$$1 - H(P, \tilde{P}) = \rho^2(P, \tilde{P}) = \tfrac{1}{2}E_Q[\sqrt{z} - \sqrt{2 - z}]^2 \leq E_Q|z - 1| = \tfrac{1}{2}\|P - \tilde{P}\|.$$

□

Remark. It can be shown in a similar way that, for every $\alpha \in (0, 1)$,

$$2[1 - H(\alpha; P, \tilde{P})] \leq \|P - \tilde{P}\| \leq \sqrt{c_\alpha(1 - H(\alpha; P, \tilde{P}))}, \tag{24}$$

where c_α is a constant.

Corollary 3. *Let P and P^n, $n \geq 1$, be probability measures on $(\Omega, \mathscr{F})$. Then (as $n \to \infty$)*

$$\|P^n - P\| \to 0 \Leftrightarrow H(P^n, P) \to 1 \Leftrightarrow \rho(P^n, P) \to 0,$$
$$\|P^n - P\| \to 2 \Leftrightarrow H(P^n, P) \to 0 \Leftrightarrow \rho(P^n, P) \to 1.$$

Corollary 4. *Since by (5)*

$$\mathscr{E}r(P, \tilde{P}) = 1 - \tfrac{1}{2}\|P - \tilde{P}\|,$$

we have, by (21) and (22),

$$\tfrac{1}{2}H^2(P, \tilde{P}) \leq 1 - \sqrt{1 - H^2(P, \tilde{P})} \leq \mathscr{E}r(P, \tilde{P}) \leq H(P, \tilde{P}). \tag{25}$$

In particular, let

$$P^n = \underbrace{P \times \cdots \times P}_{n}, \quad \tilde{P}^n = \underbrace{\tilde{P} \times \cdots \times \tilde{P}}_{n},$$

be direct products of measures. Then, since $H(P^n, \tilde{P}^n) = [H(P, \tilde{P})]^n = e^{-\lambda n}$ with $\lambda = -\log H(P, \tilde{P}) \geq \rho^2(P, \tilde{P})$, we obtain from (25) the inequalities

$$\tfrac{1}{2} e^{-2\lambda n} \leq \mathscr{E}r(P^n, \tilde{P}^n) \leq e^{-\lambda n} \leq e^{-n\rho^2(P,\tilde{P})}. \tag{26}$$

In connection with the problem, considered above, of *discrimination between two statistical hypotheses*, from these inequalities we have the following result.

Let $\xi_1, \xi_2, \ldots$ be independent identically distributed random elements, that have either the probability distribution P (Hypothesis H) or $\tilde{P}$ (Hypothesis $\tilde{H}$), with $\tilde{P} \neq P$, and therefore, $\rho^2(P, \tilde{P}) > 0$. Then, when $n \to \infty$, the function $\mathscr{E}r(P^n, \tilde{P}^n)$, which describes the quality of optimal discrimination between the hypotheses H and $\tilde{H}$ from observations of $\xi_1, \xi_2, \ldots$, decreases exponentially to zero.

3. The Hellinger integrals of order α are well suited for stating conditions of *absolute continuity* and *singularity* of probability measures.

Let P and $\tilde{P}$ be two probability measures defined on a measurable space $(\Omega, \mathscr{F})$. We say that $\tilde{P}$ is *absolutely continuous* with respect to P (notation: $\tilde{P} \ll P$) if $\tilde{P}(A) = 0$ whenever $P(A) = 0$ for $A \in \mathscr{F}$. If $\tilde{P} \ll P$ and $P \ll \tilde{P}$, we say that P and $\tilde{P}$ are *equivalent* ($\tilde{P} \sim P$). The measures P and $\tilde{P}$ are called *singular* or *orthogonal* ($\tilde{P} \perp P$), if there is an $A \in \mathscr{F}$ for which $P(A) = 1$ and $\tilde{P}(\overline{A}) = 1$ (i.e., P and $\tilde{P}$ "sit" on different sets).

Let Q be a probability measure, with $P \ll Q, \tilde{P} \ll Q$, $z = dP/dQ, \tilde{z} = d\tilde{P}/dQ$.

Theorem 2. *The following conditions are equivalent*:

(a) $\tilde{P} \ll P$,
(b) $\tilde{P}(z > 0) = 1$,
(c) $H(\alpha; P, \tilde{P}) \to 1, \alpha \downarrow 0$.

Theorem 3. *The following conditions are equivalent*:

(a) $\tilde{P} \perp P$,
(b) $\tilde{P}(z > 0) = 0$,
(c) $H(\alpha; P, \tilde{P}) \to 0, \alpha \downarrow 0$,
(d) $H(\alpha; P, \tilde{P}) = 0$ *for all* $\alpha \in (0, 1)$,
(e) $H(\alpha; P, \tilde{P}) = 0$ *for some* $\alpha \in (0, 1)$.

PROOF. The proofs of these theorems will be given simultaneously. By the definitions of z and $\tilde{z}$,

$$P(z = 0) = E_Q[zI(z = 0)] = 0, \tag{27}$$

$$\tilde{P}(A \cap \{z > 0\}) = E_Q[\tilde{z}I(A \cap \{z > 0\})]$$

$$= E_Q\left[\tilde{z}\frac{z}{z}I(A \cap \{z > 0\})\right] = E_P\left[\frac{\tilde{z}}{z}I(A \cap \{z > 0\})\right]$$

$$= E_P\left[\frac{\tilde{z}}{z}I(A)\right]. \tag{28}$$

Consequently, we have the *Lebesgue decomposition*

$$\tilde{P}(A) = E_P\left[\frac{\tilde{z}}{z}I(A)\right] + \tilde{P}(A \cap \{z = 0\}), \quad A \in \mathscr{F}, \tag{29}$$

in which $Z = \tilde{z}/z$ is called the *Lebesgue derivative of $\tilde{P}$ with respect to P* and denoted by $d\tilde{P}/dP$ (compare the remark on the Radon–Nikodým theorem, Sect. 6, Chap. 2).

Hence, we immediately obtain the equivalence of (a) and (b) in both theorems. Moreover, since

$$z^{\alpha}\tilde{z}^{1-\alpha} \to \tilde{z}I(z > 0), \quad \alpha \downarrow 0,$$

and for $\alpha \in (0, 1)$

$$0 \leq z^{\alpha}\tilde{z}^{1-\alpha} \leq \alpha z + (1 - \alpha)\tilde{z} \leq z + \tilde{z}$$

with $E_Q(z + \tilde{z}) = 2$, we have, by Lebesgue's dominated convergence theorem,

$$\lim_{\alpha \downarrow 0} H(\alpha; P, \tilde{P}) = E_Q \tilde{z}I(z > 0) = \tilde{P}(z > 0)$$

and therefore, (b) $\Leftrightarrow$ (c) in both theorems.

Finally, let us show that in the second theorem (c) $\Leftrightarrow$ (d) $\Leftrightarrow$ (e). For this, we need only note that $H(\alpha; P, \tilde{P}) = \tilde{E}(z/\tilde{z})^{\alpha}I(\tilde{z} > 0)$ and $\tilde{P}(\tilde{z} > 0) = 1$. Hence, for each $\alpha \in (0, 1)$ we have $\tilde{P}(z > 0) = 0 \Leftrightarrow H(\alpha; P, \tilde{P}) = 0$, from which there follows the implication (c) $\Leftrightarrow$ (d) $\Leftrightarrow$ (e).

□

Example 1. Let $P = P_1 \times P_2 \times \ldots$, $\tilde{P} = \tilde{P}_1 \times \tilde{P}_2 \ldots$, where P_k and $\tilde{P}_k$ are Gaussian measures on $(R, \mathscr{B}(R))$ with densities

$$p_k(x) = \frac{1}{\sqrt{2\pi}}e^{(x-a_k)^2/2}, \quad \tilde{p}_k(x) = \frac{1}{\sqrt{2\pi}}e^{-(x-\tilde{a}_k)^2/2}.$$

Since

$$H(\alpha; P, \tilde{P}) = \prod_{k=1}^{\infty} H(\alpha; P_k, \tilde{P}_k),$$

where a simple calculation shows that

$$H(\alpha; P_k, \tilde{P}_k) = \int_{-\infty}^{\infty} p_k^{\alpha}(x)\tilde{p}_k^{1-\alpha}(x)\,dx = e^{-(\alpha(1-\alpha)/2)(a_k-\tilde{a}_k)^2},$$

we have
$$H(\alpha; P, \tilde{P}) = e^{-(\alpha(1-\alpha)/2)\sum_{k=1}^{\infty}(a_k - \tilde{a}_k)^2}.$$

From Theorems 2 and 3, we find that

$$\tilde{P} \ll P \Leftrightarrow P \ll \tilde{P} \Leftrightarrow \tilde{P} \sim P \Leftrightarrow \sum_{k=1}^{\infty}(a_k - \tilde{a}_k)^2 < \infty,$$

$$\tilde{P} \perp P \Leftrightarrow \sum_{k=1}^{\infty}(a_k - \tilde{a}_k)^2 = \infty.$$

Example 2. Again let $P = P_1 \times P_2 \times \ldots$, $\tilde{P} = \tilde{P}_1 \times \tilde{P}_2 \times \ldots$, where P_k and $\tilde{P}_k$ are Poisson distributions with respective parameters $\lambda_k > 0$ and $\tilde{\lambda}_k > 0$. Then it is easily shown that

$$\tilde{P} \ll P \Leftrightarrow P \ll \tilde{P} \Leftrightarrow \tilde{P} \sim P \Leftrightarrow \sum_{k=1}^{\infty}(\sqrt{\lambda_k} - \sqrt{\tilde{\lambda}_k})^2 < \infty,$$

$$\tilde{P} \perp P \Leftrightarrow \sum_{k=1}^{\infty}(\sqrt{\lambda_k} - \sqrt{\tilde{\lambda}_k})^2 = \infty. \tag{30}$$

5. PROBLEMS

1. In the notation of Lemma 2, set

$$P \wedge \tilde{P} = E_Q(z \wedge \tilde{z}),$$

   where $z \wedge \tilde{z} = \min(z, \tilde{z})$. Show that

   $$\|P - \tilde{P}\| = 2(1 - P \wedge \tilde{P})$$

   (and consequently, $\mathscr{E}r(P, \tilde{P}) = P \wedge \tilde{P}$).

2. Let P, P_n, $n \geq 1$, be probability measures on $(R, \mathscr{B}(R))$ with densities (with respect to Lebesgue measure) $p(x)$, $p_n(x)$, $n \geq 1$. Let $p_n(x) \to p(x)$ for almost all x (with respect to Lebesgue measure). Show that then

   $$\|P - P_n\| = \int_{-\infty}^{\infty} |p(x) - p_n(x)|\, dx \to 0, \quad n \to \infty$$

   (compare Problem 17 in Sect. 6, Chap. 2).

3. Let P and $\tilde{P}$ be two probability measures. We define *Kullback information* $K(P, \tilde{P})$ (in favor of P against $\tilde{P}$) by the equation

   $$K(P, \tilde{P}) = \begin{cases} E \log(dP/d\tilde{P}) & \text{if } P \ll \tilde{P}, \\ \infty & \text{otherwise.} \end{cases}$$

   Show that
   $$K(P, \tilde{P}) \geq -2\log(1 - \rho^2(P, \tilde{P})) \geq 2\rho^2(P, \tilde{P}).$$

4. Establish formulas (11) and (12).
5. Prove inequalities (24).
6. Let P, $\tilde{P}$, and Q be probability measures on $(R, \mathscr{B}(R))$; $P * Q$ and $\tilde{P} * Q$, their convolutions (see Subsection 4 in Sect. 8, Chap. 2). Then

$$\|P * Q - \tilde{P} * Q\| \leq \|P - \tilde{P}\|.$$

7. Prove (30).
8. Let ξ and η be random elements on $(\Omega, \mathscr{F}, \mathsf{P})$ with values in a measurable space $(E, \mathscr{E})$. Show that

$$|\mathsf{P}\{\xi \in A\} - \mathsf{P}\{\eta \in A\}| \leq \mathsf{P}(\xi \neq \eta), \quad A \in \mathscr{E}.$$

10 Contiguity and Entire Asymptotic Separation of Probability Measures

1. These concepts play a fundamental role in the asymptotic theory of mathematical statistics, being natural extensions of the concepts of absolute continuity and singularity of two measures in the case of *sequences* of pairs of measures.

Let us begin with definitions.

Let $(\Omega^n, \mathscr{F}^n)_{n \geq 1}$ be a sequence of measurable spaces; let $(P^n)_{n \geq 1}$ and $(\tilde{P}^n)_{n \geq 1}$ be sequences of probability measures with P^n and $\tilde{P}^n$ defined on $(\Omega^n, \mathscr{F}^n)$, $n \geq 1$.

Definition 1. We say that a sequence $(\tilde{P}^n)$ of measures is *contiguous* to the sequence (P^n) (notation: $(\tilde{P}^n) \lhd (P^n)$) if, for all $A^n \in \mathscr{F}^n$ such that $P^n(A^n) \to 0$ as $n \to \infty$, we have $\tilde{P}^n(A^n) \to 0$, $n \to \infty$.

Definition 2. We say that sequences $(\tilde{P}^n)$ and (P^n) of measures are *entirely (asymptotically) separated* (or for short: $(\tilde{P}^n) \vartriangle (P^n)$), if there is a subsequence $n_k \uparrow \infty$, $k \to \infty$, and sets $A^{n_k} \in \mathscr{F}^{n_k}$ such that

$$P^{n_k}(A^{n_k}) \to 1 \quad \text{and} \quad \tilde{P}^{n_k}(A^{n_k}) \to 0, \quad k \to \infty.$$

We notice immediately that entire separation is a *symmetric* concept in the sense that $(\tilde{P}^n) \vartriangle (P^n) \Leftrightarrow (P^n) \vartriangle (\tilde{P}^n)$. Contiguity does not possess this property. If $(\tilde{P}^n) \lhd (P^n)$ and $(P^n) \lhd (\tilde{P}^n)$, we write $(\tilde{P}^n) \lhd\rhd (P^n)$ and say that the sequences (P^n) and $(\tilde{P}^n)$ of measures are *mutually contiguous*.

We notice that in the case when $(\Omega^n, \mathscr{F}^n) = (\Omega, \mathscr{F})$, $P^n = P$, $\tilde{P}^n = \tilde{P}$ for all $n \geq 1$, we have

$$(\tilde{P}^n) \lhd (P^n) \Leftrightarrow \tilde{P} \ll P, \tag{1}$$

$$(\tilde{P}^n) \lhd\rhd (P^n) \Leftrightarrow \tilde{P} \sim P, \tag{2}$$

$$(\tilde{P}^n) \vartriangle (P^n) \Leftrightarrow \tilde{P} \perp P. \tag{3}$$

These properties and the definitions given above explain why contiguity and entire asymptotic separation are often thought of as "asymptotic absolute continuity" and "asymptotic singularity" for sequences $(\tilde{P}^n)$ and (P^n).

2. Theorems 1 and 2 presented below are natural extensions of Theorems 2 and 3 of Sect. 9 to sequences of measures.

Let $(\Omega^n, \mathscr{F}^n)_{n \geq 1}$ be a sequence of measurable spaces; Q^n, a probability measure on $(\Omega^n, \mathscr{F}^n)$; and ξ^n a random variable (generally speaking, extended; see Sect. 4, Chap. 2) on $(\Omega^n, \mathscr{F}^n)$, $n \geq 1$.

Definition 3. A sequence (ξ^n) of random variables is *tight* with respect to a sequence of measures (Q^n) (notation: $(\xi^n \mid Q^n)$ is *tight*) if

$$\lim_{N \uparrow \infty} \limsup_n Q^n(|\xi^n| > N) = 0. \tag{4}$$

(Compare the corresponding definition of tightness of a family of probability measures in Sect. 2.)

We shall always set

$$Q^n = \frac{P^n + \tilde{P}^n}{2}, \quad z^n = \frac{dP^n}{dQ^n}, \quad \tilde{z}^n = \frac{d\tilde{P}^n}{dQ^n}.$$

We shall also use the notation

$$Z^n = \tilde{z}^n / z^n \tag{5}$$

for the Lebesgue derivative of $\tilde{P}^n$ with respect to P^n (see (29) in Sect. 9), taking $2/0 = \infty$. We note that if $\tilde{P}^n \ll P^n$, Z^n is precisely one of the versions of the density $d\tilde{P}^n/dP^n$ of the measure $\tilde{P}^n$ with respect to P^n (see Sect. 6, Chap. 2).

For later use it is convenient to note that since

$$P^n\left(z^n \leq \frac{1}{N}\right) = E_{Q^n}\left(z^n I\left(z^n \leq \frac{1}{N}\right)\right) \leq \frac{1}{N} \tag{6}$$

and $Z^n \leq 2/z^n$, we have

$$((1/z^n) \mid P^n) \quad \text{is tight}, \quad (Z^n \mid P^n) \quad \text{is tight}. \tag{7}$$

Theorem 1. *The following statements are equivalent:*

(a) $(\tilde{P}^n) \lhd (P^n)$,
(b) $(1/z^n \mid \tilde{P}^n)$ *is tight,*
(b') $(Z^n \mid \tilde{P}^n)$ *is tight,*
(c) $\lim_{\alpha \downarrow 0} \liminf_n H(\alpha; P^n, \tilde{P}^n) = 1$.

Theorem 2. *The following statements are equivalent:*

(a) $(\tilde{P}^n) \triangle (P^n)$,
(b) $\liminf_n \tilde{P}^n(z^n \geq \varepsilon) = 0$ *for every* $\varepsilon > 0$,

(b') $\quad \limsup_n \tilde{P}^n(Z^n \le N) = 0$ *for every* $N > 0$,
(c) $\quad \lim_{\alpha \downarrow 0} \liminf_n H(\alpha; P^n, \tilde{P}^n) = 0$,
(d) $\quad \liminf_n H(\alpha; P^n, \tilde{P}^n) = 0$ *for all* $\alpha \in (0,1)$,
(e) $\quad \liminf_n H(\alpha; P^n, \tilde{P}^n) = 0$ *for some* $\alpha \in (0,1)$.

PROOF OF THEOREM 1.

(a) $\Rightarrow$ (b). If (b) is not satisfied, there are an $\varepsilon > 0$ and a sequence $n_k \uparrow \infty$ such that $\tilde{P}^{n_k}(z^{n_k} < 1/n_k) \ge \varepsilon$. But by (6), $P^{n_k}(z^{n_k} < 1/n_k) \le 1/n_k$, $k \to \infty$, which contradicts the assumption that $(\tilde{P}^n) \lhd (P^n)$.

(b) $\Leftrightarrow$ (b'). We have only to note that $Z^n = (2/z^n) - 1$.
(b) $\Rightarrow$ (a). Let $A^n \in \mathscr{F}^n$ and $P^n(A^n) \to 0$, $n \to \infty$. We have

$$\tilde{P}^n(A^n) \le \tilde{P}^n(z^n \le \varepsilon) + E_{Q^n}(\tilde{z}^n I(A^n \cap \{z^n > \varepsilon\}))$$
$$\le \tilde{P}^n(z^n \le \varepsilon) + \frac{2}{\varepsilon} E_{Q^n}(z^n I(A^n)) = \tilde{P}^n(z^n \le \varepsilon) + \frac{2}{\varepsilon} P^n(A^n).$$

Therefore,

$$\limsup_n \tilde{P}^n(A^n) \le \limsup_n \tilde{P}^n(z^n \le \varepsilon), \quad \varepsilon > 0.$$

Proposition (b) is equivalent to saying that $\lim_{\varepsilon \downarrow 0} \limsup_n \tilde{P}^n(z^n \le \varepsilon) = 0$. Therefore, $\tilde{P}^n(A^n) \to 0$, i.e., (b) $\Rightarrow$ (a).

(b) $\Rightarrow$ (c). Let $\varepsilon > 0$. Then

$$H(\alpha; P^n, \tilde{P}^n) = E_{Q^n}[(z^n)^\alpha (\tilde{z}^n)^{1-\alpha}] \ge E_{Q^n}\left[\left(\frac{z^n}{\tilde{z}^n}\right)^\alpha I(z^n \ge \varepsilon) I(\tilde{z}^n > 0) \tilde{z}^n\right]$$
$$= E_{\tilde{P}^n}\left[\left(\frac{z^n}{\tilde{z}^n}\right)^\alpha I(z^n \ge \varepsilon)\right] \ge \left(\frac{\varepsilon}{2}\right)^\alpha \tilde{P}^n(z^n \ge \varepsilon), \quad (8)$$

since $z^n + \tilde{z}^n = 2$. Therefore, for $\varepsilon > 0$,

$$\liminf_{\alpha \downarrow 0} \liminf_n H(\alpha; P^n, \tilde{P}^n)$$
$$\ge \liminf_{\alpha \downarrow 0} \left(\frac{\varepsilon}{2}\right)^\alpha \liminf_n \tilde{P}^n(z^n \ge \varepsilon) = \liminf_n \tilde{P}^n(z^n \ge \varepsilon). \quad (9)$$

By (b), $\liminf_{\varepsilon \downarrow 0} \liminf_n \tilde{P}^n(z^n \ge \varepsilon) = 1$. Hence, (c) follows from (9) and the fact that $H(\alpha; P^n, \tilde{P}^n) \le 1$.

(c) $\Rightarrow$ (b). Let $\delta \in (0,1)$. Then

$$H(\alpha; P^n, \tilde{P}^n) = E_{Q^n}[(z^n)^\alpha (\tilde{z}^n)^{1-\alpha} I(z^n < \varepsilon)]$$
$$+ E_{Q^n}[(z^n)^\alpha (\tilde{z}^n)^{1-\alpha} I(z^n \ge \varepsilon, \tilde{z}^n \le \delta)]$$
$$+ E_{Q^n}[(z^n)^\alpha (\tilde{z}^n)^{1-\alpha} I(z^n \ge \varepsilon, \tilde{z}^n > \delta)]$$
$$\le 2\varepsilon^\alpha + 2\delta^{1-\alpha} + E_{Q^n}\left[\tilde{z}^n \left(\frac{z^n}{\tilde{z}^n}\right)^\alpha I(z^n \ge \varepsilon, \tilde{z}^n > \delta)\right]$$
$$\le 2\varepsilon^\alpha + 2\delta^{1-\alpha} + \left(\frac{2}{\delta}\right)^\alpha \tilde{P}^n(z^n \ge \varepsilon). \quad (10)$$

Consequently,

$$\liminf_{\varepsilon \downarrow 0} \liminf_n \tilde{P}^n(z^n \geq \varepsilon) \geq \left(\frac{\delta}{2}\right)^\alpha \liminf_n H(\alpha; P^n, \tilde{P}^n) - \frac{2}{2^\alpha}\delta$$

for all $\alpha \in (0,1)$, $\delta \in (0,1)$. If we first let $\alpha \downarrow 0$, use (c), and then let $\delta \downarrow 0$, we obtain

$$\liminf_{\varepsilon \downarrow 0} \liminf_n \tilde{P}^n(z^n \geq \varepsilon) \geq 1,$$

from which (b) follows.

$\square$

PROOF OF THEOREM 2. (a) $\Rightarrow$ (b). Let $(\tilde{P}^n) \; \triangle \; (P^n)$, $n_k \uparrow \infty$, and let $A^{n_k} \in \mathscr{F}^{n_k}$ have the property that $P^{n_k}(A^{n_k}) \to 1$ and $\tilde{P}^{n_k}(A^{n_k}) \to 0$. Then, since $z^n + \tilde{z}^n = 2$, we have

$$\tilde{P}^{n_k}(z^{n_k} \geq \varepsilon) \leq \tilde{P}^{n_k}(A^{n_k}) + E_{Q^{n_k}}\left\{z^{n_k} \cdot \frac{\tilde{z}^{n_k}}{z^{n_k}} I(\overline{A}^{n_k})I(z^{n_k} \geq \varepsilon)\right\}$$

$$= \tilde{P}^{n_k}(A^{n_k}) + E_{P^{n_k}}\left\{\frac{\tilde{z}^{n_k}}{z^{n_k}} I(\overline{A}^{n_k})I(z^{n_k} \geq \varepsilon)\right\} \leq \tilde{P}^{n_k}(A^{n_k}) + \frac{2}{\varepsilon} P^{n_k}(\overline{A}^{n_k}).$$

Consequently, $\tilde{P}^{n_k}(z^{n_k} \geq \varepsilon) \to 0$ and therefore, (b) is satisfied.

(b) $\Rightarrow$ (a). If (b) is satisfied, there is a sequence $n_k \uparrow \infty$ such that

$$\tilde{P}^{n_k}\left(z^{n_k} \geq \frac{1}{k}\right) \leq \frac{1}{k} \to 0, \quad k \to \infty.$$

Hence, having observed (see (6)) that $P^{n_k}(z^{n_k} \geq 1/k) \geq 1 - (1/k)$, we obtain (a).

(b) $\Rightarrow$ (b'). We have only to observe that $Z^n = (2/z^n) - 1$.

(b) $\Rightarrow$ (d). By (10) and (b),

$$\liminf_n H(\alpha; P^n, \tilde{P}^n) \leq 2\varepsilon^\alpha + 2\delta^{1-\alpha}$$

for arbitrary ε and δ on the interval $(0,1)$. Therefore, (d) is satisfied.

(d) $\Rightarrow$ (c) and (d) $\Rightarrow$ (e) are evident.

Finally, from (8) we have

$$\liminf_n \tilde{P}^n(z^n \geq \varepsilon) \leq \left(\frac{2}{\varepsilon}\right)^\alpha \liminf_n H(\alpha; P^n, \tilde{P}^n).$$

Therefore, (c) $\Rightarrow$ (b) and (e) $\Rightarrow$ (b), since $(2/\varepsilon)^\alpha \to 1$, $\alpha \downarrow 0$.

$\square$

3. We now consider a particular case of *independent* observations, where the calculation of the integrals $H(\alpha; P^n, \tilde{P}^n)$ and application of Theorems 1 and 2 do not present much difficulty.

Let us suppose that the measures P^n and $\tilde{P}^n$ are *direct products* of measures:

$$P^n = P_1 \times \cdots \times P_n, \quad \tilde{P}^n = \tilde{P}_1 \times \cdots \times \tilde{P}_n, \quad n \geq 1,$$

where P_k and $\tilde{P}_k$ are given on $(\Omega_k, \mathscr{F}_k)$, $k \geq 1$.

Since in this case

$$H(\alpha; P^n, \tilde{P}^n) = \prod_{k=1}^{n} H(\alpha; P_k, \tilde{P}_k) = e^{\sum_{k=1}^{n} \log[1 - (1 - H(\alpha; P_k, \tilde{P}_k))]},$$

we obtain the following result from Theorems 1 and 2:

$$(\tilde{P}^n) \lhd (P^n) \Leftrightarrow \lim_{\alpha \downarrow 0} \limsup_{n} \sum_{k=1}^{n} [1 - H(\alpha; P_k, \tilde{P}_k)] = 0, \tag{11}$$

$$(\tilde{P}^n) \vartriangle (P^n) \Leftrightarrow \limsup_{n} \sum_{k=1}^{n} [1 - H(\alpha; P_k, \tilde{P}_k)] = \infty. \tag{12}$$

Example. Let $(\Omega_k, \mathscr{F}_k) = (R, \mathscr{B}(R))$, $a_k \in [0, 1)$,

$$P_k(dx) = I_{[0,1]}(x)\, dx, \quad \tilde{P}_k(dx) = \frac{1}{1 - a_k} I_{[a_k, 1]}(x)\, dx.$$

Since here $H(\alpha; P_k, \tilde{P}_k) = (1 - a_k)^\alpha$, $\alpha \in (0, 1)$, from (11) and the fact that $H(\alpha; P_k, \tilde{P}_k) = H(1 - \alpha; \tilde{P}_k, P_k)$, we obtain

$$(\tilde{P}^n) \lhd (P^n) \Leftrightarrow \limsup_{n} n a_n < \infty, \quad \text{i.e., } a_n = O\left(\frac{1}{n}\right),$$

$$(P^n) \lhd (\tilde{P}^n) \Leftrightarrow \limsup_{n} n a_n = 0, \quad \text{i.e., } a_n = o\left(\frac{1}{n}\right),$$

$$(\tilde{P}^n) \vartriangle (P^n) \Leftrightarrow \limsup_{n} n a_n = \infty.$$

4. PROBLEMS

1. Let $P^n = P_1^n \times \cdots \times P_n^n$, $\tilde{P}^n = \tilde{P}_1^n \times \cdots \times \tilde{P}_n^n$, $n \geq 1$, where P_k^n and $\tilde{P}_k^n$ are *Gaussian measures* with parameters $(a_k^n, 1)$ and $(\tilde{a}_k^n, 1)$. Find conditions on (a_k^n) and $(\tilde{a}_k^n)$ under which $(\tilde{P}^n) \lhd (P^n)$ and $(\tilde{P}^n) \vartriangle (P^n)$.

2. Let $P^n = P_1^n \times \cdots \times P_n^n$ and $\tilde{P}^n = \tilde{P}_1^n \times \cdots \times \tilde{P}_n^n$, where P_k^n and $\tilde{P}_k^n$ are probability measures on $(R, \mathscr{B}(R))$ for which $P_k^n(dx) = I_{[0,1]}(x)\, dx$ and $\tilde{P}_k^n(dx) = I_{[a_n, 1+a_n]}(x)\, dx$, $0 \leq a_n \leq 1$. Show that $H(\alpha; P_k^n, \tilde{P}_k^n) = 1 - a_n$ and

$$(\tilde{P}^n) \lhd (P^n) \Leftrightarrow (P^n) \lhd (\tilde{P}^n) \Leftrightarrow \limsup_{n} n a_n = 0,$$

$$(\tilde{P}^n) \vartriangle (P^n) \Leftrightarrow \limsup_{n} n a_n = \infty.$$

3. Let $(\Omega, \mathscr{F}, (\mathscr{F}_n)_{n \geq 0})$ be a filtered measurable space, i.e., a measurable space $(\Omega, \mathscr{F})$ with a flow of σ-algebras $(\mathscr{F}_n)_{n \geq 0}$ such that $\mathscr{F}_0 \subseteq \mathscr{F}_1 \subseteq \cdots \subseteq \mathscr{F}$.

Assume that $\mathscr{F} = \sigma(\bigcup_n \mathscr{F}_n)$. Let P and $\tilde{P}$ be two probability measures on $(\Omega, \mathscr{F})$ and $P_n = P \mid \mathscr{F}_n$, $\tilde{P}_n = \tilde{P} \mid \mathscr{F}_n$ be their restrictions to $\mathscr{F}_n$. Show that

$$(\tilde{P}_n) \lhd (P_n) \Leftrightarrow \tilde{P} \ll P,$$
$$(\tilde{P}_n) \lhd\rhd (P_n) \Leftrightarrow \tilde{P} \sim P,$$
$$(\tilde{P}_n) \vartriangle (P_n) \Leftrightarrow \tilde{P} \perp P.$$

11 Rate of Convergence in the Central Limit Theorem

1. Let $\xi_{n1}, \ldots, \xi_{nn}$ be independent random variables, $S_n = \xi_{n1} + \cdots + \xi_{nn}$, $F_n(x) = P(S_n \leq x)$. If $S_n \overset{d}{\to} \mathscr{N}(0,1)$, then $F_n(x) \to \Phi(x)$ for every $x \in R$. Since $\Phi(x)$ is continuous, the convergence here is actually uniform (Problem 5 in Sect. 1):

$$\sup_x |F_n(x) - \Phi(x)| \to 0, \quad n \to \infty. \tag{1}$$

It is natural to ask *how rapid* the convergence in (1) is. We shall establish a result for the case when

$$S_n = \frac{\xi_1 + \cdots + \xi_n}{\sigma\sqrt{n}}, \quad n \geq 1,$$

where $\xi_1, \xi_2, \ldots$ is a sequence of *independent identically distributed* random variables with $\mathsf{E}\,\xi_k = 0$, $\mathrm{Var}\,\xi_k = \sigma^2$ and $\mathsf{E}\,|\xi_1|^3 < \infty$.

Theorem (Berry–Esseen). *We have the bound*

$$\sup_x |F_n(x) - \Phi(x)| \leq \frac{C\,\mathsf{E}\,|\xi_1|^3}{\sigma^3\sqrt{n}}, \tag{2}$$

where C is an absolute constant.

PROOF. For simplicity, let $\sigma^2 = 1$ and $\beta_3 = \mathsf{E}\,|\xi_1|^3$. By Esseen's inequality (Subsection 10 of Sect. 12, Chap. 2)

$$\sup_x |F_n(x) - \Phi(x)| \leq \frac{2}{\pi}\int_0^T \left|\frac{f_n(t) - \varphi(t)}{t}\right|\, dt + \frac{24}{\pi T}\frac{1}{\sqrt{2\pi}}, \tag{3}$$

where $\varphi(t) = e^{-t^2/2}$ and

$$f_n(t) = [f(t/\sqrt{n})]^n$$

with $f(t) = \mathsf{E}\,e^{it\xi_1}$.

In (3) we may take T *arbitrarily*. Let us choose

$$T = \sqrt{n}/(5\beta_3).$$

We are going to show that for this T,

$$|f_n(t) - \varphi(t)| \leq \frac{7}{6}\frac{\beta_3}{\sqrt{n}}|t|^3 e^{-t^2/4}, \quad |t| \leq T. \tag{4}$$

The required estimate (2), with C an absolute constant, will follow immediately from (3) by means of (4). (A more sophisticated analysis gives $0.4097 < C < 0.469$, see Remark 2 in Sect. 4.)

We now turn to the proof of (4). By formula (18) in Sect. 2, Chap. 2 ($n = 3$, $E\,\xi_1 = 0$, $E\,\xi_1^2 = 1$, $E\,|\xi_1|^3 < \infty$) we obtain

$$f(t) = E\,e^{it\xi_1} = 1 - \frac{t^2}{2} + \frac{(it)^3}{6}[E\,\xi_1^3(\cos\theta_1 t\xi_1 + i\sin\theta_2 t\xi_1)], \tag{5}$$

where $|\theta_1| \leq 1$, $|\theta_2| \leq 1$. Consequently,

$$f\left(\frac{t}{\sqrt{n}}\right) = 1 - \frac{t^2}{2n} + \frac{(it)^3}{6n^{3/2}}\left[E\,\xi_1^3\left(\cos\theta_1\frac{t}{\sqrt{n}}\xi_1 + i\sin\theta_2\frac{t}{\sqrt{n}}\xi_1\right)\right].$$

If $|t| \leq T = \sqrt{n}/5\beta_3$, we find, by using the inequality $\beta_3 \geq \sigma^3 = 1$ (see (28), Sect. 6, Chap. 2), that

$$1 - \left|f\left(\frac{t}{\sqrt{n}}\right)\right| \leq \left|1 - f\left(\frac{t}{\sqrt{n}}\right)\right| \leq \frac{t^2}{2n} + \frac{|t|^3\beta_3}{3n^{3/2}} \leq \frac{1}{25}.$$

Consequently, for $|t| \leq T$ it is possible to have the representation

$$\left[f\left(\frac{t}{\sqrt{n}}\right)\right]^n = e^{n\log f(t/\sqrt{n})}, \tag{6}$$

where $\log z$ means the principal value of the logarithm of the complex number z ($\log z = \log|z| + i\arg z$, $-\pi < \arg z \leq \pi$).

Since $\beta_3 < \infty$, we obtain from Taylor's theorem with the Lagrange remainder (compare (35) in Sect. 12, Chap. 2)

$$\log f\left(\frac{t}{\sqrt{n}}\right) = \frac{it}{\sqrt{n}}s_{\xi_1}^{(1)} + \frac{(it)^2}{2n}s_{\xi_1}^{(2)} + \frac{(it)^3}{6n^{3/2}}(\log f)'''\left(\theta\frac{t}{\sqrt{n}}\right)$$

$$= -\frac{t^2}{2n} + \frac{(it)^3}{6n^{3/2}}(\log f)'''\left(\theta\frac{t}{\sqrt{n}}\right), \quad |\theta| \leq 1, \tag{7}$$

since the semi-invariants are $s_{\xi_1}^{(1)} = E\,\xi_1 = 0$, $s_{\xi_1}^{(2)} = \sigma^2 = 1$.

In addition,

$$(\log f(s))''' = \frac{f'''(s) \cdot f^2(s) - 3f''(s)f'(s)f(s) + 2(f'(s))^3}{f^3(s)}$$

$$= \frac{\mathsf{E}[(i\xi_1)^3 e^{i\xi_1 s}]f^2(s) - 3\,\mathsf{E}[(i\xi_1)^2 e^{i\xi_1 s}]\,\mathsf{E}[(i\xi_1)e^{i\xi_1 s}]f(s) + 2\,\mathsf{E}[(i\xi_1)e^{i\xi_1 s}]^3}{f^3(s)}.$$

From this, taking into account that $|f(t/\sqrt{n})| \geq 24/25$ for $|t| \leq T$ and $|f(s)| \leq 1$, we obtain

$$\left|(\log f)'''\left(\theta\frac{t}{\sqrt{n}}\right)\right| \leq \frac{\beta_3 + 3\beta_1 \cdot \beta_2 + 2\beta_1^3}{(\frac{24}{25})^3} \leq 7\beta_3 \qquad (8)$$

($\beta_k = \mathsf{E}\,|\xi_1|^k$, $k = 1, 2, 3$; $\beta_1 \leq \beta_2^{1/2} \leq \beta_3^{1/3}$; see (28), Sect. 6, Chap. 2).

From (6)–(8), using the inequality $|e^z - 1| \leq |z|e^{|z|}$, we find for $|t| \leq T = \sqrt{n}/5\beta_3$ that

$$\left|\left[f\left(\frac{t}{\sqrt{n}}\right)\right]^n - e^{-t^2/2}\right| = |e^{n\log f(t/\sqrt{n})} - e^{-t^2/2}|$$

$$\leq \left(\frac{7}{6}\right)\frac{\beta_3|t|^3}{\sqrt{n}}\exp\left\{-\frac{t^2}{2} + \left(\frac{7}{6}\right)|t|^3\frac{\beta_3}{\sqrt{n}}\right\} \leq \frac{7}{6}\frac{\beta_3|t|^3}{\sqrt{n}}e^{-t^2/4}.$$

This completes the proof of the theorem.

□

Remark. We observe that unless we make some supplementary hypothesis about the behavior of the random variables that are added, (2) cannot be improved. In fact, let $\xi_1, \xi_2, \ldots$ be independent identically distributed Bernoulli random variables with

$$\mathsf{P}(\xi_k = 1) = \mathsf{P}(\xi_k = -1) = \tfrac{1}{2}.$$

It is evident by symmetry that

$$2\,\mathsf{P}\left(\sum_{k=1}^{2n}\xi_k < 0\right) + \mathsf{P}\left(\sum_{k=1}^{2n}\xi_k = 0\right) = 1,$$

and hence, by Stirling's formula ((6), Sect. 2, Chap. 1)

$$\left|\mathsf{P}\left(\sum_{k=1}^{2n}\xi_k < 0\right) - \frac{1}{2}\right| = \frac{1}{2}\mathsf{P}\left(\sum_{k=1}^{2n}\xi_k = 0\right)$$

$$= \frac{1}{2}C_{2n}^n \cdot 2^{-2n} \sim \frac{1}{2\sqrt{\pi n}} = \frac{1}{\sqrt{(2\pi)(2n)}}.$$

It follows, in particular, that the constant C in (2) cannot be less than $(2\pi)^{-1/2}$ and that $\mathsf{P}(\sum_{k=1}^{2n}\xi_k = 0) \sim (\pi n)^{-1/2}$.

2. PROBLEMS

1. Prove (8).
2. Let $\xi_1, \xi_2, \ldots$ be independent identically distributed random variables with $\mathsf{E}\,\xi_1 = 0$, $\mathrm{Var}\,\xi_1 = \sigma^2$ and $\mathsf{E}\,|\xi_1|^3 < \infty$. It is known that the following *nonuniform inequality* holds: for all $x \in R$,

$$|F_n(x) - \Phi(x)| \leq \frac{C\,\mathsf{E}\,|\xi_1|^3}{\sigma^3 \sqrt{n}} \cdot \frac{1}{(1 + |x|)^3}.$$

   Prove this, at least for Bernoulli random variables.

3. Let $(\xi_k)_{k \geq 1}$ be a sequence of independent identically distributed random variables, taking values ± 1 with probabilities $1/2$, and let $S_k = \xi_1 + \cdots + \xi_k$. Let $\varphi_2(t) = \mathsf{E}\,e^{it\xi_1} = \frac{1}{2}(e^{it} + e^{-it})$. Show (in accordance with Laplace), that

$$\mathsf{P}\{S_{2n} = 0\} = \frac{1}{\pi} \int_0^\pi \varphi_2^n(t)\,dt \sim \frac{1}{\sqrt{\pi n}}, \quad n \to \infty.$$

4. Let $(\xi_k)_{k \geq 0}$ be a sequence of independent identically distributed random variables, taking $2a + 1$ integer values $0, \pm 1, \ldots, \pm a$ with equal probabilities. Let $\varphi_{2a+1}(t) = \mathsf{E}\,e^{it\xi_1} = \frac{1}{1+2a}\left(1 + 2\sum_{k=1}^a \cos tk\right)$.
   Show (again in accordance with Laplace) that

$$\mathsf{P}\{S_n = 0\} = \frac{1}{\pi} \int_0^\pi \varphi_{2a+1}^n(t)\,dt \sim \frac{\sqrt{3}}{\sqrt{2\pi(a+1)n}}, \quad n \to \infty.$$

   In particular, for $a = 1$, i.e., for the case, where ξ_k's take three values $-1, 0, 1$,

$$\mathsf{P}\{S_n = 0\} \sim \frac{\sqrt{3}}{2\sqrt{\pi n}}, \quad n \to \infty.$$

12 Rate of Convergence in Poisson's Theorem

1. Let $\xi_1, \xi_2, \ldots, \xi_n$ be independent Bernoulli random variables that take the values 1 and 0 with probabilities

$$\mathsf{P}(\xi_k = 1) = p_k, \quad \mathsf{P}(\xi_k = 0) = q_k (= 1 - p_k), \quad 1 \leq k \leq n.$$

We set $S = \xi_1 + \cdots + \xi_n$; let $B = (B_0, B_1, \ldots, B_n)$ be the probability distribution of the sum S, where $B_k = \mathsf{P}(S = k)$. Also let $\Pi = (\pi_0, \pi_1, \ldots)$ be the Poisson distribution with parameter λ, where

$$\pi_k = \frac{e^{-\lambda}\lambda^k}{k!}, \quad k \geq 0.$$

We noticed in Subsection 4 of Sect. 6, Chap. 1 that if

$$p_1 = \cdots = p_n, \quad \lambda = np, \tag{1}$$

there is the following estimate (Prokhorov [75]) for the *distance in variation* between the measures B and Π $(B_{n+1} = B_{n+2} = \cdots = 0)$:

$$\|B - \Pi\| = \sum_{k=0}^{\infty} |B_k - \pi_k| \le C_1(\lambda)p = C_1(\lambda) \cdot \frac{\lambda}{n}, \tag{2}$$

where

$$C_1(\lambda) = 2\min(2, \lambda). \tag{3}$$

For the case when p_k are not necessarily equal, but satisfy $\sum_{k=1}^{n} p_k = \lambda$, Le Cam [57] showed that

$$\|B - \Pi\| = \sum_{k=0}^{\infty} |B_k - \pi_k| \le C_2(\lambda) \max_{1 \le k \le n} p_k, \tag{4}$$

where

$$C_2(\lambda) = 2\min(9, \lambda). \tag{5}$$

A theorem to be presented below will imply the estimate

$$\|B - \Pi\| \le C_3(\lambda) \max_{1 \le k \le n} p_k, \tag{6}$$

in which

$$C_3(\lambda) = 2\lambda. \tag{7}$$

Although $C_2(\lambda) < C_3(\lambda)$ for $\lambda > 9$, i.e., (6) is worse than (4), we nevertheless have preferred to give a proof of (6), since this proof is essentially elementary, whereas an emphasis on obtaining a "good" constant $C_2(\lambda)$ in (4) greatly complicates the proof.

2. Theorem. *Let* $\lambda = \sum_{k=1}^{n} p_k$. *Then*

$$\|B - \Pi\| = \sum_{k=0}^{\infty} |B_k - \pi_k| \le 2 \sum_{k=1}^{n} p_k^2. \tag{8}$$

PROOF. We use the fact that each of the distributions B and Π is a *convolution* of distributions:

$$\begin{aligned} B &= B(p_1) * B(p_2) * \cdots * B(p_n), \\ \Pi &= \Pi(p_1) * \Pi(p_2) * \cdots * \Pi(p_n), \end{aligned} \tag{9}$$

understood as a convolution of the corresponding distribution functions (see Subsection 4, Sect. 8, Chap. 2), where $B(p_k) = (1 - p_k, p_k)$ is the Bernoulli distribution on the points 0 and 1, and $\Pi(p_k)$ is the Poisson distribution with parameter p_k supported on the points $0, 1, \ldots$.

It is easy to show that the difference $B - \Pi$ can be represented in the form

$$B - \Pi = R_1 + \cdots + R_n, \tag{10}$$

where

$$R_k = (B(p_k) - \Pi(p_k)) * F_k \tag{11}$$

with

$$
\begin{aligned}
F_1 &= \Pi(p_2) * \cdots * \Pi(p_n), \\
F_k &= B(p_1) * \cdots * B(p_{k-1}) * \Pi(p_{k+1}) * \cdots * \Pi(p_n), \quad 2 \le k \le n - 1, \\
F_n &= B(p_1) * \cdots * B(p_{n-1}).
\end{aligned}
$$

By problem 6 in Sect. 9, we have $\|R_k\| \le \|B(p_k) - \Pi(p_k)\|$. Consequently, we see immediately from (10) that

$$\|B - \Pi\| \le \sum_{k=1}^{n} \|B(p_k) - \Pi(p_k)\|. \tag{12}$$

By formula (12) in Sect. 9, we see that there is no difficulty in calculating the variation $\|B(p_k) - \Pi(p_k)\|$:

$$
\begin{aligned}
&\|B(p_k) - \Pi(p_k)\| \\
&= |(1 - p_k) - e^{-p_k}| + |p_k - p_k e^{-p_k}| + \sum_{j \ge 2} \frac{p_k^j e^{-p_k}}{j!} \\
&- |(1 - p_k) - e^{-p_k}| + |p_k - p_k e^{-p_k}| + 1 - e^{-p_k} - p_k e^{-p_k} \\
&= 2p_k(1 - e^{-p_k}) \le 2p_k^2.
\end{aligned}
$$

From this, together with (12), we obtain the required inequality (8).

This completes the proof of the theorem.

$\square$

Corollary. *Since $\sum_{k=1}^{n} p_k^2 \le \lambda \max_{1 \le k \le n} p_k$, we obtain* (6).

3. PROBLEMS

1. Show that, if $\lambda_k = -\log(1 - p_k)$,

$$\|B(p_k) - \Pi(\lambda_k)\| = 2(1 - e^{-\lambda_k} - \lambda_k e^{-\lambda_k}) \le \lambda_k^2$$

and consequently, $\|B - \Pi\| \le \sum_{k=1}^{n} \lambda_k^2$.

2. Establish the representations (9) and (10).

13 Fundamental Theorems of Mathematical Statistics

1. In Sect. 7, Chap. 1, we considered some problems of estimation and constructing confidence intervals for the probability of "success" from observations of random variables in Bernoulli trials. These are typical problems of *mathematical statistics*, which deals with in a certain sense *inverse* problems of probability theory. Indeed, whereas probability theory is mainly interested in computation, for a given probability model, of some probabilistic quantities (probabilities of events, probability distributions of random elements and their characteristics, and so on), in mathematical statistics we are interested to reveal (with certain degree of reliability), based on available statistical data, the probabilistic model for which the *statistical* properties of the empirical data agree best of all with *probabilistic* properties of the random mechanism generating these data.

The results given below (due to Glivenko, Cantelli, Kolmogorov, and Smirnov) may be rightly named *fundamental theorems of mathematical statistics* because they not only establish the *principal possibility* of extracting probabilistic information (about the distribution function of the observed random variables) from statistical raw material, but also make it possible to *estimate the goodness of fit* between statistical data and one or another probability model.

2. Let $\xi_1, \xi_2, \ldots$ be a sequence of independent identically distributed random variables defined on a probability space $(\Omega, \mathscr{F}, \mathsf{P})$ and let $F = F(x)$, $x \in R$, be their distribution function, $F(x) = \mathsf{P}\{\xi_k \leq x\}$. Define for any $N \geq 1$ the *empirical distribution function*

$$F_N(x; \omega) = \frac{1}{N} \sum_{k=1}^{N} I(\xi_k(\omega) \leq x), \quad x \in R. \tag{1}$$

By the law of large numbers (Sect. 3, Theorem 2) for *any* $x \in R$

$$F_N(x; \omega) \xrightarrow{\mathsf{P}} F(x), \quad N \to \infty, \tag{2}$$

i.e., $F_N(x)$ converges to $F(x)$ in P-probability.

Moreover, it follows from Theorems 1 and 2 of Sect. 3, Chap. 4, Vol. 2 that for *any* $x \in R$ this convergence holds *with probability one*: as $N \to \infty$,

$$F_N(x; \omega) \to F(x) \quad \text{(P-a. s.)}. \tag{3}$$

Remarkably, a stronger result on *uniform* convergence in (3) also holds.

Theorem 1 (Glivenko–Cantelli). *Under the above conditions the random variables*

$$D_N(\omega) = \sup_{x \in R} |F_N(x; \omega) - F(x)| \tag{4}$$

converge to zero with probability one:

$$\mathsf{P}(\lim_N D_N(\omega) = 0) = 1. \tag{5}$$

PROOF. Let Q be the set of rational numbers in R. Clearly,

$$\sup_{r \in Q} |F_N(r; \omega) - F(r)|$$

is a random variable. And since

$$D_N(\omega) = \sup_{x \in R} |F_N(x; \omega) - F(x)| = \sup_{r \in Q} |F_N(r; \omega) - F(r)|,$$

the statistic $D_N(\omega)$ is also a *random variable*, so that we can speak of its distribution. Let $M \geq 2$ be an integer and $k = 1, 2, \ldots, M-1$. Define the sequence of numbers

$$x_{M,k} = \min\{x \in R \colon k/M \leq F(x)\},$$

setting also $x_{M,0} = -\infty$, $x_{M,M} = +\infty$.

Take an interval $[x_{M,k}, x_{M,k+1}) \neq \varnothing$ and let x belong to this interval. Then obviously

$$
\begin{aligned}
F_N(x; \omega) - F(x) &\leq F_N(x_{M,k+1} - 0) - F(x_{M,k}) \\
&= [F_N(x_{M,k+1} - 0) - F(x_{M,k+1} - 0)] + [F(x_{M,k+1} - 0) - F(x_{M,k})] \\
&\leq F_N(x_{M,k+1} - 0) - F(x_{M,k+1} - 0) + 1/M.
\end{aligned}
$$

In a similar way, assuming again that $x \in [x_{M,k}, x_{M,k+1})$, we find that

$$F_N(x; \omega) - F(x) \geq F_N(x_{M,k}; \omega) - F(x_{M,k}) - 1/M.$$

Therefore, for any $x \in R$

$$
\begin{aligned}
&|F_N(x; \omega) - F(x)| \\
&\leq \max_{\substack{1 \leq k \leq M-1 \\ 1 \leq l \leq M-1}} \{|F_N(x_{M,k}; \omega) \quad F(x_{M,k})|, |F_N(x_{M,l} - 0; \omega) - F(x_{M,l} - 0)|\} + 1/M,
\end{aligned}
$$

hence

$$\lim_{n \to \infty} \sup_x |F_N(x; \omega) - F(x)| \leq 1/M \quad \text{(P-a. s.)}.$$

Since M is arbitrary, this implies (5).

$\square$

The Glivenko–Cantelli theorem, which is one of the *fundamental* theorems of mathematical statistics, states, as we pointed out, the principal possibility to verify, based on observations of (*independent identically distributed*) random variables $\xi_1, \xi_2, \ldots$ that the distribution function of these variables is precisely $F = F(x)$. In other words, this theorem guarantees the possibility to establish an agreement between the "theory and experiment."

3. As is seen from (1), $F_N(x)$ for each $x \in R$ is the relative frequency of events $\{\xi_i \leq x\}$, $i = 1, \ldots, N$, in N Bernoulli trials. This implies (2) and (3) by the Law of Large Numbers (LLN) and the Strong Law of Large Numbers (SLLN) respectively. But a much deeper information about the distribution of the frequency in the Bernoulli scheme is given by the de Moivre–Laplace theorem, which in this case states that for any *fixed* $x \in R$

$$\sqrt{N}(F_N(x) - F(x)) \xrightarrow{law} \mathscr{N}(0, F(x)[1 - F(x)]), \tag{6}$$

i.e., that the distribution of $\sqrt{N}\,(F_N(x) - F(x))$ converges to the normal one with zero mean and variance $\sigma^2(x) = F(x)\,(1 - F(x))$. (See Sect. 6, Chap. 1.) From this, one can easily derive the limit distribution of $\sqrt{N}|F_N(x) - F(x)|$ using the fact that $P(|\xi| > x) = 2\,P(\xi > x)$ for any $\xi \sim \mathscr{N}(0, \sigma^2)$ due to symmetry of the normal distribution. Namely, restating (6) as

$$P\left(\frac{\sqrt{N}(F_N(x) - F(x))}{\sigma(x)} > x\right) \to 1 - \Phi(x), \quad x \in R,$$

we can write our statement about $\sqrt{N}|F_N(x) - F(x)|$ as

$$P\left(\frac{\sqrt{N}|F_N(x) - F(x)|}{\sigma(x)} > x\right) \to 2(1 - \Phi(x)), \quad x \ge 0,$$

where $\Phi(\cdot)$ is the standard normal distribution function.

However, like in the case of the Glivenko–Cantelli theorem, we will be interested in the *maximal* deviation of $F_N(x)$ from $F(x)$, more precisely, in the distribution of D_N defined by (4) and

$$D_N^+ = \sup_x (F_N(x) - F(x)) \tag{7}$$

describing the maximum of the one-sided deviation of $F_N(x)$ from $F(x)$.

Now we will formulate a theorem on the limit distributions of D_N and D_N^+. This theorem shows, in particular, that these limit distributions hold with the same normalization as in (6), i.e., with multiplying these quantities by $\sqrt{N}$ (which is by no means obvious a priori), and that these limit distributions are essentially different, in contrast to the simple relation between those of $\sqrt{N}\,(F_N(x) - F(x))$ and $\sqrt{N}|F_N(x) - F(x)|$ for a fixed $x \in R$. The result (8) is due to Smirnov [94] and the result (9), (10) to Kolmogorov [48].

Theorem 2. *Assume that* $F(x)$ *is continuous. With the above notation, we have*

$$P(\sqrt{N}D_N^+ \le y) \to 1 - e^{-2y^2}, \quad y \ge 0, \tag{8}$$

$$P(\sqrt{N}D_N \le y) \to K(y), \tag{9}$$

where

$$K(y) = \sum_{k=-\infty}^{\infty} (-1)^k e^{-2k^2 y^2}, \quad y \ge 0. \tag{10}$$

The rigorous proof of this theorem goes beyond the scope of this book. It can be found in Billingsley [9], Sect. 13. Here we give an outline of the proof and a heuristic derivation of the formulas in (8) and (10).

The following observation (A. N. Kolmogorov) is of key importance for obtaining the limit distributions of these statistics.

Lemma 1. *Let* $\mathbb{F}$ *be the class of continuous distribution functions* $F = F(x)$. *For any* $N \geq 1$ *the probability distribution of* $D_N(\omega)$ *is the same for all* $F \in \mathbb{F}$. *The same is true also for* $D_N^+(\omega)$.

PROOF. Let $\eta_1, \eta_2, \ldots$ be a sequence of independent identically distributed random variables with uniform distribution on $[0, 1]$, i.e., having distribution function $U(x) = \mathsf{P}\{\eta_1 \leq x\} = x, 0 \leq x \leq 1$.

To prove the lemma we will show that for any *continuous* function $F = F(x)$ the distribution of the statistic $\sup_x |F_N(x; \omega) - F(x)|$ coincides with the distribution of $\sup_x |U_N(x; \omega) - U(x)|$, where

$$U_N(x; \omega) = N^{-1} \sum_{k=1}^{N} I(\eta_k(\omega) \leq x)$$

is the empirical distribution function of the variables $\eta_1, \ldots, \eta_N$.

Denote by A the union of intervals $I = [a, b]$, $-\infty < a < b < \infty$, on which the distribution function $F = F(x)$ is constant, so that $\mathsf{P}\{\xi_1 \in I\} = 0$. Then

$$D_N(\omega) = \sup_{x \in R} |F_N(x; \omega) - F(x)| = \sup_{x \in \bar{A}} |F_N(x; \omega) - F(x)|.$$

Introduce the variables $\tilde{\eta}_k = F(\xi_k)$ and empirical distribution functions

$$U_N(x; \omega) = \frac{1}{N} \sum_{k=1}^{N} I(\tilde{\eta}_k(\omega) \leq x).$$

Then we find that for $x \in \bar{A}$

$$U_N(F(x); \omega) = \frac{1}{N} \sum_{k=1}^{N} I(F(\xi_k(\omega)) \leq F(x)) = \frac{1}{N} \sum_{k=1}^{N} I(\xi_k(\omega) \leq x) = F_N(x; \omega),$$

since for such x we have $\{\omega : \xi_k(\omega) \leq x\} = \{\omega : F(\xi_k(\omega)) \leq F(x)\}$. Thus

$$D_N(\omega) = \sup_{x \in \bar{A}} |F_N(x; \omega) - F(x)| = \sup_{x \in \bar{A}} |U_N(F(x); \omega) - F(x)|$$

$$= \sup_{x \in R} |U_N(F(x); \omega) - F(x)| = \sup_{y \in (0,1)} |U_N(y; \omega) - y| \overset{\text{P-a. s.}}{=} \sup_{y \in [0,1]} |U_N(y; \omega) - y|,$$

where the last equality ($\overset{\text{P-a. s.}}{=}$) is a consequence of

$$P\{\tilde{\eta}_1 = 0\} = P\{\tilde{\eta}_1 = 1\} = 0. \tag{11}$$

Now we will show that the random variables $\tilde{\eta}_k$ are uniformly distributed on $[0, 1]$. To this end denote (for $y \in (0, 1)$)

$$x(y) = \inf\{x \in R \colon F(x) \geq y\}.$$

Then $F(x(y)) = y, y \in (0, 1)$, and

$$P\{\tilde{\eta}_1 \leq y\} = P\{F(\xi_1) \leq y\}$$
$$= P\{F(\xi_1) \leq F(x(y))\} = P\{\xi_1 \leq x(y)\} = F(x(y)) = y.$$

Combined with (11) this proves that the random variable $\tilde{\eta}_1$ (and hence each of $\tilde{\eta}_2, \tilde{\eta}_3, \dots$) is uniformly distributed on $[0, 1]$.

□

4*. This lemma shows that for obtaining limit distributions of D_N and D_N^+ (based on independent observations $\xi_1, \xi_2, \dots$ with a continuous distribution function $F = F(x) \in \mathbb{F}$) we may assume from the outset that $\xi_1, \xi_2, \dots$ are independent random variables *uniformly* distributed on $[0, 1]$.

Setting

$$U_N(t) = \frac{1}{N} \sum_{i=1}^{N} I(\xi_i \leq t) \tag{12}$$

and

$$w_N(t) = \sqrt{N}(U_N(t) - t), \tag{13}$$

we have

$$D_N^+ = \sup_{t \in (0,1)} w_N(t), \quad D_N = \sup_{t \in (0,1)} |w_N(t)|. \tag{14}$$

The proof of Theorem 2 consists of two steps:
(i) The proof of weak convergence

$$w_N^*(\cdot) \overset{w}{\to} w^0(\cdot), \tag{15}$$

where $w_N^*(\cdot)$ is a continuous random function approximating $w_N(\cdot)$ (e.g., by linear interpolation) so that $\sup_{t \in (0,1)} |w_N(t) - w_N^*(t)| \to 0$ as $N \to \infty$, and w^0 is the conditional Wiener process (see Subsection 7 of Sect. 13). For ease of notation, we henceforth often write sup without indicating that $t \in (0, 1)$. The convergence (15) is understood as weak convergence of the corresponding distributions in the space C of continuous functions on $[0, 1]$. (See Subsection 6 of Sect. 2, Chap. 2. See also Sect. 1, Chap. 3 for the concept of weak convergence. Recall that C is endowed with distance $\rho(x_1(\cdot), x_2(\cdot)) = \sup |x_1(t) - x_2(t)|, x_1(\cdot), x_2(\cdot) \in C$.)

---

* Subsections 4 and 5 are written by D.M. Chibisov.

Then (15) implies that the distributions of ρ-continuous functionals of w_N^* (and hence of the same functionals of w_N) converge to the distributions of these functionals of w^0. It is easily seen that if we change a function $x(t) \in C$ less than by ε uniformly in $t \in [0, 1]$ then $\sup x(t)$ and $\sup |x(t)|$ will change less than by ε. This means that these are ρ-continuous functionals of $x(\cdot) \in C$ and therefore by (15)

$$\sup_{t \in (0,1)} w_N(t) \xrightarrow{d} \sup_{t \in (0,1)} w^0(t), \qquad \sup_{t \in (0,1)} |w_N(t)| \xrightarrow{d} \sup_{t \in (0,1)} |w^0(t)|. \tag{16}$$

(ii) The proof that the distributions of $\sup_{t \in (0,1)} w^0(t)$ and $\sup_{t \in (0,1)} |w^0(t)|$ are given by the right-hand sides of (8) and (9).

For the proof of (i) we have to establish convergence of the finite-dimensional distributions of $w_N^*(\cdot)$ to those of w^0 and tightness of the sequence of distributions of $w_N^*(\cdot)$ in the space C. For the proof of tightness the reader is referred to [9].

Here we will only show the convergence of finite-dimensional distributions. The aim of introducing w_N^* was to replace w_N by a random element in the space C of continuous functions. (Or else w_N could be treated as an element of D, see [9]. See also [9] for difficulties which arise when trying to consider w_N in the space of discontinuous functions with metric ρ.) The fact that the functions $w_N^*(\cdot)$ and $w_N(\cdot)$ approach each other implies that their finite-dimensional distributions converge to the same limits, so that in this part of the proof we may deal directly with w_N instead of w_N^*.

Recall that $w^0(t)$, $t \in [0, 1]$, is the Gaussian process with covariance function

$$r^0(s, t) = \min(s, t) - st, \quad s, t \in [0, 1], \tag{17}$$

see Example 2 in Sect. 13, Chapter 2. Therefore to prove the convergence of interest we have to prove that for any $0 < t_1 < \cdots < t_k < 1$ the joint distribution of $(w_N(t_1), \ldots, w_N(t_k))$ converges to the k-dimensional normal distribution with zero mean and covariance matrix $\| \min(t_i, t_j) - t_i t_j \|_{i,j=1}^k$.

Notice that for $k = 1$ this is just the statement of the de Moivre–Laplace theorem. Indeed, by (12) $U_N(t)$ is the relative frequency of the events $(\xi_i \le t)$, $i = 1, \ldots, N$, having the probability t of occurrence, in N Bernoulli trials, and by (13) $w_N(t)$ is this frequency properly centered and normalized to obey the integral de Moivre–Laplace theorem saying that its distribution converges to $\mathcal{N}(0, t(1 - t))$, where $t(1 - t)$ is the variance of a single Bernoulli variable $I(\xi_i < t)$. Since $\mathrm{Var}\, w^0(t) = r^0(t, t) = t(1 - t)$ by (17), this conforms with the above statement on convergence $w_N(t) \xrightarrow{d} w^0(t)$ for a single $t \in (0, 1)$.

To prove a similar result for arbitrary $0 < t_1 < \cdots < t_k < 1$, we will use the following multidimensional version of Theorem 3, Sect. 3, Chap. 3 (it is stated as Problem 5 therein; the proof can be found, e.g., in [15]). Let $X_i = (X_{i1}, \ldots, X_{ik})$, $i = 1, 2, \ldots$, be independent identically distributed k-dimensional random vectors.

Theorem. *Suppose X_i's have a (common) finite covariance matrix $\mathbb{R}$. Denote by P_N the distribution of $S_N = (S_{N1}, \ldots, S_{Nk})$, where*

$$S_{Nj} = \frac{1}{\sqrt{N}} \sum_{i=1}^{N} (X_{ij} - \mathsf{E}X_{1j}), \quad j = 1, \dots, k. \tag{18}$$

Then

$$P_N \xrightarrow{w} \mathcal{N}(0, \mathbb{R}) \quad as \quad N \to \infty, \tag{19}$$

where $\mathcal{N}(0, \mathbb{R})$ is the k-dimensional normal distribution with zero mean and co-variance matrix $\mathbb{R}$.

To apply this theorem to $(w_N(t_1), \dots, w_N(t_k))$, let $X_{ij} = I(\xi_i \leq t_j)$, $i = 1, \dots, N$, $j = 1, \dots, k$. Then by (12), (13) $w_N(t_j) = S_{Nj}$ as in (18), and to obtain the required convergence, we only need to check that

$$\mathrm{Cov}(I(\xi_1 \leq s), I(\xi_1 \leq t)) = \min(s, t) - st \quad \text{for any} \quad 0 < s, t < 1. \tag{20}$$

Using the formulas $\mathrm{Cov}(\xi, \eta) = \mathsf{E}(\xi\eta) - \mathsf{E}\xi\mathsf{E}\eta$ and $\mathsf{E}I(\xi_1 \leq t) = t$, we see that (20) follows from the fact that $I(\xi_1 \leq s)I(\xi_1 \leq t) = I(\xi_1 \leq \min(s, t))$.

Thus we have established the "finite-dimensional convergence" part of the proof of (15) (or of step (i)). As we said above, for the remaining part, the proof of tightness, the reader is referred to [9].

5. Now we turn to the statement (ii). Reversing the inequalities in (8) and (9) this statement can be written as

$$\mathsf{P}\left(\sup_{t \in (0,1)} w^0(t) > y \right) = e^{-2y^2}, \quad y \geq 0, \tag{21}$$

$$\mathsf{P}\left(\sup_{t \in (0,1)} |w^0(t)| > y \right) = 2 \sum_{k=1}^{\infty} (-1)^{k+1} e^{-2k^2 y^2}, \quad y \geq 0. \tag{22}$$

For the detailed rigorous proof of (21) and (22) we again refer to [9]. Here we give a heuristic version of that proof.

First, let us check that the conditional Wiener process $w^0(\cdot)$ introduced in Example 2, Sect. 13, Chapter 2, is indeed the Wiener process $w(\cdot)$ conditioned on $\{w(1) = 0\}$. For that we show that the conditional finite-dimensional distributions of $w(\cdot)$ given that $w(1) = 0$ are the same as the finite-dimensional distributions of w^0. We will check this for $w(t)$ at a single point t, $0 < t < 1$. A general statement for $0 < t_1 < \dots < t_k < 1$, $k \geq 1$, can be proved in a similar way. Recall that the Wiener process $w(\cdot)$ is the Gaussian process with covariance function $r(s, t) = \min(s, t)$, $s, t \geq 0$. Let $\xi = w(1)$ and $\eta = w(t)$. Then $\mathsf{E}\xi = \mathsf{E}\eta = 0$, $\mathrm{Var}\,\xi = 1$, $\mathrm{Var}\,\eta = t$, and $\mathrm{Cov}(\xi, \eta) = \mathsf{E}\xi\eta = t$, hence $\mathrm{Cov}(\xi, \eta - t\xi) = 0$. (This holds because $\mathsf{E}(\eta\,|\,\xi) = t\xi$, see (12) in Sect. 13, Chap. 2, and in general $\mathsf{E}(\xi(\eta - \mathsf{E}(\eta\,|\,\xi))) = 0$.) Since (ξ, η) are jointly normally distributed, this implies that $\xi = w(1)$ and $\eta - t\xi = w(t) - tw(1)$ are independent. Now $\mathrm{Cov}(w(s) - sw(1), w(t) - tw(1)) = \min(s, t) - 2st + st = r^0(s, t)$ as in (17), hence we may set $w^0(t) = w(t) - tw(1)$, $0 \leq t \leq 1$. Then the conditional distribution of $(w(t)\,|\,w(1) = 0)$ equals that of $(w(t) - tw(1)\,|\,w(1) = 0)$, which is equal to the unconditional distribution of $w^0(t)$ due to independence of $w^0(t)$ and $w(1)$ shown above.

Now the left-hand sides of (21) and (22) can be replaced by

$$P\left(\sup_{t\in(0,1)} w(t) > y \,\middle|\, w(1) = 0\right) \quad \text{and} \quad P\left(\sup_{t\in(0,1)} |w(t)| > y \,\middle|\, w(1) = 0\right) \qquad (23)$$

respectively. Rewrite the first of these probabilities as

$$\lim_{\delta\to 0} \frac{P\left(\{\sup_{t\in(0,1)} w(t) > y\} \cap \{w(1) \in U_\delta\}\right)}{P(w(1) \in U_\delta)}, \qquad (24)$$

where $U_\delta = (-\delta, +\delta)$. (Of course, it is a heuristic step, but the formula (24) for the above conditional probability is in fact correct, see [9].) Let $\tau = \min\{t : w(t) = y\}$. This is a stopping time defined on the first event in the numerator of (24). For a fixed τ the event under P in the numerator occurs if the increment of $w(t)$ on $(\tau, 1)$ equals $-y$ up to $\pm\delta$, i.e., lies within $(-y - \delta, -y + \delta)$. By symmetry, the probability of this event equals the probability that this increment lies in $(y - \delta, y + \delta)$. Therefore the probability in the numerator of (24) equals

$$P\left(\{\sup_{t\in(0,1)} w(t) > y\} \cap \{w(1) \in 2y + U_\delta\}\right).$$

But for small δ the second event implies the first, so that this probability is $P(w(1) \in 2y + U_\delta)$. Taking the limit as $\delta \to 0$, we obtain that

$$P\left(\sup_{t\in(0,1)} w(t) > y \,\middle|\, w(1) = 0\right) = \frac{\varphi(2y)}{\varphi(0)} = e^{-2y^2},$$

where $\varphi(y)$ is the density of $\mathcal{N}(0, 1)$, which proves (21).

To prove (22) we represent its left-hand side as the second probability in (23) and use the arguments in the above derivation of (21). The event in the numerator of (24) with $w(t)$ replaced by $|w(t)|$ describes two possibilities for the path of $w(\cdot)$ to leave the stripe between the $\pm y$ lines, viz. by crossing the upper $(+y)$ or the lower $(-y)$ boundary. This apparently implies that the probability in (22) is twice the one in (21). But in this way some paths are counted twice, namely, those which cross both boundaries, so that the probability of crossing two boundaries has to be subtracted. This reasoning continuous by the principle of inclusion–exclusion to give the formula in (22). A detailed derivation of this formula can again be found in [9].

6. Let us consider how the knowledge, say, of relation (9), where $K(y)$ is given by (10), enables us to provide *a test for agreement between experiment and theory* or a *goodness-of-fit test*. To this end we first give a short table of the distribution function $K(y)$:

y	$K(y)$	y	$K(y)$	y	$K(y)$
		1.10	0.822282	2.10	0.999705
0.28	0.000001	1.20	0.887750	2.20	0.999874
0.30	0.000009	1.30	0.931908	2.30	0.999949
0.40	0.002808	1.40	0.960318	2.40	0.999980
0.50	0.036055	1.50	0.977782	2.50	0.9999925
0.60	0.135718	1.60	0.988048	2.60	0.9999974
0.70	0.288765	1.70	0.993828	2.70	0.9999990
0.80	0.455857	1.80	0.996932	2.80	0.9999997
0.90	0.607270	1.90	0.998536	2.90	0.99999990
1.00	0.730000	2.00	0.999329	3.00	0.99999997

If N is sufficiently large, we may assume that $K(y)$ provides an accurate enough approximation for $\mathsf{P}\{\sqrt{N}D_N(\omega) \leq y\}$.

Naturally, if the value $\sqrt{N}D_N(\omega)$ computed on the basis of empirical data $\xi_1(\omega), \ldots, \xi_N(\omega)$ turns out to be *large*, then the hypothesis that the (hypothesized) distribution function of these variables is just the (continuous) function $F = F(x)$ *has to be rejected.*

The above table enables us to get an idea about the *degree of reliability* of our conclusions. If, say, $\sqrt{N}D_N(\omega) > 1.80$, then (since $K(1.80) = 0.996932$) we know that this event has probability approximately equal to $0.0031\ (= 1.0000 - 0.9969)$. If we think of events with such a small probability $(= 0.0031)$ as practically almost impossible, we conclude that *the hypothesis* that the distribution function $\mathsf{P}\{\xi_1 \leq x\}$ is $F(x)$, where $F(x)$ is the function used in the formula for $D_N(\omega)$, should be *rejected.* On the contrary, if, say, $\sqrt{N}D_N(\omega) \leq 1.80$, then we can say (invoking the law of large numbers) that agreement between "experiment and theory" will hold in 9 969 such cases out of 10 000.

Remark. It is important to stress that when applying goodness-of-fit tests using Kolmogorov's or Smirnov's distributions, the distribution function $F = F(x)$ to be tested has to be *completely specified.* These tests "do not work" if the hypothesis assumes only that the distribution function $F = F(x)$ belongs to a *parametric family* $\mathbb{G} = \{G = G(x; \theta); \theta \in \Theta\}$ of distribution functions $G(x; \theta)$ depending on a parameter $\theta \in \Theta$. (Although for each θ the function $G(x; \theta)$ is supposed to be uniquely determined.) In this case the following way of testing the agreement between empirical data and the hypothesis $F \in \mathbb{G}$ comes to mind: first, to estimate θ based on N observations by means of some estimator $\hat{\theta}_N = \hat{\theta}_N(\omega)$ and then to make a decision using the quantity $\sqrt{N} \sup_{x \in R} |F_N(x; \omega) - G(x; \hat{\theta}_N(\omega))|$ as it was done in the above example. Unfortunately, the distribution function $G(x; \hat{\theta}_N(\omega))$ will be *random* and the distribution of the statistic $\sqrt{N} \sup_{x \in R} |F_N(x; \omega) - G(x; \hat{\theta}_N(\omega))|$ will not be, in general, given by Kolmogorov's distribution $K = K(y)$.

Concerning the problem of testing the hypothesis that $F \in \mathbb{G}$ see, e.g., [45].

Historical and Bibliographical Notes

Chapter 1: Introduction

The history of probability theory up to the time of Laplace is described by Todhunter [96]. The period from Laplace to the end of the nineteenth century is covered by Gnedenko and Sheinin in [53]. Stigler [95] provides very detailed exposition of the history of probability theory and mathematical statistics up to 1900. Maistrov [66] discusses the history of probability theory from the beginning to the thirties of the twentieth century. There is a brief survey of the history of probability theory in Gnedenko [32]. For the origin of much of the terminology of the subject see Aleksandrova [2].

For the basic concepts see Kolmogorov [51], Gnedenko [32], Borovkov [12], Gnedenko and Khinchin [33], A. M. and I. M. Yaglom [97], Prokhorov and Rozanov [77], handbook [54], Feller [30, 31], Neyman [70], Loève [64], and Doob [22]. We also mention [38, 90] and [91] which contain a large number of problems on probability theory.

In putting this text together, the author has consulted a wide range of sources. We mention particularly the books by Breiman [14], Billingsley [10], Ash [3, 4], Ash and Gardner [5], Durrett [24, 25], and Lamperti [56], which (in the author's opinion) contain an excellent selection and presentation of material.

The reader can find useful reference material in Great Soviet Encyclopedia, Encyclopedia of Mathematics [40] and Encyclopedia of Probability Theory and Mathematical Statistics [78].

The basic journal on probability theory and mathematical statistics in our country is *Teoriya Veroyatnostei i ee Primeneniya* published since 1956 (translated as *Theory of Probability and its Applications*).

Referativny Zhournal published in Moscow as well as *Mathematical Reviews* and *Zentralblatt für Mathematik* contain abstracts of current papers on probability and mathematical statistics from all over the world.

© Springer Science+Business Media New York 2016
A.N. Shiryaev, *Probability-1*, Graduate Texts
in Mathematics 95, DOI 10.1007/978-0-387-72206-1

A useful source for many applications, where statistical tables are needed, is *Tablicy Matematicheskoy Statistiki* (*Tables of Mathematical Statistics*) by Bol'shev and Smirnov [11]. Nowadays statistical computations are mostly performed using computer packages.

Chapter 2

Section 1. Concerning the construction of probabilistic models see Kolmogorov [49] and Gnedenko [32]. For further material on problems of distributing objects among boxes see, e.g., Kolchin, Sevastyanov, and Chistyakov [47].

Section 2. For other probabilistic models (in particular, the one-dimensional Ising model) that are used in statistical physics, see Isihara [42].

Section 3. Bayes's formula and theorem form the basis for the "Bayesian approach" to mathematical statistics. See, for example, De Groot [20] and Zacks [98].

Section 4. A variety of problems about random variables and their probabilistic description can be found in Meshalkin [68], Shiryayev [90], Shiryayev, Erlich and Yaskov [91], Grimmet and Stirzaker [38].

Section 6. For sharper forms of the local and integral theorems, and of Poisson's theorem, see Borovkov [12] and Prokhorov [75].

Section 7. The examples of Bernoulli schemes illustrate some of the basic concepts and methods of mathematical statistics. For more detailed treatment of mathematical statistics see, for example, Lehmann [59] and Lehmann and Romano [60] among many others.

Section 8. Conditional probability and conditional expectation with respect to a decomposition will help the reader understand the concepts of conditional probability and conditional expectation with respect to σ-algebras, which will be introduced later.

Section 9. The ruin problem was considered in essentially the present form by Laplace. See Gnedenko and Sheinin in [53]. Feller [30] contains extensive material from the same circle of ideas.

Section 10. Our presentation essentially follows Feller [30]. The method of proving (10) and (11) is taken from Doherty [21].

Section 11. Martingale theory is thoroughly covered in Doob [22]. A different proof of the ballot theorem is given, for instance, in Feller [30].

Section 12. There is extensive material on Markov chains in the books by Feller [30], Dynkin [26], Dynkin and Yushkevich [27], Chung [18, 19], Revuz [81], Kemeny and Snell [44], Sarymsakov [84], and Sirazhdinov [93]. The theory of branching processes is discussed by Sevastyanov [85].

Chapter 2

Section 1. Kolmogorov's axioms are presented in his book [51].

Section 2. Further material on algebras and σ-algebras can be found in, for example, Kolmogorov and Fomin [52], Neveu [69], Breiman [14], and Ash [4].

Section 3. For a proof of Carathéodory's theorem see Loève [64] or Halmos [39].

Sections 4–5. More material on measurable functions is available in Halmos [39].

Section 6. See also Kolmogorov and Fomin [51], Halmos [39], and Ash [4]. The Radon–Nikodým theorem is proved in these books.

Inequality (23) was first stated without proof by Bienaymé [8] in 1853 and proved by Chebyshev [16] in 1867. Inequality (21) and the proof given here are due to Markov [67] (1884). This inequality together with its corollaries (22), (23) is usually referred to as Chebyshev's inequality. However sometimes inequality (21) is called Markov's inequality, whereas Chebyshev's name is attributed to inequality (23).

Section 7. The definitions of conditional probability and conditional expectation with respect to a σ-algebra were given by Kolmogorov [51]. For additional material see Breiman [14] and Ash [4].

Section 8. See also Borovkov [12], Ash [4], Cramér [17], and Gnedenko [32].

Section 9. Kolmogorov's theorem on the existence of a process with given finite-dimensional distributions is in his book [51]. For Ionescu-Tulcea's theorem see also Neveu [69] and Ash [4]. The proof in the text follows [4].

Sections 10–11. See also Kolmogorov and Fomin [52], Ash [4], Doob [22], and Loève [64].

Section 12. The theory of characteristic functions is presented in many books. See, for example, Gnedenko [32], Gnedenko and Kolmogorov [34], and Ramachandran [79]. Our presentation of the connection between moments and semi-invariants follows Leonov and Shiryaev [61].

Section 13. See also Ibragimov and Rozanov [41], Breiman [14], Liptser and Shiryaev [62], Grimmet and Stirzaker [37], and Lamperti [56].

Chapter 3

Section 1. Detailed investigations of problems on weak convergence of probability measures are given in Gnedenko and Kolmogorov [34] and Billingsley [9].

Section 2. Prokhorov's theorem appears in his paper [76].

Section 3. The monograph [34] by Gnedenko and Kolmogorov studies the limit theorems of probability theory by the method of characteristic functions. See also Billingsley [9]. Problem 2 includes both Bernoulli's law of large numbers and Poisson's law of large numbers (which assumes that $\xi_1, \xi_2, \ldots$ are independent and take only two values (1 and 0), but in general are differently distributed: $\mathsf{P}(\xi_i = 1) = p_i$, $\mathsf{P}(\xi_i = 0) = 1 - p_i$, $i \geq 1$).

Section 4. Here we give the standard proof of the central limit theorem for sums of independent random variables under the Lindeberg condition. Compare [34] and [72].

Section 5. Questions of the validity of the central limit theorem without the hypothesis of asymptotic negligibility have already attracted the attention of P. Lévy. A detailed account of the current state of the theory of limit theorems in the *nonclassical setting* is contained in Zolotarev [99]. The statement and proof of Theorem 1 were given by Rotar [82].

Section 6. The presentation uses material from Gnedenko and Kolmogorov [34], Ash [4], and Petrov [71, 72].

Section 7. The Lévy–Prohorov metric was introduced in the well-known paper by Prohorov [76], to whom the results on metrizability of weak convergence of measures given on metric spaces are also due. Concerning the metric $\|P - \tilde{P}\|_{BL}^*$, see Dudley [23] and Pollard [73].

Section 8. Theorem 1 is due to Skorokhod. Useful material on the method of a single probability space may be found in Borovkov [12] and in Pollard [73].

Sections 9–10. A number of books contain a great deal of material touching on these questions: Jacod and Shiryaev [43], LeCam [58], Greenwood and Shiryaev [36].

Section 11. Petrov [72] contains a lot of material on estimates of the rate of convergence in the central limit theorem. The proof of the Berry–Esseen theorem given here is contained in Gnedenko and Kolmogorov [34].

Section 12. The proof follows Presman [74].

Section 13. For additional material on fundamental theorems of mathematical statistics, see Breiman [14], Cramér [17], Rényi [80], Billingsley [10], and Borovkov [13].

References

[1] M. Abramowitz and I. A. Stegun. *Handbook of Mathematical Functions: with Formulas, Graphs, and Mathematical Tables*. Courier Dover Publications, New York, 1972.

[2] N. V. Aleksandrova. *Mathematical Terms* [*Matematicheskie terminy*]. Vysshaia Shkola, Moscow, 1978.

[3] R. B. Ash. *Basic Probability Theory*. Wiley, New York, 1970.

[4] R. B. Ash. *Real Analysis and Probability*. Academic Press, New York, 1972.

[5] R. B. Ash and M. F. Gardner. *Topics in Stochastic Processes*. Academic Press, New York, 1975.

[6] K. B. Athreya and P. E. Ney. *Branching Processes*. Springer, New York, 1972.

[7] S. N. Bernshtein. Chebyshev's work on the theory of probability (in Russian), in *The Scientific Legacy of P. L. Chebyshev* [*Nauchnoe nasledie P. L. Chebysheva*], pp. 43–68, Akademiya Nauk SSSR, Moscow-Leningrad, 1945.

[8] I.-J. Bienaymé. Considérations à l'appui de la découverte de Laplace sur la loi de probabilité das la méthode de moindres carrés. *C. R. Acad. Sci. Paris*, **37** (1853), 309–324.

[9] P. Billingsley. *Convergence of Probability Measures*. Wiley, New York, 1968.

[10] P. Billingsley. *Probability and Measure*. 3rd ed. New York, Wiley, 1995.

[11] L. N. Bol'shev and N. V. Smirnov. *Tables of Mathematical Statistics* [*Tablicy Matematicheskoĭ Statistiki*]. Nauka, Moscow, 1983.

[12] A. A. Borovkov. *Wahrscheinlichkeitstheorie: eine Einführung*, first edition Birkhäuser, Basel–Stuttgart, 1976; *Theory of Probability*, 3rd edition [*Teoriya veroyatnosteĭ*]. Moscow, URSS, 1999.

[13] A. A. Borovkov. *Mathematical Statistics* [*Matematicheskaya Statistika*]. Nauka, Moscow, 1984.

[14] L. Breiman. *Probability*. Addison-Wesley, Reading, MA, 1968.

[15] A. V. Bulinsky and A. N. Shiryayev. *Theory of Random Processes* [*Teoriya Sluchaĭnykh Processov*]. Fizmatlit, Moscow, 2005.

© Springer Science+Business Media New York 2016
A.N. Shiryaev, *Probability-1*, Graduate Texts
in Mathematics 95, DOI 10.1007/978-0-387-72206-1

[16] P. L. Chebyshev. On mean values. [O srednikh velichinakh] *Matem. Sbornik*, II (1867), 1–9 (in Russian). Also: *J. de math. pures et appl.*, II série, XII (1867), 177–184. Reproduced in: Complete works of Chebyshev [Polnoye sobraniye sochineniǐ P. L. Chebysheva], Vol. 2, pp. 431–437, Acad. Nauk USSR, Moscow–Leningrad, 1947.

[17] H. Cramér, *Mathematical Methods of Statistics*. Princeton University Press, Princeton, NJ, 1957.

[18] K. L. Chung. *Markov Chains with Stationary Transition Probabilities*. Springer-Verlag, New York, 1967.

[19] K. L. Chung. *Elementary Probability Theory wigh Stochastic Processes*. 3rd ed., Springer-Verlag, 1979.

[20] M. H. De Groot. *Optimal Statistical Decisions*. McGraw-Hill, New York, 1970.

[21] M. Doherty. An amusing proof in fluctuation theory. *Lecture Notes in Mathematics*, no. 452, 101–104, Springer-Verlag, Berlin, 1975.

[22] J. L. Doob. *Stochastic Processes*. Wiley, New York, 1953.

[23] R. M. Dudley. Distances of Probability Measures and Random Variables. *Ann. Math. Statist.* **39**, 5 (1968), 1563–1572.

[24] R. Durrett. *Probability: Theory and Examples*. Pacific Grove, CA. Wadsworth & Brooks/Cole, 1991.

[25] R. Durrett. *Brownian Motion and Martingales in Analysis*. Belmont, CA. Wadsworth International Group, 1984.

[26] E. B. Dynkin. *Markov Processes*. Plenum, New York, 1963.

[27] E. B. Dynkin and A. A. Yushkevich. *Theorems and problems on Markov processes* [*Teoremy i Zadachi o Processakh Markova*]. Nauka, Moscow, 1967.

[28] N. Dunford and J. T. Schwartz, *Linear Operators*, Part 1, *General Theory*. Wiley, New York, 1988.

[29] C.-G. Esseen. A moment inequality with an application to the central limit theorem. *Skand. Aktuarietidskr.*, **39** (1956), 160–170.

[30] W. Feller. *An Introduction to Probability Theory and Its Applications*, vol. 1, 3rd ed. Wiley, New York, 1968.

[31] W. Feller. *An Introduction to Probability Theory and Its Applications*, vol. 2, 2nd ed. Wiley, New York, 1966.

[32] B. V. Gnedenko. *The Theory of Probability* [*Teoriya Voroyatnosteǐ*]. Mir, Moscow, 1988.

[33] B. V. Gnedenko and A. Ya. Khinchin. *An Elementary Introduction to the Theory of Probability*. Freeman, San Francisco, 1961; ninth edition [*Elementarnoe Vvedenie v Teoriyu Veroyatnosteǐ*]. "Nauka", Moscow, 1982.

[34] B. V. Gnedenko and A. N. Kolmogorov. *Limit Distributions for Sums of Independent Random Variables*, revised edition. Addison-Wesley, Reading, MA, 1968.

[35] I. S. Gradshteyn and I. M. Ryzhik, *Table of Integrals, Series, and Products*, 5th ed. Academic Press, New York, 1994.

[36] P. E. Greenwood and A. N. Shiryaev. *Contiguity and the Statistical Invariance Principle*. Gordon and Breach, New York, 1985.

[37] G. R. Grimmet and D. R. Stirzaker. *Probability and Random Processes*. Oxford: Clarendon Press, 1983.

[38] G. R. Grimmet and D. R. Stirzaker. *One Thousand Exercises in Probability*. Oxford Univ. Press, Oxford, 2004.

[39] P. R. Halmos. *Measure Theory*. Van Nostrand, New York, 1950.

[40] M. Hazewinkel, editor. *Encyclopaedia of Mathematics*, Vols. 1–10 + Supplement I–III. Kluwer, 1987–2002. [Engl. transl. (extended) of: I. M. Vinogradov, editor. *Matematicheskaya Entsiklopediya*, in 5 Vols.], Moscow, Sov. Entsiklopediya, 1977–1985.

[41] I. A. Ibragimov and Yu. A. Rozanov. *Gaussian Random Processes*. Springer-Verlag, New York, 1978.

[42] A. Isihara. *Statistical Physics*. Academic Press, New York, 1971.

[43] J. Jacod and A. N. Shiryaev. *Limit Theorems for Stochastic Processes*. Springer-Verlag, Berlin–Heidelberg, 1987.

[44] J. Kemeny and L. J. Snell. *Finite Markov Chains*. Van Nostrand, Princeton, 1960.

[45] E. V. Khmaladze, Martingale approach in the theory of nonparametric goodness-of-fit tests. *Probability Theory and its Applications*, **26**, 2 (1981), 240–257.

[46] A. Khrennikov. *Interpretations of Probability*. VSP, Utrecht, 1999.

[47] V. F. Kolchin, B. A. Sevastyanov, and V. P. Chistyakov. *Random Allocations*. Halsted, New York, 1978.

[48] A. Kolmogoroff, Sulla determinazione empirica di una leggi di distribuzione. *Giornale dell'Istituto degli Attuari*, **IV** (1933), 83–91.

[49] A. N. Kolmogorov, *Probability theory* (in Russian), in *Mathematics: Its Contents, Methods, and Value [Matematika, ee soderzhanie, metody i znachenie]*. Akad. Nauk SSSR, vol. 2, 1956.

[50] A. N. Kolmogorov, The contribution of Russian science to the development of probability theory. *Uchen. Zap. Moskov. Univ.* **1947** (91), 53–64. (in Russian).

[51] A. N. Kolmogorov. *Foundations of the Theory of Probability*. Chelsea, New York, 1956; second edition [*Osnovnye poniatiya Teorii Veroyatnosteĭ*]. "Nauka", Moscow, 1974.

[52] A. N. Kolmogorov and S. V. Fomin. *Elements of the Theory of Functions and Functionals Analysis*. Graylok, Rochester, 1957 (vol. 1), 1961 (vol. 2); sixth edition [*Elementy teorii funktsiĭ i funktsional'nogo analiza*]. "Nauka" Moscow, 1989.

[53] A. N. Kolmogorov and A. P. Yushkevich, editors. *Mathematics of the Nineteenth Century [Matematika XIX veka]*. Nauka, Moscow, 1978.

[54] V. S. Korolyuk, editor. *Handbook of probability theory and mathematical statistics [Spravochnik po teorii veroyatnosteĭ i matematicheskoĭ statistike]*. Kiev, Naukova Dumka, 1978.

[55] Ky Fan, Entfernung zweier zufälliger Grössen and die Konvergenz nach Wahrscheinlichkeit. *Math. Zeitschr.* **49**, 681–683.

[56] J. Lamperti. *Stochastic Processes*. Aarhus Univ. (Lecture Notes Series, no. 38), 1974.

[57] L. Le Cam. An approximation theorem for the Poisson binomial distribution. *Pacif. J. Math.*, **19**, 3 (1956), 1181–1197.

[58] L. Le Cam. *Asymptotic Methods in Statistical Decision Theory*. Springer-Verlag, Berlin etc., 1986.

[59] E. L. Lehmann. *Theory of Point Estimation*. Wiley, New York, 1983.

[60] E. L. Lehmann and J. P. Romano. *Testing Statistical Hypotheses*. 3rd Ed., Springer-Verlag, New York, 2006.

[61] V. P. Leonov and A. N. Shiryaev. On a method of calculation of semi-invariants. *Theory Probab. Appl.* **4** (1959), 319–329.

[62] R. S. Liptser and A. N. Shiryaev. *Statistics of Random Processes*. Springer-Verlag, New York, 1977.

[63] R. Sh. Liptser and A. N. Shiryaev. *Theory of Martingales*. Kluwer, Dordrecht, Boston, 1989.

[64] M. Loève. *Probability Theory*. Springer-Verlag, New York, 1977–78.

[65] E. Lukacs. *Characteristic functions*. 2nd edition. Briffin, London, 1970.

[66] D. E. Maistrov. *Probability Theory: A Historical Sketch*. Academic Press, New York, 1974.

[67] A.A. Markov *On certain applications of algebraic continued fractions*, Ph.D. thesis, St. Petersburg, 1884.

[68] L. D. Meshalkin. *Collection of Problems on Probability Theory [Sbornik zadach po teorii veroyatnostĕ]*. Moscow University Press, 1963.

[69] J. Neveu. *Mathematical Foundations of the Calculus of Probability*. Holden-Day, San Francisco, 1965.

[70] J. Neyman. *First Course in Probability and Statistics*. Holt, New York, 1950.

[71] V. V. Petrov. *Sums of Independent Random Variables*. Springer-Verlag, Berlin, 1975.

[72] V. V. Petrov. *Limit Theorems of Probability Theory*. Clarendon Press, Oxford, 1995.

[73] D. Pollard. *Convergence of Stochastic Processes*. Springer-Verlag, Berlin, 1984.

[74] E. L. Presman. Approximation in variation of the distribution of a sum of independent Bernoulli variables with a Poisson law. *Theory of Probability and Its Applications* **30** (1985), no. 2, 417–422.

[75] Yu. V. Prohorov [Prokhorov]. Asymptotic behavior of the binomial distribution. *Uspekhi Mat. Nauk* **8**, no. 3(55) (1953), 135–142 (in Russian).

[76] Yu. V. Prohorov. Convergence of random processes and limit theorems in probability theory. *Theory Probab. Appl.* **1** (1956), 157–214.

[77] Yu. V. Prokhorov and Yu. A. Rozanov. *Probability theory*. Springer-Verlag, Berlin–New York, 1969; second edition [*Teoriia veroiatnostĕ*]. "Nauka", Moscow, 1973.

[78] Yu. V. Prohorov, editor. *Probability Theory and Mathematical Statistics, Encyclopedia [Teoriya Veroyatnostĕ i Matematicheskaya Statistika, Encyclopedia]* (in Russian). Sov. Entsiklopediya, Moscow, 1999.

[79] B. Ramachandran. *Advanced Theory of Characteristic Functions*. Statistical Publishing Society, Calcutta, 1967.

[80] A. Rényi. *Probability Theory*, North-Holland, Amsterdam, 1970.

[81] D. Revuz. *Markov Chains*. 2nd ed., North-Holland, Amsterdam, 1984.

[82] V. I. Rotar. An extension of the Lindeberg–Feller theorem. *Math. Notes* 18 (1975), 660–663.

[83] Yu. A. Rozanov. *The Theory of Probability, Stochastic Processes, and Mathematical Statistics [Teoriya Veroyatnosteĭ, Sluchaĭnye Protsessy i Matematicheskaya Statistika]*. Nauka, Moscow, 1985.

[84] T. A. Sarymsakov. *Foundations of the Theory of Markov Processes [Osnovy teorii protsessov Markova]*. GITTL, Moscow, 1954.

[85] B. A. Sevastyanov [Sewastjanow]. *Verzweigungsprozesse*. Oldenbourg, Munich-Vienna, 1975.

[86] B. A. Sevastyanov. *A Course in the Theory of Probability and Mathematical Statistics [Kurs Teorii Veroyatnosteĭ i Matematicheskoĭ Statistiki]*. Nauka, Moscow, 1982.

[87] I. G. Shevtsova. On absolute constants in the Berry-Esseen inequality and its structural and nonuniform refinements. *Informatika i ee primeneniya*. **7**, 1 (2013), 124–125 (in Russian).

[88] A. N. Shiryaev. *Probability*. 2nd Ed. Springer-Verlag, Berlin etc. 1995.

[89] A. N. Shiryaev. *Wahrscheinlichkeit*. VEB Deutscher Verlag der Wissenschafter, Berlin, 1988.

[90] A. N. Shiryaev. *Problems in Probability Theory [Zadachi po Teorii Veroyatnostey]*. MCCME, Moscow, 2011.

[91] A. N. Shiryaev, I. G. Erlich, and P. A. Yaskov. *Probability in Theorems and Problems [Veroyatnost' v Teoremah i Zadachah]*. Vol. 1, MCCME, Moscow, 2013.

[92] Ya. G. Sinai. *A Course in Probability Theory [Kurs Teorii Veroyatnosteĭ]*. Moscow University Press, Moscow, 1985; 2nd ed. 1986.

[93] S. H. Sirazhdinov. *Limit Theorems for Stationary Markov Chains [Predelnye Teoremy dlya Odnorodnyh Tsepeĭ Markova]*. Akad. Nauk Uzbek. SSR, Tashkent, 1955.

[94] N. V. Smirnov. On the deviations of the empirical distribution curve [Ob ukloneniyakh empiricheskoĭ krivoĭ raspredeleniya] *Matem. Sbornik*, **6**, (48), no. 1 (1939), 3–24.

[95] S. M. Stigler. *The History of Statistics: The Measurement of Uncertainty Before 1900*. Cambridge: Belknap Press of Harvard Univ. Press, 1986.

[96] I. Todhunter. *A History of the Mathematical Theory of Probability from the Time of Pascal to that of Laplace*. Macmillan, London, 1865.

[97] A. M. Yaglom and I. M. Yaglom. *Probability and Information*. Reidel, Dordrecht, 1983.

[98] S. Zacks. *The Theory of Statistical Inference*. Wiley, New York, 1971.

[99] V. M. Zolotarev. *Modern Theory of Summation of Random Variables [Sovremennaya Teoriya Summirovaniya Nezavisimyh Sluchaĭnyh Velichin]*. Nauka, Moscow, 1986.

Keyword Index

© Springer Science+Business Media New York 2016
A.N. Shiryaev, *Probability-1*, Graduate Texts
in Mathematics 95, DOI 10.1007/978-0-387-72206-1

Symbol Index

© Springer Science+Business Media New York 2016
A.N. Shiryaev, *Probability-1*, Graduate Texts
in Mathematics 95, DOI 10.1007/978-0-387-72206-1

Printed in the United States
By Bookmasters